FUNDAMENTALS OF
ELECTRIC POWER ENGINEERING

FUNDAMENTALS OF ELECTRIC POWER ENGINEERING

Isaak D. Mayergoyz • Patrick McAvoy
University of Maryland, USA

World Scientific

NEW JERSEY • LONDON • SINGAPORE • BEIJING • SHANGHAI • HONG KONG • TAIPEI • CHENNAI

Published by

World Scientific Publishing Co. Pte. Ltd.
5 Toh Tuck Link, Singapore 596224
USA office: 27 Warren Street, Suite 401-402, Hackensack, NJ 07601
UK office: 57 Shelton Street, Covent Garden, London WC2H 9HE

Library of Congress Cataloging-in-Publication Data
Mayergoyz, I. D.
Fundamentals of electric power engineering / Isaak D. Mayergoyz, University of Maryland, USA, Patrick McAvoy, University of Maryland, USA.
pages cm
Includes bibliographical references and index.
ISBN 978-9814616584 (hardbound : alk. paper) -- ISBN 9814616583 (hardbound : alk. paper) --
ISBN 978-9814616591 (e-book) -- ISBN 9814616591 (e-book) --
ISBN 978-9814616607 (mobile) -- ISBN 9814616605 (mobile)
1. Electric power systems. I. McAvoy, Patrick. II. Title.
TK1001.M34 2014
621.31--dc23

2014028094

British Library Cataloguing-in-Publication Data
A catalogue record for this book is available from the British Library.

Cover illustration: High voltage switch gear for three phases, by Angie (Sawara, Japan) under CC BY 2.0, Wikimedia Commons

Printed in Singapore

TO OUR STUDENTS

Preface

You have in your hands an undergraduate text on fundamentals of electric power engineering. This text reflects the experience of the first author in teaching electric power engineering courses in the Electrical and Computer Engineering Department of the University of Maryland College Park during the past thirty-four years. These courses have constituted the educational core of the electric power engineering program. This program was originally established (in the early 1980s) with the financial support and sponsorship of Baltimore Gas and Electric (BGE) Company, Potomac Electric Power Company (PEPCO), Virginia Electric Power Company (VEPCO), Bechtel Corporation and General Electric (GE) Foundation. This program has been designed as a sequence of senior elective courses in the area of electric power engineering. This design has two main advantages. First, students in such elective courses usually have strong interest in electric power and are really motivated to learn the related material. Second, senior students have already been exposed to fundamentals in electric and electronic circuits, electromagnetics and control theory. This opens the opportunity to cover the material in power courses at sufficiently high level and with the same mathematical and physical rigor which is now practiced in courses on communication, control and electromagnetics. This text on fundamentals of electric power engineering reflects this approach to teaching power courses.

Electric power engineering has always been an integral part of electrical engineering education. This is especially true nowadays in view of renewed emphasis in the area of energy in general and electric power engineering in particular. This textbook may provide a viable alternative to existing textbooks on the market by covering in one volume in a concise and rigorous manner such topics as power systems, electrical machines and power electronics. For this reason, this book can be used for teaching three different courses such as Power Systems, Electrical Machines and Power Electronics. These power courses form the mainstream of electric power engineering curriculum at most universities worldwide.

The book consists of three parts. The first part of the book deals with the review of electric and magnetic circuits. This review stresses the topics which nowadays are usually deemphasized (or ignored) in required circuits and electromagnetics courses. Namely, the phasor diagrams for ac circuits and analysis of electric circuits with periodic non-sinusoidal sources are stressed. Phasor diagrams have practically disappeared from circuit courses and textbooks, while these diagrams are still very instrumental in electric power engineering. Analysis of electric circuits with periodic non-sinusoidal sources is very important in the study of steady-state operation of power electronics converters. The frequency-domain and time-domain techniques for such analysis are presented in the book. In the review of magnetic circuits, a special emphasis is made on the analysis of magnetic circuits with permanent magnets. This is justified, on the one hand, by the proliferation of permanent magnets in power devices and, on the other hand, by the insufficient discussion of this topic in the existing undergraduate textbooks. Furthermore, the analysis of nonlinear magnetic circuits and eddy current losses for circularly (or elliptically) polarized magnetic fields are presented. The former is important because magnetic saturation of ferromagnetic cores often occurs in power devices. The latter is of interest because ferromagnetic cores of ac electric machines are subject to rotating (not linearly polarized) magnetic fields.

The second part of the book can be used for teaching courses on power systems and electrical machines. This part starts with a brief review of the structure of power systems, analysis of three-phase circuits and the discussion of ac power and power factor. Next, the analysis of faults in power systems is presented. This analysis is first performed by using the Thevenin theorem. Then, the concept of symmetrical components is introduced and the sequence networks are derived. Finally, the analysis of faults based on sequence networks is discussed. The next chapter deals with transformers. Here, the design and principle of operation of transformers are first considered along with the study of the ideal transformer. Then, the equivalent circuit for a single-phase transformer is derived on the basis of equivalent mathematical transformation of coupled circuit equations and the importance of leakage inductances (leakage reactances) is stressed. Next, open-circuit and short-circuit tests are described as the experimental means of determining parameters of equivalent circuits. The chapter is concluded by the discussion of three-phase transformers. The following chapter deals with synchronous generators. Here, the design and principle of operation of synchronous generators with cylindrical rotors and salient pole rotors are first considered. Then, the mathematical analysis of armature reaction magnetic fields of ideal cylindrical rotor generators is carried out. The results of this analysis are used in the discussion of the design of stator windings and in the computation of their synchronous reactance. It is stressed that stator windings are designed as filters of spatial and temporal harmonics. This is achieved due to their distributed nature, their two-layer structure and the use of fractional pitch. Next, the two-reactance theory of salient pole synchronous generators is presented. The chapter is concluded by the derivation of formulas for the power of synchronous generators and by the discussion of static stability of these generators as well as of their performance when connected to an infinite bus. The next chapter deals with the power flow analysis and dynamic (transient) stability of power systems. Here, the nonlinear power flow equations are first derived and then their numerical solutions by using Newton-Raphson and continuation techniques are discussed. The analysis of the transient stability is carried out by using the "swing" equation for mechanical motion of rotors of synchronous generators. This equation is presented in the Hamiltonian form, which leads to the phase portrait of rotor dynamics. This phase portrait results in a simple algebraic criterion for transient stability of rotor dynamics which contains as a particular case the celebrated equal area stability criterion. The last chapter of the second part deals with induction machines. In the past, induction machines have mostly been used as motors. However, recently induction machines have found applications as generators in wind energy systems. First, the design and principle of operation of induction machines is discussed. Then, by using the fact that in the induction machines the electromagnetic coupling between the rotor and stator windings is mostly realized through rotating magnetic fields, the coupled circuit equations are derived. The equivalent mathematical transformation of these coupled circuit equations leads to the equivalent electric circuits for induction machines. The chapter is concluded by the discussion of mechanical (torque-speed) characteristics of induction machines, which reveal the possibility of frequency control of speed of induction motors.

The third part of the book deals with power electronics and it consists of four chapters. The first chapter covers the material related to power semiconductor devices. It starts with a brief review of the scope and nature of power electronics and then provides the summary of basic facts of semiconductor physics. This is followed by the discussion of *p-n* junctions and diodes, where all the basic relations are derived and characteristic features of power diodes are outlined. Then, the principles of operation and designs of bipolar junction transistors (BJTs) and thyristors (SCRs) are presented and the special emphasis is placed on the understanding of operation of these devices as switches. Next, such devices as the MOSFET, power MOSFET and IGBT are discussed and their main advantages as power electronics switches are articulated. The chapter is concluded with brief descriptions of snubber circuits and resonant switches. The principles of zero-current switching (ZCS) and zero-voltage switching (ZVS) are outlined and it is stressed that the design of resonant (quasi-resonant) power converters is a very active and promising area of research. The next chapter deals with rectifiers. Here, single-phase bridge rectifiers with RL, RC and RLC loads are first discussed along with center-tapped transformer rectifiers. The time-domain technique is extensively used to derive analytical expressions for currents and voltages in such rectifiers. Then, three-phase diode rectifiers are studied and various circuit topologies of these rectifiers are presented along with derivation of analytical expressions for voltages and currents. The chapter is concluded with the discussion of phase-controlled rectifiers, and single-phase as well as three-phase versions of such rectifiers are studied in detail. The following chapter deals with inverters. It starts with the discussion of single-phase bridge inverters, and the design of bidirectional (bilateral) switches needed for the operation of the inverters is motivated and described. This is followed by the detailed study of pulse-width modulation (PWM) and analytical expressions for spectra of PWM voltages are derived. The chapter is concluded with the discussion of three-phase inverters. The design of ac-to-ac converters by cascading three-phase rectifiers with three-phase inverters is then briefly outlined along with their applications in ac motor drives. The last chapter of the third part of the book deals with dc-to-dc converters (choppers). Here, the buck converter, boost converter and buck-boost converter are first covered, and continuous and discontinuous modes of their operation are studied in detail. The chapter is concluded with the discussion of "flyback" and "forward" (indirect) converters, and physical aspects of their operation are carefully described along with all pertinent analytical formulas.

Each part of the book is supplemented by a list of problems of varying difficulty. Many of these problems are pointed questions related to the theoretical aspects discussed in the text. It is hoped that this may motivate readers to go through the appropriate parts of the text again and again, which may eventually result in the better comprehension of the material.

We have made an effort to produce a relatively short book that covers the fundamentals of the very broad area of electric power engineering. Naturally, this can only be achieved by omitting some topics. We have omitted the discussion of transmission lines because this subject is usually covered in courses on electromagnetics. We have also omitted the discussion of power system protection and economic operation (optimal dispatch) of generating resources. In our view, these topics are more suited for more advanced courses. In the power electronics portion of the book, we have limited our discussion to the most basic facts related to the circuit topology and principles of operation of power converters. This book has a strong theoretical flavor with emphasis on physical and mathematical aspects of electric power engineering fundamentals. It clearly reveals the multidisciplinary nature of power engineering and it stresses its connections with other areas of electrical engineering. It is believed that this approach can be educationally beneficial.

In undertaking this project, we wanted to produce a student-friendly textbook. We have come to the conclusion that students' interests are best served when the discussion of complicated concepts is not avoided. We have tried to introduce these concepts in a straightforward way and strived to achieve clarity and precision in exposition. We believe that material which is carefully and rigorously presented is better absorbed. It is for students to judge to what extent we have succeeded.

Contents

Part II: Power Systems 129

PART I

Review of Electric and Magnetic Circuit Theory

Chapter 1

Basic Electric Circuit Theory

1.1 Review of Basic Equations of Electric Circuit Theory

In circuit theory, there are two distinct types of basic mathematical relations. In the first type, these relations depend on the physical nature of circuit elements and they are called terminal relations. The second type of relations reflects the connectivity of the electric circuit, namely, how the circuit elements are interconnected. For this reason, they are sometimes called topological relations. These relations are based on the Kirchhoff Current Law (KCL) and Kirchhoff Voltage Law (KVL).

We start with terminal relations and consider five basic two-terminal elements: resistor, inductor, capacitor and ideal (independent) voltage and current sources. These circuit elements are ubiquitous in power engineering applications. However, in power electronics, multi-terminal circuit elements are used as well. These multi-terminal circuit elements are models of power semiconductor devices and they will be discussed in Chapter 1 of Part III.

A two-terminal element is schematically represented in Figure 1.1. Each two-terminal element is characterized by the voltage $v(t)$ across the terminals and by the current $i(t)$ through the element. In order to write the meaningful equations relating the voltages and currents in electric circuits, it is necessary to assign a polarity to the voltage and a direction to the current. These assigned (not actual) directions and polarities are called *reference directions* and *reference polarities.* These reference directions and polarities are assigned arbitrarily. The actual current directions and voltage polarities are not known beforehand and they may change with time. The reference direction for a current is indicated by an arrow, while the reference polarity is specified by placing plus and minus signs next to element terminals (see Figure 1.1). The reference directions and polarities are

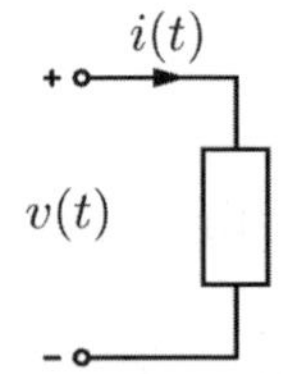

Fig. 1.1

usually coordinated by choosing the reference direction of the current from the positive reference terminal to the negative reference terminal.

As discussed below, the reference directions and polarities are used in writing KCL and KVL equations. These equations are then solved and the signs of currents and voltages are found at any instant of time. If at time t a current is positive,

$$i(t) > 0, \tag{1.1}$$

then the actual direction of this current coincides with its reference direction. If, on the other hand, the found current is negative at time t,

$$i(t) < 0, \tag{1.2}$$

then the actual direction of this current at time t is opposite to its reference direction.

Similarly, if a voltage is found to be positive,

$$v(t) > 0, \tag{1.3}$$

then the actual polarity coincides with its reference polarity. On the other hand, if the voltage is found to be negative,

$$v(t) < 0, \tag{1.4}$$

then the actual voltage polarity is opposite to its reference polarity. It is clear from the above discussion that the reference directions and polarities allow one to write KCL and KVL equations, then solve them and eventually find actual directions and polarities of circuit variables.

Now, we consider a resistor. Its circuit notation is shown in Figure 1.2. The terminal relation for the resistor is given by Ohm's law,

$$\boxed{v(t) = Ri(t),} \tag{1.5}$$

where R is the resistance of the resistor. By using the terminal relation (1.5), we find the expression for instantaneous power $p(t)$ for the resistor,

$$p(t) = v(t)i(t) = Ri^2(t), \tag{1.6}$$

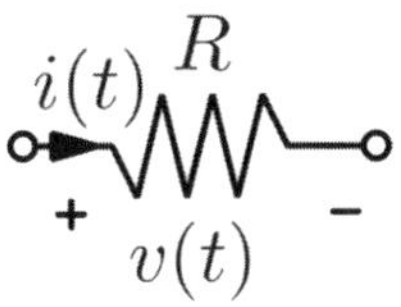

Fig. 1.2

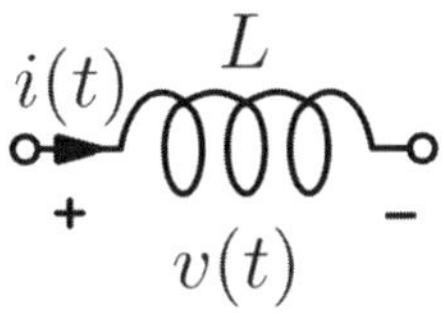

Fig. 1.3

or

$$p(t) = \frac{v^2(t)}{R}. \tag{1.7}$$

It is clear from the last two equations that in the case of the resistor the instantaneous power is always positive, which means that the resistor always consumes electric power. This is the reason why resistors are often used in electric power engineering to model irreversible losses of electric energy or its conversion into other forms of energy.

Next, we discuss an inductor. Its circuit notation is shown in Figure 1.3. The inductor is characterized by inductance L and its terminal relation is given by the formula

$$\boxed{v(t) = L\frac{di(t)}{dt}.} \tag{1.8}$$

The last equation implies that the current through an inductor is differentiable and, consequently, a continuous function of time. The latter means that for any instant of time t_0 the value of the current $i(t_{0-})$ immediately before t_0 is equal to the value of the current $i(t_{0+})$ immediately after t_0:

$$\boxed{i(t_{0-}) = i(t_{0+}).} \tag{1.9}$$

It is clear from equation (1.8) that in the case of dc currents ($i(t) = const$) the voltage across the inductor is equal to zero and the inductor can be replaced by a "short-circuit" branch. On the other hand, if the inductor (as a part of a circuit) is connected by some switch to a source and the initial current through the inductor before switching is zero, then according to the continuity condition (1.9) the current through the inductor will remain zero immediately after switching. This means that immediately after switching the inductor is an "open-circuit" branch.

It is apparent from equation (1.8) that the instantaneous power $p(t)$ for the inductor is given by the formula

$$p(t) = v(t)i(t) = Li(t)\frac{di(t)}{dt} = \frac{L}{2}\frac{d\left(i^2(t)\right)}{dt}. \tag{1.10}$$

It follows from the last equation that power is positive if the absolute value of inductor current increases with time and it is negative if the absolute value of the current decreases with time. This implies that when the absolute value of the inductor current is increasing with time, the electric power is being consumed and stored in the inductor's magnetic field. On the other hand, when the absolute value of the inductor current is decreasing with time, then the electric power is "given back" at the expense of the energy previously stored in the magnetic field. This clearly suggests that the inductor is an energy storage element and it finds many applications as such in electric power engineering.

Another important application of the inductor is for "ripple suppression" in power electronics. Power electronics converters are switching-mode converters in which semiconductor devices are used as switches that are repeatedly (periodically) switched "on" and "off" to achieve the desired performance of the converters. This periodic switching results in periodic components of electric currents which manifest themselves as ripple in output converter voltages. These ripples can be suppressed by using inductors. Indeed, from equation (1.8) we find

$$i(t) = i(0) + \frac{1}{L}\int_0^t v(\tau)d\tau. \tag{1.11}$$

If the current is a periodic function of time with period T, then

$$i(0) = i(T) \tag{1.12}$$

and

$$\int_0^T v(\tau)d\tau = 0. \tag{1.13}$$

It is apparent that the second term in the right-hand side of equation (1.11) can be construed as a ripple in electric current. It is clear from formula (1.11) that this ripple can be suppressed by increasing the value of inductance L. It is also clear from formulas (1.11) and (1.13) that the same

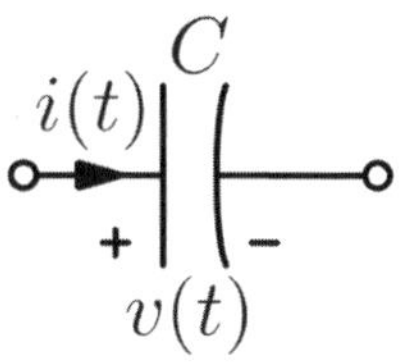

Fig. 1.4

ripple can be suppressed by decreasing T, i.e., by increasing the frequency of switching. This leads to the "trade-off" that is practiced in power electronics: the higher the frequency of switching of power semiconductor devices, the smaller the value of inductance that is needed for the ripple suppression.

Now, we proceed to the discussion of a capacitor. Its circuit notation is shown in Figure 1.4. The capacitor is characterized by capacitance C and its terminal relation is given by the formula

$$\boxed{i(t) = C\frac{dv(t)}{dt}.} \tag{1.14}$$

The last equation implies that the voltage across a capacitor is differentiable and, consequently, a continuous function of time. The latter means that for any instant of time t_0 the value of the voltage $v(t_{0-})$ immediately before t_0 is equal to the value of the voltage $v(t_{0+})$ immediately after t_0:

$$\boxed{v(t_{0-}) = v(t_{0+}).} \tag{1.15}$$

It is apparent from equation (1.14) that in the case of dc voltages ($v(t) = const$) the current through the capacitor is equal to zero and the capacitor is an "open-circuit" branch. This is consistent with the fact that the current through the capacitor is a displacement current, which exists only when the electric field in the capacitor varies with time.

In the case when an uncharged capacitor with zero voltage is connected through some switch to a source, then according to the continuity condition (1.15) the voltage across the capacitor will remain zero immediately after switching. This means that immediately after switching the capacitor is a "short-circuit" branch. This implies that capacitors connected in parallel with power equipment may protect this equipment from large initial transient currents, which mostly flow through these capacitors.

It is apparent from equation (1.14) that the instantaneous power $p(t)$ for the capacitor is given by the formula

$$p(t) = v(t)i(t) = Cv(t)\frac{dv(t)}{dt} = \frac{C}{2}\frac{d\left(v^2(t)\right)}{dt}. \tag{1.16}$$

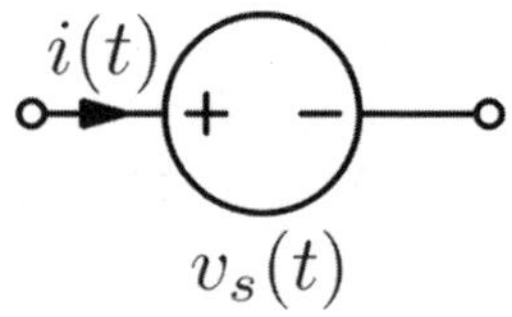

Fig. 1.5

It follows from the last equation that power is positive if the absolute value of capacitor voltage increases with time and it is negative if the absolute value of capacitor voltage decreases with time. This implies that, when the absolute value of the capacitor voltage is increasing with time, the electric power is being consumed and stored in the electric field within the capacitor. On the other hand, when the absolute value of the capacitor voltage is decreasing with time, then the electric power is "given back" at the expense of energy stored in the electric field. This clearly reveals that the capacitor is an energy storage element and it is used as such in many power-related applications.

Another important application of the capacitor is for "ripple suppression" in power electronics. Indeed, from equation (1.14) we derive

$$v(t) = v(0) + \frac{1}{C}\int_0^t i(\tau)d\tau. \tag{1.17}$$

If the capacitor voltage is a periodic function of time with period T, then

$$v(0) = v(T) \tag{1.18}$$

and

$$\int_0^T i(\tau)d\tau = 0. \tag{1.19}$$

It is clear that the second term in the right-hand side of formula (1.17) can be construed as a ripple in capacitor voltage. It is apparent from formula (1.17) that this ripple can be suppressed by increasing the value of capacitance C. It is also apparent from formulas (1.17) and (1.19) that the same ripple can be suppressed by decreasing T, i.e., by increasing the frequency of switching. Hence, the trade-off: the higher the frequency of switching of semiconductor devices in power converters, the smaller the value of the capacitance that is needed for ripple suppression.

Finally, we shall discuss ideal (independent) voltage and current sources, whose circuit notations are shown in Figures 1.5 and 1.6, respectively. An

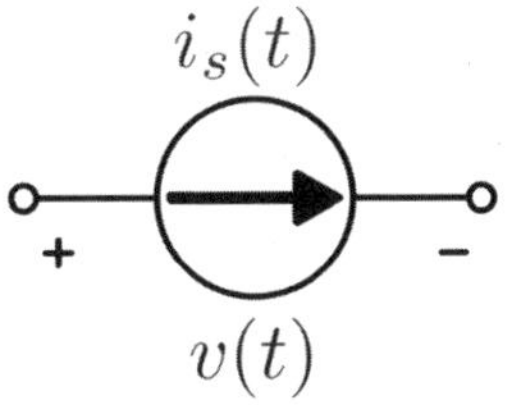

Fig. 1.6

ideal voltage source is a two-terminal element with the property that the voltage across its terminals is specified at every instant of time,

$$\boxed{v(t) = v_s(t),} \tag{1.20}$$

where $v_s(t)$ is a given (known) function of time. It is apparent from the given definition that the terminal voltage does not depend on the current through the voltage source, which is reflected in the terminology *independent* voltage source.

An ideal current source is by definition a two-terminal element with the property that the current through this element is specified at every instant of time,

$$\boxed{i(t) = i_s(t),} \tag{1.21}$$

where $i_s(t)$ is a given (known) function of time. It is clear from the given definition that the terminal current does not depend on the voltage across the current source; in this sense, this is an ideal (or independent) current source.

Previously we discussed the terminal relations, which are determined by the physical nature of the circuit elements. Now, we proceed to the brief discussion of the relations which are due to the connectivity of elements in an electric circuit. There are two types of such relations. We begin with the Kirchhoff Current Law (KCL). KCL equations are written for nodes of electric circuits. A node of an electric circuit is a "point" where three or more elements are connected together. KCL states that the algebraic sum of electric currents at any node of an electric circuit is equal to zero at every instant of time. This is mathematically expressed as follows:

$$\boxed{\sum_k i_k(t) = 0.} \tag{1.22}$$

The term "algebraic sum" implies that some currents are taken with positive signs while others are taken with negative signs. Two equivalent rules can be used for sign assignments. One rule is that positive signs are assigned to currents with reference directions toward the node, while negative signs are assigned to currents with reference directions from the node. KCL equations can be written for any node. However, only $(n - 1)$ equations will be linearly independent, where n is the number of nodes in a given circuit. The $(n-1)$ nodes for which KCL equations are written can be chosen arbitrarily. The KCL equation for the last (n-th) node can be obtained by summing up the previously written KCL equations. This clearly suggests that the equation for the last (n-th) node is not linearly independent. A "point" in an electric circuit where only two elements are connected is not qualified as a node because of the triviality of the KCL equation in this case, which simply suggests that the same current flows through both circuit elements, i.e., these two circuit elements are connected in series.

Next, we discuss equations written by using the Kirchhoff Voltage Law (KVL). These equations are written for loops. A loop is defined as a set of branches that form a closed path with the property that each node is encountered only once as the loop is traced. A branch is defined as a single two-terminal element or several two-terminal elements connected in series. KVL states that the algebraic sum of branch voltages around any loop of an electric circuit is equal to zero at every instant of time. This is mathematically expressed as follows:

$$\boxed{\sum_k v_k(t) = 0.} \tag{1.23}$$

The term "algebraic sum" implies that some branch voltages are taken with positive signs while others are taken with negative signs. The following rule can be used for sign assignment. If the tracing direction of the branch coincides with the reference direction of the branch current then the positive sign is assigned to the branch voltage, otherwise the negative sign is assigned to the branch voltage.

Since there may be many possible loops for any given circuit, determining which loops must be traced in order to write linearly independent KVL equations is not entirely obvious. One method which will always produce the correct number of linearly independent KVL equations is based on the use of a graph tree. By definition, a graph tree of an electric circuit is a subset of branches with the property that all nodes of the circuit are connected together, but there are no closed loops formed by these branches. It is clear that any graph tree of an electric circuit with n nodes contains $n-1$ branches. By adding a new branch to the graph tree we create a new loop and a new KVL equation can be written for the created loop. This KVL equation will contain a new variable – branch voltage for the added branch. KVL equations written in this way will be linearly independent because each new equation contains a new variable. It is easy to see that the total number of linearly independent KVL equations written in the described way is equal to $b-(n-1)$, where b is the total number of branches in the circuit. This is so because $b-(n-1)$ branches should be removed to form a graph tree and, consequently, $b-(n-1)$ loops will be formed by adding one by one the removed branches.

It is clear that the total number of linearly independent equations written by using KCL and KVL (i.e., the total number of equations (1.22) and (1.23)) is equal to the number b of branches. An additional b equations are obtained by using terminal relations (1.5), (1.8), (1.14), (1.20) and (1.21). Thus, the total number of equations is $2b$, which is the total number of circuit variables, i.e., the total number of branch currents and branch voltages. These are ordinary differential equations for which the initial conditions can be found by using the continuity conditions (1.9) and (1.15). Thus, the framed equations (1.5), (1.8), (1.9), (1.14), (1.15), (1.20), (1.21), (1.22) and (1.23) form the foundation of electric circuit theory. In this theory, various analysis techniques have been developed which exploit the connectivity of electric circuits as well as the particular nature of excitation of these circuits. Some of these techniques will be discussed in this first part of the book due to their wide use in electric power engineering.

It is worthwhile to note that in electric circuit theory the basic (framed) relations (1.5), (1.8), (1.9), (1.14), (1.15), (1.20), (1.21), (1.22) and (1.23) are treated as axioms (postulates) which are fully consistent with experimental facts. However, within the framework of electromagnetic field theory, all these fundamental circuit relations can be derived by using approximations relevant to the notion of electric circuits with lumped parameters.

It is worthwhile to stress in the conclusion of this section that electric circuits are models for actual devices – models which are based on some simplifications and approximations. This may lead in some cases to intrinsic (logical) contradictions between the basic circuit relations. We shall illustrate such possible contradictions for two cases of very simple electric circuits shown in Figures 1.7 and 1.8. In Figure 1.7, at time $t_0 = 0$ a dc current source is connected by switch (SW) to an inductor and a resistor connected in series. It is clear that prior to the switching the current

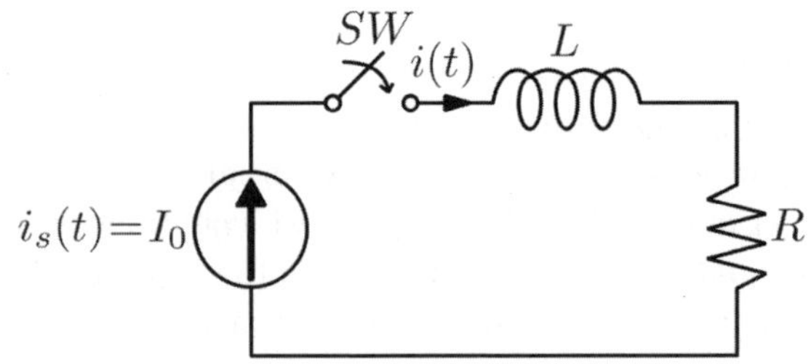

Fig. 1.7

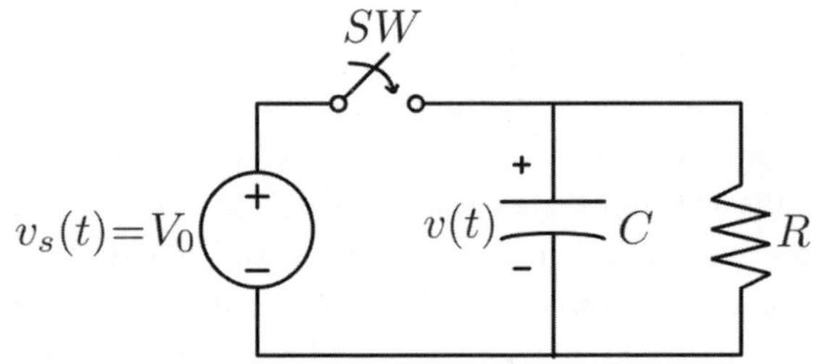

Fig. 1.8

through the inductor is equal to zero,

$$i(0_-) = 0. \tag{1.24}$$

According to the continuity condition (1.9), the current $i(0_+)$ immediately after switching must be equal to zero,

$$i(0_+) = 0. \tag{1.25}$$

However, according to KCL the same current must be equal to the current I_0 of the current source,

$$i(0_+) = I_0 \neq 0, \tag{1.26}$$

which is a contradiction.

Similarly, for the circuit shown in Figure 1.8, according to the continuity condition (1.15) the voltage $v(0_+)$ across the capacitor is equal to zero,

$$v(0_+) = 0, \tag{1.27}$$

if the capacitor was not charged before switching. However, according to KVL the same voltage must be equal to the voltage V_0 of the dc voltage source,

$$v(0_+) = V_0 \neq 0, \tag{1.28}$$

which is a contradiction.

The above contradictions appear because in the circuits shown in Figures 1.7 and 1.8 some small parameters have been neglected. For instance, for an actual (real) inductor there is always small parallel to L capacitance between the turns of the inductor. The presence of this capacitance will remove the contradiction between equations (1.25) and (1.26) because immediately after switching the current I_0 will flow through this capacitance. Similarly, the presence of a small resistance of connecting wires in the circuit shown in Fig. 1.8 will remove the contradiction between equations (1.27) and (1.28), because immediately after switching the voltage V_0 will be applied across such resistance. The presented discussion suggests that in the actual devices represented by the circuits shown in Figures 1.7 and 1.8 the initial stages of transients will be controlled by neglected small parameters. This implies that small parameters can be important for proper modeling of the performance of actual devices. It is demonstrated later in this text that the importance of small parameters is typical for such power devices as transformers and induction machines as well as boost and buck-boost choppers.

1.2 Phasor Analysis of AC Electric Circuits

In many electric power applications, electric circuits are excited by ac (sinusoidal) sources. Under steady-state conditions, all voltages and currents in such circuits will be sinusoidal. A special and very useful circuit analysis technique exists which exploits this fact. It is known as the phasor technique and it uses complex numbers to represent time-harmonic sinusoidal quantities. The main advantage of the phasor technique is that it reduces calculus operations on time-harmonic sinusoidal quantities to algebraic operations on complex numbers (phasors). As a result, the basic differential equations of electric circuits discussed in the previous section are reduced to linear algebraic equations with respect to phasors. This significantly simplifies the analysis of electric circuits under ac steady-state conditions.

The central idea of the phasor technique can be described as follows. Every time-harmonic sinusoidal quantity is fully characterized by three numbers: its frequency, peak value and initial phase. In ac steady-state analysis, the frequency of sinusoidal quantities is fixed and known. Consequently, every time-harmonic quantity of known frequency is fully characterized by two numbers: its peak value and initial phase. The same is true for complex numbers. In the polar form, each complex number is characterized by its magnitude (absolute value) and polar angle. This suggests to represent any time-harmonic quantity by a phasor, which is a complex number whose magnitude is equal to the peak value of the time-harmonic quantity and whose polar angle is equal to the initial phase of this time-harmonic quantity. This definition of the phasor is illustrated by the following two formulas for time-harmonic (sinusoidal) voltage and current, respectively,

$$\boxed{v(t) = V_m \cos(\omega t + \varphi_V) \leftrightarrow \hat{V} = V_m e^{j\varphi_V},} \tag{1.29}$$

$$\boxed{i(t) = I_m \cos(\omega t + \varphi_I) \leftrightarrow \hat{I} = I_m e^{j\varphi_I},} \tag{1.30}$$

where $\hat{V}$ and $\hat{I}$ are the phasors of voltage and current, respectively. It is clear from the above formulas that if sinusoidal quantities are given then it is easy to write their phasors. On the other hand, if the phasor is known and represented in the polar form, then it is easy to write the corresponding time-harmonic quantity.

Now, it is easy to see that in the case of ac steady state all terminal relations can be represented in the phasor form. Consider first a resistor. Then, by substituting sinusoidal voltage and current in formula (1.5), we find

$$V_m \cos(\omega t + \varphi_V) = RI_m \cos(\omega t + \varphi_I). \tag{1.31}$$

The last equality implies that

$$V_m = RI_m, \tag{1.32}$$

$$\varphi_V = \varphi_I. \tag{1.33}$$

The last formula reveals that in the case of resistors, their time-harmonic voltages and currents are in phase. From the last two formulas, we also derive

$$\hat{V} = V_m e^{j\varphi_V} = RI_m e^{j\varphi_I} = R\hat{I}. \tag{1.34}$$

Thus, it is established that the terminal relation for the resistor can be written in the phasor form as follows:

$$\boxed{\hat{V} = R\hat{I}.} \tag{1.35}$$

Next, we consider an inductor. Then, by substituting sinusoidal voltage and current in formula (1.8), we find

$$V_m \cos(\omega t + \varphi_V) = -\omega L I_m \sin(\omega t + \varphi_I), \tag{1.36}$$

or

$$V_m \cos(\omega t + \varphi_V) = \omega L I_m \cos\left(\omega t + \varphi_I + \frac{\pi}{2}\right). \tag{1.37}$$

The last equality implies that

$$V_m = \omega L I_m, \tag{1.38}$$

$$\varphi_V = \varphi_I + \frac{\pi}{2}. \tag{1.39}$$

The last formula reveals that in the case of the inductor, its time-harmonic voltage leads its time-harmonic current by $\frac{\pi}{2}$. Equivalently, the current lags behind the voltage by $\frac{\pi}{2}$. From the last two formulas we also derive

$$\hat{V} = V_m e^{j\varphi_V} = \omega L I_m e^{j(\varphi_I + \pi/2)} = j\omega L I_m e^{j\varphi_I} = j\omega L \hat{I}. \tag{1.40}$$

Thus, it is established that the terminal relation for the inductor can be written in the phasor form as follows:

$$\boxed{\hat{V} = j\omega L \hat{I}.} \tag{1.41}$$

Finally, we consider a capacitor. By substituting sinusoidal voltage and current in formula (1.14), we find

$$I_m \cos(\omega t + \varphi_I) = -\omega C V_m \sin(\omega t + \varphi_V), \tag{1.42}$$

or

$$I_m \cos(\omega t + \varphi_I) = \omega C V_m \cos\left(\omega t + \varphi_V + \frac{\pi}{2}\right). \tag{1.43}$$

The last equality implies that

$$I_m = \omega C V_m, \tag{1.44}$$

$$\varphi_I = \varphi_V + \frac{\pi}{2}. \tag{1.45}$$

The last formula reveals that in the case of the capacitor, its time-harmonic current leads its time-harmonic voltage by $\frac{\pi}{2}$. Equivalently, the voltage lags behind the current by $\frac{\pi}{2}$. From the last two formulas we also derive

$$\hat{I} = I_m e^{j\varphi_I} = \omega C V_m e^{j(\varphi_V + \pi/2)} = j\omega C V_m e^{j\varphi_V} = j\omega C \hat{V}. \tag{1.46}$$

Thus, it is established that the terminal relation for the capacitor can be written in the phasor form as follows:

$$\boxed{\hat{V} = -\frac{j}{\omega C}\hat{I}.} \tag{1.47}$$

It is also clear that given sinusoidal voltage and current sources, they can be written in phasor forms:

$$v_s(t) = V_{ms}\cos(\omega t + \varphi_{V_s}) \leftrightarrow \hat{V}_s = V_{ms}e^{j\varphi_{V_s}}, \tag{1.48}$$

$$i_s(t) = I_{ms}\cos(\omega t + \varphi_{I_s}) \leftrightarrow \hat{I}_s = I_{ms}e^{j\varphi_{I_s}}. \tag{1.49}$$

It can be concluded that in the case of ac steady state all terminal relations can be written in the algebraic phasor form.

Now, we turn to KCL and KVL equations and write them in the phasor form. To do this we shall use the fact that the phasor of the sum of sinusoidal quantities is equal to the sum of the phasors of sinusoidal quantities being summed. The proof of this fact is based on the following mathematical relation between sinusoidal quantities and phasors:

$$v(t) = V_m\cos(\omega t + \varphi_V) = \mathrm{Re}\left[V_m e^{j(\omega t+\varphi_V)}\right] = \mathrm{Re}\left[\hat{V}e^{j\omega t}\right]. \tag{1.50}$$

Indeed, consider a sum of sinusoidal quantities of arbitrary physical nature

$$g(t) = \sum_k G_{mk}\cos(\omega t + \varphi_k) \tag{1.51}$$

and we want to prove that

$$\hat{G} = \sum_k \hat{G}_k. \tag{1.52}$$

By using the Euler formula and formula (1.50), we find

$$\begin{aligned} g(t) &= \sum_k G_{mk}\mathrm{Re}\left[e^{j(\omega t+\varphi_k)}\right] = \sum_k \mathrm{Re}\left[G_{mk}e^{j\varphi_k}e^{j\omega t}\right] \\ &= \sum_k \mathrm{Re}\left[\hat{G}_k e^{j\omega t}\right] = \mathrm{Re}\left[\left(\sum_k \hat{G}_k\right)e^{j\omega t}\right] = \mathrm{Re}\left[\hat{G}e^{j\omega t}\right], \end{aligned} \tag{1.53}$$

which proves the equality (1.52).

Having established the last fact, we can write KCL equations (1.22) and KVL equations (1.23) in the phasor form as follows:

$$\boxed{\sum_k \hat{I}_k = 0,} \quad [(n-1) \text{ lin. ind. eqs.}], \tag{1.54}$$

$$\boxed{\sum_k \hat{V}_k = 0,} \quad [b-(n-1) \text{ lin. ind. eqs.}]. \tag{1.55}$$

Next, we shall discuss the very important concept of impedance. To this end, consider a branch where a resistor, an inductor and a capacitor are connected in series (see Figure 1.9). According to KVL, we have

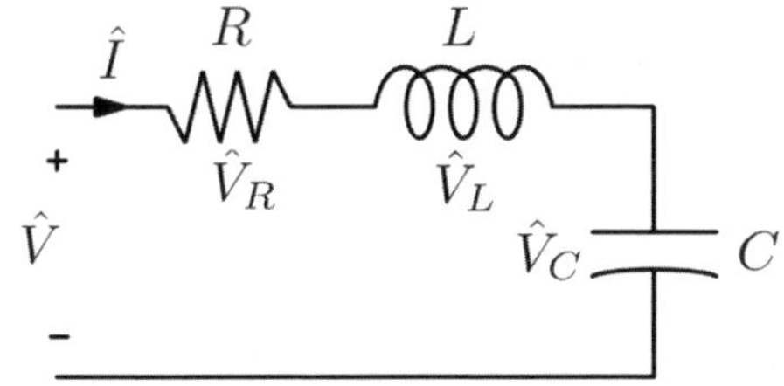

Fig. 1.9

$$\hat{V} = \hat{V}_R + \hat{V}_L + \hat{V}_C. \tag{1.56}$$

By using phasor relations (1.35), (1.41) and (1.47), we transform the last equation as follows:

$$\hat{V} = R\hat{I} + j\omega L\hat{I} - \frac{j}{\omega C}\hat{I} = \hat{I}\left[R + j\left(\omega L - \frac{1}{\omega C}\right)\right]. \tag{1.57}$$

Now, the impedance Z of the RLC branch can be naturally introduced by the formula

$$\boxed{Z = R + j\left(\omega L - \frac{1}{\omega C}\right) = R + jX,} \tag{1.58}$$

where

$$X = \omega L - \frac{1}{\omega C} \tag{1.59}$$

is called the reactance of the branch and it is determined by the energy storage elements in the branch.

By using the definition (1.58) of impedance, we shall write the terminal relation for the RLC branch in the form

$$\boxed{\hat{V} = \hat{I}Z.} \tag{1.60}$$

In the polar form, the impedance can be written as

$$\boxed{Z = |Z|e^{j\varphi}.} \tag{1.61}$$

By substituting the last formula into equation (1.60), we find

$$V_m e^{j\varphi_V} = I_m|Z|e^{j(\varphi_I + \varphi)}, \tag{1.62}$$

which implies that

$$V_m = I_m|Z|, \tag{1.63}$$

$$\varphi_V - \varphi_I = \varphi. \tag{1.64}$$

Thus, the magnitude of branch impedance relates peak values of branch voltage and current, while the polar angle of the impedance is equal to the phase shift in time between branch voltage and current.

From equations (1.58) and (1.61) the following useful expressions for $|Z|$ and φ can be obtained:

$$|Z| = \sqrt{R^2 + X^2} = \sqrt{R^2 + \left(\omega L - \frac{1}{\omega C}\right)^2}, \tag{1.65}$$

$$\tan\varphi = \frac{X}{R} = \frac{\omega L - \frac{1}{\omega C}}{R}. \tag{1.66}$$

The branch shown in Figure 1.9 has the most general composition when three physically distinct two-terminal elements are connected in series. The expression (1.58) for the impedance of such a branch is naturally simplified when only one or two distinct two-terminal elements are present in the branch. These simplifications are given by the following formulas:

$$Z = R, \tag{1.67}$$

if a branch contains only a resistor;

$$Z = j\omega L, \tag{1.68}$$

if a branch contains only an inductor;

$$Z = -\frac{j}{\omega C}, \tag{1.69}$$

if a branch contains only a capacitor;

$$Z = R + j\omega L \tag{1.70}$$

for an RL branch;

$$Z = R - \frac{j}{\omega C} \tag{1.71}$$

for an RC branch and

$$Z = j\left(\omega L - \frac{1}{\omega C}\right) \tag{1.72}$$

for an LC branch.

The last formula suggests that the impedance is equal to zero if

$$\omega = \frac{1}{\sqrt{LC}}. \tag{1.73}$$

It is also clear that under the condition (1.73), the impedance of the *RLC* branch is given by the formula

$$Z = R. \tag{1.74}$$

In other words, under the condition (1.73) an *RLC* branch acts as a pure resistor and branch voltage and current are in phase. This phenomenon is called resonance and may have important implications.

The presented discussion can be summarized as follows. At ac steady state, each branch can be characterized by impedance and the expression for the branch impedance is determined by the composition of the branch. Let us consider a branch number k and let Z_k be its impedance. Then, the phasor branch voltage $\hat{V}_k$ and the phasor branch current $\hat{I}_k$ are related by the formula (see (1.60))

$$\hat{V}_k = \hat{I}_k Z_k. \tag{1.75}$$

Formula (1.75) can be used in the phasor form of KVL equations (1.55). In these equations (as well as in equations (1.54)) we can also identify known phasors of voltage and current sources and move them with appropriate signs to the right-hand sides. The described transformations eventually result in the following ac steady-state equations for the phasors of branch currents:

$$\boxed{\begin{aligned} \sum_k \hat{I}_k &= -\sum_k \hat{I}_{sk}, & (1.76)\\ \sum_k \hat{I}_k Z_k &= -\sum_k \hat{V}_{sk}. & (1.77) \end{aligned}}$$

These are simultaneous linear algebraic equations and it is apparent that the total number of these equations is equal to the total number of passive (i.e., without sources) branches. By solving these equations, $\hat{I}_k$ can be found and then by using formula (1.75) the phasors of branch voltages $\hat{V}_k$ can be determined. As soon as this is done, instantaneous voltages and branch currents can be computed by using formulas similar to (1.50). In the circuit theory, various techniques have been developed which exploit connectivity of electric circuits. One example of such a technique is the method of equivalent transformations which exploits series and parallel connections of various branches to achieve the overall simplification of electric circuits.

It is interesting to mention that from the mathematical point of view the phasor technique allows to find particular periodic solutions of ordinary differential equations (ODEs) which describe the performance of electric circuits in the case of their time-harmonic excitation. This phasor technique is more efficient and more powerful than the technique of undetermined coefficients used in courses on ordinary differential equations. In this text, the phasor technique will be frequently used to derive simple expressions for particular periodic solutions of ODEs in cases when actual regimes are not ac steady states.

As has been emphasized in our discussion, the main idea of the phasor technique is to reduce the operations of calculus on sinusoidal quantities to algebraic operations on their phasors. It turns out that this idea can be extended to a broader class of voltages and currents.

Consider the following voltage:

$$v(t) = V_m e^{\sigma t} \cos(\omega t + \varphi_V). \tag{1.78}$$

By using the Euler formula, the last formula can be transformed as follows:

$$v(t) = V_m e^{\sigma t} \mathrm{Re}\left[e^{j(\omega t + \varphi_V)}\right] = \mathrm{Re}\left[V_m e^{j\varphi_V} e^{(\sigma + j\omega)t}\right]. \tag{1.79}$$

Now, we introduce the voltage phasor $\hat{V}$,

$$\hat{V} = V_m e^{j\varphi_V}, \tag{1.80}$$

as well as the complex frequency

$$s = \sigma + j\omega. \tag{1.81}$$

By using the last two formulas in equation (1.79), we find

$$\boxed{v(t) = \mathrm{Re}\left[\hat{V} e^{st}\right].} \tag{1.82}$$

The last formula extends the notion of phasors to voltages (1.78) which are characterized by complex frequency s. Similarly, for a current of complex frequency s we have

$$i(t) = I_m e^{\sigma t} \cos(\omega t + \varphi_I) = \mathrm{Re}\left[\hat{I} e^{st}\right], \tag{1.83}$$

where as before $\hat{I} = I_m e^{j\varphi_I}$. Now, it can be shown that the phasor terminal relations for resistors, inductors and capacitors in the case of voltages and currents of the same complex frequency s are similar to formulas (1.35), (1.41) and (1.47). Indeed, in the case of a resistor we have

$$V_m e^{\sigma t} \cos(\omega t + \varphi_V) = R I_m e^{\sigma t} \cos(\omega t + \varphi_I). \tag{1.84}$$

By using formulas (1.79)-(1.83), we derive

$$\mathrm{Re}\left[\hat{V} e^{st}\right] = \mathrm{Re}\left[R\hat{I} e^{st}\right], \tag{1.85}$$

which implies that

$$\hat{V} = R\hat{I}. \tag{1.86}$$

In the case of an inductor, we find

$$V_m e^{\sigma t}\cos(\omega t + \varphi_V) = \sigma L I_m e^{\sigma t}\cos(\omega t + \varphi_I) + \omega L I_m e^{\sigma t}\cos\left(\omega t + \varphi_I + \frac{\pi}{2}\right). \tag{1.87}$$

The last equality can be transformed as follows:

$$\mathrm{Re}\left[\hat{V}e^{st}\right] = \mathrm{Re}\left[(\sigma + j\omega)L\hat{I}e^{st}\right], \tag{1.88}$$

or

$$\mathrm{Re}\left[\hat{V}e^{st}\right] = \mathrm{Re}\left[sL\hat{I}e^{st}\right], \tag{1.89}$$

which implies that

$$\hat{V} = sL\hat{I}. \tag{1.90}$$

By using the same line of reasoning, it can be shown that in the case of the capacitor we have

$$\hat{V} = \frac{1}{sC}\hat{I}. \tag{1.91}$$

By using this algebraization of terminal relations, it can be demonstrated that in the case of an RLC branch subject to a voltage (1.78) of complex frequency s we have the branch current of the same complex frequency and their phasors are related by the impedance which is a function of the same complex frequency. Namely,

$$\hat{V} = \hat{I}Z(s), \tag{1.92}$$

where

$$Z(s) = R + sL + \frac{1}{sC}. \tag{1.93}$$

In applications, it is quite rare that electric circuits are excited by sources of complex frequency. However, voltages and currents of complex frequency regularly appear in the case of transients in electric circuits. For this reason, the notion of complex frequency as well as the notion of impedance $Z(s)$ as a function of complex frequency s can be useful in the analysis of transients in electric circuits. Indeed, it can be shown that the complex frequencies of transient response in RLC circuits are zeros of impedance $Z(s) = 0$. The detailed discussion of this matter is beyond the scope of this section.

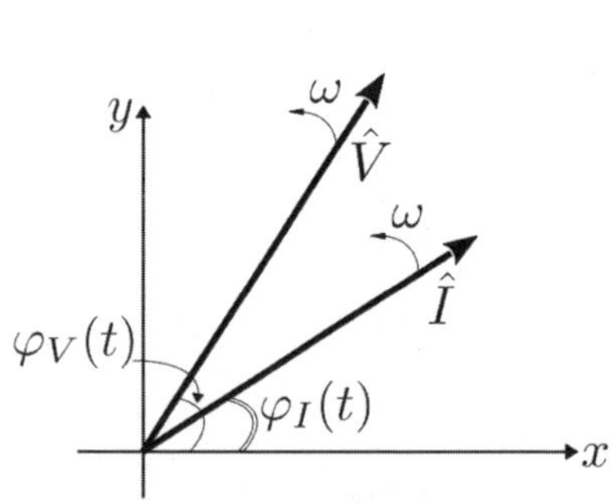

Fig. 1.10

1.3 Phasor Diagrams

Phasor diagrams are ubiquitous in electric power engineering. They provide geometric visualization for time phase shifts between different sinusoidal quantities and for their peak values. The starting point in the discussion of phasor diagrams is the representation of a sinusoidal quantity by a uniformly rotating vector. Consider a time-harmonic voltage

$$v(t) = V_m \cos(\omega t + \varphi_V). \tag{1.94}$$

We can represent this voltage by a vector $\hat{V}$ (called a phasor) whose length is equal to the peak value V_m of $v(t)$ and whose initial angle with the x-axis of the Cartesian coordinate system is equal to the initial phase φ_V of $v(t)$. This angle is called "initial" because the vector $\hat{V}$ is uniformly rotating in the counterclockwise direction with constant angular velocity equal to the angular frequency ω of $v(t)$. Thus, at time t the angle $\varphi_V(t)$ between the vector $\hat{V}$ and the x-axis (see Figure 1.10) is equal to

$$\varphi_V(t) = \omega t + \varphi_V, \tag{1.95}$$

i.e., angle $\varphi_V(t)$ is the sum of the initial angle φ_V and the angle ωt through which the vector has rotated over the time t. If at any instant of time we consider the projection of this vector onto the x-axis, it is clear that this projection is equal to the instantaneous value of voltage $v(t)$. In this sense, $v(t)$ is represented by the uniformly rotating vector $\hat{V}$. Now consider a time-harmonic current of the same angular frequency ω,

$$i(t) = I_m \cos(\omega t + \varphi_I). \tag{1.96}$$

It can also be represented by a uniformly rotating vector (phasor) $\hat{I}$ whose length is equal to the peak value I_m of $i(t)$ and whose initial angle with the x-axis is equal to the initial phase φ_I of $i(t)$. Since this vector is uniformly

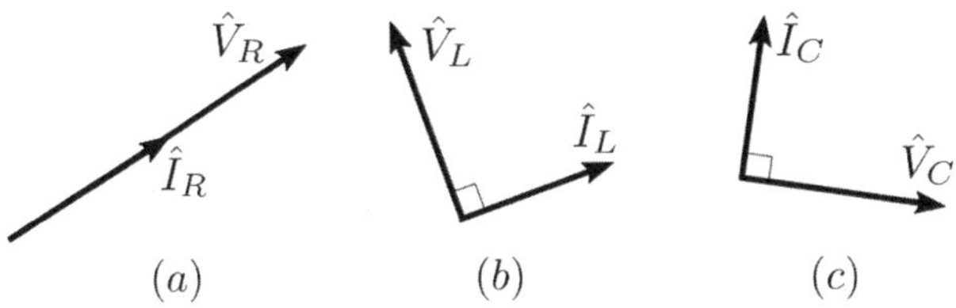

Fig. 1.11

rotating in the counterclockwise direction with angular velocity ω, the angle $\varphi_I(t)$ between $\hat{I}$ and the x-axis at time t (see Figure 1.10) is equal to

$$\varphi_I(t) = \omega t + \varphi_I. \tag{1.97}$$

Again, it is apparent that at any instant of time t the projection of vector $\hat{I}$ onto the x-axis is equal to the instantaneous value of current $i(t)$. Now, the important fact emerges. If we consider the angle φ between the two rotating vectors $\hat{V}$ and $\hat{I}$, we find according to formulas (1.95) and (1.97) that

$$\varphi = \varphi_V(t) - \varphi_I(t) = \varphi_V - \varphi_I = const. \tag{1.98}$$

Thus, this angle φ does not change with time and it is equal to the phase shift in time between sinusoidal voltage $v(t)$ and sinusoidal current $i(t)$. In this sense, one may say that the rotation of the vectors does not matter because it does not change the lengths of the vectors and the angle between them. These lengths and the angle are the most important because they represent the peak values of the sinusoidal quantities and their phase shift in time. For this reason, the rotation of vectors (phasors) can be completely ignored and we represent sinusoidal quantities by vectors (phasors) whose lengths are equal to the peak values of the sinusoidal quantities and the angles between the vectors are equal to the phase shifts in time between the sinusoidal quantities. This is the central idea of the phasor diagrams, i.e., to represent the phase shifts in time by geometric angles between the vectors (phasors). This idea helps to visualize different relations between sinusoidal quantities and to use geometry in calculations.

Phasor diagrams can be constructed for complicated electric circuits. These constructions are based on generic phasor diagrams for the three basic two-terminal elements (resistor, inductor and capacitor). These generic phasor diagrams are shown in Figure 1.11. Consider first the resistor. According to formula (1.33), time-harmonic voltage across the resistor and time-harmonic current through the resistor are in phase, i.e., the phase shift in time between the voltage and current is equal to zero. That is

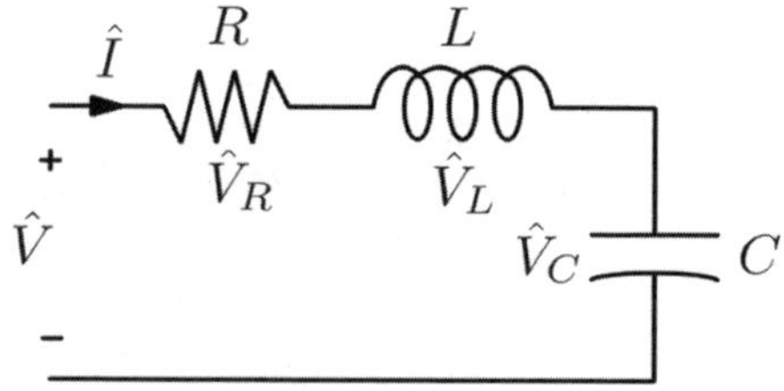

Fig. 1.12

why a generic phasor diagram for the resistor has the form shown in Figure 1.11a. This diagram is called generic because it reflects the main feature, zero phase shift, while the lengths of phasors $\hat{V}_R$ and $\hat{I}_R$ may vary from problem to problem. Now consider the inductor. According to formula (1.39), time-harmonic voltage across the inductor leads the time-harmonic current through the inductor by $\frac{\pi}{2}$. For this reason, in the generic phasor diagram the geometric angle between vectors $\hat{V}_L$ and $\hat{I}_L$ is equal to $\frac{\pi}{2}$ (as shown in Figure 1.11b) and vector $\hat{V}_L$ leads vector $\hat{I}_L$ in the sense of counterclockwise rotation. Finally, consider the capacitor. According to formula (1.45), time-harmonic current through the capacitor leads the time-harmonic voltage across the capacitor by $\frac{\pi}{2}$. This means that in the generic phasor diagram the geometric angle between vectors $\hat{V}_C$ and $\hat{I}_C$ is equal to $\frac{\pi}{2}$ (as shown in Figure 1.11c) and vector $\hat{I}_C$ leads vector $\hat{V}_C$ in the sense of counterclockwise rotation.

Now, through several examples we shall demonstrate how the phasor diagrams can be constructed for actual circuits.

Example 1. Consider the *RLC* circuit shown in Figure 1.12 excited by time-harmonic voltage. First, we shall write the KVL equation for this circuit,

$$\hat{V} = \hat{V}_R + \hat{V}_L + \hat{V}_C. \tag{1.99}$$

This equation implies that in the phasor diagram vector $\hat{V}$ is the vectorial sum of vectors $\hat{V}_R$, $\hat{V}_L$ and $\hat{V}_C$. As a general rule, we start the construction of the phasor diagram by identifying the quantity which is common to the resistor, inductor and capacitor. This quantity is the current, so we shall first draw vector $\hat{I}$. Then, according to the generic diagram 1.11a, the vector $\hat{V}_R$ has the same direction as vector $\hat{I}$ (see Figure 1.13), while according to the generic diagrams 1.11b and 1.11c vectors $\hat{V}_L$ and $\hat{V}_C$ are shifted from vector $\hat{I}$ by $\frac{\pi}{2}$ in counterclockwise and clockwise directions, respectively. By performing the vector addition of the three vectors $\hat{V}_R$, $\hat{V}_L$ and $\hat{V}_C$, we

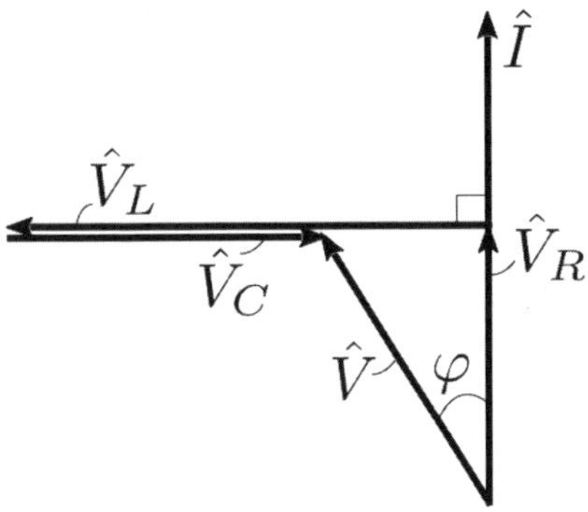

Fig. 1.13

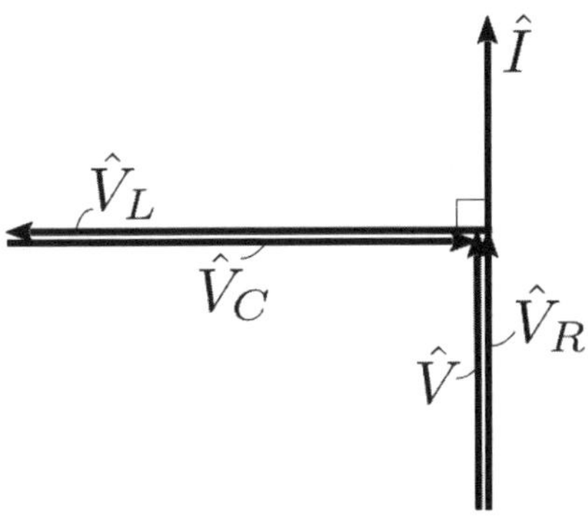

Fig. 1.14

arrive at the final form of the phasor diagram shown in Figure 1.13. In this diagram, the geometric angle φ between vectors $\hat{V}$ and $\hat{I}$ represents the phase shift in time between the input voltage and input current.

It is of interest to consider a particular form of this phasor diagram corresponding to the case of resonance. The resonance occurs under the condition specified by formula (1.73). At resonance, the input voltage and current are in phase. This leads to the phasor diagram shown in Figure 1.14. It is apparent from this figure that the voltages across the inductor and capacitor have the same peak values but opposite phases (i.e., shifted in phase by π). For this reason, they compensate one another at any instant of time. It is also apparent from Figure 1.14 that the peak values of voltages across the inductor and capacitor may be much higher than the peak value of the input voltage. This may never happen in dc circuits.

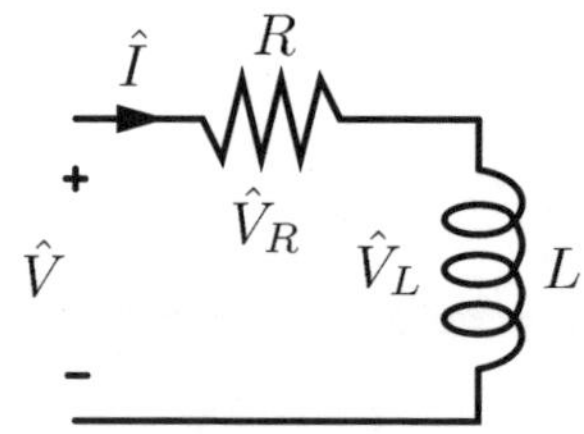

Fig. 1.15

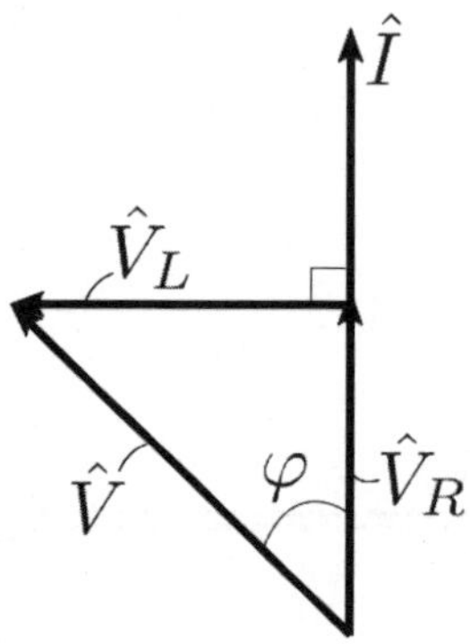

Fig. 1.16

Example 2. Consider the RL circuit shown in Figure 1.15. Suppose that by using a voltmeter the peak values of the input voltage and the voltage across the resistor have been measured and found to be 50 V and 30 V, respectively. The question is what the peak value of the voltage across the inductor and the phase shift in time between the input voltage and input current are.

The visualization of the phasor diagram for the above circuit appreciably facilitates and simplifies the solution of this problem. A well-versed person will visualize this phasor diagram in mind (without actually drawing it) to come up with the immediate answers that the peak value of the voltage across the inductor is 40 V, while the phase shift in time between the input voltage and input current is arctan 4/3. For pedagogical reasons, we draw this phasor diagram shown in Figure 1.16, which is a particular case ($\hat{V}_C = 0$) of the phasor diagram shown in Figure 1.13. From the right triangle in Figure 1.16, the above-stated solution of the problem becomes immediately apparent.

Example 3. Consider a more complicated circuit shown in Figure 1.17.

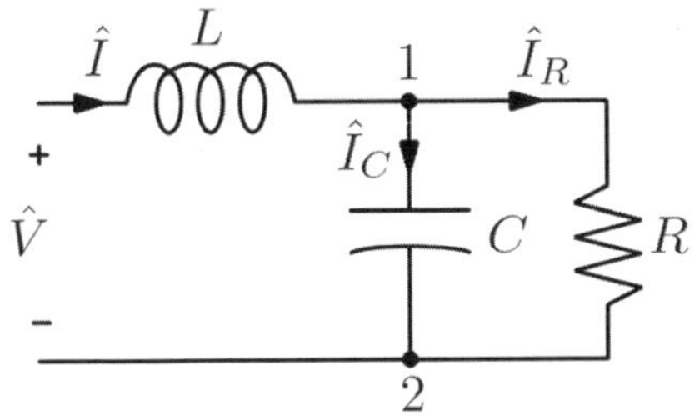

Fig. 1.17

First, we shall write the relevant KCL and KVL equations:

$$\hat{I} = \hat{I}_C + \hat{I}_R, \tag{1.100}$$

$$\hat{V} = \hat{V}_L + \hat{V}_{12}, \tag{1.101}$$

where $\hat{V}_{12}$ is the phasor of the voltage across the nodes 1 and 2. These equations imply that vectorial sums of $\hat{I}_C$ with $\hat{I}_R$ and $\hat{V}_L$ with $\hat{V}_{12}$ will result in $\hat{I}$ and $\hat{V}$, respectively. Second, as a general rule, we start the construction of the phasor diagram from the "end" of the circuit where a resistor and a capacitor are connected in parallel and we identify the voltage $\hat{V}_{12}$ as the common quantity for R and C. So, we start the construction of the phasor diagram by drawing the vector $\hat{V}_{12}$. According to the generic phasor diagrams shown in Figures 1.11a and 1.11c, vector $\hat{I}_R$ has the same direction as $\hat{V}_{12}$, while vector $\hat{I}_C$ is perpendicular to $\hat{V}_{12}$ and "leads" it as far as counterclockwise rotation is concerned. According to (1.100), the vectorial sum of $\hat{I}_R$ and $\hat{I}_C$ results in $\hat{I}$ as shown in Figure 1.18. According to the generic diagram shown in Figure 1.11b, vector $\hat{V}_L$ is perpendicular to $\hat{I}$ and leads it in the sense of counterclockwise rotation. According to equation (1.101), the vectorial sum of $\hat{V}_{12}$ and $\hat{V}_L$ results in $\hat{V}$. This completes the construction of the phasor diagram and the geometric angle φ is equal to the phase shift in time between the input voltage and input current.

Example 4. Consider a circuit shown in Figure 1.19. First, we write relevant KCL and KVL equations

$$\hat{V}_{12} = \hat{V}_R + \hat{V}_{L_2}, \tag{1.102}$$

$$\hat{I} = \hat{I}_R + \hat{I}_C, \tag{1.103}$$

$$\hat{V} = \hat{V}_{L_1} + \hat{V}_{12}, \tag{1.104}$$

where, as before, $\hat{V}_{12}$ is the phasor of the voltage across the nodes 1 and 2. Second, as a general rule, we start the construction of the phasor diagram

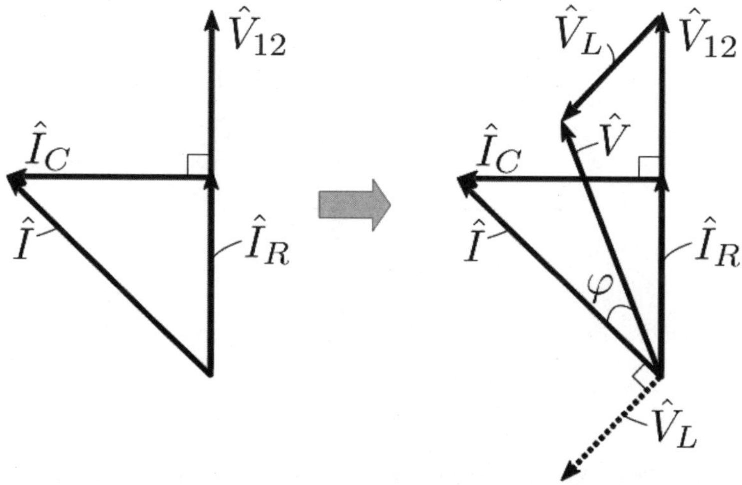

Fig. 1.18

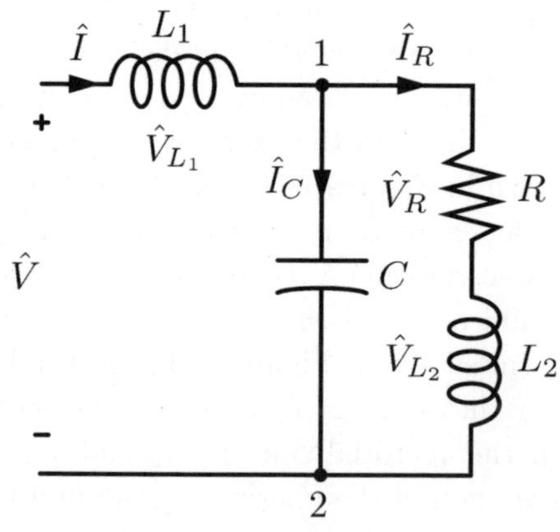

Fig. 1.19

from the "end" of the circuit, i.e., from the branch RL_2 and we identify the current $\hat{I}_R$ as the quantity common to R and L_2. So, we draw vector $\hat{I}_R$. According to the generic phasor diagrams shown in Figure 1.11a and 1.11b, vector $\hat{V}_R$ has the same direction as the vector $\hat{I}_R$, while vector $\hat{V}_{L_2}$ is perpendicular to the vector $\hat{I}_R$ and its orientation reflects that the voltage across the inductor leads the current by $\frac{\pi}{2}$. According to equation (1.102), the vectorial sum of $\hat{V}_R$ and $\hat{V}_{L_2}$ results in $\hat{V}_{12}$ (see Figure 1.20). According to the generic diagram shown in Figure 1.11c, vector $\hat{I}_C$ is per-

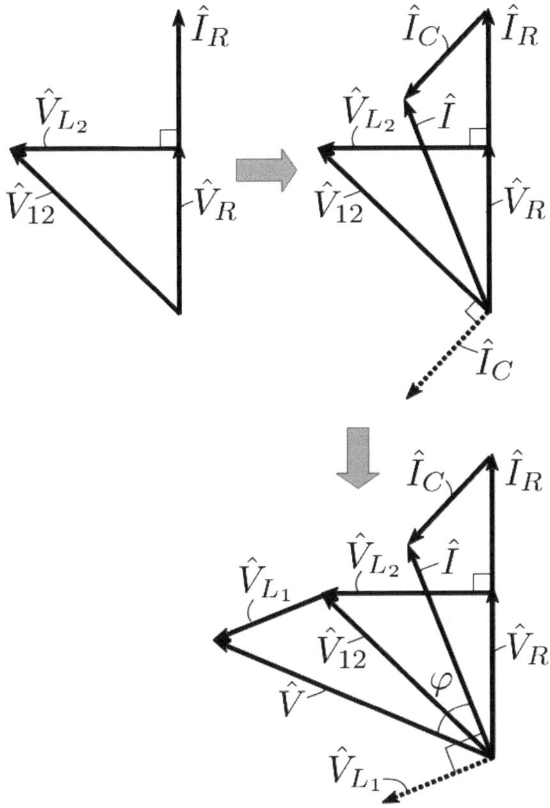

Fig. 1.20

pendicular to vector $\hat{V}_{12}$ and "leads" it in the sense of counterclockwise rotation. The vectorial sum of $\hat{I}_C$ and $\hat{I}_R$ results according to equation (1.103) in $\hat{I}$. Finally, according to the generic diagram shown in Figure 1.11b, vector $\hat{V}_{L_1}$ is perpendicular to vector $\hat{I}$ and "leads" it. According to equation (1.104), the vectorial sum of $\hat{V}_{L_1}$ and $\hat{V}_{12}$ results in $\hat{V}$, and this completes the construction of the phasor diagram.

This concludes the discussion of the phasor diagrams, which will be extensively used throughout this text.

Chapter 2

Analysis of Electric Circuits with Periodic Non-sinusoidal Sources

2.1 Fourier Series Analysis

In power electronics, analysis of steady-state performance of switching-mode power converters is reduced to the analysis of electric circuits excited by time-periodic non-sinusoidal sources. There are two analytical techniques that will be extensively used in this text for the analysis of such circuits. The first one is the frequency-domain technique, which is based on the Fourier series expansions of time-periodic functions (sources) and subsequent use of the phasor technique. The second one is the time-domain technique, which is based on the formulation of the steady-state circuit analysis as a boundary value problem for ordinary differential equations with periodic boundary conditions.

We begin with the frequency-domain technique and we first review in this section the basic facts related to the Fourier series. These series are used for periodic functions. Below, we consider periodic functions of time, which is relevant to circuit analysis. However, the Fourier series are also very instrumental in the design of windings of synchronous and induction machines discussed in the second part of this book. In that case, Fourier series are used for the expansion of periodic functions in space rather than in time.

Function $f(t)$ is said to be periodic with period T (see Figure 2.1) if

$$f(t+T) = f(t). \tag{2.1}$$

It is apparent that a multiple of T by any natural number is a period of $f(t)$ as well. It will be assumed in the following that T is the smallest period of $f(t)$. The fundamental angular frequency

$$\omega = \frac{2\pi}{T} \tag{2.2}$$

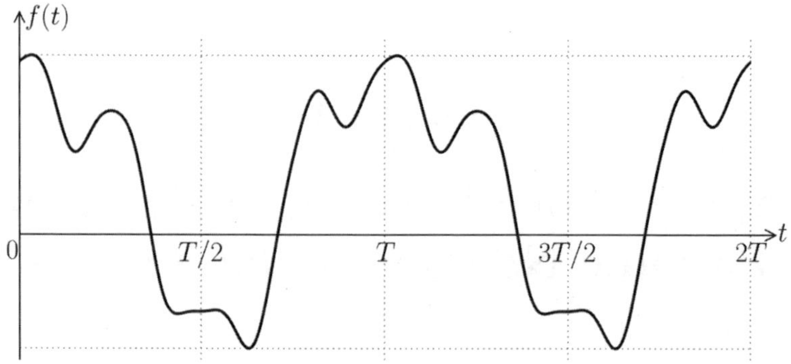

Fig. 2.1

will be associated with this period. It turns out that under some general conditions periodic functions can be expanded into trigonometric Fourier series

$$\boxed{f(t) = c_0 + \sum_{n=1}^{\infty} [a_n \cos n\omega t + b_n \sin n\omega t]\,,} \tag{2.3}$$

where numbers c_0, a_n and b_n are called the Fourier expansion coefficients. The next step is to find the expression for these expansion coefficients in terms of function $f(t)$. This can be accomplished by using the following "orthogonality" relations for trigonometric functions:

$$\int_0^T \cos n\omega t dt = 0, \tag{2.4}$$

$$\int_0^T \sin n\omega t dt = 0, \tag{2.5}$$

$$\int_0^T \cos n\omega t \sin m\omega t dt = 0, \tag{2.6}$$

$$\int_0^T \cos n\omega t \cos m\omega t dt = \frac{T}{2}\delta_{nm}, \tag{2.7}$$

$$\int_0^T \sin n\omega t \sin m\omega t dt = \frac{T}{2}\delta_{nm}, \tag{2.8}$$

where the symbol δ_{nm} is the so-called Kronecker delta defined as follows:

$$\delta_{nm} = \begin{cases} 1, & \text{if } n = m, \\ 0, & \text{if } n \neq m. \end{cases} \tag{2.9}$$

These orthogonality conditions are easy to prove. Formulas (2.4) and (2.5) are immediately obvious because functions $\cos n\omega t$ and $\sin n\omega t$ are periodic with period $\frac{T}{n}$, and the integrals of sine and cosine functions over any number of periods are equal to zero. The proof of formulas (2.6), (2.7) and (2.8) is only slightly more complicated. This proof is based on the well-known trigonometric identities which reduce products of sine and cosine functions to sums of these functions with modified arguments, and then formulas similar to (2.4) and (2.5) can be used. This line of reasoning can be followed when $n \neq m$. In the case when $n = m$, the trigonometric identities relating squares of cosine and sine functions to cosine functions of double argument can be used to arrive at formulas (2.7) and (2.8). We omit the details of the outlined derivations and encourage the reader to perform these derivations, which will be beneficial for proper understanding of the material.

In calling formulas (2.4)-(2.8) orthogonality conditions, we use geometric language which implies the analogy between formulas (2.4)-(2.8) and orthogonality of vectors. In this sense, the functional set consisting of a constant function and trigonometric functions is an orthogonal functional set, and formulas (2.4) and (2.5) can be understood as orthogonality relations between a constant function and cosine and sine functions. Orthogonal vectors are linearly independent and may be used as bases for expansions of an arbitrary vector. Similarly, the functional set consisting of a constant function and cosine and sine functions can be used as the functional basis for the expansion of an arbitrary function $f(t)$, and formula (2.3) can be understood as such an expansion. The Fourier expansion coefficients can be interpreted as "projections" of $f(t)$ on the "axes" identified with a constant function and cosine and sine functions. This interpretation is consistent with the following formulas for the expansion coefficients:

$$c_0 = \frac{1}{T}\int_0^T f(t)dt, \tag{2.10}$$

$$a_n = \frac{2}{T}\int_0^T f(t)\cos n\omega t dt, \tag{2.11}$$

$$b_n = \frac{2}{T}\int_0^T f(t)\sin n\omega t dt. \tag{2.12}$$

These formulas can be derived by using the orthogonality relations. Indeed, by integrating both sides of formula (2.3) over T and by using formulas (2.4) and (2.5) we arrive at equation (2.10). In geometric language, it can be said that c_0 is the projection of $f(t)$ on the constant (unity) function. Similarly, by multiplying both sides of formula (2.3) by $\cos n\omega t$, subsequently integrating both sides over T and using the orthogonality relations (2.4), (2.6) and (2.7), we arrive at the formula (2.11). In geometric language, it can be said that a_n is the projection of $f(t)$ on $\cos n\omega t$. Finally, by multiplying both sides of formula (2.3) by $\sin n\omega t$, subsequently integrating both sides over T and using orthogonality relations (2.5), (2.6) and (2.8), we arrive at the formula (2.12). In geometric language, it can be said that b_n is the projection of $f(t)$ on $\sin n\omega t$.

Next, the following two remarks are in order.

Remark 1. Integrands in formulas (2.10), (2.11) and (2.12) are periodic functions of period T. It is apparent from the geometric point of view that integrals of such functions over period T do not depend on the location of the time interval of integration. Consequently, the last three formulas can also be written in the following form:

$$c_0 = \frac{1}{T}\int_{-\frac{T}{2}}^{\frac{T}{2}} f(t)dt, \tag{2.13}$$

$$a_n = \frac{2}{T}\int_{-\frac{T}{2}}^{\frac{T}{2}} f(t)\cos n\omega t dt, \tag{2.14}$$

$$b_n = \frac{2}{T}\int_{-\frac{T}{2}}^{\frac{T}{2}} f(t)\sin n\omega t dt. \tag{2.15}$$

Remark 2. Under some general conditions, expansion coefficients a_n and b_n tend to zero with the increase of n, i.e.,

$$\lim_{n\to\infty} a_n = 0, \qquad \lim_{n\to\infty} b_n = 0. \tag{2.16}$$

It is interesting to note that formulas (2.16) are valid despite the fact that the integrands in formulas (2.11) and (2.12) do not tend to zero. This raises the question of the generic intuitive explanation for the validity of relations (2.16). The reason is that the functions $\cos n\omega t$ and $\sin n\omega t$ become very fast oscillating as n is increased. In other words, the period $\frac{T}{n}$ of these functions becomes smaller and smaller. A sufficiently regular (normal) function $f(t)$ can be accurately approximated by a piecewise constant function with constant values in each time interval of $\frac{T}{n}$-duration. According to (2.4) and (2.5), for such piecewise constant functions, the integrals in (2.11) and (2.12) are equal to zero. A simple and rigorous proof of the validity of relations (2.16) can be given by assuming that $f(t)$ is differentiable and using

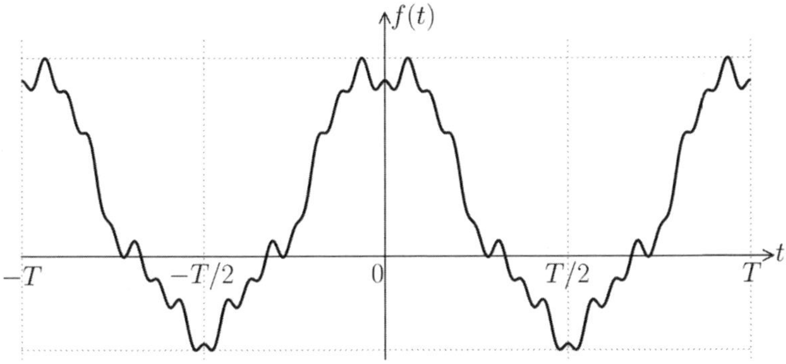

Fig. 2.2

integration by parts in formulas (2.11) and (2.12). The reader is encouraged to carry out this proof.

A periodic function $f(t)$ may have some symmetry properties. These symmetry properties may result in substantial simplifications of Fourier series (2.3). Below, we shall discuss three types of symmetries: even symmetry, odd symmetry and half-wave symmetry.

a) Even symmetry

A function $f(t)$ is called even (or even symmetric) if for any t we have

$$f(t) = f(-t). \tag{2.17}$$

An example of the graph of such function is shown in Figure 2.2, and it is clear that this graph exhibits mirror symmetry with respect to the vertical axis. It is apparent from formula (2.17) as well as from Figure 2.2 that for any even function we have

$$\int_{-\frac{T}{2}}^{\frac{T}{2}} f(t)dt = 2\int_{0}^{\frac{T}{2}} f(t)dt. \tag{2.18}$$

b) Odd symmetry

A function $f(t)$ is called odd (or odd symmetric) if for any t we have

$$f(t) = -f(-t). \tag{2.19}$$

An example of the graph of such function is shown in Figure 2.3, and it is clear that this graph exhibits rotational symmetry with respect to the origin. It is apparent from formula (2.19) as well as from Figure 2.3 that for any odd function we have

$$\int_{-\frac{T}{2}}^{\frac{T}{2}} f(t)dt = 0. \tag{2.20}$$

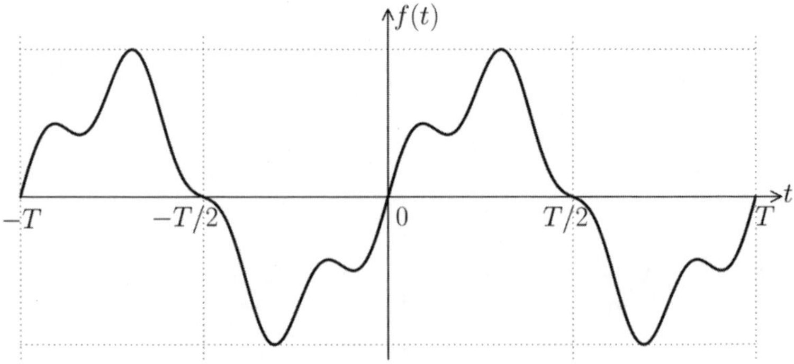

Fig. 2.3

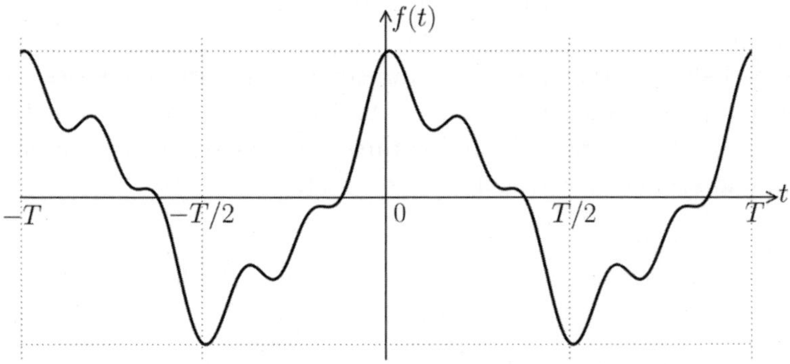

Fig. 2.4

Remark 3. It follows from the definitions (2.17) and (2.19) that the product of two even or two odd functions is an even function, while the product of an even function and an odd function is an odd function.

c) Half-wave symmetry

A periodic function $f(t)$ is called a half-wave symmetric function if for any t we have

$$f\left(t+\frac{T}{2}\right)=-f(t), \tag{2.21}$$

where T is the period of $f(t)$.

An example of the graph of such function is shown in Figure 2.4. It is apparent from the definition (2.21) and Figure 2.4 that a half-wave symmetric function has two identical but of opposite sign half-cycles.

For this reason,

$$\int_0^T f(t)dt = 0. \tag{2.22}$$

Remark 4. Half-wave symmetric functions are related to functions of period $\frac{T}{2}$. Indeed, it is easy to see that the product of two half-wave symmetric functions is a function of period $\frac{T}{2}$, while the product of a half-wave symmetric function and a function of period $\frac{T}{2}$ is a half-wave symmetric function.

Now, we demonstrate symmetry-related simplifications of Fourier series (2.3). We start with the case when $f(t)$ is an even function. We recall that $\cos n\omega t$ are even functions, while $\sin n\omega t$ are odd functions. From the last observation and Remark 3, we find that the integrands in formulas (2.13) and (2.14) are even functions, while integrands in formula (2.15) are odd functions. From these facts and formulas (2.18) and (2.20), we conclude that

$$b_n = 0, \tag{2.23}$$

$$f(t) = c_0 + \sum_{n=1}^{\infty} a_n \cos n\omega t, \tag{2.24}$$

$$c_0 = \frac{2}{T}\int_0^{\frac{T}{2}} f(t)dt, \tag{2.25}$$

$$a_n = \frac{4}{T}\int_0^{\frac{T}{2}} f(t)\cos n\omega t dt. \tag{2.26}$$

The achieved simplification is twofold. First, only a constant term and cosine terms are present in the Fourier series expressions. Second, the calculation of c_0 and a_n requires the evaluation of integrals over $\frac{T}{2}$ rather than over T.

Next, we consider the case when $f(t)$ is an odd function. In this case, according to Remark 3 the integrands in formulas (2.13) and (2.14) are odd functions, while the integrands in formulas (2.15) are even functions. From these facts and formulas (2.18) and (2.20), we conclude respectively

$$c_0 = 0, \tag{2.27}$$

$$a_n = 0, \tag{2.28}$$

$$f(t) = \sum_{n=1}^{\infty} b_n \sin n\omega t, \tag{2.29}$$

$$b_n = \frac{4}{T}\int_0^{\frac{T}{2}} f(t)\sin n\omega t dt. \tag{2.30}$$

Finally, we consider the case when $f(t)$ is a half-wave symmetric function. According to formulas (2.22) and (2.10), we immediately find

$$c_0 = 0. \tag{2.31}$$

To achieve further simplification in Fourier series expansions, we observe that for odd n functions $\cos n\omega t$ and $\sin n\omega t$ are half-wave symmetric functions, while for even n functions $\cos n\omega t$ and $\sin n\omega t$ are periodic with period $\frac{T}{2}$. From the last observation and Remark 4, we find that the integrands in formulas (2.11) and (2.12) are of half-wave symmetry for even n, while these integrands are periodic functions of period $\frac{T}{2}$ for odd n. Thus, according to (2.22) all even Fourier coefficients are equal to zero, while odd Fourier coefficients are not equal to zero and integration over T can be reduced to double the integration over $\frac{T}{2}$. Thus, we arrive at the following simplification of Fourier series:

$$f(t) = \sum_{n=0}^{\infty} \left[a_{2n+1}\cos(2n+1)\omega t + b_{2n+1}\sin(2n+1)\omega t\right], \tag{2.32}$$

$$a_{2n+1} = \frac{4}{T}\int_0^{\frac{T}{2}} f(t)\cos(2n+1)\omega t dt, \tag{2.33}$$

$$b_{2n+1} = \frac{4}{T}\int_0^{\frac{T}{2}} f(t)\sin(2n+1)\omega t dt. \tag{2.34}$$

It is worthwhile to mention that the case of half-wave symmetric functions is encountered quite often in various electric power-related applications.

We conclude this section by the discussion of an alternative form of Fourier series. This form is very convenient for the coupling of the Fourier series expansion with the phasor technique. This coupling is the foundation of the frequency-domain technique for the analysis of electric circuits with periodic non-sinusoidal sources.

Consider one term of the infinite sum in Fourier series expansion (2.3) and perform the following transformations:

$$a_n \cos n\omega t + b_n \sin n\omega t = \sqrt{a_n^2 + b_n^2}\left(\frac{a_n}{\sqrt{a_n^2 + b_n^2}} \cos n\omega t + \frac{b_n}{\sqrt{a_n^2 + b_n^2}} \sin n\omega t\right). \quad (2.35)$$

Next, we introduce φ_n by the formulas

$$\cos \varphi_n = \frac{a_n}{\sqrt{a_n^2 + b_n^2}}, \quad (2.36)$$

$$\sin \varphi_n = -\frac{b_n}{\sqrt{a_n^2 + b_n^2}}, \quad (2.37)$$

which means that

$$\tan \varphi_n = -\frac{b_n}{a_n}. \quad (2.38)$$

It is easy to see that this introduction is consistent with trigonometric identity

$$\cos^2 \varphi_n + \sin^2 \varphi_n = 1. \quad (2.39)$$

We shall also introduce the notation

$$c_n = \sqrt{a_n^2 + b_n^2}. \quad (2.40)$$

By substituting formulas (2.36), (2.37) and (2.39) into equation (2.35), we obtain

$$\begin{aligned} a_n \cos n\omega t + b_n \sin n\omega t &= c_n(\cos \varphi_n \cos n\omega t - \sin \varphi_n \sin n\omega t) \\ &= c_n \cos(n\omega t + \varphi_n). \end{aligned} \quad (2.41)$$

By using the last relation in formula (2.3), we arrive at the following alternative form of the Fourier series:

$$\boxed{f(t) = c_0 + \sum_{n=1}^{\infty} c_n \cos(n\omega t + \varphi_n),} \quad (2.42)$$

where c_0, c_n and φ_n can be computed by using the following formulas:

$$c_0 = \frac{1}{T}\int_0^T f(t)dt, \quad (2.43)$$

$$a_n = \frac{2}{T}\int_0^T f(t) \cos n\omega t dt, \quad (2.44)$$

$$b_n = \frac{2}{T}\int_0^T f(t) \sin n\omega t dt, \quad (2.45)$$

$$c_n = \sqrt{a_n^2 + b_n^2}, \quad (2.46)$$

$$\tan \varphi_n = -\frac{b_n}{a_n}. \quad (2.47)$$

This concludes the review of the Fourier series.

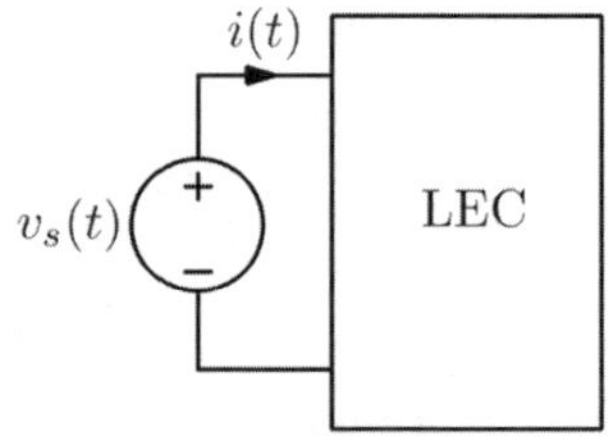

Fig. 2.5

2.2 Frequency-Domain Technique

In this section, we consider the frequency-domain technique for the analysis of steady-state regimes of electric circuits excited by periodic non-sinusoidal sources. This technique is based on the combined usage of Fourier series expansions and phasors. We first present the general description and justification of the frequency domain technique and then we illustrate this technique by two examples.

Consider an electric circuit shown in Figure 2.5. Here, $v_s(t)$ is a given periodic non-sinusoidal voltage source

$$v_s(t+T) = v_s(t), \tag{2.48}$$

while LEC is the abbreviation for a generic linear electric circuit with given lumped parameters. It is required to find the input electric current $i(t)$ in the steady-state regime, i.e.,

$$i(t+T) = i(t). \tag{2.49}$$

The frequency-domain technique for the solution of the stated problem consists of the following three steps.

Step 1. The given periodic function $v_s(t)$ is expanded into Fourier series by using formulas (2.42)-(2.47). These formulas in the notation relevant to our problem can be written as follows:

$$v_s(t) = V_{s0} + \sum_{n=1}^{\infty} V_{sn}\cos(n\omega t + \varphi_{sn}), \tag{2.50}$$

$$V_{s0} = \frac{1}{T}\int_0^T v_s(t)dt, \tag{2.51}$$

$$a_n = \frac{2}{T}\int_0^T v_s(t)\cos n\omega t dt, \tag{2.52}$$

$$b_n = \frac{2}{T}\int_0^T v_s(t)\sin n\omega t dt, \tag{2.53}$$

$$V_{sn} = \sqrt{a_n^2 + b_n^2}, \tag{2.54}$$

$$\tan\varphi_{sn} = -\frac{b_n}{a_n}, \tag{2.55}$$

where

$$\omega = \frac{2\pi}{T}. \tag{2.56}$$

Each term in the expansion (2.50) can be interpreted as a voltage source:

$$v_{sn}(t) = V_{sn}\cos(n\omega t + \varphi_{sn}), \tag{2.57}$$

$$v_s(t) = V_{s0} + \sum_{n=1}^{\infty} v_{sn}(t), \tag{2.58}$$

and the given voltage source can be interpreted as the series connection of these voltage sources (see Figure 2.6).

Step 2. Next, we shall use the superposition principle illustrated in Figure 2.7 and consider the current $i(t)$ as the sum of currents I_0 and $i_n(t)$ excited in LEC when only one of the voltage sources V_{s0} or $v_{sn}(t)$, respectively, is active:

$$i(t) = I_0 + \sum_{n=1}^{\infty} i_n(t). \tag{2.59}$$

The calculation of I_0 requires dc analysis of LEC subject to dc voltage V_{s0}. In this analysis, inductors in LEC are replaced by short-circuit branches, while capacitors are replaced by open-circuit branches. Thus, the determination of I_0 is reduced to the dc analysis of the resistive electric circuit corresponding to LEC (see Figure 2.8).

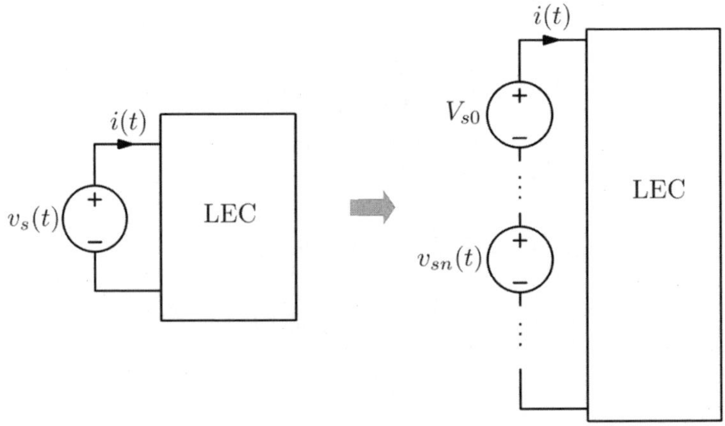

Fig. 2.6

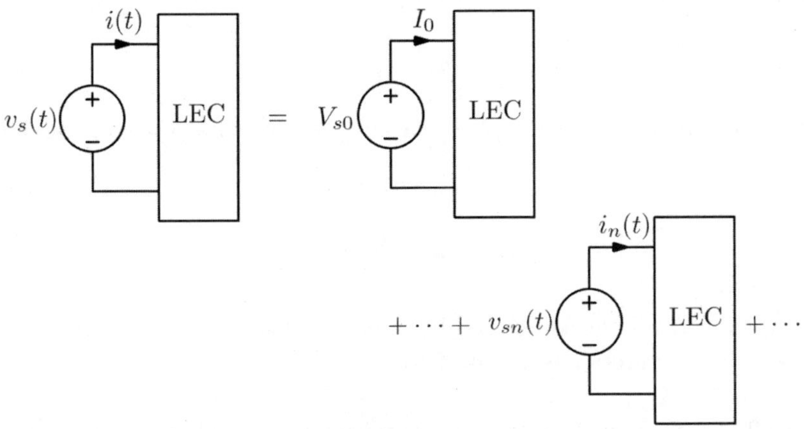

Fig. 2.7

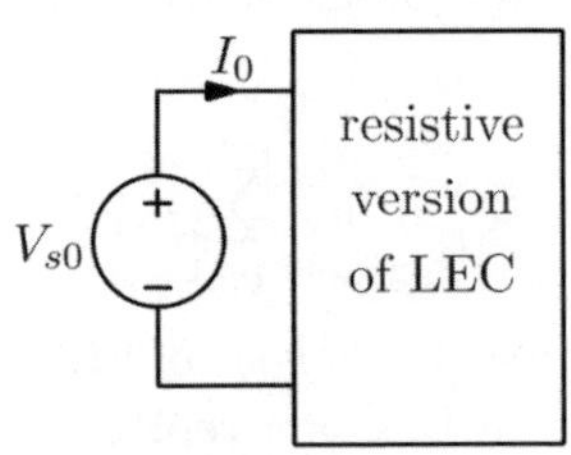

Fig. 2.8

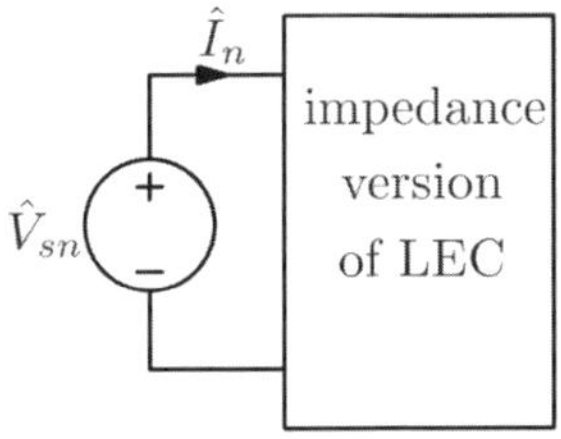

Fig. 2.9

The calculation of $i_n(t)$ can be carried out by using the phasor technique. Namely, the time-harmonic voltage source $v_{sn}(t)$ is represented by the phasor

$$v_{sn}(t) = V_{sn}\cos(n\omega t + \varphi_{sn}) \to \hat{V}_{sn} = V_{sn}e^{j\varphi_{sn}}, \tag{2.60}$$

each branch of LEC is represented by its impedance evaluated at the angular frequency $n\omega$ and the phasor analysis technique is used to find the phasor $\hat{I}_n$ in the impedance version of LEC (see Figure 2.9). Having found the phasor $\hat{I}_n = I_{mn}e^{j\varphi_{I_n}}$, the current $i_n(t)$ can be written as follows:

$$\hat{I}_n = I_{mn}e^{j\varphi_{I_n}} \to i_n(t) = I_{mn}\cos(n\omega t + \varphi_{I_n}). \tag{2.61}$$

Step 3. By using formulas (2.59) and (2.61), the final expression for $i(t)$ can be represented in the form

$$i(t) = I_0 + \sum_{n=1}^{\infty} I_{mn}\cos(n\omega t + \varphi_{I_n}). \tag{2.62}$$

We conclude the general description of the frequency-domain technique with the following two remarks.

Remark 1. In many power electronics-related applications, the first term in the right-hand side of formula (2.62) can be interpreted as the main desired signal, while the infinite sum in (2.62) can be interpreted as undesirable "ripple." Thus, the frequency-domain technique leads to the clear separation between the main desired component of the signal and its ripple.

Remark 2. In the presented general description of the frequency-domain technique, the calculation of the input current $i(t)$ was discussed. It is easy to see that the same three steps can be applied to the calculation of any branch current or any branch voltage of LEC.

Now, we shall illustrate the frequency-domain technique by two examples.

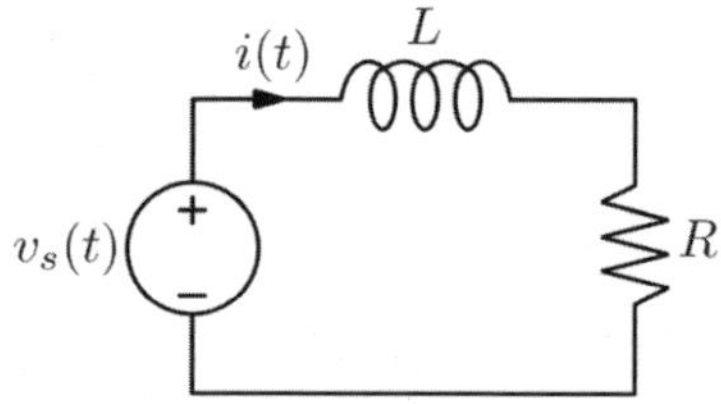

Fig. 2.10

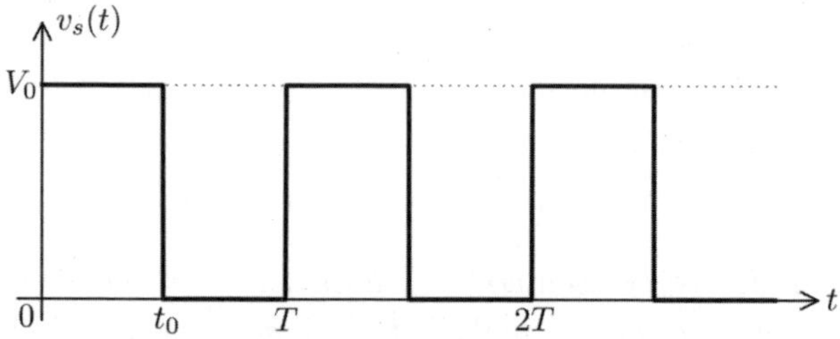

Fig. 2.11

Example 1. Consider the electric circuit shown in Figure 2.10, where the voltage source $v_s(t)$ is a periodic sequence (train) of identical rectangular pulses (see Figure 2.11). Thus, V_0, T, t_0, L and R are given, and it is required to find $i(t)$ and $v_R(t)$ in the steady-state regime, i.e.,

$$i(t+T) = i(t), \qquad v_R(t+T) = v_R(t). \tag{2.63}$$

Step 1. We represent $v_s(t)$ as the following Fourier series:

$$v_s(t) = V_{s0} + \sum_{n=1}^{\infty} V_{sn} \cos(n\omega t + \varphi_{sn}), \tag{2.64}$$

where

$$\omega = \frac{2\pi}{T}. \tag{2.65}$$

To find V_{s0}, V_{sn} and φ_{sn}, we sequentially use the formulas (2.51)-(2.55).

$$V_{s0} = \frac{1}{T}\int_0^T v_s(t)dt = \frac{1}{T}\int_0^{t_0} V_0 dt = \frac{t_0}{T}V_0. \tag{2.66}$$

By introducing the notation D for so-called duty factor,

$$D = \frac{t_0}{T}, \tag{2.67}$$

we find

$$\boxed{V_{s0} = DV_0.} \tag{2.68}$$

Next,

$$a_n = \frac{2}{T}\int_0^T v_s(t)\cos n\omega t dt = \frac{2V_0}{T}\int_0^{t_0} \cos n\omega t dt$$
$$= \frac{2V_0}{T}\left(\frac{1}{n\omega}\sin n\omega t\right)\Big|_0^{t_0} = \frac{2V_0}{n\omega T}\sin n\omega t_0. \tag{2.69}$$

Taking into account in the last formula the relation (2.65), we find

$$a_n = \frac{V_0}{\pi n}\sin n\omega t_0. \tag{2.70}$$

Similarly,

$$b_n = \frac{2}{T}\int_0^T v_s(t)\sin n\omega t dt = \frac{2V_0}{T}\int_0^{t_0} \sin n\omega t dt$$
$$= \frac{2V_0}{T}\left(-\frac{1}{n\omega}\cos n\omega t\right)\Big|_0^{t_0} = \frac{2V_0}{n\omega T}(1 - \cos n\omega t_0). \tag{2.71}$$

Invoking again the relation (2.65), the last formula can be written as follows:

$$b_n = \frac{V_0}{\pi n}(1 - \cos n\omega t_0). \tag{2.72}$$

Next,

$$V_{sn} = \sqrt{a_n^2 + b_n^2} = \frac{V_0}{\pi n}\sqrt{\sin^2 n\omega t_0 + (1 - \cos n\omega t_0)^2}$$
$$= \frac{V_0}{\pi n}\sqrt{2(1 - \cos n\omega t_0)} = \frac{V_0}{\pi n}\sqrt{4\sin^2\frac{n\omega t_0}{2}}, \tag{2.73}$$

which implies that

$$\boxed{V_{sn} = \frac{2V_0}{\pi n}\left|\sin\frac{n\omega t_0}{2}\right|.} \tag{2.74}$$

Finally,

$$\boxed{\tan\varphi_{sn} = \frac{\cos n\omega t_0 - 1}{\sin n\omega t_0}.} \tag{2.75}$$

Thus, the explicit analytical expressions are found for V_{s0}, V_{sn} and φ_{sn}. This concludes this first step.

Step 2. First consider the dc analysis of the resistive version of the electric circuit shown in Figure 2.10. This is illustrated by Figure 2.12. This analysis is trivial and results in

$$I_0 = \frac{V_{s0}}{R} = \frac{DV_0}{R}, \qquad V_{R0} = DV_0. \tag{2.76}$$

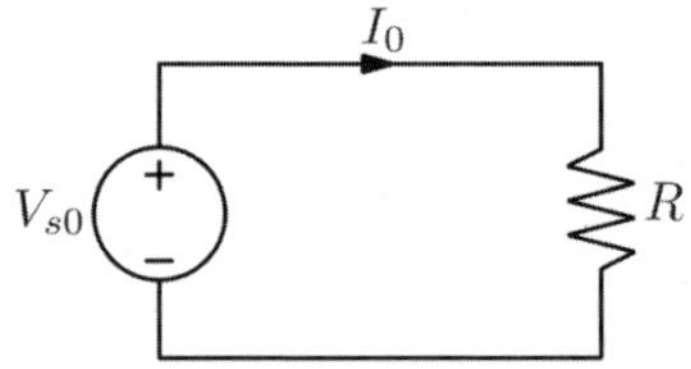

Fig. 2.12

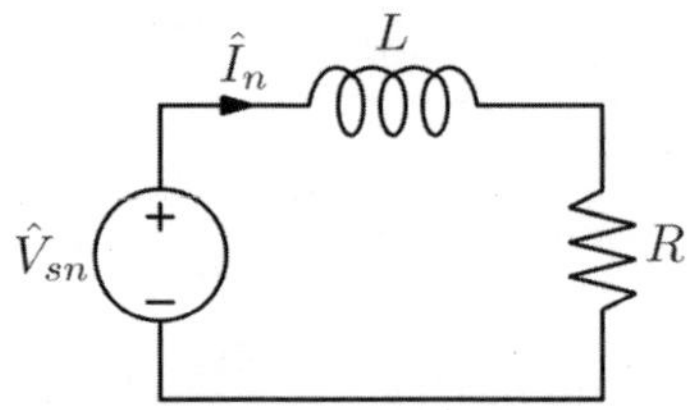

Fig. 2.13

Second, consider the phasor analysis of ac steady state in the circuit shown in Figure 2.13 at the frequency $n\omega$. Here, we find

$$\hat{V}_{sn} = V_{sn} e^{j\varphi_{sn}}, \tag{2.77}$$

and V_{sn} and φ_{sn} are given by formulas (2.74) and (2.75). Furthermore,

$$\hat{I}_n = \frac{\hat{V}_{sn}}{Z_n}, \tag{2.78}$$

where

$$Z_n = R + jn\omega L = \sqrt{R^2 + n^2\omega^2 L^2} e^{j\varphi_n} \tag{2.79}$$

and

$$\tan\varphi_n = \frac{n\omega L}{R}. \tag{2.80}$$

By substituting formulas (2.77) and (2.79) into equation (2.78), we end up with

$$\hat{I}_n = \frac{V_{sn}}{\sqrt{R^2 + n^2\omega^2 L^2}} e^{j(\varphi_{sn} - \varphi_n)}. \tag{2.81}$$

From the last formula, we find

$$i_n(t) = \frac{V_{sn}}{\sqrt{R^2 + n^2\omega^2 L^2}} \cos(n\omega t + \varphi_{sn} - \varphi_n). \tag{2.82}$$

Step 3. The input current $i(t)$ is the sum of currents I_0 and $i_n(t)$ for all ac steady-state regimes:

$$i(t) = I_0 + \sum_{n=1}^{\infty} i_n(t), \tag{2.83}$$

or

$$\boxed{i(t) = \frac{DV_0}{R} + \frac{2V_0}{\pi} \sum_{n=1}^{\infty} \frac{\left|\sin \frac{n\omega t_0}{2}\right|}{n\sqrt{R^2 + n^2\omega^2 L^2}} \cos(n\omega t + \varphi_{sn} - \varphi_n),} \tag{2.84}$$

where we have used formula (2.74) for V_{sn} in the equation (2.82). By multiplying the last formula by R we find the voltage across the resistor

$$\boxed{v_R(t) = DV_0 + \frac{2V_0 R}{\pi} \sum_{n=1}^{\infty} \frac{\left|\sin \frac{n\omega t_0}{2}\right|}{n\sqrt{R^2 + n^2\omega^2 L^2}} \cos(n\omega t + \varphi_{sn} - \varphi_n).} \tag{2.85}$$

In some applications

$$\omega L \gg R, \tag{2.86}$$

and

$$\sqrt{R^2 + n^2\omega^2 L^2} \approx n\omega L. \tag{2.87}$$

This leads to the following simplification of formula (2.85):

$$\boxed{v_R(t) \approx DV_0 + \frac{2V_0 R}{\pi\omega L} \sum_{n=1}^{\infty} \frac{\left|\sin \frac{n\omega t_0}{2}\right|}{n^2} \cos(n\omega t + \varphi_{sn} - \varphi_n).} \tag{2.88}$$

It is clear that due to the inequality (2.86) the second term (the ripple) in the right-hand side of formula (2.88) is small. It is also clear from the last formula that the ripple suppression is controlled by the product ωL. This implies the trade-off for ripple suppression between the value of L and the frequency of switching used to produce the train of rectangular pulses shown in Figure 2.11. We point out that this trade-off has already been discussed in general terms in section 1 of Chapter 1.

Example 2. Consider the electric circuit shown in Figure 2.14, where the voltage source $v_s(t)$ is a periodic function of time shown in Figure 2.11. It is assumed that V_0, T, t_0, R, L and C are given, and it is required to find voltage $v_R(t)$ across the resistor in the steady-state regime, i.e.,

$$v_R(t + T) = v_R(t). \tag{2.89}$$

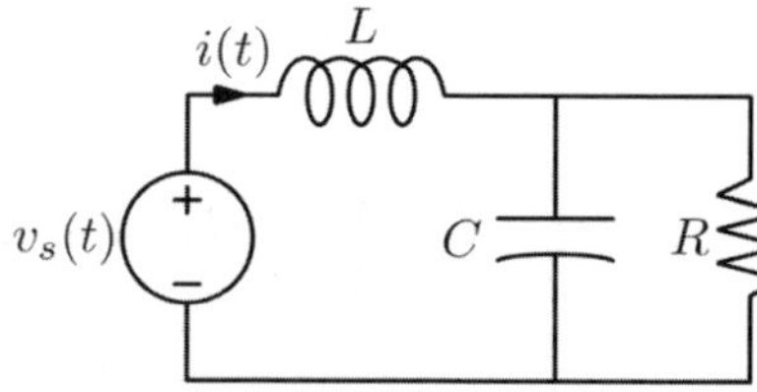

Fig. 2.14

Step 1. As before, we represent $v_s(t)$ by the Fourier series

$$v_s(t) = V_{s0} + \sum_{n=1}^{\infty} V_{sn} \cos(n\omega t + \varphi_{sn}), \tag{2.90}$$

where $\omega = \frac{2\pi}{T}$ and V_{s0}, V_{sn} and φ_{sn} can be computed by using formulas (2.68), (2.74) and (2.75). In other words, the first step in this example is identical to the first step in the first example, because the electric circuits in these examples are excited by identical voltage sources.

Step 2. The dc analysis of the electric circuit shown in Figure 2.14 and subject to dc voltage source V_{s0} (instead of voltage source $v_s(t)$) leads to the circuit shown in Figure 2.12. The result of this analysis is obvious:

$$V_{R0} = V_{s0} = DV_0. \tag{2.91}$$

Next, we consider the phasor analysis of ac steady state in the electric circuit shown in Figure 2.15 at the frequency $n\omega$. Here, as before,

$$\hat{V}_{sn} = V_{sn} e^{j\varphi_{sn}} \tag{2.92}$$

and V_{sn} and φ_{sn} are found in the first step (see formulas (2.74) and (2.75)). It is apparent that the phasor $\hat{I}_n$ of the input current is equal to

$$\hat{I}_n = \frac{\hat{V}_{sn}}{Z_n}, \tag{2.93}$$

where Z_n is the input impedance of the electric circuit shown in Figure 2.15 at the frequency $n\omega$. It is clear that this input impedance is found as follows:

$$Z_n = jn\omega L + \frac{-\frac{j}{n\omega C} R}{-\frac{j}{n\omega C} + R}. \tag{2.94}$$

By using simple algebraic transformations, we find

$$Z_n = jn\omega L + \frac{R}{1 + jn\omega CR} = \frac{R - n^2\omega^2 LCR + jn\omega L}{1 + jn\omega CR}. \tag{2.95}$$

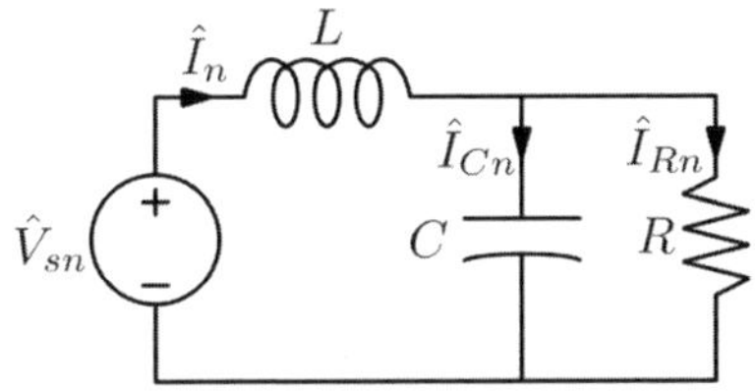

Fig. 2.15

By substituting the last formula into (2.93), we arrive at

$$\hat{I}_n = \hat{V}_{sn} \frac{1 + jn\omega CR}{R - n^2\omega^2 LCR + jn\omega L}. \tag{2.96}$$

Now, by using the current divider rule, we obtain

$$\hat{I}_{Rn} = \hat{I}_n \frac{\frac{-j}{n\omega C}}{\frac{-j}{n\omega C} + R} = \hat{I}_n \frac{1}{1 + jn\omega CR}. \tag{2.97}$$

By substituting formula (2.96) into the last equation, we find

$$\hat{I}_{Rn} = \frac{\hat{V}_{sn}}{R - n^2\omega^2 LCR + jn\omega L}, \tag{2.98}$$

which leads to

$$\hat{V}_{Rn} = \hat{V}_{sn} \frac{R}{R - n^2\omega^2 LCR + jn\omega L}. \tag{2.99}$$

By using formula (2.92) and simple transformations, the last equation can be written as follows:

$$\hat{V}_{Rn} = \frac{V_{sn} R}{\sqrt{\left(n^2\omega^2 LCR - R\right)^2 + n^2\omega^2 L^2}} e^{j(\varphi_{sn} - \varphi_n)}, \tag{2.100}$$

where

$$\tan \varphi_n = \frac{n\omega L}{R - n^2\omega^2 LCR}. \tag{2.101}$$

From formula (2.100) we obtain

$$v_{Rn}(t) = \frac{V_{sn} R}{\sqrt{\left(n^2\omega^2 LCR - R\right)^2 + n^2\omega^2 L^2}} \cos(n\omega t + \varphi_{sn} - \varphi_n). \tag{2.102}$$

Step 3. Now, by using the superposition principle, we arrive at

$$v_R(t) = V_{R0} + \sum_{n=1}^{\infty} v_{Rn}(t). \tag{2.103}$$

By using in the last equation formula (2.91) for V_{R0} and formula (2.102) for $v_{Rn}(t)$ as well as formula (2.74) for V_{sn}, we obtain the final expression for $v_R(t)$:

$$\boxed{v_R(t) = DV_0 + \frac{2V_0 R}{\pi} \sum_{n=1}^{\infty} \frac{\left|\sin \frac{n\omega t_0}{2}\right|}{n\sqrt{\left(n^2\omega^2 LCR - R\right)^2 + n^2\omega^2 L^2}} \cos(n\omega t + \varphi_{sn} - \varphi_n).} \tag{2.104}$$

In certain applications, the second term in the right-hand side of the last equation can be construed as a ripple. This ripple will be effectively suppressed if the lumped parameters of the circuit shown in Figure 2.14 are chosen in such a way that

$$\omega^2 LC \gg 1 \text{ and } \omega^2 C^2 R^2 \gg 1. \tag{2.105}$$

The last inequalities imply, respectively, that

$$n^2\omega^2 LCR \gg R \tag{2.106}$$

and

$$n^4\omega^4 L^2 C^2 R^2 \gg n^2\omega^2 L^2. \tag{2.107}$$

Formulas (2.106) and (2.107) mean that

$$\sqrt{\left(n^2\omega^2 LCR - R\right)^2 + n^2\omega^2 L^2} \approx n^2\omega^2 LCR. \tag{2.108}$$

This leads to the following simplification of equation (2.104):

$$\boxed{v_R(t) \approx DV_0 + \frac{2V_0}{\pi\omega^2 LC} \sum_{n=1}^{\infty} \frac{\left|\sin \frac{n\omega t_0}{2}\right|}{n^3} \cos(n\omega t + \varphi_{sn} - \varphi_n).} \tag{2.109}$$

Now, it is apparent that more efficient suppression of ripple can be achieved in the circuit shown in Figure 2.14 in comparison with the circuit shown in Figure 2.10. Indeed, each term in the infinite sum of formula (2.109) decays as $1/n^3$ rather than $1/n^2$ as in formula (2.88). In addition, the suppression of the ripple in formula (2.109) is controlled by the product $\omega^2 LC$ rather than by the product ωL. This means that the increase in switching frequency suppresses the ripple more efficiently in the circuit shown in Figure 2.14 as compared with the circuit shown in Figure 2.10. Finally, the dependence of ripple suppression on $\omega^2 LC$ reveals, as before, the trade-off between the values of energy storage elements L and C and the frequency of switching.

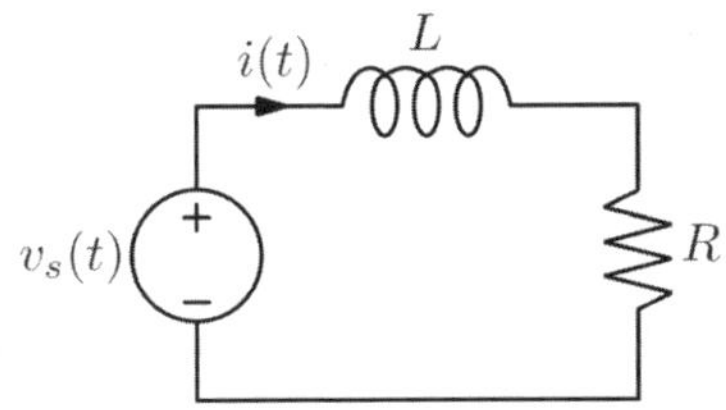

Fig. 2.16

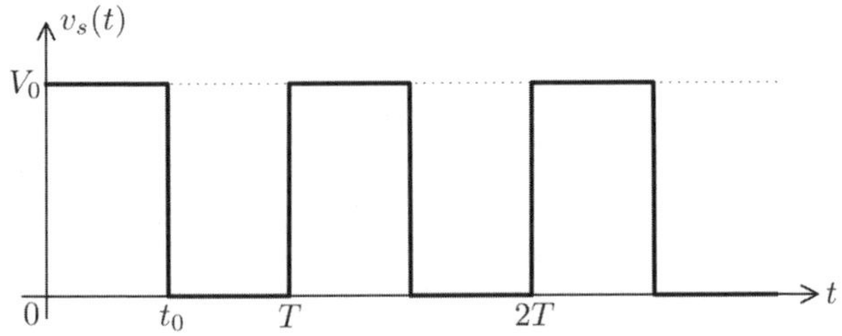

Fig. 2.17

2.3 Time-Domain Technique

Now, we proceed to the discussion of the time-domain technique for the analysis of steady-state regimes of linear electric circuits excited by periodic non-sinusoidal sources. This technique is based on the formulation of steady-state analysis as a boundary value problem for ordinary differential equations with periodic boundary conditions. We illustrate this technique by the following two examples.

Example 1. Consider the electric circuit shown in Figure 2.16 excited by the voltage source $v_s(t)$, where $v_s(t)$ is the periodic function of time shown in Figure 2.17. Here, V_0, T, t_0, L and R are given, and it is required to find $i(t)$ and $v_R(t)$ which are periodic with period T:

$$i(t+T) = i(t), \qquad v_R(t+T) = v_R(t). \tag{2.110}$$

It is apparent that this example is identical to Example 1 from the previous section. This is done on purpose in order for the reader to compare advantages and disadvantages of the time-domain and frequency-domain techniques.

The time-domain technique can be presented as a sequence of three distinct steps.

Step 1. is to formulate the problem of steady-state analysis as a boundary value problem with periodic boundary conditions. To this end, we first write the KVL for the circuit shown in Figure 2.16:

$$L\frac{di(t)}{dt} + Ri(t) = v_s(t). \tag{2.111}$$

It is apparent from Figure 2.17 that the last equation can be written as two distinct equations for two time intervals:

$$\boxed{\begin{aligned} &L\frac{di(t)}{dt} + Ri(t) = V_0, \quad \text{if } 0 < t < t_0, \\ &L\frac{di(t)}{dt} + Ri(t) = 0, \quad \text{if } t_0 < t < T. \end{aligned}} \tag{2.112, 2.113}$$

These two equations can be complemented by two conditions

$$\boxed{\begin{aligned} i(0) &= i(T), \\ i(t_{0-}) &= i(t_{0+}). \end{aligned}} \tag{2.114, 2.115}$$

The equation (2.114) is the periodic boundary condition which follows from (2.110) for $t = 0$, while the equation (2.115) is the interface condition which expresses the continuity of electric current through the inductor.

The last four equations constitute the boundary value problem for differential equations (2.112) and (2.113) with periodic boundary and interface conditions (2.114) and (2.115), respectively. As soon as the solution of this boundary value problem is found, the value of $i(t)$ can be found at any time (i.e., not only in time interval $[0, T]$) by using the first equation in (2.110). Indeed, by using this equation, we can extend the solution from the time interval $[0, T]$ to the time interval $[T, 2T]$, and then from the time interval $[T, 2T]$ to the time interval $[2T, 3T]$ and so on. In other words, by using the first equation in (2.110), the solution of the boundary value problem (2.112)-(2.115) with periodic boundary condition can be periodically extended to the infinite time interval. It is clear that this periodically extended solution has the physical meaning of the steady state in the electric circuit shown in Figure 2.16.

Step 2. is to find general solutions of differential equations (2.112) and (2.113). We start with equation (2.112). This is a linear inhomogeneous equation of first order with constant coefficients. Its general solution has two distinct components: a particular solution $i_p(t)$ of inhomogeneous equation (2.112) and a general solution $i_h(t)$ of the corresponding homogeneous equation. Namely,

$$i(t) = i_p(t) + i_h(t), \tag{2.116}$$

where

$$L\frac{di_p(t)}{dt} + Ri_p(t) = V_0, \tag{2.117}$$

while

$$L\frac{di_h(t)}{dt} + Ri_h(t) = 0. \tag{2.118}$$

In mathematics, the particular solution of inhomogeneous differential equation (2.117) is sought in the same form as the right-hand side of the equation:

$$i_p(t) = B = const. \tag{2.119}$$

By substituting formula (2.119) into equation (2.117) we find

$$RB = V_0, \qquad B = \frac{V_0}{R} \tag{2.120}$$

and

$$i_p(t) = \frac{V_0}{R}. \tag{2.121}$$

It is apparent that $i_p(t)$ is identical to the dc steady state in the electric circuit shown in Figure 2.16 excited by the dc voltage V_0. This observation is very helpful and can be used for the calculation of particular solutions of differential equations for more complicated circuits excited by dc or ac voltage sources. In the latter case, the phasor technique can be used for the calculation of particular solutions.

Now, we proceed to the calculation of $i_h(t)$ by using equation (2.118). We look for a solution of this equation in the form

$$i_h(t) = A_1 e^{st}, \tag{2.122}$$

where A_1 and s are some constants. By substituting the last formula into equation (2.118), we arrive at

$$sLA_1e^{st} + RA_1e^{st} = 0, \tag{2.123}$$

which leads to

$$sL + R = 0 \tag{2.124}$$

and

$$s = -\frac{R}{L}. \tag{2.125}$$

Thus, the general solution $i_h(t)$ has the form

$$i_h(t) = A_1 e^{-\frac{R}{L}t}. \tag{2.126}$$

By substituting formulas (2.121) and (2.126) into equation (2.116), we find the general solution of equation (2.112):

$$i(t) = \frac{V_0}{R} + A_1 e^{-\frac{R}{L}t}, \qquad \text{if } 0 < t < t_0. \tag{2.127}$$

Next, we consider the general solution of equation (2.113). This equation is identical in structure to equation (2.118). Consequently, its general solution has the form

$$i(t) = A_2 e^{-\frac{R}{L}t}, \qquad \text{if } t_0 < t < T, \tag{2.128}$$

where A_2 is some constant.

Step 3. is to find constants A_1 and A_2 from the periodic boundary condition (2.114) and interface boundary condition (2.115). These conditions lead to the following equations:

$$\begin{cases} \dfrac{V_0}{R} + A_1 = A_2 e^{-\frac{RT}{L}}, & (2.129) \\ \dfrac{V_0}{R} + A_1 e^{-\frac{Rt_0}{L}} = A_2 e^{-\frac{Rt_0}{L}}. & (2.130) \end{cases}$$

Indeed, by using formula (2.127) for the evaluation of $i(0)$ and formula (2.128) for the evaluation of $i(T)$, we end up with equation (2.129). Similarly, by using equation (2.127) for the evaluation of $i(t_{0-})$ and formula (2.128) for the evaluation of $i(t_{0+})$, we arrive at equation (2.130). The last two equations can be easily solved. Indeed, these equations can be written as follows:

$$\begin{cases} A_1 - A_2 e^{-\frac{RT}{L}} = -\dfrac{V_0}{R}, & (2.131) \\ A_1 - A_2 = -\dfrac{V_0}{R} e^{\frac{Rt_0}{L}}. & (2.132) \end{cases}$$

By subtracting the second equation from the first, we find

$$A_2 \left(1 - e^{-\frac{RT}{L}}\right) = \frac{V_0}{R} \left(e^{\frac{Rt_0}{L}} - 1\right) \tag{2.133}$$

and

$$A_2 = \frac{V_0}{R} \frac{e^{\frac{Rt_0}{L}} - 1}{1 - e^{-\frac{RT}{L}}}. \tag{2.134}$$

Now, by substituting the last formula into equation (2.132), after simple transformations we obtain

$$A_1 = \frac{V_0}{R} \frac{e^{\frac{R(t_0 - T)}{L}} - 1}{1 - e^{-\frac{RT}{L}}}. \tag{2.135}$$

By using the last two formulas in equations (2.127) and (2.128) we arrive at the final expression

$$i(t) = \begin{cases} \dfrac{V_0}{R}\left[1 + \dfrac{e^{\frac{R(t_0-T)}{L}} - 1}{1 - e^{-\frac{RT}{L}}} e^{-\frac{Rt}{L}}\right], & \text{if } 0 < t < t_0, \\ \dfrac{V_0}{R} \dfrac{e^{\frac{Rt_0}{L}} - 1}{1 - e^{-\frac{RT}{L}}} e^{-\frac{Rt}{L}}, & \text{if } t_0 < t < T. \end{cases} \tag{2.136}$$

By taking into account that

$$v_R(t) = i(t)R, \tag{2.137}$$

we find

$$v_R(t) = \begin{cases} V_0\left[1 + \dfrac{e^{\frac{R(t_0-T)}{L}} - 1}{1 - e^{-\frac{RT}{L}}} e^{-\frac{Rt}{L}}\right], & \text{if } 0 < t < t_0, \\ V_0 \dfrac{e^{\frac{Rt_0}{L}} - 1}{1 - e^{-\frac{RT}{L}}} e^{-\frac{Rt}{L}}, & \text{if } t_0 < t < T. \end{cases} \tag{2.138}$$

Formulas (2.136) and (2.138) are the final results of our analysis.

It is interesting to deduce from the last formula the expression for $v_R(t)$ in the case when

$$\frac{RT}{L} \ll 1. \tag{2.139}$$

By using inequality (2.139), we conclude that

$$e^{-\frac{R}{L}t} \approx 1, \qquad \text{if } 0 < t < T, \tag{2.140}$$

$$e^{-\frac{R}{L}(T-t_0)} - 1 \approx 1 - \frac{R}{L}(T - t_0) - 1 = -\frac{R}{L}(T - t_0), \tag{2.141}$$

$$1 - e^{-\frac{R}{L}T} \approx 1 - 1 + \frac{R}{L}T = \frac{R}{L}T, \tag{2.142}$$

$$e^{\frac{R}{L}t_0} - 1 \approx 1 + \frac{R}{L}t_0 - 1 = \frac{R}{L}t_0. \tag{2.143}$$

Consequently,

$$v_R(t) \approx \begin{cases} V_0\left[1 + \dfrac{-\frac{R}{L}(T - t_0)}{\frac{R}{L}T}\right] = V_0\dfrac{t_0}{T} = DV_0, & \text{if } 0 < t < t_0, \\ V_0\dfrac{\frac{R}{L}t_0}{\frac{R}{L}T} = V_0\dfrac{t_0}{T} = DV_0, & \text{if } t_0 < t < T. \end{cases} \tag{2.144}$$

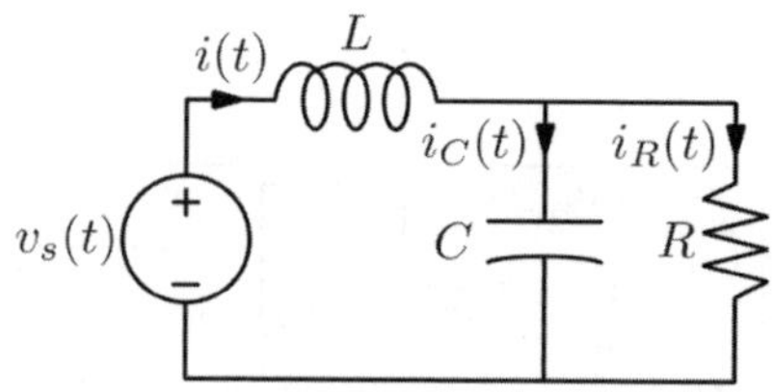

Fig. 2.18

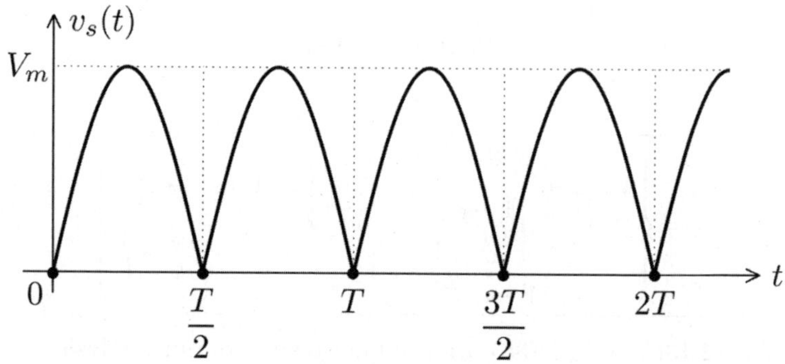

Fig. 2.19

Thus, as expected,

$$v_R(t) \approx DV_0. \tag{2.145}$$

Example 2. Consider the electric circuit shown in Figure 2.18 excited by voltage source $v_s(t)$, where $v_s(t)$ is a periodic function of time given by the formula

$$v_s(t) = V_m|\sin \omega t| \tag{2.146}$$

and shown in Figure 2.19. Here, V_m, $\omega = \frac{2\pi}{T}$, L, C and R are given, and it is required to find $v_R(t)$ at the steady state, i.e.,

$$v_R(t + T/2) = v_R(t). \tag{2.147}$$

This problem is encountered in the analysis of rectifiers in power electronics.

Step 1. First, it is apparent that

$$v_R(t) = v_C(t). \tag{2.148}$$

Then, by using KVL and KCL, we respectively find

$$L\frac{di(t)}{dt} + v_C(t) = v_s(t), \tag{2.149}$$

$$i(t) = i_C(t) + i_R(t) = C\frac{dv_C(t)}{dt} + \frac{v_C(t)}{R}. \tag{2.150}$$

By substituting the last formula into equation (2.149), we obtain

$$LC\frac{d^2v_C(t)}{dt^2} + \frac{L}{R}\frac{dv_C(t)}{dt} + v_C(t) = v_s(t). \tag{2.151}$$

Now we consider the last equation for the time interval between 0 and $\frac{T}{2}$, which is the period of $v_s(t)$. It is clear from formula (2.146) as well as Figure 2.19 that for this time interval

$$v_s(t) = V_m \sin \omega t. \tag{2.152}$$

By substituting this formula into equation (2.151) we end up with

$$\boxed{LC\frac{d^2v_C(t)}{dt^2} + \frac{L}{R}\frac{dv_C(t)}{dt} + v_C(t) = V_m \sin \omega t.} \tag{2.153}$$

This is a second-order differential equation, which is consistent with the fact that the circuit being discussed has two energy storage elements L and C. At the steady state, current $i(t)$ and voltage $v_C(t)$ satisfy the periodic boundary conditions

$$i(0) = i(T/2), \tag{2.154}$$

$$\boxed{v_C(0) = v_C(T/2).} \tag{2.155}$$

From the last two formulas and formula (2.150) we find

$$\boxed{\frac{dv_C}{dt}(0) = \frac{dv_C}{dt}(T/2).} \tag{2.156}$$

Thus, the problem of the analysis of the steady state is reduced to the boundary value problem for the second-order differential equation (2.153) with two periodic boundary conditions (2.155) and (2.156). This concludes the first step.

Step 2. A general solution of differential equation (2.153) is the sum of a particular solution $v_C^{(p)}(t)$ of this equation and a general solution $v_C^{(h)}(t)$ of the corresponding homogeneous equation

$$LC\frac{d^2v_C^{(h)}(t)}{dt^2} + \frac{L}{R}\frac{dv_C^{(h)}(t)}{dt} + v_C^{(h)}(t) = 0. \tag{2.157}$$

We consider the specific particular solution of equation (2.153) which corresponds to the ac steady state of the circuit in Figure 2.18 excited by the voltage source $V_m \sin \omega t$. This particular solution can be found by using the phasor technique. Actually, this has been already done in the previous section when we used the phasor technique to analyze the circuit shown in Figure 2.15. There are only two minor differences. First, the phasor of the voltage source $\hat{V}_s$ in our case is given not by formula (2.92) but by the formula

$$\hat{V}_s = -V_m e^{j\frac{\pi}{2}}. \tag{2.158}$$

Second, in formulas (2.102) and (2.101) n must be omitted. Thus,

$$v_C^{(p)}(t) = -\frac{V_m R}{\sqrt{(\omega^2 LCR - R)^2 + \omega^2 L^2}} \cos\left(\omega t + \frac{\pi}{2} - \varphi\right), \tag{2.159}$$

$$\tan \varphi = \frac{\omega L}{R - \omega^2 LCR}. \tag{2.160}$$

We look for a solution of equation (2.157) in the form

$$v_C^{(h)}(t) = Ae^{st}. \tag{2.161}$$

By substituting the last formula into equation (2.157), we find after simple transformations that

$$LCs^2 + \frac{L}{R}s + 1 = 0. \tag{2.162}$$

This quadratic equation has two roots

$$s_1 = -\frac{1}{2RC} + \sqrt{\frac{1}{4R^2C^2} - \frac{1}{LC}}, \tag{2.163}$$

$$s_2 = -\frac{1}{2RC} - \sqrt{\frac{1}{4R^2C^2} - \frac{1}{LC}}. \tag{2.164}$$

For simplicity consider the case when s_1 and s_2 are real and distinct. Other cases can be treated in a similar way. Then, a general solution of homogeneous equation (2.157) is

$$v_C^{(h)}(t) = A_1 e^{s_1 t} + A_2 e^{s_2 t}, \tag{2.165}$$

and a general solution of equation (2.153) is given by the formula

$$v_C(t) = -\frac{V_m R \cos(\omega t + \frac{\pi}{2} - \varphi)}{\sqrt{(\omega^2 LCR - R)^2 + \omega^2 L^2}} + A_1 e^{s_1 t} + A_2 e^{s_2 t}. \tag{2.166}$$

This concludes the second step.

Step 3. To find constants A_1 and A_2 we use periodic boundary conditions (2.155) and (2.156). This leads, after simple transformations, to the following simultaneous equations:

$$A_1\left(1-e^{\frac{s_1T}{2}}\right)+A_2\left(1-e^{\frac{s_2T}{2}}\right)=\frac{2RV_m\sin\varphi}{\sqrt{\left(\omega^2LCR-R\right)^2+\omega^2L^2}}, \tag{2.167}$$

$$A_1s_1\left(1-e^{\frac{s_1T}{2}}\right)+A_2s_2\left(1-e^{\frac{s_2T}{2}}\right)=-\frac{2\omega RV_m\cos\varphi}{\sqrt{\left(\omega^2LCR-R\right)^2+\omega^2L^2}}. \tag{2.168}$$

The solution of these equations is given by the formulas

$$A_1=\frac{2RV_m(s_2\sin\varphi+\omega\cos\varphi)}{(s_2-s_1)\left(1-e^{\frac{s_1T}{2}}\right)\sqrt{\left(\omega^2LCR-R\right)^2+\omega^2L^2}}, \tag{2.169}$$

$$A_2=\frac{2RV_m(s_1\sin\varphi+\omega\cos\varphi)}{(s_1-s_2)\left(1-e^{\frac{s_2T}{2}}\right)\sqrt{\left(\omega^2LCR-R\right)^2+\omega^2L^2}}. \tag{2.170}$$

By substituting the last two formulas into equation (2.166), we obtain the analytical solution for the steady state of the electric circuit shown in Figure 2.18. This analytical solution is valid for the time interval $\left[0, \frac{T}{2}\right]$. By using formula (2.147), it can be periodically extended for any time interval.

Chapter 3

Magnetic Circuit Theory

3.1 Basic Equations of Magnetic Circuit Theory

Many power devices (such as transformers, generators, motors, relays, etc.) contain magnetic systems consisting of ferromagnetic cores with coils wound around them. A schematic representation of such magnetic systems is shown in Figure 3.1. Ferromagnetic cores are used because their magnetic permeability μ_c is much larger than the magnetic permeability of free space μ_0. For this reason, ferromagnetic cores can be utilized for guiding of magnetic flux. In other words, most of the magnetic field lines are confined to the ferromagnetic core and form the core flux Φ_c, while only a small portion of magnetic field lines leak out and form the leakage flux Φ_ℓ. The rigorous analysis of magnetic systems with ferromagnetic cores is quite complicated and it requires the solution of Maxwell equations of electromagnetic field theory. However, there exists an approximate but rather accurate approach to the analysis of magnetic systems which is called magnetic circuit theory. This approach is justified when the two inequalities below are valid:

$$\mu_c \gg \mu_0, \tag{3.1}$$

$$\Phi_c \gg \Phi_\ell, \tag{3.2}$$

and it is based on the following assumptions (approximations).

- **Assumption 1**. Leakage flux Φ_ℓ is completely neglected.
- **Assumption 2**. Magnetic field is assumed to be spatially uniform within each leg of the iron core and parallel to the sides of the leg. Here, a leg is understood as a part of the iron core with the same normal cross-sectional area. Symbols 1, 2, 3, etc. in Figure 3.1 mark the legs of the core.

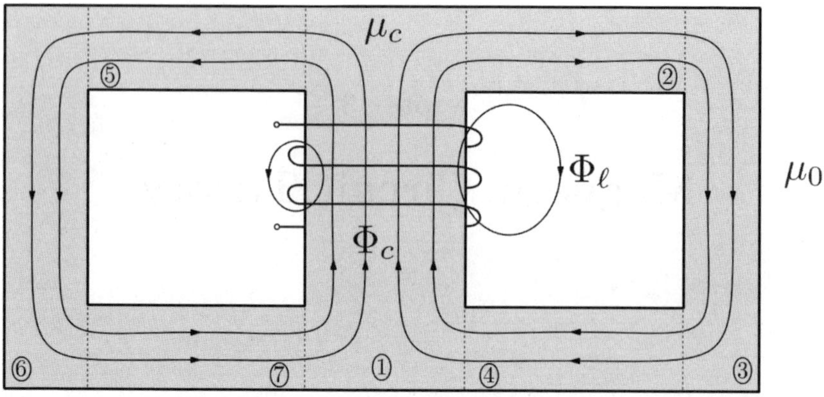

Fig. 3.1

Magnetic circuit theory equations are derived from the integral form of Maxwell equations for stationary magnetic fields and the above Assumptions 1 and 2. The equations of stationary (static) magnetic field in integral form can be stated as follows:

$$\oint_L \mathbf{H} \cdot d\boldsymbol{\ell} = I_{total}, \tag{3.3}$$

$$\oint_S \mathbf{B} \cdot \mathbf{ds} = 0, \tag{3.4}$$

$$\mathbf{B} = \mu \mathbf{H}, \tag{3.5}$$

where $\mathbf{H}$ and $\mathbf{B}$ are magnetic field and magnetic flux density, respectively, while μ is magnetic permeability.

The first equation (3.3) is Ampere's Law, which states that for any closed path L the line integral of magnetic field is equal to the total current (I_{total}) enclosed by this path. This total current is the algebraic sum of all current enclosed by the path of integration. The term "algebraic" implies that some of these currents are taken with positive signs while others are taken with negative signs. The proper signs are assigned by using the right-hand rule, which specifies that an enclosed current is taken with positive sign if its direction coincides with the direction of progress of a right-hand screw turn in the direction of traverse (tracing) of path L; otherwise, an enclosed current is taken with negative sign.

The second equation (3.4) is the principle of continuity of magnetic flux, which states that the magnetic flux through any closed surface S is equal to zero. In formula (3.4), $\mathbf{ds}$ is a vector whose direction coincides with the direction of outward normal to ds.

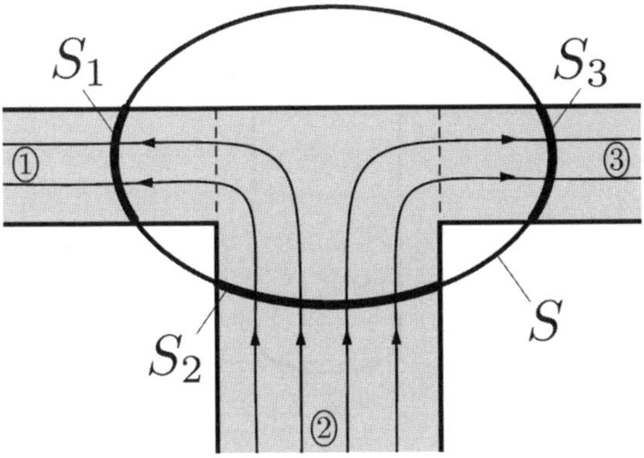

Fig. 3.2

Finally, it must be remarked that equation (3.5) is the simplest form of the constitutive relation, which is valid when ferromagnetic media can be characterized by their magnetic permeability. It will be discussed later on in this chapter that this is not always the case.

Now, we proceed with the derivation of basic equations of magnetic circuit theory. We start with the discussion of the first Kirchhoff's Law for magnetic circuits. This law deals with nodes of magnetic circuits. Here, a node is defined as a place in a ferromagnetic core where three or more legs are connected together (see Figure 3.2). Consider an arbitrary closed surface S which contains the node and apply equation (3.4) to this surface. According to the first assumption, all magnetic flux leakage is neglected. This is tantamount to the assumption that all magnetic field lines are confined to the ferromagnetic core. For this reason, equation (3.4) can be written as follows:

$$\int_{S_1} \mathbf{B}_1 \cdot \mathbf{ds} + \int_{S_2} \mathbf{B}_2 \cdot \mathbf{ds} + \int_{S_3} \mathbf{B}_3 \cdot \mathbf{ds} = 0, \tag{3.6}$$

where S_1, S_2 and S_3 are the parts of S inside of the first, second and third legs, respectively, while $\mathbf{B}_1$, $\mathbf{B}_2$ and $\mathbf{B}_3$ are magnetic flux densities within these legs (see Figure 3.2). It is clear that the integrals in (3.6) are related

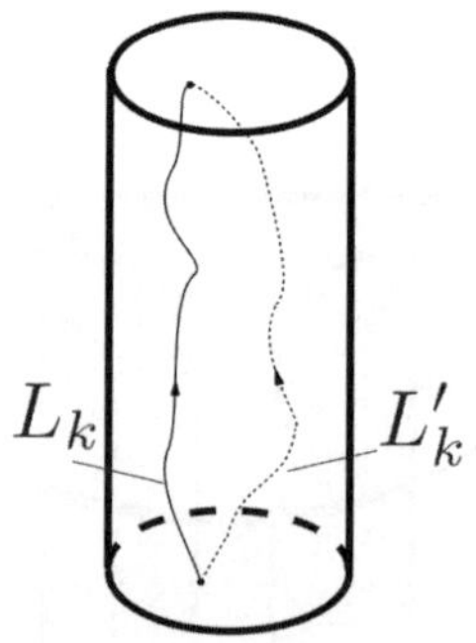

Fig. 3.3

to magnetic fluxes through the legs:

$$\int_{S_1} \mathbf{B}_1 \cdot \mathbf{ds} = \Phi_1, \tag{3.7}$$

$$\int_{S_2} \mathbf{B}_2 \cdot \mathbf{ds} = -\Phi_2, \tag{3.8}$$

$$\int_{S_3} \mathbf{B}_3 \cdot \mathbf{ds} = \Phi_3. \tag{3.9}$$

It is understandable that the negative sign in formula (3.8) appears because the direction of $\mathbf{B}_2$ is opposite to the direction of $\mathbf{ds}$ in the integral over S_2. By substituting formulas (3.7), (3.8) and (3.9) into equation (3.6), we end up with

$$\Phi_1 - \Phi_2 + \Phi_3 = 0. \tag{3.10}$$

The last formula implies that *the algebraic sum of all fluxes at the node of the ferromagnetic core is equal to zero.* It is apparent that the last statement as well as the derivation of (3.10) are very general in nature and applicable to any node. Thus, equation (3.10) can be written in general form as

$$\boxed{\sum_k \Phi_k = 0,} \tag{3.11}$$

and it constitutes the first Kirchhoff's Law of magnetic circuits.

Next, we introduce the important concept of a drop of magnetic potential across a leg of ferromagnetic core. Consider an arbitrary leg (i.e., leg number k) of ferromagnetic core and a path L_k confined to this leg and connecting its two ends (see Figure 3.3). The drop of magnetic potential

u_{mk} across the leg is defined as

$$u_{mk} = \int_{L_k} \mathbf{H}_k \cdot d\boldsymbol{\ell}. \tag{3.12}$$

We next demonstrate that this quantity is well defined, i.e., that it does not depend on the choice of L_k. Indeed, consider another path L'_k confined to the leg and connecting its ends (see Figure 3.3). For this path, the drop of magnetic potential u'_{mk} is defined as

$$u'_{mk} = \int_{L'_k} \mathbf{H}_k \cdot d\boldsymbol{\ell}. \tag{3.13}$$

Now,

$$u_{mk} - u'_{mk} = \int_{L_k} \mathbf{H}_k \cdot d\boldsymbol{\ell} - \int_{L'_k} \mathbf{H}_k \cdot d\boldsymbol{\ell} = \oint_{L_k + L'_k} \mathbf{H}_k \cdot d\boldsymbol{\ell} = 0. \tag{3.14}$$

In the derivation performed in the last formula, the direction of traverse of L'_k was first reversed, the two integrals were combined into one integral over closed path $L_k + L'_k$ and according to Ampere's Law this integral is equal to zero because the closed path $L_k + L'_k$ is confined to the leg and, therefore, it cannot enclose any current.

From the last formula we find

$$u_{mk} = u'_{mk}, \tag{3.15}$$

which proves that u_{mk} does not depend on the choice of L_k.

Now, we can proceed to the discussion of the second Kirchhoff's Law of magnetic circuits. Consider a "loop" in a ferromagnetic core (see Figure 3.4) created by four legs. Consider a closed path L within this loop and apply Ampere's Law to this path. Since each turn of the coil with current I is enclosed by this path, we find that

$$I_{total} = NI \tag{3.16}$$

and

$$\oint_L \mathbf{H} \cdot d\boldsymbol{\ell} = NI, \tag{3.17}$$

where N is the number of turns of the coil.

The integral in the last formula can be represented as

$$\oint_L \mathbf{H} \cdot d\boldsymbol{\ell} = \int_{L_1} \mathbf{H}_1 \cdot d\boldsymbol{\ell} + \int_{L_2} \mathbf{H}_2 \cdot d\boldsymbol{\ell} + \int_{L_3} \mathbf{H}_3 \cdot d\boldsymbol{\ell} + \int_{L_4} \mathbf{H}_4 \cdot d\boldsymbol{\ell}. \tag{3.18}$$

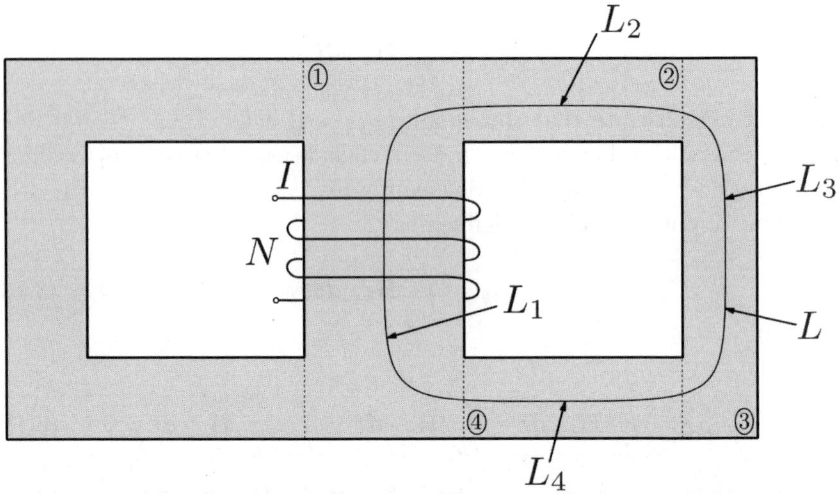

Fig. 3.4

Each integral in the right-hand side of formula (3.18) can be identified as the drop of magnetic potential across the corresponding leg,

$$\int_{L_k} \mathbf{H}_k \cdot d\boldsymbol{\ell} = u_{mk}, \quad (k = 1, 2, 3, 4). \tag{3.19}$$

It is customary in magnetic circuit theory to call the quantity in the right-hand side of formula (3.17) magnetomotive force and to use the abbreviation mmf for this term:

$$NI = \text{mmf}. \tag{3.20}$$

Now, by substituting formula (3.19) into equation (3.18), which is then substituted into relation (3.17), and taking into account the notation (3.20), we arrive at

$$u_{m1} + u_{m2} + u_{m3} + u_{m4} = \text{mmf}. \tag{3.21}$$

It is apparent that the presented derivation is quite general in nature and valid for any number of legs in a loop. Thus, it can be stated that for any loop of ferromagnetic core the following relation is valid:

$$\boxed{\sum_k u_{mk} = \text{mmf},} \tag{3.22}$$

where mmf is the magnetomotive force acting in the loop. This is the second Kirchhoff's Law of magnetic circuits.

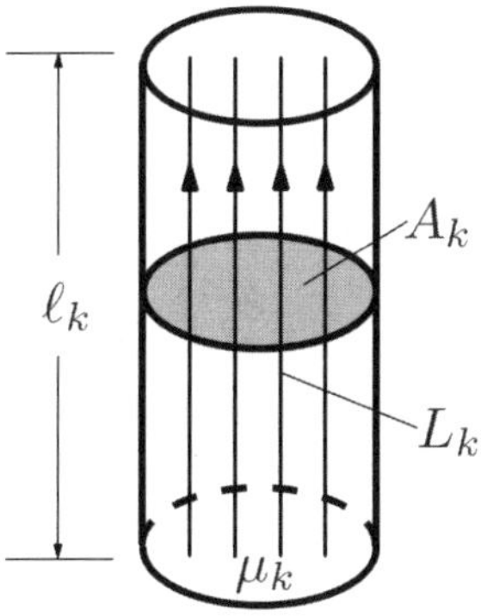

Fig. 3.5

We next turn to the discussion of Ohm's Law of magnetic circuits. This Ohm's Law relates the drop of magnetic potential u_{mk} across any leg of ferromagnetic core to the flux Φ_k through this leg. The derivation of this law proceeds as follows. Consider an arbitrary leg (i.e., leg number k) of length ℓ_k, cross-sectional area A_k and permeability μ_k (see Figure 3.5). According to the definition (3.12) of the drop of magnetic potential u_{mk} we have

$$u_{mk} = \int_{L_k} \mathbf{H}_k \cdot \boldsymbol{d\ell}. \tag{3.23}$$

Since the integration path L_k can be arbitrarily chosen within the leg, we shall use the choice when L_k coincides with one of the magnetic field lines. Due to this choice and also using Assumption 2, the integral in the last formula can be simplified as follows:

$$u_{mk} = \int_{L_k} \mathbf{H}_k \cdot \boldsymbol{d\ell} = \int_{L_k} H_k d\ell = H_k \int_{L_k} d\ell = H_k \ell_k. \tag{3.24}$$

The first simplification occurs because $\mathbf{H}_k$ and $\boldsymbol{d\ell}$ have the same directions since L_k coincides with the magnetic field line. The second simplification occurs because the magnetic field is spatially uniform within the leg according to Assumption 2.

Next, formula (3.24) can be written as follows:

$$u_{mk} = \mu_k H_k \frac{\ell_k}{\mu_k}. \tag{3.25}$$

By taking into account that

$$B_k = \mu_k H_k, \tag{3.26}$$

we find

$$u_{mk} = B_k \frac{\ell_k}{\mu_k}, \tag{3.27}$$

or

$$u_{mk} = B_k A_k \frac{\ell_k}{\mu_k A_k}. \tag{3.28}$$

By definition, magnetic flux Φ_k through the leg is given by the formula

$$\Phi_k = \int_{A_k} \mathbf{B}_k \cdot \mathbf{ds}. \tag{3.29}$$

According to Assumption 2, magnetic flux density $\mathbf{B}_k$ is normal to A_k and, consequently, has the same direction as $\mathbf{ds}$. Furthermore, according to the same assumption, B_k is spatially uniform in the leg. Consequently,

$$\Phi_k = \int_{A_k} \mathbf{B}_k \cdot \mathbf{ds} = \int_{A_k} B_k ds = B_k \int_{A_k} ds = B_k A_k. \tag{3.30}$$

By using the last formula in equation (3.28) we arrive at

$$u_{mk} = \Phi_k \frac{\ell_k}{\mu_k A_k}. \tag{3.31}$$

Finally, by introducing the reluctance $\mathcal{R}_{mk}$ of the leg as

$$\boxed{\mathcal{R}_{mk} = \frac{\ell_k}{\mu_k A_k},} \tag{3.32}$$

we arrive at Ohm's Law

$$\boxed{u_{mk} = \Phi_k \mathcal{R}_{mk}.} \tag{3.33}$$

This completes the derivation of the basic equations of magnetic circuit theory. This theory can be summarized as follows. Each leg of ferromagnetic core is characterized by three quantities: drop of magnetic potential u_{mk} across it, magnetic flux Φ_k through it and magnetic reluctance $\mathcal{R}_{mk}$. These quantities are related to one another and to magnetomotive forces (i.e., to currents in coils) by the basic equations (3.11), (3.22), (3.32) and (3.33). The close examination of these equations reveals that their mathematical structure is identical to the mathematical structure of basic equations of resistive electric circuits. This is the principle of mathematical similarity between the basic equations of magnetic and electric circuits. This principle is illustrated in the following table.

Table

Magnetic Circuits		Electric Circuits	
$\sum_k \Phi_k = 0$	$\Longleftrightarrow$	$\sum_k i_k = 0$	(3.34)
$\sum_k u_{mk} = \text{mmf}$	$\Longleftrightarrow$	$\sum_k v_k = \text{emf}$	(3.35)
$\text{mmf} = NI$	$\Longleftrightarrow$	$\text{emf} = v_s$	(3.36)
$u_{mk} = \Phi_k \mathcal{R}_{mk}$	$\Longleftrightarrow$	$v_k = i_k R_k$	(3.37)
$\mathcal{R}_{mk} = \dfrac{\ell_k}{\mu_k A_k}$	$\Longleftrightarrow$	$R_k = \dfrac{\ell_k}{\sigma_k A_k}$	(3.38)

It is apparent that the mathematical structure of the equations in the right-hand column of the table is identical to the mathematical structure of the equations in the left-hand column of the table. From the purely mathematical point of view, the only difference in these two sets of equations is in the notation of the variables. Indeed, if we replace Φ_k by i_k, u_{mk} by v_k, mmf by emf, $\mathcal{R}_{mk}$ by R_k, μ_k by σ_k (and the other way around), then one set of equations is transformed into the other. The question can be immediately asked what the utility of this principle of mathematical similarity is. The answer is that all techniques that have been developed for the analysis of electric circuits can be used for the analysis of magnetic circuits. This is so because the analysis techniques for electric circuits have been developed by using the mathematical structure of the basic equations for these circuits and this structure is the same as for magnetic circuits. There is another utility of this principle of mathematical similarity that was used in the past, i.e., before the advent of powerful computers. Namely, electric circuits can be used for analog modeling of magnetic circuits. The advantage of such modeling is that electric circuits are easier and cheaper to assemble and electric circuit measurements are usually more accurate than magnetic measurements.

It must be stressed that the similarity between magnetic and electric circuits is purely mathematical in nature. There is no physical similarity, and quantities which describe magnetic and electric circuits have different physical nature and, hence, different physical dimensions.

Now, we shall discuss how the magnetic circuit theory can be used for the analysis of magnetic systems. In this analysis, the geometry of ferromagnetic cores and currents in coils are given and it is required to find

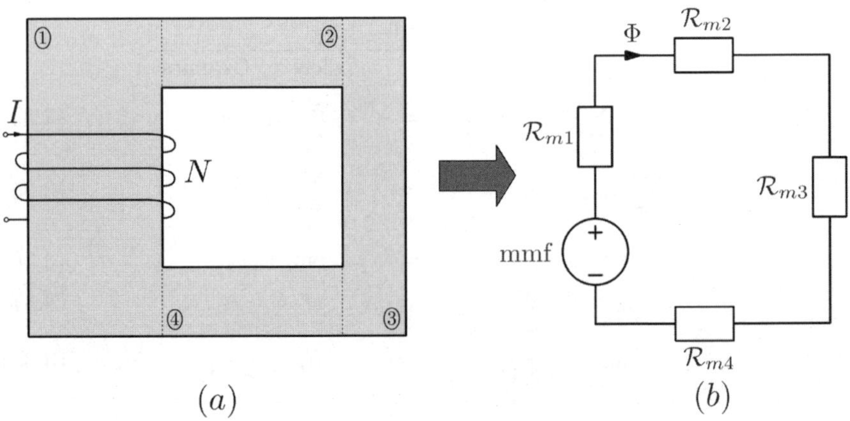

Fig. 3.6

magnetic fluxes through core legs. We illustrate below this analysis by two simple examples.

Example 1. Consider the magnetic system shown in Figure 3.6a. The current I through the coil and its number of turns N are given as well as geometry and magnetic permeability of each leg, i.e., ℓ_k, A_k, μ_k, $(k = 1, 2, 3, 4)$. It is required to find the flux Φ through the legs. This flux is the same for all legs because they are connected in series; this also immediately follows from the basic equation (3.11).

Step 1. In this step, we replace the actual magnetic system by the equivalent magnetic circuit. In doing so, we replace the coil with current I by the magnetomotive force

$$\text{mmf} = NI, \tag{3.39}$$

and each leg of the ferromagnetic core by the magnetic reluctance

$$\mathcal{R}_{mk} = \frac{\ell_k}{\mu_k A_k}, \tag{3.40}$$

and connect these reluctances in the magnetic circuit in the same way as the corresponding legs are connected in the ferromagnetic core (see Figure 3.6b). It is apparent that mmf and $\mathcal{R}_{mk}$ can be easily computed because N, I, ℓ_k, A_k and μ_k are given.

Step 2. By using the principle of mathematical similarity between magnetic and electric circuits, we can use the same technique for the analysis of the magnetic circuit in Figure 3.6 as for the corresponding electric circuit.

In the case being discussed, all magnetic reluctances are connected in series; consequently, the flux Φ can be found as follows:

$$\Phi = \frac{\text{mmf}}{\sum_{k=1}^{4} \mathcal{R}_{mk}}. \tag{3.41}$$

Now, by using formulas (3.39) and (3.40) in the last equation, we end up with the following explicit formula for the magnetic flux:

$$\Phi = \frac{NI}{\sum_{k=1}^{4} \frac{\ell_k}{\mu_k A_k}}. \tag{3.42}$$

In some applications, it may be of interest to find the current I which will guarantee the desired flux Φ. From formula (3.42), the answer is immediate:

$$I = \frac{1}{N}\Phi \sum_{k=1}^{4} \frac{\ell_k}{\mu_k A_k}. \tag{3.43}$$

Example 2. Consider the magnetic system shown in Figure 3.7. The current I and the number of turns N are given as well as geometry and magnetic permeability of each leg ℓ_k, A_k, μ_k, $(k = 1, 2, 3)$. It is required to find all fluxes Φ_k, $(k = 1, 2, 3)$.

Step 1. In this step, as before, we replace the actual magnetic system by the equivalent magnetic circuit. In doing so, we replace the coil by the magnetomotive force

$$\text{mmf} = NI, \tag{3.44}$$

and each leg of the ferromagnetic core by its magnetic reluctance

$$\mathcal{R}_{mk} = \frac{\ell_k}{\mu_k A_k}, \tag{3.45}$$

and connect these reluctances in the magnetic circuit in the same way as the corresponding legs are connected in the ferromagnetic core (see Figure 3.7b). It is apparent that mmf and $\mathcal{R}_{mk}$ can be easily computed by using the given data.

Step 2. We shall analyze the magnetic circuit in Figure 3.7b in exactly the same way as we would analyze the similar electric circuit. Namely, we identify that reluctances $\mathcal{R}_{m2}$ and $\mathcal{R}_{m3}$ are connected in parallel and can be replaced by the equivalent reluctance. This equivalent reluctance is connected in series with reluctance $\mathcal{R}_{m1}$ and, consequently, they can be

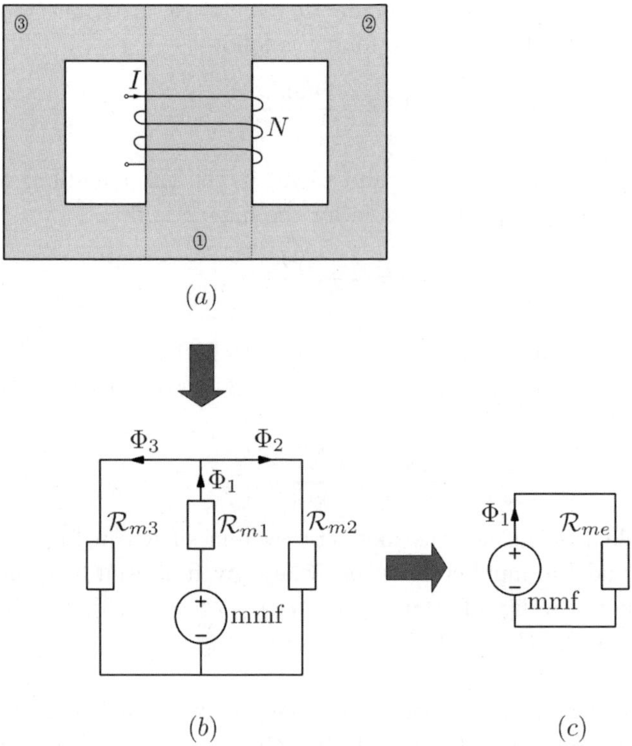

Fig. 3.7

replaced by the overall equivalent reluctance $\mathcal{R}_{me}$ (see Figure 3.7c), which is given by the formula

$$\mathcal{R}_{me} = \mathcal{R}_{m1} + \frac{\mathcal{R}_{m2}\mathcal{R}_{m3}}{\mathcal{R}_{m2} + \mathcal{R}_{m3}} \tag{3.46}$$

or

$$\mathcal{R}_{me} = \frac{\mathcal{R}_{m1}\mathcal{R}_{m2} + \mathcal{R}_{m1}\mathcal{R}_{m3} + \mathcal{R}_{m2}\mathcal{R}_{m3}}{\mathcal{R}_{m2} + \mathcal{R}_{m3}}. \tag{3.47}$$

Step 3. From the magnetic circuit shown in Figure 3.7c, we find

$$\Phi_1 = \frac{\text{mmf}}{\mathcal{R}_{me}} \tag{3.48}$$

or

$$\Phi_1 = \frac{\text{mmf}\,(\mathcal{R}_{m2} + \mathcal{R}_{m3})}{\mathcal{R}_{m1}\mathcal{R}_{m2} + \mathcal{R}_{m1}\mathcal{R}_{m3} + \mathcal{R}_{m2}\mathcal{R}_{m3}}. \tag{3.49}$$

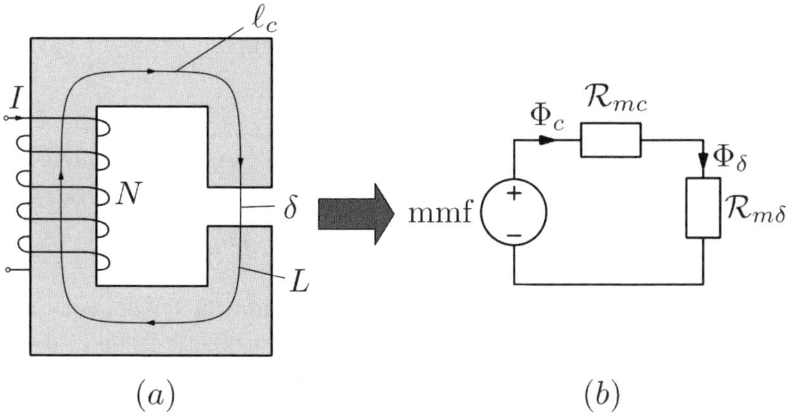

Fig. 3.8

Having found Φ_1, we turn to the circuit shown in Figure 3.7b and find Φ_2 and Φ_3 by using the flux divider rule, which is identical to the current divider rule for parallel electric circuits. This leads to formulas

$$\Phi_2 = \frac{\text{mmf}\,\mathcal{R}_{m3}}{\mathcal{R}_{m1}\mathcal{R}_{m2} + \mathcal{R}_{m1}\mathcal{R}_{m3} + \mathcal{R}_{m2}\mathcal{R}_{m3}}, \tag{3.50}$$

$$\Phi_3 = \frac{\text{mmf}\,\mathcal{R}_{m2}}{\mathcal{R}_{m1}\mathcal{R}_{m2} + \mathcal{R}_{m1}\mathcal{R}_{m3} + \mathcal{R}_{m2}\mathcal{R}_{m3}}. \tag{3.51}$$

By substituting formulas (3.44) and (3.45) into equations (3.49), (3.50) and (3.51), we find explicit expressions for fluxes in terms of the given data.

It is apparent from the above two examples that the magnetic circuit theory is a quite simple and powerful tool for computing fluxes in ferromagnetic cores of magnetic systems. Now, we shall extend the magnetic circuit theory to the case when ferromagnetic cores may have air gaps. Such air gaps are typical for electric power devices. In such air gaps, electromagnetic interaction between currents and magnetic fields occurs, and as a result, mechanical energy is converted into electric energy or the other way around. Thus, air gaps are the regions where energy conversion occurs.

We consider the simplest magnetic system with an air gap shown in Figure 3.8a, and shall demonstrate that in the framework of the magnetic circuit theory the air gap can be treated as another leg of the magnetic system and can be represented by the appropriate magnetic reluctance (see Figure 3.8b). This demonstration is of general nature and applicable to more complicated magnetic circuits.

To start the derivation, we consider a magnetic field line L which goes through the core and the gap. We next apply Ampere's Law for path L:

$$\oint_L \mathbf{H} \cdot d\boldsymbol{\ell} = NI. \tag{3.52}$$

It is apparent that the integral in formula (3.52) can be represented as the sum of two integrals,

$$\oint_L \mathbf{H} \cdot d\boldsymbol{\ell} = \int_{\ell_c} \mathbf{H}_c \cdot d\boldsymbol{\ell} + \int_{\delta} \mathbf{H}_\delta \cdot d\boldsymbol{\ell} = NI, \tag{3.53}$$

where ℓ_c and δ are the parts of L which are within the ferromagnetic core and the gap, respectively, while $\mathbf{H}_c$ and $\mathbf{H}_\delta$ are magnetic fields in the core and the gap, respectively. By taking into account that ℓ_c and δ are parts of the magnetic field line L, we conclude that the directions of $d\boldsymbol{\ell}$ and magnetic fields along ℓ_c and δ coincide. Furthermore, according to Assumption 2 used in magnetic circuit theory, it can be assumed that magnetic field is spatially uniform in the ferromagnetic core as well as in the air gap. These facts lead to the following simplification of formula (3.53):

$$H_c \ell_c + H_\delta \delta = \text{mmf}, \tag{3.54}$$

where ℓ_c and δ in the last equation can be construed as the average length of the core and the length of the gap, respectively.

Now, we shall invoke the continuity of the normal component of the magnetic flux density at the interface between the air gap and the ferromagnetic core,

$$B_{cn} = B_{\delta n}. \tag{3.55}$$

The last relation can also be written as

$$\mu_c H_c = \mu_0 H_\delta, \tag{3.56}$$

because it is assumed that the magnetic field line is perpendicular to the core-gap interface. By using the last formula in equation (3.54), we find

$$H_c \left(\ell_c + \frac{\mu_c}{\mu_0} \delta \right) = \text{mmf}, \tag{3.57}$$

which leads to

$$H_c = \frac{\text{mmf}}{\ell_c + \frac{\mu_c}{\mu_0} \delta}. \tag{3.58}$$

Within the framework of the two assumptions of the magnetic circuit theory, we have

$$\Phi_c = \Phi_\delta = A_c B_c = A_c \mu_c H_c. \tag{3.59}$$

By substituting formula (3.58) into the last equation, we derive

$$\Phi_c = \Phi_\delta = \frac{A_c \mu_c \,\mathrm{mmf}}{\ell_c + \frac{\mu_c \delta}{\mu_0}} = \frac{\mathrm{mmf}}{\frac{\ell_c}{\mu_c A_c} + \frac{\delta}{\mu_0 A_c}}. \tag{3.60}$$

It is apparent that the reluctance of the core is

$$\mathcal{R}_{mc} = \frac{\ell_c}{\mu_c A_c}. \tag{3.61}$$

As mentioned before, we shall treat the air gap as another leg of the magnetic system. This leg has length δ, permeability μ_0 and cross-sectional area A_c. This implies that the magnetic reluctance $\mathcal{R}_{m\delta}$ of this leg is given by the formula

$$\boxed{\mathcal{R}_{m\delta} = \frac{\delta}{\mu_0 A_c}.} \tag{3.62}$$

From formulas (3.60), (3.61) and (3.62) we conclude

$$\Phi_\delta = \Phi_c = \frac{\mathrm{mmf}}{\mathcal{R}_{mc} + \mathcal{R}_{m\delta}}. \tag{3.63}$$

The last equation implies the validity of the magnetic circuit representation (see Figure 3.8b) of the magnetic system shown in Figure 3.8a.

It is interesting and instructive to compare the values of magnetic reluctances of the ferromagnetic core and the air gap. To this end, consider the ratio of $\mathcal{R}_{mc}$ to $\mathcal{R}_{m\delta}$, which according to formulas (3.61) and (3.62) is given by

$$\frac{\mathcal{R}_{mc}}{\mathcal{R}_{m\delta}} = \frac{\ell_c}{\delta}\frac{\mu_0}{\mu_c}. \tag{3.64}$$

In typical designs,

$$\ell_c \approx 50\delta, \qquad \mu_c > 10^3 \mu_0, \tag{3.65}$$

which leads to the conclusion that

$$\frac{\mathcal{R}_{mc}}{\mathcal{R}_{m\delta}} < 0.05 \qquad \text{and} \qquad \mathcal{R}_{mc} \ll \mathcal{R}_{m\delta}. \tag{3.66}$$

According to formula (3.63), this means that

$$\Phi_\delta \approx \frac{\mathrm{mmf}}{\mathcal{R}_{m\delta}} = \frac{\mu_0 A N I}{\delta}. \tag{3.67}$$

Thus, the smaller the air gap, the stronger the magnetic flux and magnetic fields in the air gap. This fact immediately reveals that ferromagnetic cores serve two useful purposes as illustrated by Figure 3.9. First, they reduce stray (useless) magnetic fields and, second, they concentrate and

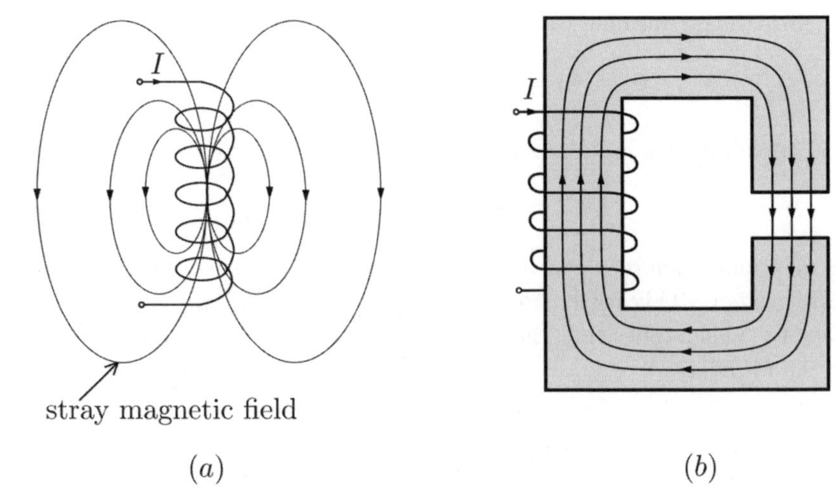

Fig. 3.9

appreciably enhance magnetic fields in desired regions identified as air gaps. Indeed, according to Figure 3.9a, when the ferromagnetic core is not used, the magnetic field created by the coil with current I is spread out. As a result, only a small portion of the magnetic field may reach the desired region, while most of the magnetic field manifests itself as a stray field. On the contrary, when a ferromagnetic core is used (see Figure 3.9b), this core due to its high magnetic permeability guides practically all magnetic field lines through the air gap resulting in the focusing and enhancement (also due to high μ_c) of magnetic field in the desired region of the air gap. In addition, since practically all magnetic field lines are confined to the core, undesired stray magnetic field is dramatically reduced.

It has been assumed in our discussion that all of the magnetic field lines in the air gap are confined to the same cross-sectional area as in the ferromagnetic core (see formula (3.62)). This assumption ignores the fringing effect shown in Figure 3.10, when magnetic field lines appear around the sides of the air gap. It is intuitively apparent that the smaller the air gap δ, the smaller the fringing effect. For this reason, the fringing effect is often taken into account by using in formula (3.62) effective cross-sectional area A_e instead of A_c. The effective cross-sectional area is often obtained by adding δ to each side of A_c (see Figure 3.11).

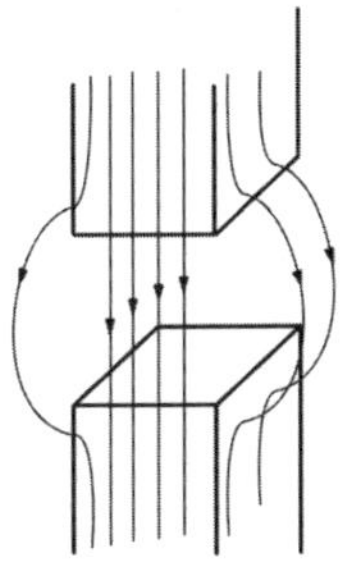

Fig. 3.10

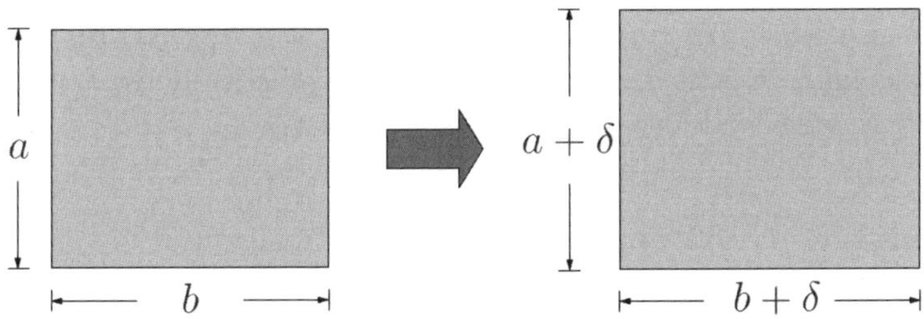

Fig. 3.11

3.2 Application of Magnetic Circuit Theory to the Calculation of Inductance and Mutual Inductance

In this section, we demonstrate that the magnetic circuit theory can be effectively used for the calculation of inductance and mutual inductance of coils wound around legs of ferromagnetic cores. We start with the definition of inductance. Consider a coil energized with current I (see Figure 3.12). This current creates a magnetic field that can be represented by magnetic field lines. In general, these field lines may link a different number of turns of the coil. This may result in different magnetic fluxes linking a different number of turns. Suppose that magnetic flux Φ_k links N_k turns of the coil. Then, the total flux linkage of the coil is defined as follows:

$$\psi = \sum_k N_k \Phi_k. \tag{3.68}$$

Now, the inductance can be defined as the following ratio:

$$L = \frac{\psi}{I}. \tag{3.69}$$

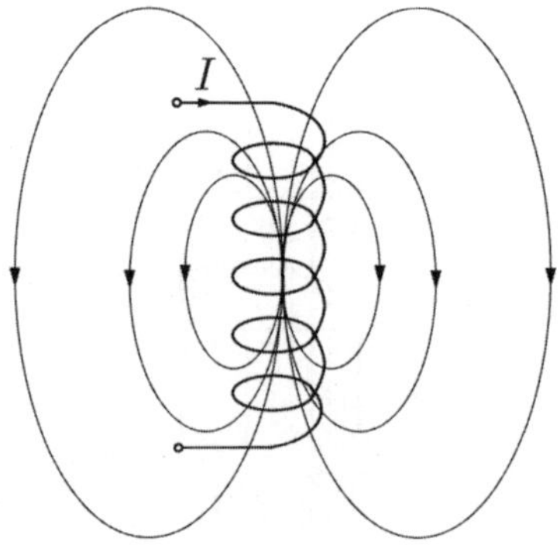

Fig. 3.12

In the case when the turns of the coil are closely spaced, the last two formulas can be modified. Namely, in this case practically the same flux Φ links all N turns and formula (3.68) can be written as

$$\psi = N\Phi, \tag{3.70}$$

which leads to the following expression for the inductance:

$$L = \frac{N\Phi}{I}. \tag{3.71}$$

It is important to stress that, although the inductance is defined as the ratio of the flux linkage to the current, it does not depend on the flux linkage or the current. The inductance depends on the number N of turns of the coil and its geometry. In the case of closely spaced turns, that is, when formula (3.71) is valid, the inductance is proportional to the square of the number of turns (N^2). This suggests that the inductance can be increased by increasing the number of turns. Another efficient way to increase the inductance is to use ferromagnetic cores with high magnetic permeability. In fact, this approach is widely used in electric power engineering. Figure 3.13 presents one possible realization of this approach. In this case, the flux linkage ψ of the coil has two distinct components: the main (core) component ψ_c which is due to the magnetic field lines entirely confined to the ferromagnetic core and the "leakage" component ψ_ℓ which is due to the magnetic field lines that leak out,

$$\psi = \psi_c + \psi_\ell. \tag{3.72}$$

By substituting the last relation into formula (3.69), we find

$$L = \frac{\psi_c}{I} + \frac{\psi_\ell}{I}. \tag{3.73}$$

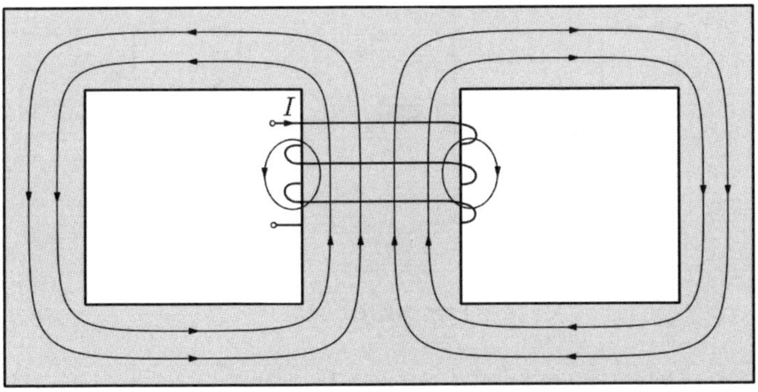

Fig. 3.13

This means that the inductance also has two distinct components,

$$L = L_m + L_\ell, \tag{3.74}$$

where L_m is the main inductance defined as

$$L_m = \frac{\psi_c}{I}, \tag{3.75}$$

while L_ℓ is the leakage inductance specified as

$$L_\ell = \frac{\psi_\ell}{I}. \tag{3.76}$$

It is apparent that for ferromagnetic cores with high magnetic permeability

$$\psi_c \gg \psi_\ell, \tag{3.77}$$

which implies that

$$L_m \gg L_\ell, \tag{3.78}$$

and, consequently,

$$L \approx L_m. \tag{3.79}$$

It turns out that the main inductance can be effectively computed by using the magnetic circuit theory. Below, we shall first derive the general formula for L_m that will be next illustrated by two examples.

Suppose that we want to compute the inductance of the coil wound around some leg of a ferromagnetic core. Since inductance does not depend on a specific value of the current through the coil, we will assume that the coil is excited by some current I. Now, we replace the actual magnetic

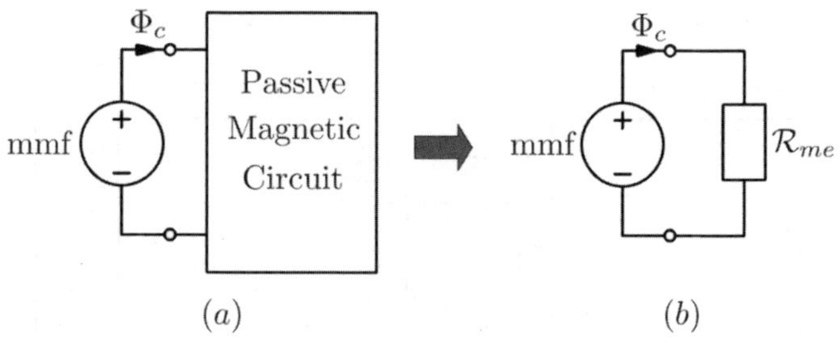

Fig. 3.14

system by the equivalent magnetic circuit. In doing so, we replace the coil with current I by the magnetomotive force

$$\text{mmf} = NI \tag{3.80}$$

and legs of the ferromagnetic core by the appropriate magnetic reluctances which are connected in the magnetic circuit in the same way as the corresponding legs are connected in the ferromagnetic core. A general case of such magnetic circuit is shown in Figure 3.14a, where Φ_c is the core magnetic flux through the leg around which the coil is wound, while the rest of the magnetic circuit is passive (i.e., without any mmf sources). This passive magnetic circuit is represented by the box to emphasize the general nature of our discussion. Any passive circuit can be represented by the equivalent magnetic reluctance with respect to the mmf terminals (see Figure 3.14b). Consequently,

$$\Phi_c = \frac{\text{mmf}}{\mathcal{R}_{me}} = \frac{NI}{\mathcal{R}_{me}}. \tag{3.81}$$

It is apparent that the core magnetic flux Φ_c links all N turns of the coil. This implies that

$$\psi_c = N\Phi_c = \frac{N^2 I}{\mathcal{R}_{me}}. \tag{3.82}$$

Now, by using formulas (3.75), (3.79) and (3.82), we derive

$$\boxed{L \approx L_m = \frac{N^2}{\mathcal{R}_{me}}.} \tag{3.83}$$

The last formula clearly reveals that the inductance is indeed proportional to the square of the number of coil turns and depends on geometry and magnetic properties of the ferromagnetic core through the equivalent magnetic reluctance $\mathcal{R}_{me}$. The last formula is very general because it is applicable

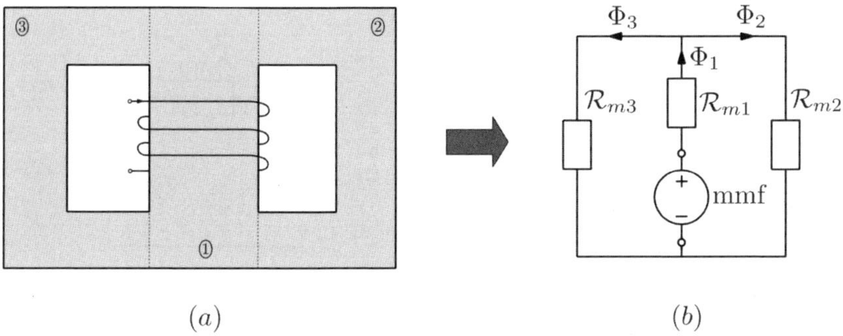

Fig. 3.15

to any geometry of the ferromagnetic core. The usefulness of this general formula is illustrated below by the following two examples.

Example 1. Consider the magnetic system shown in Figure 3.15a and assume that the number N of coil turns and magnetic permeability μ_c of the ferromagnetic core are given along with geometry of each leg: ℓ_k, A_k, $(k = 1, 2, 3)$. It is required to find the inductance of the coil. The equivalent magnetic circuit is shown in Figure 3.15b. It is clear from this figure that the equivalent magnetic reluctance with respect to the terminals of the mmf is given by the formula (see Example 2 and equation (3.47) from the previous section)

$$\mathcal{R}_{me} = \frac{\mathcal{R}_{m1}\mathcal{R}_{m2} + \mathcal{R}_{m1}\mathcal{R}_{m3} + \mathcal{R}_{m2}\mathcal{R}_{m3}}{\mathcal{R}_{m2} + \mathcal{R}_{m3}}. \tag{3.84}$$

By using the last relation in formula (3.83), we find

$$L \approx L_m = N^2 \frac{\mathcal{R}_{m2} + \mathcal{R}_{m3}}{\mathcal{R}_{m1}\mathcal{R}_{m2} + \mathcal{R}_{m1}\mathcal{R}_{m3} + \mathcal{R}_{m2}\mathcal{R}_{m3}}. \tag{3.85}$$

This expression can be appreciably simplified in the practical (symmetrical) case when legs 2 and 3 are identical, have the same permeability μ_c as leg 1 and

$$A_2 = \frac{A_1}{2}. \tag{3.86}$$

In this case,

$$\mathcal{R}_{m2} = \mathcal{R}_{m3} = \frac{2\ell_2}{\mu_c A_1}, \tag{3.87}$$

$$\mathcal{R}_{m1} = \frac{\ell_1}{\mu_c A_1}. \tag{3.88}$$

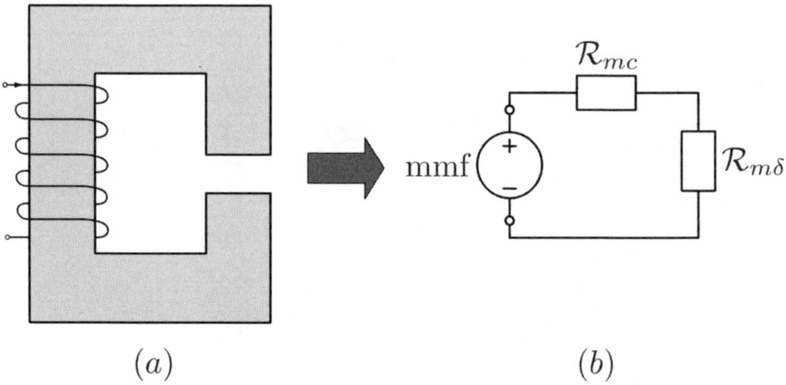

Fig. 3.16

By using the last two equations in formula (3.85), we end up with

$$L \approx L_m = \frac{\mu_c A_1 N^2}{\ell_1 + \ell_2}. \tag{3.89}$$

The last expression clearly reveals that the inductance can be substantially enhanced by using high magnetic permeability ferromagnetic cores.

Example 2. Consider the magnetic system shown in Figure 3.16a and assume that the number N of coil turns is given along with geometry and permeability of the core as well as the gap: ℓ_c, A_c, μ_c, δ. It is required to find the inductance of the coil. Figure 3.16b represents the equivalent magnetic circuit for the magnetic system shown in Figure 3.16a. From this equivalent magnetic circuit and formula (3.83), we find

$$L \approx L_m = \frac{N^2}{\mathcal{R}_{mc} + \mathcal{R}_{m\delta}} = \frac{\mu_c A_c N^2}{\ell + \frac{\mu_c}{\mu_0}\delta}. \tag{3.90}$$

In the typical case when $\mathcal{R}_{mc} \ll \mathcal{R}_{m\delta}$, the last formula can be simplified as

$$L \approx L_m \approx \frac{N^2}{\mathcal{R}_{m\delta}} = \frac{\mu_0 A_c N^2}{\delta}. \tag{3.91}$$

This equation reveals the important fact that the inductance can be effectively controlled by changing the air gap length. This fact is used in many applications.

It is clear from the presented discussion that the general approximate formula (3.83) for the inductance L has been derived by neglecting leakage inductance L_ℓ. The leakage inductance cannot be accounted for within the framework of the magnetic circuit theory because this theory completely ignores leakage fluxes. At first, it may be thought that it is only natural

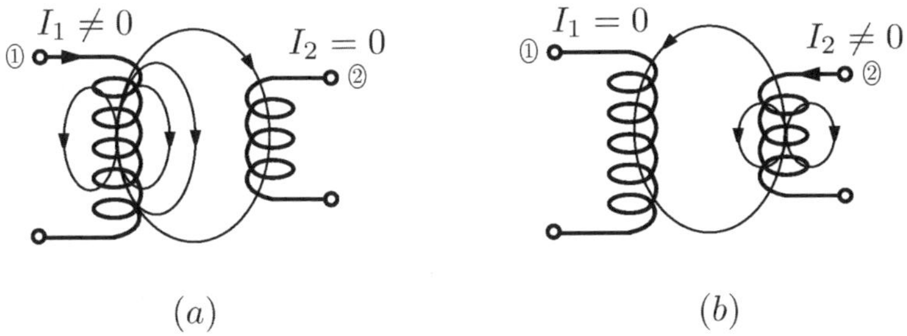

Fig. 3.17

to neglect leakage inductance because it is quite small. However, this is often not the case and in many engineering applications leakage inductances play an important and even crucial role. This is true when the performance of devices is based on *strong* electromagnetic coupling between coils (windings). This strong coupling is typical for transformers and induction motors and leakage inductances play important roles in the theory of these devices. The fact that the magnetic circuit theory is helpless in calculation of leakage inductances implies that more sophisticated magnetic field computation techniques must be employed for this purpose. These techniques are beyond the scope of our current discussion.

Next, we proceed with the discussion of the calculation of mutual inductance and we start this discussion with the definition of mutual inductance. Consider two coils and two distinct cases: a) the first coil is energized ($I_1 \neq 0$), while the second coil is not (see Figure 3.17a); and b) the second coil is energized ($I_2 \neq 0$), while the first coil is not (see Figure 3.17b). In case a), the current I_1 through the first coil creates the magnetic field which is represented by field lines. Some of these field lines may link the turns of the second coil, resulting in the flux linkage ψ_{21} (see Figure 3.17a). The subscripts in ψ_{21} indicate that this is the flux linkage of the second coil due to the current through the first coil. The mutual inductance M_{12} between the first and second coils is defined as the ratio

$$M_{12} = \frac{\psi_{21}}{I_1}. \tag{3.92}$$

In case b), some field lines of the magnetic field created by the current I_2 through the second coil link the turns of the first coil resulting in the flux linkage ψ_{12} (see Figure 3.17b). The mutual inductance M_{21} between the second and the first coils is defined as the ratio

$$M_{21} = \frac{\psi_{12}}{I_2}. \tag{3.93}$$

It is proved in the electromagnetic field theory that the following equality (often called reciprocity principle) is valid:

$$M_{12} = M_{21} = M. \tag{3.94}$$

Mutual inductances are defined above as the ratios of flux linkages to currents. However, the mutual inductance M does not depend on flux linkages or current; rather, it depends on the number of turns N_1 and N_2 of the first and the second coils (namely, their product $N_1 N_2$), as well as their geometry and mutual location.

It is apparent from the definition of mutual inductance that it can be viewed as a measure of electromagnetic coupling between two coils. Indeed, the larger the mutual inductance, the larger the flux linkage of one coil created by the same current through another coil, and, consequently, the larger the electromagnetic coupling between the two coils. It is important to stress that the mutual inductance as the measure of electromagnetic coupling between two coils depends on the number of turns. To illustrate this, it is instructive to consider the case when the second coil is quite remote from the first coil and, consequently, all turns of the second coil are linked by small magnetic fluxes created by the current through the first coil. If the turns of the second coil are closely spaced, then practically the same small magnetic flux Φ_{21} links all the turns of the second coil and the flux linkage of the second coil is equal to $\psi_{21} = N_2 \Phi_{21}$. Thus, by using a very large number N_2 of turns of the second coil, ψ_{21} as well as $M = M_{12}$ can be appreciably enhanced despite the fact that only a very small number of field lines link the second coil. There is another effective way to enhance electromagnetic coupling between two coils, that is, by using a ferromagnetic core (see Figure 3.18). Indeed, a ferromagnetic core with high magnetic permeability is very good for guiding magnetic field lines from the location of the first coil to the location of the second coil (Figure 3.18b). This clearly suggests that ferromagnetic cores can be used for the enhancement and control of mutual inductance. This property of ferromagnetic cores is utilized in the design of transformers and electric machines. It turns out that the magnetic circuit theory can be effectively used for the calculation of mutual inductance of coils whose electromagnetic coupling is assisted by ferromagnetic cores. We present the general algorithm for such calculations by considering the following example.

Example. Consider a magnetic system with two coils shown in Figure 3.19a. It is assumed that the number of turns N_1 and N_2 of the first coil and second coil respectively are given as well as geometry and magnetic

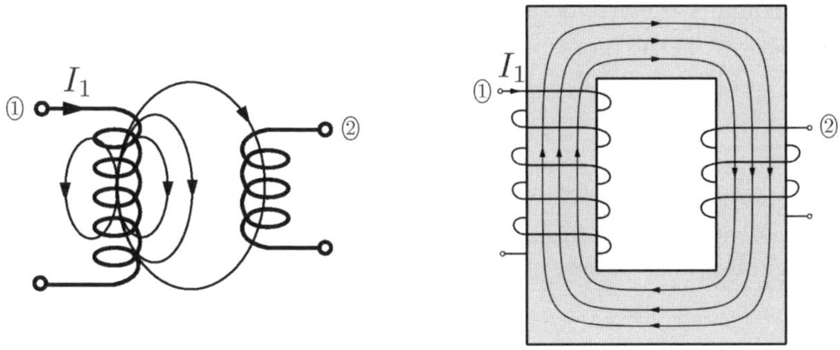

(a) (b)

Fig. 3.18

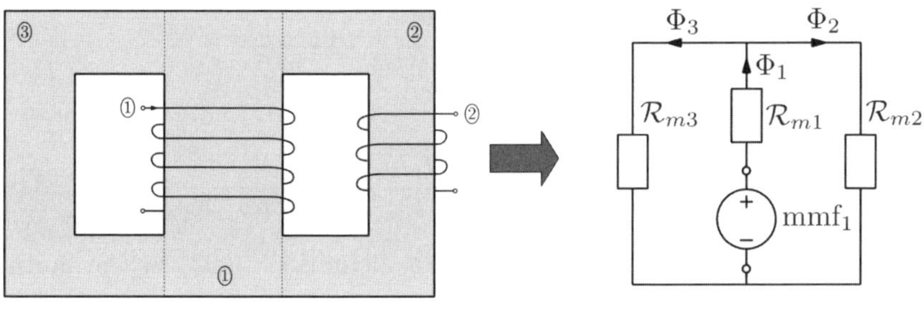

(a) (b)

Fig. 3.19

permeability of each leg of the ferromagnetic core. It is required to derive the formula for the mutual inductance. The general algorithm for the calculation of the mutual inductance consists of the following three steps.

Step 1. We assume that the first coil is energized by an arbitrary current I_1, while the second coil is not energized ($I_2 = 0$). Then we replace such excited magnetic system by the equivalent magnetic circuit. In doing so, we replace the first coil by the magnetomotive force

$$\mathrm{mmf}_1 = N_1 I_1 \tag{3.95}$$

and each leg of the ferromagnetic core by its magnetic reluctance

$$\mathcal{R}_{mk} = \frac{\ell_k}{\mu_k A_k}, \qquad (k = 1, 2, 3), \tag{3.96}$$

and we connect these reluctances in the magnetic circuit in Figure 3.19b in the same way as the corresponding legs are connected in the ferromagnetic core. It is clear that all $\mathcal{R}_{mk}$ can be computed on the basis of the given data.

Step 2. The purpose of this step is to find the flux Φ_2 through the second leg as a linear function of I_1. This has been already done in Example 2 of the previous section where the following formula was derived (see equation (3.50)):

$$\Phi_2 = \frac{\text{mmf}_1 \, \mathcal{R}_{m3}}{\mathcal{R}_{m1}\mathcal{R}_{m2} + \mathcal{R}_{m1}\mathcal{R}_{m3} + \mathcal{R}_{m2}\mathcal{R}_{m3}}, \tag{3.97}$$

or, by using equation (3.95),

$$\Phi_2 = \frac{N_1 I_1 \mathcal{R}_{m3}}{\mathcal{R}_{m1}\mathcal{R}_{m2} + \mathcal{R}_{m1}\mathcal{R}_{m3} + \mathcal{R}_{m2}\mathcal{R}_{m3}}. \tag{3.98}$$

Step 3. It is clear that

$$\psi_{21} = N_2 \Phi_2 = \frac{N_1 N_2 I_1 \mathcal{R}_{m3}}{\mathcal{R}_{m1}\mathcal{R}_{m2} + \mathcal{R}_{m1}\mathcal{R}_{m3} + \mathcal{R}_{m2}\mathcal{R}_{m3}}. \tag{3.99}$$

Finally, by using the last formula in the definition (3.92) of the mutual inductance, we find

$$\boxed{M = M_{12} = \frac{N_1 N_2 \mathcal{R}_{m3}}{\mathcal{R}_{m1}\mathcal{R}_{m2} + \mathcal{R}_{m1}\mathcal{R}_{m3} + \mathcal{R}_{m2}\mathcal{R}_{m3}}.} \tag{3.100}$$

Thus, as discussed before, the mutual inductance depends on the product of turns of the two coils as well as geometry and magnetic permeability of the legs of the ferromagnetic core.

The last formula can be appreciably simplified in the case when legs 2 and 3 are identical, have the same permeability μ_c as leg 1 and each of their cross-sectional areas is half the cross-sectional area of the first leg. Under these conditions, relations (3.87) and (3.88) are valid and the last formula can be transformed to arrive at

$$M = \frac{\mu_c A_1 N_1 N_2}{2(\ell_1 + \ell_2)}. \tag{3.101}$$

The three-step algorithm used in this example is of general nature and can be used for the calculation of mutual inductances of coils with ferromagnetic cores of any geometry and number of legs.

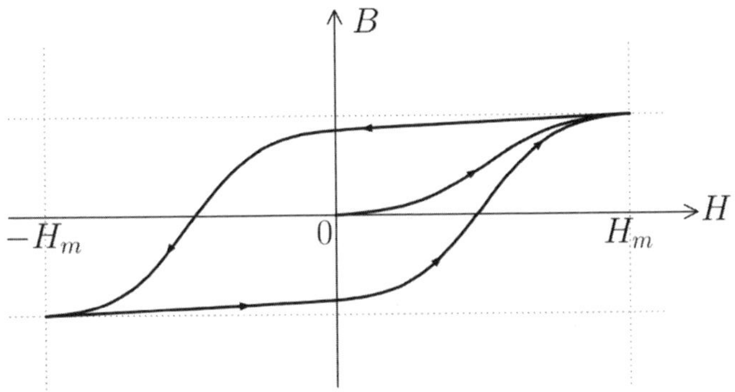

Fig. 3.20

3.3 Magnetic Circuits with Permanent Magnets

In our previous discussion, we have considered magnetic systems excited by coils with currents. This type of excitation requires power supplies. There exists another way to excite magnetic systems by using permanent magnets. This type of excitation can be realized without any power supplies and, for this reason, it is very attractive in many power engineering applications. In this section, we shall further develop the magnetic circuit theory to make it applicable to the analysis of magnetic systems excited (energized) by permanent magnets.

We shall first discuss what a permanent magnet is. To start this discussion, we need some simple facts related to the phenomenon of magnetic hysteresis. This phenomenon is exhibited by ferromagnetic materials and one of its simplest manifestations is the formation of hysteresis loops. This is illustrated by Figure 3.20, where a symmetric hysteresis loop is presented. This loop is formed as a result of periodic in time variations of magnetic field H between two extremum values H_m and $-H_m$. As far as the shape of hysteresis loop is concerned, there are two types of ferromagnetic materials which are important in power applications. They are soft magnetic materials and hard magnetic materials. Soft magnetic materials are characterized by narrow hysteresis loops (see Figure 3.21a) and these materials are used for magnetic cores of power devices. Hard magnetic materials have wide hysteresis loops (see Figure 3.21b) and these materials are used for permanent magnets as well as in magnetic data storage (hard drives). It is apparent that the notion of magnetic permeability is not applicable to hysteretic magnetic materials and the constitutive relation (3.5) must be

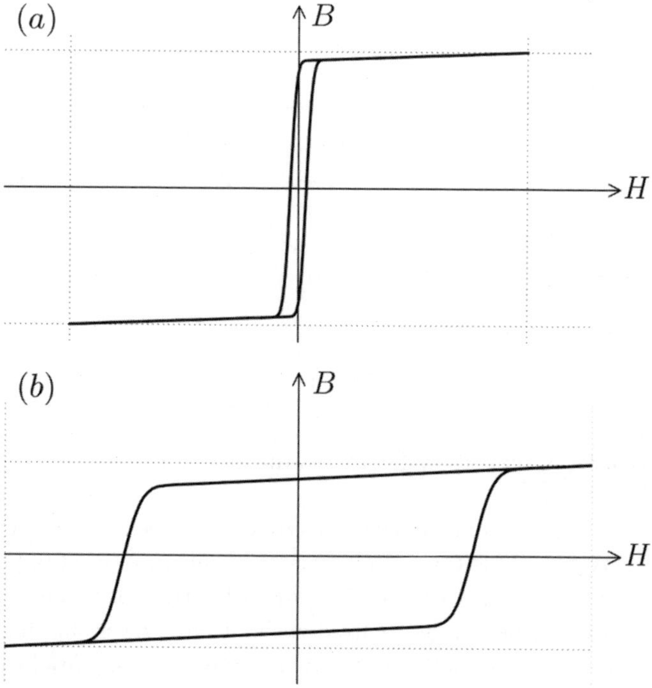

Fig. 3.21

replaced by the relation

$$\mathbf{B} = \mu_0(\mathbf{M} + \mathbf{H}), \tag{3.102}$$

where $\mathbf{M}$ is the magnetization vector.

Hysteresis loops are often measured for toroidal samples for which $\mathbf{B}$, $\mathbf{H}$ and $\mathbf{M}$ have the same directions. In this case, the vectorial equation (3.102) can be replaced by a scalar one,

$$B = \mu_0(M + H), \tag{3.103}$$

or

$$M = \frac{B}{\mu_0} - H. \tag{3.104}$$

By using the last equation for each value of magnetic field H, the B-H hysteresis loop shown in Figure 3.21b can be transformed into the M-H loop shown in Figure 3.22. In this figure, there are four points of special significance. They are $(0, M_r)$, $(0, -M_r)$, $(H_c, 0)$ and $(-H_c, 0)$, where M_r

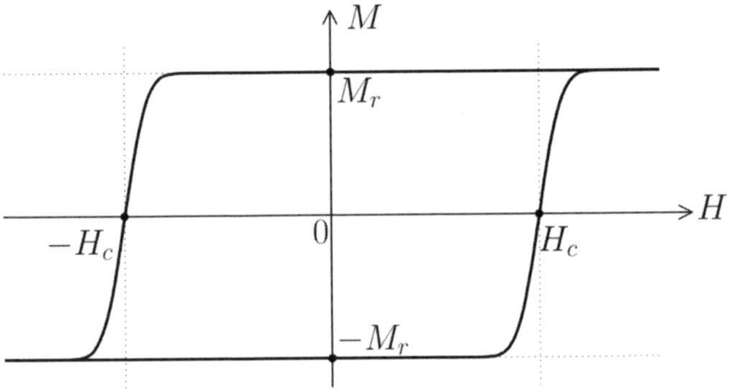

Fig. 3.22

is called remanent (or residual) magnetization, while H_c is called coercivity. The origin of this terminology is quite transparent. Magnetization M_r is called remanent or residual because this is the magnetization existing for zero magnetic field. Magnetic field H_c is called coercivity because this magnetic field is needed to be applied to coerce magnetization into changing its direction. As far as application of hard magnetic materials for permanent magnets is concerned, there are three figures of merit: remanent magnetization, coercivity and the squareness of hysteresis loop. It is demonstrated below that remanent magnetization determines the strength of permanent magnets (i.e., the strength of their magnetic field), H_c determines the stability of permanent magnets (i.e., their ability to sustain remanent magnetization under demagnetizing fields) and squareness of hysteresis loop determines the ability of permanent magnets to maintain their strength under demagnetizing fields. As far as the squareness of hysteresis loop is concerned, the ideal permanent magnet materials (and ideal permanent magnets) can be defined as materials (or magnets) for which

$$M(H) = \pm M_r = \pm M_s \qquad \text{for } -H_c < H < H_c, \tag{3.105}$$

where M_s is the saturation magnetization. There are permanent magnet materials whose hysteresis loops have almost ideal squareness. These are samarium-cobalt (SmCo) and neodymium-iron-boron (NdFeB) materials. For samarium-cobalt materials, $\mu_0 M_s$ = 0.8-1 Tesla and $H_c = 400 - 600$ kA/m, while for neodymium-iron-boron materials $\mu_0 M_s \simeq 1.2$ Tesla and $H_c \simeq 850$ kA/m. Thus, NdFeB materials have somewhat better figures of merit than SmCo materials. Furthermore, neodymium is much more abundant than samarium and, for this reason, NdFeB magnets are more

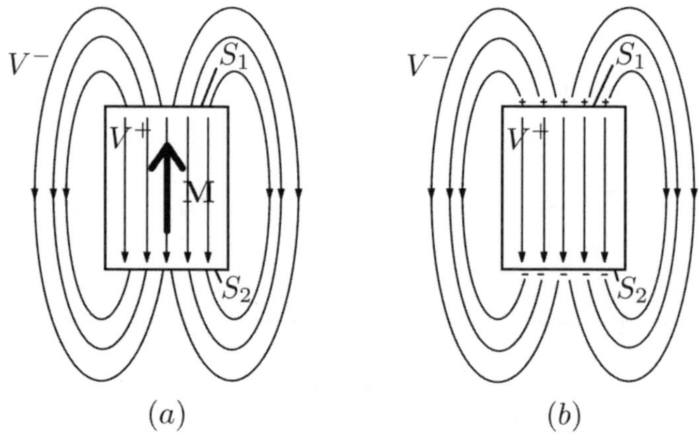

Fig. 3.23

cost effective. These magnets are made from powder through the steps of pressing, sintering and heat treatment.

Now, we can define what a permanent magnet is. A permanent magnet is a piece of hard magnetic material (usually of cylindrical shape) with remanent magnetization (see Figure 3.23). This permanent magnet creates magnetic field due to its remanent magnetization. We shall discuss the magnetic field of such permanent magnet by using the "magnetic charge" model. To arrive at this model, we assume that the magnetization **M** is parallel to the magnet sides and spatially uniform. This implies that

$$\operatorname{div} \mathbf{M} = 0. \tag{3.106}$$

By recalling that

$$\operatorname{div} \mathbf{B} = 0 \tag{3.107}$$

and that in the space around the magnet

$$\mathbf{B} = \mu_0 \mathbf{H}, \tag{3.108}$$

we find from equations (3.102), (3.106) and the last two formulas that

$$\operatorname{div} \mathbf{H} = 0 \tag{3.109}$$

inside (V^+) and outside (V^-) the permanent magnet. Furthermore,

$$\operatorname{curl} \mathbf{H} = 0, \tag{3.110}$$

because the magnetic field created only by the permanent magnet without any external current sources is being considered. It is clear that the magnetic field of an ideal permanent magnet does not have any volume sources.

We shall demonstrate next that this field has only surface sources located on the top S_1 and the bottom S_2 of the permanent magnet. To this end, we shall use the continuity of normal component of magnetic flux density on S_1 and S_2,

$$B_n^- = B_n^+, \tag{3.111}$$

where superscripts "−" and "+" indicate the values of B_n from outside and inside the permanent magnet.

By using formulas (3.102) and (3.108) the boundary condition (3.111) can be written as follows:

$$\mu_0 H_n^- = \mu_0 H_n^+ + \mu_0 M_s \quad \text{on } S_1 \tag{3.112}$$

and

$$\mu_0 H_n^- = \mu_0 H_n^+ - \mu_0 M_s \quad \text{on } S_2, \tag{3.113}$$

where $M_s = |\mathbf{M}|$. It is apparent that the difference in the form of the last two equations is due to the fact that the direction of $\mathbf{M}$ coincides with the outward normal to S_1 and it is opposite to the outward normal to S_2.

The last two equations can be written as follows:

$$H_n^- - H_n^+ = M_s \quad \text{on } S_1 \tag{3.114}$$

and

$$H_n^- - H_n^+ = -M_s \quad \text{on } S_2. \tag{3.115}$$

Thus, the normal components of magnetic field of an ideal permanent magnet are discontinuous on S_1 and S_2, and this discontinuity is the source (the origin) of the magnetic field. To visualize this magnetic field, we introduce the "magnetic charge" model. In this model (see Figure 3.23b), the remanent magnetization $\mathbf{M}$ is removed and replaced by fictitious (virtual) magnetic charges σ_m with densities

$$\sigma_m = \mu_0 M_s \quad \text{on } S_1 \tag{3.116}$$

and

$$\sigma_m = -\mu_0 M_s \quad \text{on } S_2. \tag{3.117}$$

Then, like in electrostatics, the normal component of the magnetic field created by these surface magnetic charges satisfies the boundary conditions

$$H_n^- - H_n^+ = \frac{\sigma_m}{\mu_0} = M_s \quad \text{on } S_1 \tag{3.118}$$

and

$$H_n^- - H_n^+ = \frac{\sigma_m}{\mu_0} = -M_s \quad \text{on } S_2. \tag{3.119}$$

These boundary conditions are identical to the boundary conditions (3.114) and (3.115), respectively. Furthermore, the magnetic field created by surface magnetic charges σ_m satisfies the equations (3.109) and (3.110) in V^- and V^+. Thus, the actual magnetic field of the permanent magnet and the magnetic field created by magnetic charges σ_m satisfy the same equations and boundary conditions. Consequently, these two magnetic fields are identical. The magnetic field created by σ_m is easy to visualize. The magnetic field lines of this field are directed from positive charges on S_1 to negative charges on S_2 (see Figure 3.23). Thus, these magnetic field lines are directed opposite to the remanent magnetization $\mathbf{M}$ inside of the permanent magnet. For this reason, the magnetic field inside the permanent magnet is called *demagnetizing* field. The magnetic field outside the permanent magnet can be utilized for the excitation of a magnetic system.

It is clear from the presented discussion that the larger the remanent (saturation) magnetization M_s, the larger $|\sigma_m|$ and the stronger the magnetic field created by these charges. In other words, the larger M_s, the stronger the permanent magnet. Larger M_s also results in stronger demagnetizing field. Furthermore, this demagnetizing field is quite strong for short permanent magnets because positive and negative charges σ_m are less spatially separated for short magnets. For strong permanent magnets to sustain strong demagnetizing fields, their coercivity H_c must be sufficiently large. Finally, the squareness of hysteresis M-H loop guarantees that the magnetization of a permanent magnet is not affected by the demagnetizing field and it remains equal to M_s. In other words, the strength of the permanent magnet is not degraded by its demagnetizing field. Thus, it is evident that M_s, H_c and the squareness of M-H loops are the figures of merit of the permanent magnet.

Next, we shall demonstrate that within the framework of the magnetic circuit theory the ideal permanent magnet can be represented as a *nonideal flux source*. The reasoning proceeds as follows. Consider a cylindrical ideal permanent magnet with remanent magnetization M_s, length ℓ_0 and normal cross-sectional area A_0 (see Figure 3.24a). Inside of the permanent magnet, there exists the demagnetizing field H which is assumed to be uniform within the framework of the magnetic circuit theory. Since the demagnetizing magnetic field is directed opposite to magnetization, equation (3.102)

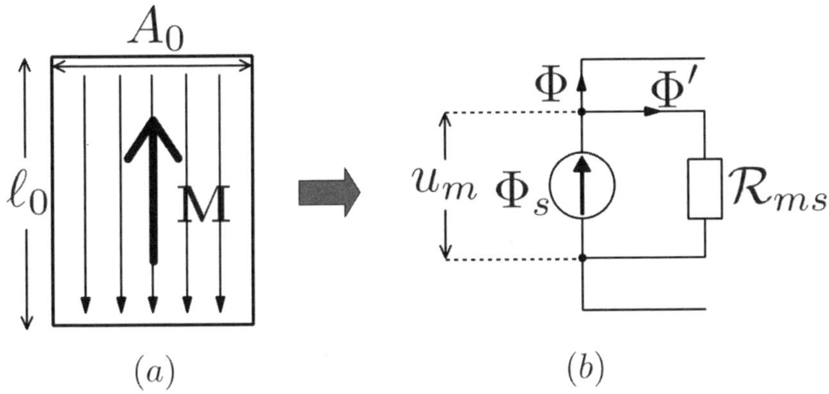

Fig. 3.24

can be written in the form

$$B = \mu_0 M_s - \mu_0 H. \quad (3.120)$$

We shall multiply both sides of the last formula by A_0,

$$BA_0 = \mu_0 M_s A_0 - \mu_0 A_0 H, \quad (3.121)$$

and shall take into account that the flux Φ through the permanent magnet is

$$\Phi = BA_0, \quad (3.122)$$

while the source flux Φ_s can be introduced as

$$\boxed{\Phi_s = \mu_0 M_s A_0.} \quad (3.123)$$

From formulas (3.121), (3.122) and (3.123) we conclude that

$$\Phi = \Phi_s - \mu_0 A_0 H. \quad (3.124)$$

Next, the last formula can be transformed as follows:

$$\Phi = \Phi_s - \frac{H\ell_0}{\dfrac{\ell_0}{\mu_0 A_0}}. \quad (3.125)$$

Finally, we introduce the drop of magnetic potential u_m across the permanent magnet as

$$u_m = H\ell_0 \quad (3.126)$$

and the permanent magnet reluctance

$$\boxed{\mathcal{R}_{ms} = \frac{\ell_0}{\mu_0 A_0}.} \tag{3.127}$$

Then, equation (3.125) can be written in the form

$$\boxed{\Phi = \Phi_s - \frac{u_m}{\mathcal{R}_{ms}}.} \tag{3.128}$$

It is apparent that the last equation can be interpreted as the equation of the nonideal flux source shown in Figure 3.24b. Indeed, according to this figure, we have

$$\Phi = \Phi_s - \Phi' \tag{3.129}$$

and

$$\Phi' = \frac{u_m}{\mathcal{R}_{ms}}. \tag{3.130}$$

By substituting the last formula into equation (3.129), we arrive at the relation (3.128).

In magnetic circuit calculations, it may be convenient to represent the ideal permanent magnet as a nonideal magnetomotive force. This representation can be obtained through the equivalent transformation of the nonideal flux source into nonideal mmf, illustrated in Figure 3.25. It is clear that the transformation shown in Figure 3.25 will be equivalent with respect to the magnetic circuit in the box if terminal relations between Φ and u_m are identical. Indeed, the equivalence is then the consequence of the fact that the box magnetic circuits in Figures 3.25a and 3.25b are described by identical mathematical equations. According to Figure 3.25a, we have

$$\Phi = \Phi_s - \frac{u_m}{\mathcal{R}_{ms}}, \tag{3.131}$$

while according to Figure 3.25b we get

$$\text{mmf} = \Phi\mathcal{R}'_{ms} + u_m, \tag{3.132}$$

or

$$\Phi = \frac{\text{mmf}}{\mathcal{R}'_{ms}} - \frac{u_m}{\mathcal{R}'_{ms}}. \tag{3.133}$$

By comparing relations (3.131) and (3.133), we conclude that they will be identical if

$$\mathcal{R}'_{ms} = \mathcal{R}_{ms} \tag{3.134}$$

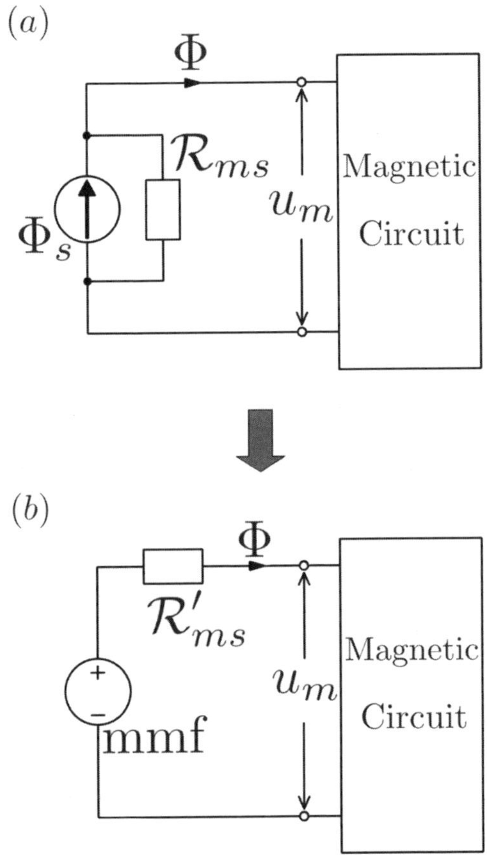

Fig. 3.25

and

$$\frac{\text{mmf}}{\mathcal{R}'_{ms}} = \Phi_s. \tag{3.135}$$

From the last two equations, we find

$$\text{mmf} = \Phi_s \mathcal{R}_{ms}. \tag{3.136}$$

Now, by using formulas (3.123) and (3.127) in the last equation, we derive

$$\boxed{\text{mmf} = M_s \ell_0.} \tag{3.137}$$

In summary, the ideal permanent magnet with remanent magnetization M_s, length ℓ_0 and normal cross-sectional area A_0 can be represented in the

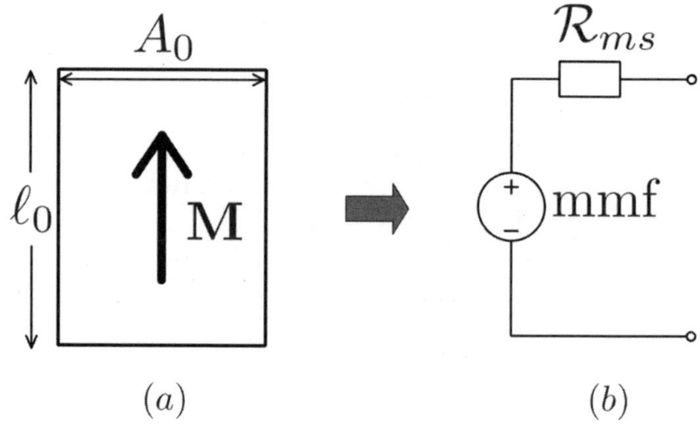

Fig. 3.26

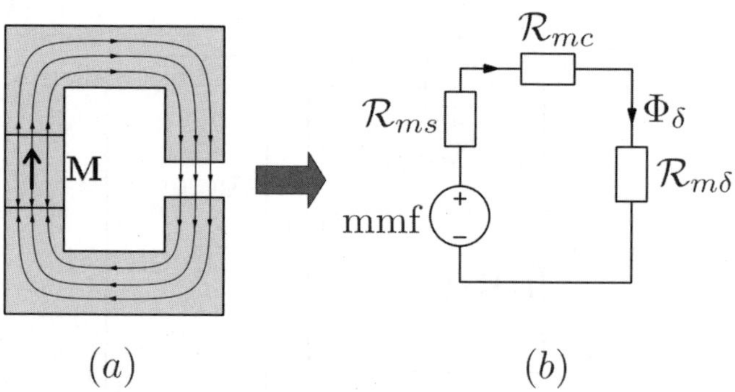

Fig. 3.27

magnetic circuit theory as a nonideal mmf (see Figure 3.26) with mmf and $\mathcal{R}_{ms}$ given by formulas (3.137) and (3.127), respectively.

As shown in Figure 3.23a, the exterior magnetic field of the permanent magnet is spread out. However, the magnetic field lines of this exterior field can be guided and focused in the desired air gap region by using a ferromagnetic core of high magnetic permeability. This type of arrangement is shown in Figure 3.27a, which represents a simple example of a magnetic system energized by a permanent magnet.

We shall next analyze this magnetic system by assuming that the remanent magnetization M_s of the ideal permanent magnet, its length ℓ_0 and normal cross-sectional area A_0 are given along with the length ℓ_c of the two legs of the ferromagnetic core, its cross-sectional area $A_c = A_0$ and permeability μ_c as well as the length δ of the air gap. It is required to find the flux through the air gap. Analysis is performed as the sequence of the following steps.

Step 1. We replace the magnetic system shown in Figure 3.27a by the equivalent magnetic circuit shown in Figure 3.27b. In this magnetic circuit, $\mathcal{R}_{mc}$ is the magnetic reluctance of the two legs of the ferromagnetic core which are connected in series. It is clear that by using the given data we find

$$\text{mmf} = M_s \ell_0, \tag{3.138}$$

$$\mathcal{R}_{ms} = \frac{\ell_0}{\mu_0 A_0}, \tag{3.139}$$

$$\mathcal{R}_{mc} = \frac{\ell_c}{\mu_c A_0}, \tag{3.140}$$

$$\mathcal{R}_{m\delta} = \frac{\delta}{\mu_0 A_0}. \tag{3.141}$$

Step 2. From Figure 3.27b, we find

$$\Phi_\delta = \frac{\text{mmf}}{\mathcal{R}_{ms} + \mathcal{R}_{mc} + \mathcal{R}_{m\delta}}. \tag{3.142}$$

Step 3. In applications, $\mu_c \gg \mu_0$ and it is usual that

$$\mathcal{R}_{mc} \ll \mathcal{R}_{m\delta} \quad \text{and} \quad \mathcal{R}_{mc} \ll \mathcal{R}_{ms}. \tag{3.143}$$

Consequently,

$$\Phi_\delta \approx \frac{\text{mmf}}{\mathcal{R}_{ms} + \mathcal{R}_{m\delta}}. \tag{3.144}$$

By substituting formulas (3.138), (3.139) and (3.141) into the last equation, we find

$$\Phi_\delta \approx \frac{M_s \ell_0}{\frac{\ell_0}{\mu_0 A_0} + \frac{\delta}{\mu_0 A_0}} = \mu_0 M_s A_0 \frac{\ell_0}{\ell_0 + \delta}. \tag{3.145}$$

Now, by recalling formula (3.123), we obtain

$$\Phi_\delta \approx \Phi_s \frac{\ell_0}{\ell_0 + \delta}. \tag{3.146}$$

It is clear from the last formula that the magnetic flux through the air gap is always smaller than the source flux Φ_s. It is also clear that the optimal design of the magnetic system is realized when

$$\ell_0 \gg \delta \tag{3.147}$$

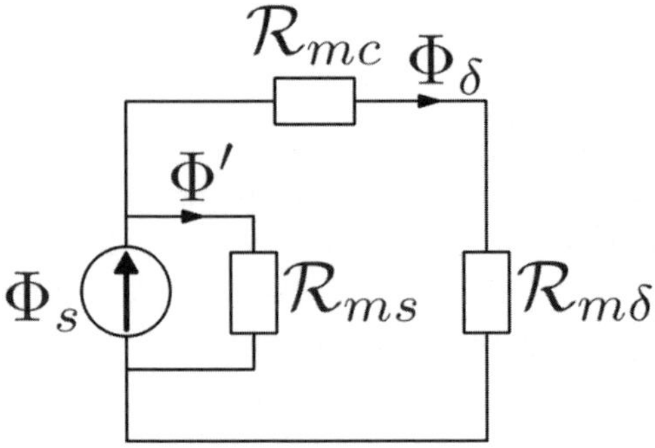

Fig. 3.28

and

$$\Phi_\delta \approx \Phi_s. \tag{3.148}$$

In this case, the magnetic reluctance of the air gap shunts the magnetic reluctance of the permanent magnet and practically all magnetic field lines go through the air gap. This not only increases the magnetic flux through the air gap, but also appreciably reduces the undesired demagnetizing field. This physical picture becomes especially transparent when the nonideal flux source representation of the ideal permanent magnet is used in computations. Indeed, in this case the equivalent magnetic circuit for the magnetic system shown in Figure 3.27 is presented by Figure 3.28. Under the condition (3.147), it follows from equations (3.139) and (3.141) that

$$\mathcal{R}_{ms} \gg \mathcal{R}_{m\delta} \tag{3.149}$$

and this leads to formula (3.148). It is apparent from the presented discussion that under the condition (3.147) the strength of the permanent magnet is almost completely utilized.

It is interesting to point out that the nonideal magnetomotive force representation of the ideal permanent magnet can be obtained from the "electric current" model of such magnets. To start the discussion of this model, we first remark that from the definition of the ideal permanent magnet it follows that

$$\operatorname{curl} \mathbf{M} = 0. \tag{3.150}$$

Since magnetic field only of the permanent magnet is being considered, we find that

$$\operatorname{curl} \mathbf{H} = 0. \tag{3.151}$$

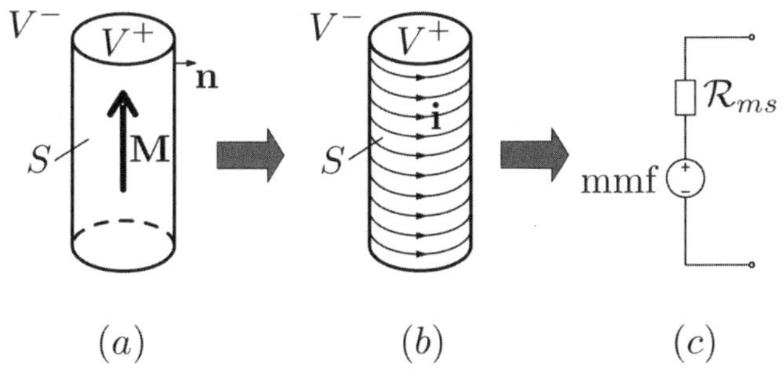

Fig. 3.29

From equations (3.102), (3.150) and (3.151) we obtain

$$\operatorname{curl} \mathbf{B} = 0. \tag{3.152}$$

In addition,

$$\operatorname{div} \mathbf{B} = 0. \tag{3.153}$$

It is clear from the last two equations that the field of magnetic flux density **B** of the ideal permanent magnet does not have any volume sources. Next, we shall demonstrate that this field has only surface sources distributed over its sides S. To this end, we shall use the continuity of tangential components of magnetic field on S,

$$\mathbf{n} \times \left(\mathbf{H}^- - \mathbf{H}^+\right) = 0, \tag{3.154}$$

which can also be written as

$$\frac{1}{\mu_0}\mathbf{n} \times \left(\mathbf{B}^- - \mathbf{B}^+\right) = \mathbf{n} \times \mathbf{M}, \tag{3.155}$$

where $\mathbf{n}$ is the unit outward normal to S (see Figure 3.29a).

Thus, the tangential component of magnetic flux density is discontinuous across S and this discontinuity is the source of the magnetic flux density field. In the "electric current" model of ideal permanent magnets (see Figure 3.29b), the remanent magnetization is removed and replaced by virtual surface currents **i** on S with density

$$\mathbf{i} = \mathbf{n} \times \mathbf{M}. \tag{3.156}$$

Then, the tangential component of the magnetic flux density field created by this surface current satisfies the boundary condition on S

$$\frac{1}{\mu_0}\mathbf{n} \times \left(\mathbf{B}^- - \mathbf{B}^+\right) = \mathbf{i} = \mathbf{n} \times \mathbf{M}. \tag{3.157}$$

This boundary condition is identical to the boundary condition (3.155). Furthermore, the magnetic flux density created by these surface currents satisfies the homogeneous equations (3.152) and (3.153) in V^- and V^+. Thus, the actual field $\mathbf{B}$ of the ideal permanent magnet and the field $\mathbf{B}$ created by surface currents $\mathbf{i}$ satisfy the same equations and boundary conditions. Consequently, these two fields of $\mathbf{B}$ are identical.

Now, if the region V^+ with surface currents $\mathbf{i}$ on S is considered as a leg of the overall magnetic system, then this leg can be characterized by its reluctance

$$\mathcal{R}_{ms} = \frac{\ell_0}{\mu_0 A_0} \tag{3.158}$$

as well as by the magnetomotive force

$$\text{mmf} = i\ell_0 \tag{3.159}$$

due to the surface electric currents $\mathbf{i}$. It is clear according to formula (3.156) that $i = |\mathbf{M}| = M_s$ and the last expression can be written in the form

$$\text{mmf} = M_s\ell_0. \tag{3.160}$$

Thus, we have arrived at the nonideal magnetomotive force model (Figure 3.29c) of the ideal permanent magnet.

In conclusion, there are two equivalent models of ideal permanent magnets: 1) the magnetic charge model that reproduces the magnetic field $\mathbf{H}$ of the ideal permanent magnet and leads to its magnetic circuit representation as a nonideal flux source; and 2) the electric current model that reproduces the magnetic flux density field $\mathbf{B}$ of the ideal permanent magnet and leads to its magnetic circuit representation as a nonideal magnetomotive force.

3.4 Nonlinear Magnetic Circuits

In our previous discussion of the magnetic circuit theory, it has been assumed that each leg of ferromagnetic core can be characterized by constant magnetic permeability and by magnetic reluctance. This is tantamount to the assumption of linearity of the magnetic properties of the ferromagnetic core when each leg can be characterized by a linear magnetization curve $B_k = \mu_k H_k$ (see Figure 3.30). However, the actual magnetic properties of ferromagnetic cores may appreciably deviate from this linearity assumption. As has been mentioned in the previous section, soft magnetic materials are used for ferromagnetic cores. These materials have very narrow hysteresis

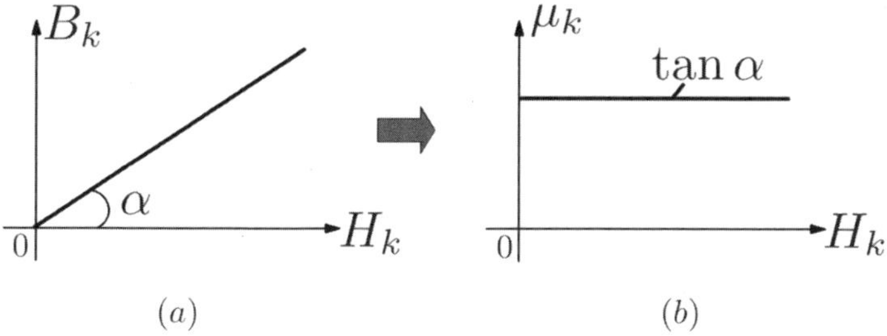

Fig. 3.30

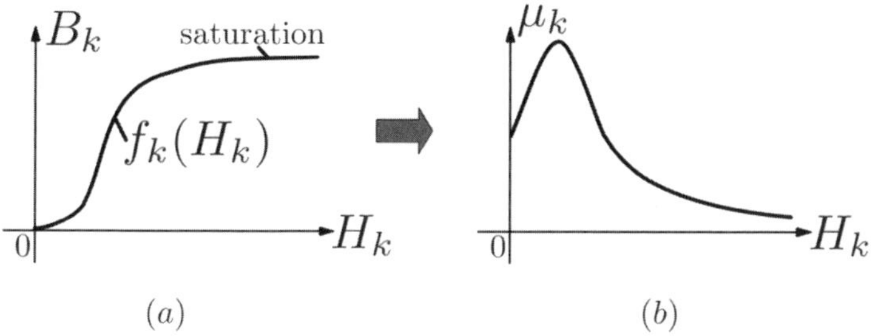

Fig. 3.31

loops (see Figure 3.21a). For this reason, as a reasonable approximation, hysteresis of these materials may be neglected and soft magnetic materials can be characterized by a nonlinear magnetization curve (see Figure 3.31a). It is clear that magnetic permeability $\mu_k(H_k) = B_k/H_k$ of such materials is not constant (see Figure 3.31b) and exhibits monotonic decrease for sufficiently large magnetic field. This phenomenon of flattening of magnetization curve and decrease in magnetic permeability is called saturation.

In this section, we shall further develop the magnetic circuit theory to account for nonlinear magnetic properties of ferromagnetic cores. It is clear that the first two fundamental equations (3.11) and (3.22) of the magnetic circuit theory do not depend on the assumption of linearity of magnetic properties of ferromagnetic cores and, consequently, these equations are valid for nonlinear magnetic circuits as well. On the other hand, the derivation of the Ohm's Law (3.33) was based on the linearity assumption and must be modified to be valid for nonlinear magnetic circuits. This modification can be performed as follows. Consider nonlinear magnetization

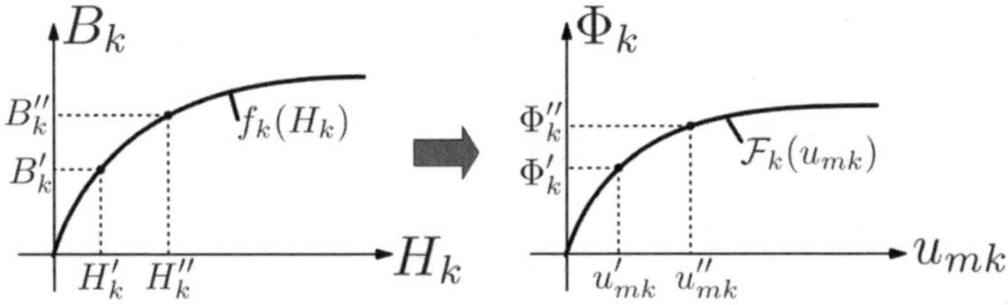

Fig. 3.32

curve

$$B_k = f_k(H_k) \tag{3.161}$$

of leg number k. We next multiply both sides of the last equation by the cross-sectional area A_k and, taking into account that $\Phi_k = B_k A_k$, we obtain

$$\Phi_k = A_k f_k(H_k). \tag{3.162}$$

Then, from equation (3.24) we find

$$H_k = \frac{u_{mk}}{\ell_k}. \tag{3.163}$$

By substituting the last relation into formula (3.162), we arrive at

$$\Phi_k = A_k f_k\left(\frac{u_{mk}}{\ell_k}\right). \tag{3.164}$$

This means that for nonlinear magnetic circuits the linear Ohm's Law (3.33) must be replaced by the nonlinear Ohm's Law

$$\boxed{\Phi_k = \mathcal{F}_k(u_{mk}),} \tag{3.165}$$

where

$$\boxed{\mathcal{F}_k(u_{mk}) = A_k f_k\left(\frac{u_{mk}}{\ell_k}\right).} \tag{3.166}$$

It is apparent from the last formula that nonlinear function $\mathcal{F}_k(u_{mk})$ is obtained by the appropriate scaling of magnetization curve $f_k(H_k)$ and this scaling is performed by using geometric parameters A_k and ℓ_k of the leg. This scaling is illustrated by Figure 3.32 and formula (3.167),

$$u'_{mk} \to H'_k = \frac{u'_{mk}}{\ell_k} \to B'_k = f_k\left(H'_k\right) \to \Phi'_k = A_k B'_k. \tag{3.167}$$

According to this formula, for an arbitrary chosen value of u'_{mk} we sequentially compute H'_k, B'_k and Φ'_k and determine a point of the graph

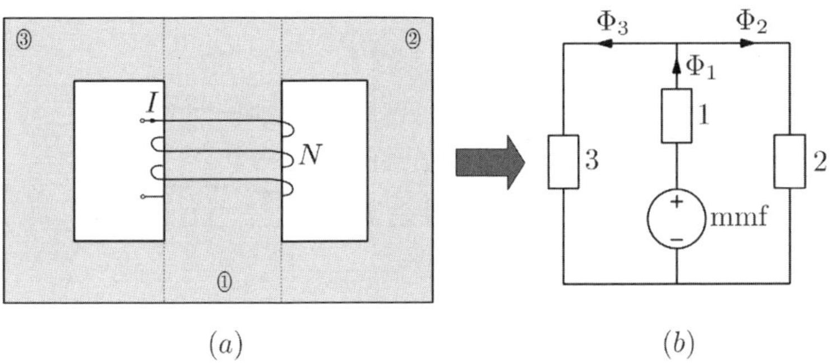

Fig. 3.33

representing the function $\mathcal{F}_k(u_{mk})$. In this way, point by point, the graph of $\mathcal{F}_k(u_{mk})$ can be constructed.

Next, we shall discuss graphical analysis of nonlinear magnetic circuits by using the nonlinear Ohm's Law defined by formulas (3.165)-(3.166). We shall illustrate this analysis by considering several examples.

Example 1. Consider a magnetic system shown in Figure 3.33a. It is assumed that the number N of turns of the coil and the current I through the coil are given along with the length ℓ_k, cross-sectional area A_k and magnetization curves $B_k = f_k(H_k)$ of each leg of the ferromagnetic core. It is required to find fluxes Φ_k, $(k = 1, 2, 3)$.

Step 1. We replace the actual magnetic system by the equivalent nonlinear magnetic circuit. In doing so, we replace the current-carrying coil with the magnetomotive force

$$\text{mmf} = NI \tag{3.168}$$

and each leg of the ferromagnetic core with nonlinear magnetic reluctance whose Φ-u_m curves can be found by using formulas

$$\Phi_k = \mathcal{F}_k(u_{mk}), \qquad (k = 1, 2, 3), \tag{3.169}$$

$$\mathcal{F}_k(u_{mk}) = A_k f_k\left(\frac{u_{mk}}{\ell_k}\right). \tag{3.170}$$

These nonlinear magnetic reluctances are connected in the equivalent magnetic circuit in the same way as the corresponding legs are connected in the ferromagnetic core (see Figure 3.33b). It is apparent that the given data can be used to perform the calculations in formulas (3.168)-(3.170).

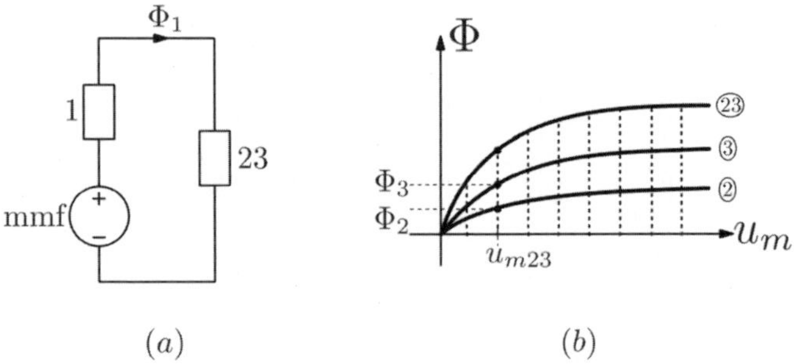

Fig. 3.34

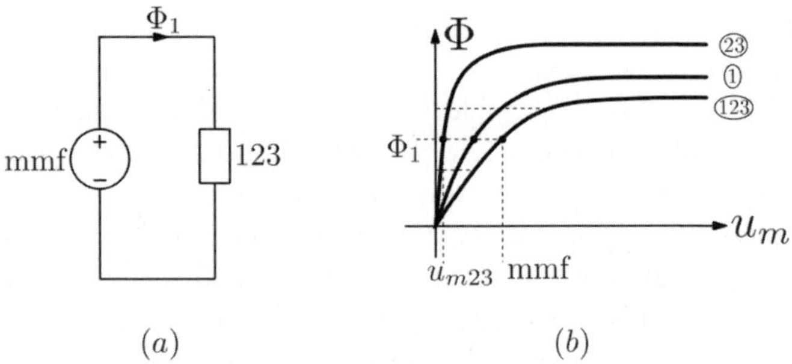

Fig. 3.35

Step 2. Next, we perform two equivalent transformations. The first transformation is to replace the two nonlinear reluctances 2 and 3 connected in parallel by one equivalent reluctance 23 (see Figure 3.34a). The latter means that we have to find the Φ-u_m graph for the reluctance 23 by using the Φ-u_m graphs for reluctances 2 and 3. This can be done by adding graphs for reluctances 2 and 3 "along the Φ-axis" as illustrated in Figure 3.34b. Namely, for any chosen value of u_m the flux Φ_{23} for the equivalent reluctance 23 is the sum of fluxes Φ_2 and Φ_3 corresponding to the chosen value of u_m. It is obvious that this graph addition along the Φ-axis is justified due to the parallel connection of nonlinear reluctances 2 and 3.

The second transformation is to replace the two nonlinear reluctances 1 and 23 connected in series by one equivalent reluctance 123 (see Figure 3.35a). The latter means to find the Φ-u_m graph for the reluctance 123 by using the Φ-u_m graphs for reluctances 1 and 23. This can be accomplished by adding graphs for reluctances 1 and 23 "along the u_m-axis" (see Figure 3.35b). Namely, for any chosen value of Φ the u_m for the equivalent reluctance 123 is equal to the sum of the u_m's for reluctances 1 and 23 for the same chosen value of Φ. This graph addition along the u_m-axis is justified due to the series connection of reluctances 1 and 23.

Step 3. It is clear from Figure 3.35a that the drop of magnetic potential across the equivalent nonlinear reluctance 123 is equal to the mmf. By using this fact and the curve 123 in Figure 3.35b we can find the flux Φ_1 as well as the drop of magnetic potential u_{m23} across the reluctance 23, which is the same as the drop of magnetic potential across reluctances 2 and 3. By using this fact and returning to Figure 3.34b, we find fluxes Φ_2 and Φ_3. This completes the solution of the problem.

It is clear that the nonlinear magnetic circuit shown in Figure 3.33b is mathematically described by nonlinear algebraic equations. The presented analysis is the graphical solution of these nonlinear equations which exploits the connectivity of the magnetic circuit. This graphical analysis can be easily programmed and run on computers. It is also worthwhile to point out that the presented technique has one important advantage. Namely, the third step can be done simultaneously for many different values of mmf (different values of current I) without any changes in the laborious step 2.

Example 2. Consider a magnetic system with two coils shown in Figure 3.36a. It is assumed that the numbers N_1 and N_2 of turns of the coils and the current I_1 through the first coil are given along with the length ℓ_k, cross-sectional area A_k and magnetization curve $B_k = f_k(H_k)$ of each leg of the ferromagnetic core. It is required to find the current I_2 through the second coil that guarantees the desired flux Φ_3 through the third leg.

Step 1. As before, we shall replace the actual magnetic system by the equivalent nonlinear magnetic circuit shown in Figure 3.36b. By using the given data, we find the magnetomotive force of the first coil

$$\text{mmf}_1 = N_1 I_1 \tag{3.171}$$

and the Φ-u_m graphs for the three nonlinear magnetic reluctances,

$$\Phi_k = \mathcal{F}_k(u_{mk}), \qquad (k = 1, 2, 3), \tag{3.172}$$

$$\mathcal{F}_k(u_{mk}) = A_k f_k\left(\frac{u_{mk}}{\ell_k}\right). \tag{3.173}$$

Step 2. By using the Φ-u_m graph for the third nonlinear reluctance and given flux Φ_3, we find the drop of magnetic potential across the third reluctance by using the procedure illustrated in Figure 3.37.

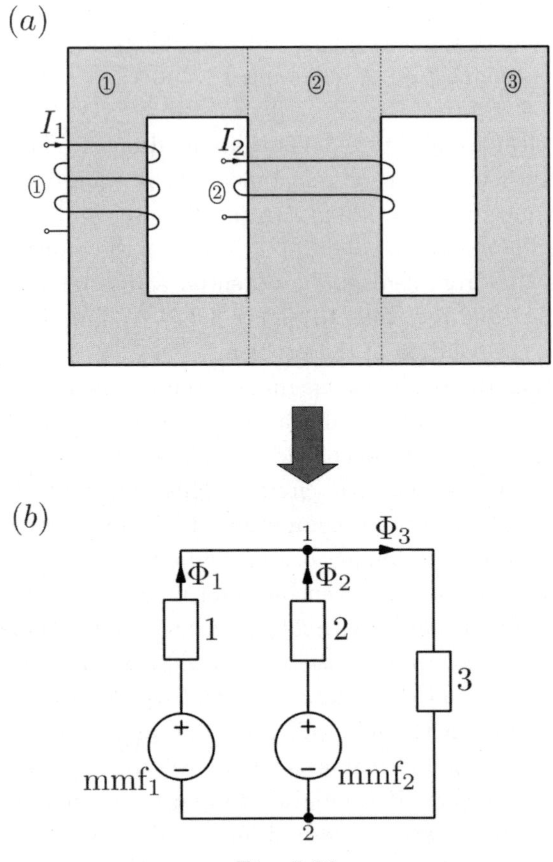

Fig. 3.36

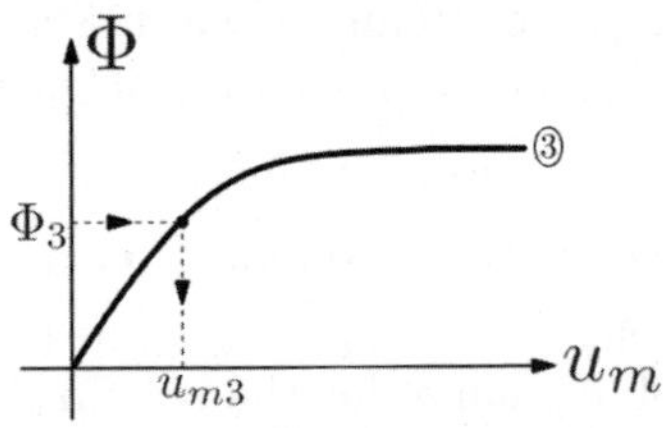

Fig. 3.37

Step 3. Then, for the loop consisting of the first and third reluctances and mmf_1 we find

$$u_{m1} + u_{m3} = \text{mmf}_1. \tag{3.174}$$

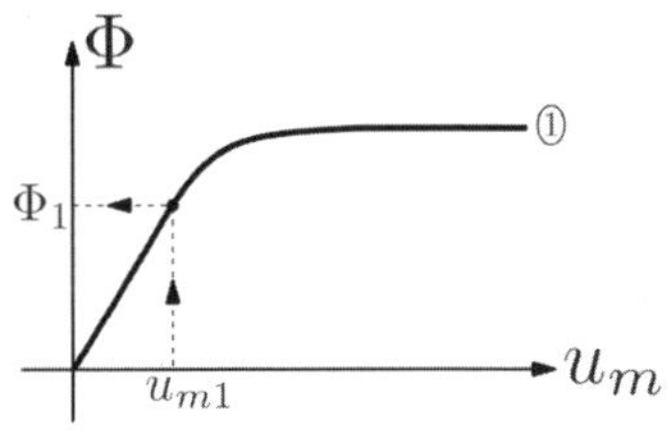

Fig. 3.38

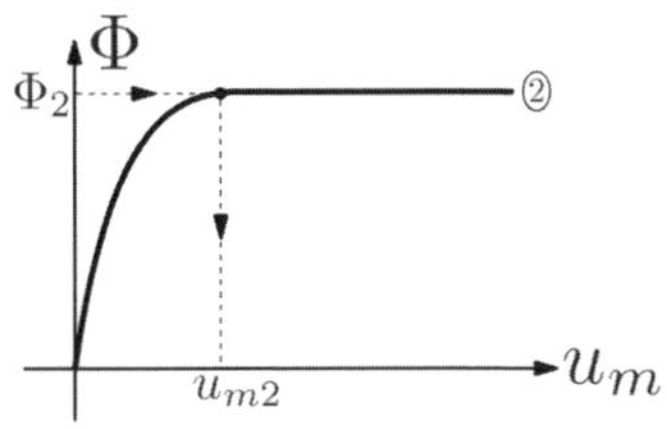

Fig. 3.39

Consequently,

$$u_{m1} = \mathrm{mmf}_1 - u_{m3} \tag{3.175}$$

and u_{m1} can be found.

Step 4. By using the Φ-u_m graph for the first nonlinear reluctance and the value of u_{m1} found in the third step, we find the flux Φ_1 through the first leg as illustrated in Figure 3.38.

Step 5. For node 1, we find

$$\Phi_1 + \Phi_2 - \Phi_3 = 0, \tag{3.176}$$

which leads to

$$\Phi_2 = \Phi_3 - \Phi_1. \tag{3.177}$$

Thus, the flux Φ_2 can be found.

Step 6. By using the Φ-u_m curve for the second nonlinear reluctance and the value of Φ_2 found in the previous step, we find u_{m2} (see Figure 3.39).

Step 7. Now, for the loop consisting of the second and third reluctances and mmf_2 we can write

$$\mathrm{mmf}_2 = u_{m2} + u_{m3}. \tag{3.178}$$

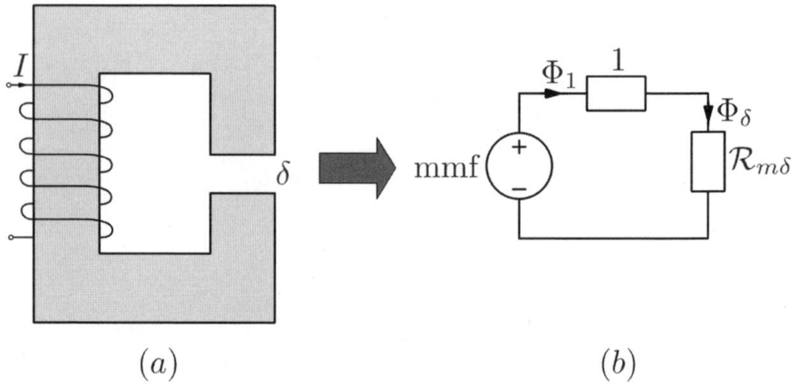

Fig. 3.40

Since

$$\mathrm{mmf}_2 = N_2 I_2, \tag{3.179}$$

from the last two formulas we find

$$I_2 = \frac{u_{m2} + u_{m3}}{N_2}, \tag{3.180}$$

which completes the solution of the problem.

Example 3. Consider a magnetic system with an air gap shown in Figure 3.40a. It is assumed that the number N of turns of the coil and the current I through the coil are given along with the air gap length δ, the length ℓ_1, cross-sectional area A_1 and the magnetization curve $B_1 = f_1(H_1)$ of the ferromagnetic core. It is required to find the flux Φ_δ through the air gap.

Step 1. As before, we replace the actual magnetic system by the equivalent nonlinear magnetic circuit shown in Figure 3.40b. By using the given data, we find

$$\mathrm{mmf} = NI, \tag{3.181}$$

$$\Phi_1 = \mathcal{F}_1(u_{m1}), \tag{3.182}$$

$$\mathcal{F}_1(u_{m1}) = A_1 f_1\left(\frac{u_{m1}}{\ell_1}\right), \tag{3.183}$$

$$\mathcal{R}_{m\delta} = \frac{\delta}{\mu_0 A_1}. \tag{3.184}$$

Step 2. From the magnetic circuit shown in Figure 3.40b, we conclude

$$\mathrm{mmf} = u_{m1}(\Phi_1) + \mathcal{R}_{m\delta}\Phi_\delta, \tag{3.185}$$

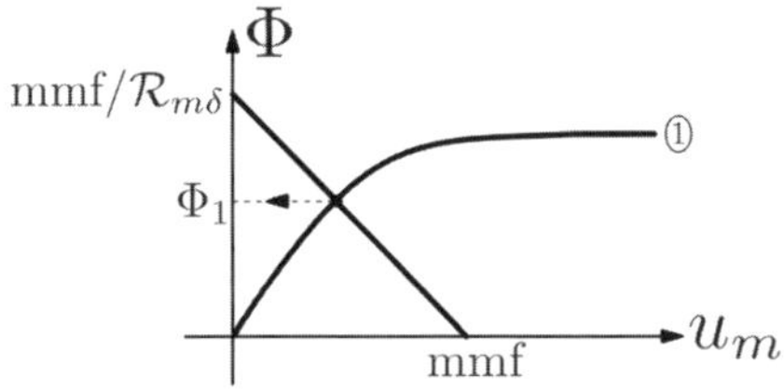

Fig. 3.41

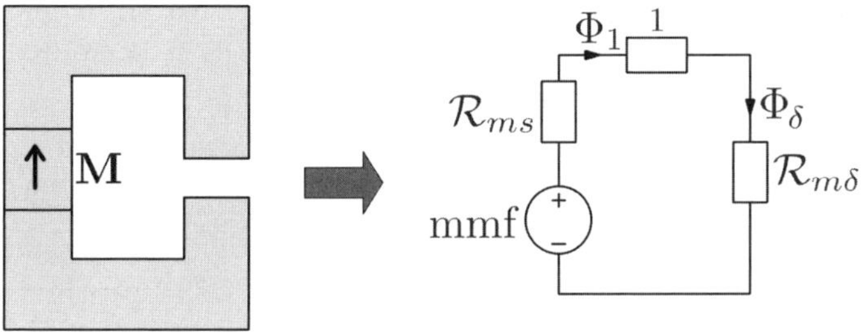

(a) (b)

Fig. 3.42

where $u_{m1}(\Phi_1)$ is the inverse function of $\Phi_1 = \mathcal{F}_1(u_{m1})$.

It is also clear that

$$\Phi_1 = \Phi_\delta, \tag{3.186}$$

and the last equation can be written as follows:

$$\text{mmf} - \mathcal{R}_{m\delta}\Phi_1 = u_{m1}(\Phi_1). \tag{3.187}$$

It is clear that the solution Φ_1 of equation (3.187) is such that the functions on both sides of this equation have the same value. This fact can be used for the graphical solution of equation (3.187), which is illustrated by Figure 3.41. In this figure, the linear function representing the left-hand side of equation (3.187) is plotted along with the Φ-u_m function for the nonlinear reluctance 1. It is apparent that these two functions of Φ_1 are equal to one another for the value of Φ_1 which corresponds to the point of intersection of their plots. This completes the solution of the problem.

Example 4. Consider a magnetic system with an ideal permanent magnet shown in Figure 3.42a. It is assumed that remanent magnetization M_s of the permanent magnet, its length ℓ_0 and its cross-sectional area A_0 are given. Furthermore, the air gap length δ, the total length ℓ_1 of the two

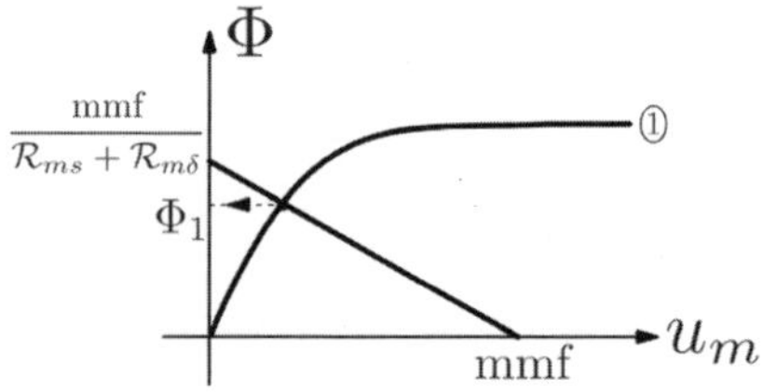

Fig. 3.43

legs of the ferromagnetic core, its cross-sectional area $A_1 = A_0$ and the magnetization curve $B_1 = f_1(H_1)$ are given as well. It is required to find the flux Φ_δ through the air gap.

Step 1. We replace the actual magnetic system by the equivalent magnetic circuit shown in Figure 3.42b. Here,

$$\text{mmf} = M_s \ell_0, \tag{3.188}$$

$$\mathcal{R}_{ms} = \frac{\ell_0}{\mu_0 A_0}, \tag{3.189}$$

$$\mathcal{R}_{m\delta} = \frac{\delta}{\mu_0 A_0}, \tag{3.190}$$

$$\Phi_1 = \mathcal{F}_1(u_{m1}), \qquad \mathcal{F}_1(u_{m1}) = A_0 f_1\left(\frac{u_{m1}}{\ell_1}\right). \tag{3.191}$$

Step 2. From the magnetic circuit shown in Figure 3.42b, we find

$$\Phi_1 = \Phi_\delta, \tag{3.192}$$

$$\text{mmf} = \mathcal{R}_{ms}\Phi_1 + u_{m1}(\Phi_1) + \mathcal{R}_{m\delta}\Phi_1. \tag{3.193}$$

The last equation can be rearranged by separating linear and nonlinear parts,

$$\text{mmf} - (\mathcal{R}_{ms} + \mathcal{R}_{m\delta})\Phi_1 = u_{m1}(\Phi_1), \tag{3.194}$$

and then solved graphically by using the same reasoning as in the previous example. This graphical solution is illustrated by Figure 3.43. This concludes the discussion of this example.

3.5 Hysteresis and Eddy Current Losses

In previous sections of this chapter, we have discussed magnetic systems excited by coils with dc currents or by permanent magnets, and we have

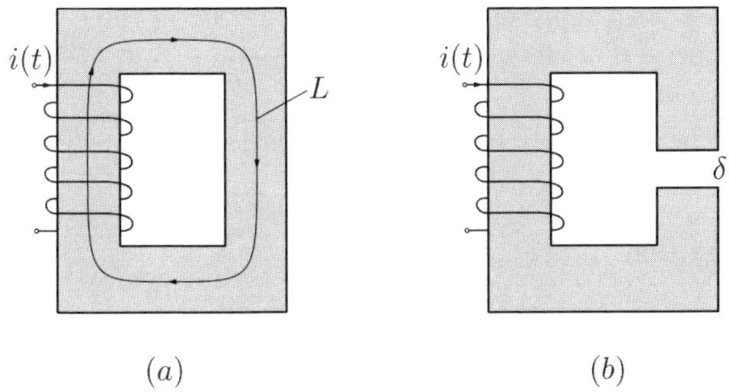

Fig. 3.44

developed the magnetic circuit theory to analyze such magnetic systems. In power engineering-related applications, magnetic systems are often excited by ac currents. The magnetic circuit theory still can be used for the analysis of magnetic systems with ac excitations by applying this theory at each instant of time. Nevertheless, the ac excitations lead to new important effects such as generation of higher-order harmonics and appearance of energy losses due to the phenomena of hysteresis and eddy currents.

As far as the generation of higher-order harmonics is concerned, it occurs due to nonlinear magnetic properties of ferromagnetic cores. As a consequence of these nonlinear properties, sinusoidal (time-harmonic) variations of magnetic field result in non-sinusoidal time variations of magnetic flux density, and the other way around: sinusoidal variations of magnetic flux density result in non-sinusoidal variations of magnetic field. This implies that if magnetic systems are excited by sinusoidal currents then magnetic fluxes and voltages will not be sinusoidal and will contain higher-order harmonics in their Fourier series expansions. Similarly, if magnetic systems are excited by sinusoidal voltages applied to their coils, then currents in these coils will not be sinusoidal and will contain higher-order harmonics.

Next, we consider the phenomena of hysteresis and eddy current losses, which are often called core losses. The existence of these losses appreciably affects the actual design of magnetic systems of power equipment. We start with hysteresis losses and shall first discuss the issue of energy stored in the magnetic field of magnetic systems. To do this, consider a simple magnetic system shown in Figure 3.44a. For the input voltage across the terminals of the coil of this magnetic system we can write the equation

$$v(t) = Ri(t) + \frac{d\psi(t)}{dt}, \tag{3.195}$$

where the last term represents the voltage induced in the coil by the time-varying magnetic flux through the ferromagnetic core. (As before, leakage flux is neglected.)

We shall next multiply both sides of equation (3.195) by $i(t)dt$:

$$v(t)i(t)dt = Ri^2(t)dt + i(t)d\psi(t). \tag{3.196}$$

It is apparent that

$$v(t)i(t)dt = p(t)dt = dW, \tag{3.197}$$

where dW stands for an infinitesimal amount of energy supplied to the magnetic system during the time dt. It is also clear that

$$Ri^2(t)dt = dW_R \tag{3.198}$$

is the amount of energy dissipated into heat during the time dt due to the ohmic resistance of the coil.

The last term in equation (3.196),

$$i(t)d\psi(t) = dW_f, \tag{3.199}$$

can be interpreted as the change in the energy stored in the magnetic field during the time dt. Thus, the equation (3.196) can be written as the following energy balance relation:

$$dW = dW_R + dW_f. \tag{3.200}$$

Next, we shall derive the expression for dW_f in terms of magnetic field and magnetic flux density. To this end, we shall apply Ampere's Law to some magnetic field line L:

$$\oint_L \mathbf{H}(t) \cdot d\boldsymbol{\ell} = Ni(t). \tag{3.201}$$

Since L is a magnetic field line and since within the framework of magnetic circuit theory magnetic field is assumed to be uniform, the integral in the last formula can be evaluated as follows:

$$\oint_L \mathbf{H}(t) \cdot d\boldsymbol{\ell} = \int_L H(t)d\ell = H(t)\ell, \tag{3.202}$$

where ℓ is the average length of the ferromagnetic core.

From formulas (3.201) and (3.202) we derive

$$i(t) = \frac{H(t)\ell}{N}. \tag{3.203}$$

On the other hand, we have

$$d\psi(t) = Nd\Phi(t) = NAdB(t), \tag{3.204}$$

where A is the normal cross-sectional area of the ferromagnetic core.

By substituting formulas (3.203) and (3.204) into equation (3.199), we derive

$$dW_f = \ell A H(t) dB(t) = V H(t) dB(t), \tag{3.205}$$

where $V = \ell A$ is the volume of the ferromagnetic core.

By integrating the last equation, we find

$$\boxed{W_f = V \int_0^B H dB.} \tag{3.206}$$

In the last equation W_f has the meaning of the change in the energy stored in the magnetic field when the magnetic flux density is monotonically increased from zero to B. If the stored magnetic energy is equal to zero at $B = 0$, then W_f in (3.206) has the meaning of the energy stored in the magnetic field when magnetic flux density is equal to B.

In the case of linear constitutive relation $B = \mu H$, from the last formula we find

$$W_f = V \frac{B^2}{2\mu}. \tag{3.207}$$

We shall apply the last formula to the case of magnetic systems with air gaps (see Figure 3.44b). In this case, the total magnetic energy consists of two distinct parts:

(1) energy W_{fc} stored in the magnetic field of the ferromagnetic core; and
(2) energy $W_{f\delta}$ stored in the magnetic field of the air gap:

$$W_f = W_{fc} + W_{f\delta}. \tag{3.208}$$

By using formula (3.207), we find

$$W_{fc} = V_c \frac{B^2}{2\mu_c}, \tag{3.209}$$

$$W_{f\delta} = V_\delta \frac{B^2}{2\mu_0}. \tag{3.210}$$

Here, V_c and V_δ are volumes of the core and the gap regions, respectively, μ_c is the magnetic permeability of the core, and the same value B of magnetic flux density is used in the last two formulas to reflect the continuity of magnetic flux. It is instructive to compare W_{fc} and $W_{f\delta}$ by evaluating the ratio

$$\frac{W_{fc}}{W_{f\delta}} = \frac{V_c}{V_\delta} \frac{\mu_0}{\mu_c} = \frac{\ell_c}{\delta} \frac{\mu_0}{\mu_c}, \tag{3.211}$$

where ℓ_c is the average length of the ferromagnetic core. It is typical that

$$\mu_c > 10^3 \mu_0 \qquad \text{and} \qquad \ell_c \approx 50\delta. \tag{3.212}$$

Then,

$$\frac{W_{fc}}{W_{f\delta}} \ll 1, \tag{3.213}$$

and

$$W_f \approx W_{f\delta}. \tag{3.214}$$

Thus, by using high permeability ferromagnetic cores, the localization of stored magnetic energy in air gap regions can be achieved. The latter is very important because air gaps are the regions where the energy conversion occurs due to electromagnetic interactions between magnetic fields and currents (as will be evident later in studying electrical machines).

Next, we demonstrate that formula (3.206) can be used for the evaluation of hysteresis energy losses. Consider a hysteresis loop shown in Figure 3.45a and assume that the magnetic field is monotonically increased from 0 to H_m, resulting in monotonic increase in magnetic flux density ($dB > 0$) from $-B_r$ to B_m. Then, according to formula (3.206), the increase in magnetic energy provided by the source (through the energized coil) can be computed as follows:

$$W_f^+ = V \int_{-B_r}^{B_m} H dB > 0. \tag{3.215}$$

It is clear that the integral in the last formula is equal to the shaded area shown in Figure 3.45a.

Next, consider the monotonic decrease of magnetic field from H_m to 0, resulting in monotonic decrease of magnetic flux density ($dB < 0$) from B_m to B_r. Then, according to formula (3.206), the decrease in magnetic energy stored in the magnetic field (and returned to the source) can be computed as follows:

$$W_f^- = V \int_{B_m}^{B_r} H dB < 0. \tag{3.216}$$

It is clear that the absolute value of the integral in the last formula is equal to the shaded area shown in Figure 3.45b. Thus, the amount of energy W_f^{hc} provided by the source to the magnetic media during one half-cycle is

$$W_f^{hc} = W_f^+ + W_f^-. \tag{3.217}$$

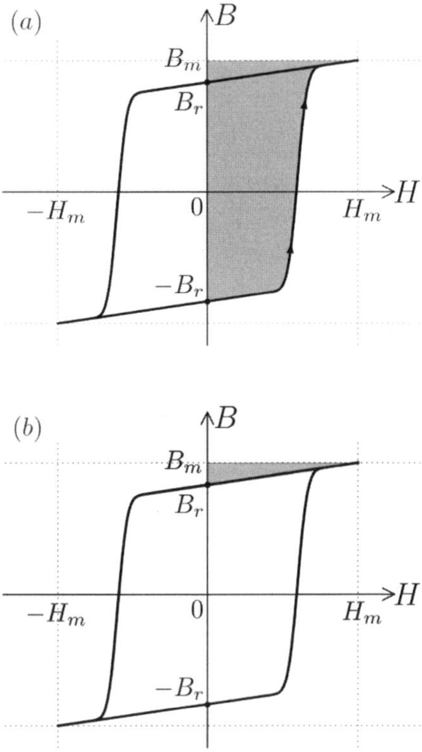

Fig. 3.45

It is clear from formulas (3.215) and (3.216) and their geometric interpretation (see Figure 3.45) that the last equation can be written in the form

$$W_f^{hc} = \frac{1}{2} V A_{\text{loop}}, \tag{3.218}$$

where A_{loop} is the area enclosed by the hysteresis loop. In formula (3.218) the coefficient $\frac{1}{2}$ is used because the subtraction of shaded areas shown in Figure 3.45 results in the shaded area shown in Figure 3.46a, which is one half of the area enclosed by the hysteresis loop.

By using the same line of reasoning as before, we can find that during the negative half-cycle of magnetic field variation (from 0 to $-H_m$ and back to 0), the amount of magnetic energy provided by the source to the ferromagnetic media will also be given by formula (3.218) with $\frac{1}{2}A_{\text{loop}}$ being the unshaded area of the hysteresis loop. Thus, the total amount of energy W_f^{fc} provided during the full cycle of variation of magnetic field is given

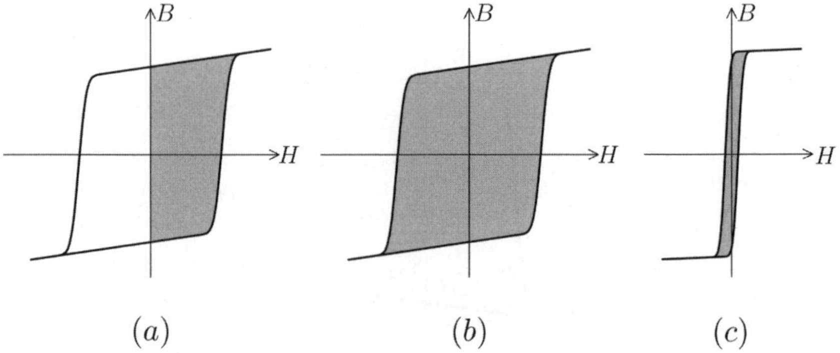

Fig. 3.46

by the formula

$$W_f^{fc} = VA_{\text{loop}}. \tag{3.219}$$

Since after one full cycle of variation of magnetic field, the magnetic field and magnetic flux density in the ferromagnetic core are the same as at the beginning of the cycle, the energy W_f^{fc} provided by the source can be interpreted only as dissipated energy. This dissipated energy is fully controlled by the shape of the hysteresis loop, namely, by the area enclosed by the loop (see Figure 3.46b). This dissipation results in energy losses which are usually referred to as hysteresis losses. The power P_h of these hysteresis losses is naturally given by the formula

$$P_h = fW_f^{fc}, \tag{3.220}$$

because the number f of full-cycle losses occur during one second. By combining formulas (3.219) and (3.220), we find

$$\boxed{P_h = fVA_{\text{loop}}.} \tag{3.221}$$

It is apparent that hysteresis losses can be appreciably reduced by using soft magnetic materials with narrow hysteresis loops (Figure 3.46c). These soft magnetic materials are used for the construction of ferromagnetic cores of many power devices (transformers, generators, motors, actuators, etc.).

Now, we switch to the discussion of eddy current losses in ferromagnetic cores. These losses exist because ferromagnetic cores have finite electric conductivity. For this reason, eddy currents are induced in these ferromagnetic cores when they are subject to time-varying magnetic fields. These induced currents result in energy losses called eddy current losses. To discuss these

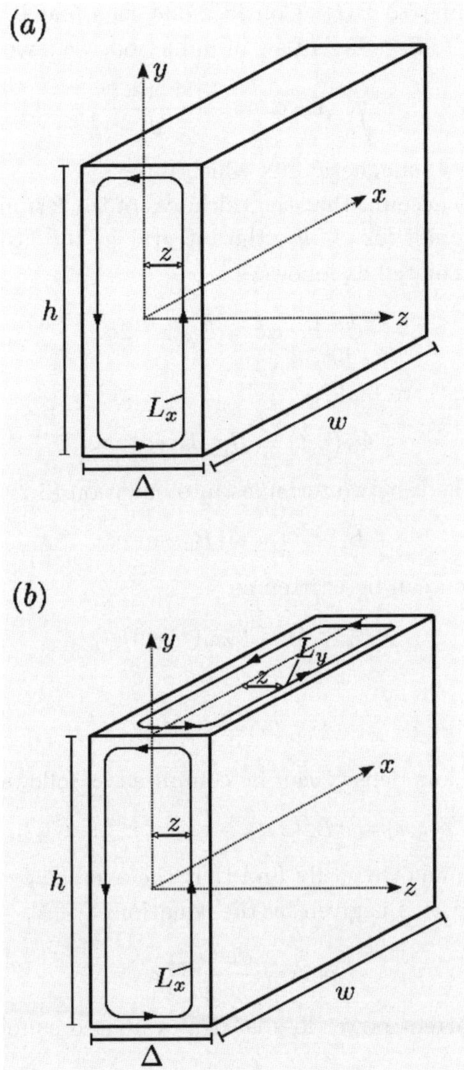

Fig. 3.47

losses, consider a piece of conducting ferromagnetic material shown in Figure 3.47a. Suppose that this lamination is subject to the time-varying magnetic flux density

$$\mathbf{B}(t) = \mathbf{e}_x B_m \cos \omega t. \tag{3.222}$$

This magnetic flux density will induce electric field whose field lines are in planes perpendicular to $\mathbf{B}(t)$. Consider one such line L_x shown in Figure 3.47a. According to Faraday's Law of induction, we have

$$\oint_{L_x} \mathbf{E} \cdot d\boldsymbol{\ell} = -\frac{d\Phi_x(z,t)}{dt}, \tag{3.223}$$

where $\Phi_x(z,t)$ is the magnetic flux which links L_x.

By taking into account that the thickness of the ferromagnetic material is usually quite small ($\Delta \ll h$), the integral in the last formula can be approximately evaluated as follows:

$$\oint_{L_x} \mathbf{E} \cdot d\boldsymbol{\ell} \approx E_y(z,t)2h. \tag{3.224}$$

For the flux $\Phi_x(z,t)$ we find

$$\Phi_x(z,t) = 2hzB_m \cos \omega t. \tag{3.225}$$

By substituting the last two formulas into equation (3.223), we derive

$$E_y(z,t) = \omega z B_m \sin \omega t. \tag{3.226}$$

The last expression can be written as

$$E_y(z,t) = E_{my}(z) \sin \omega t, \tag{3.227}$$

where

$$E_{my}(z) = \omega z B_m. \tag{3.228}$$

Now, local power loss density can be computed as follows:

$$p(z,t) = \sigma E_y^2(z,t) = \sigma \omega^2 z^2 B_m^2 \sin^2 \omega t. \tag{3.229}$$

From the last formula we easily find that the average over period $T = \frac{2\pi}{\omega}$ power loss density $\overline{p}(z)$ is given by the equation

$$\overline{p}(z) = \frac{\sigma \omega^2 B_m^2}{2} z^2. \tag{3.230}$$

The total eddy current power loss is then evaluated as follows:

$$P_{ec} = \int_0^w \left(\int_{-\frac{h}{2}}^{\frac{h}{2}} \left(\int_{-\frac{\Delta}{2}}^{\frac{\Delta}{2}} \overline{p}(z) dz \right) dy \right) dx. \tag{3.231}$$

By substituting formula (3.230) into the last equation and performing the integration, we find

$$\boxed{P_{ec} = \frac{1}{24} \sigma V \omega^2 B_m^2 \Delta^2,} \tag{3.232}$$

where $V = wh\Delta$ is the volume of the lamination.

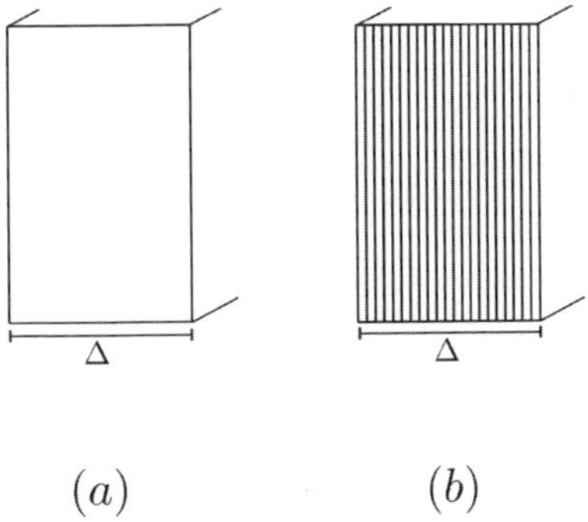

Fig. 3.48

It is clear from the last formula that the eddy current losses can be reduced by reducing conductivity σ. This is usually achieved by deliberately doping steel with silicon. This doping reduces the intrinsic conductivity of steel without appreciably affecting its high magnetic permeability. Such a silicon-doped steel is often called transformer steel, and it is used in the construction of ferromagnetic (steel) cores of many power devices. There is also another efficient way to reduce eddy current losses by using laminated structures for ferromagnetic cores. In these structures, the steel cores are assembled from a very large number of thin steel laminations which are electrically isolated from one another by very thin oxidation (or varnish) layers. To illustrate this way of reducing eddy current losses, we shall compare two designs of steel cores, a) solid core and b) laminated core (see Figure 3.48). In the case of the solid core, we have

$$P_{ec}^{(a)} = \frac{1}{24}\sigma V \omega^2 B_m^2 \Delta^2, \tag{3.233}$$

while in the case of the laminated core we derive

$$P_{ec}^{(b)} = \frac{n}{24}\sigma \left(\frac{V}{n}\right) \omega^2 B_m^2 \left(\frac{\Delta}{n}\right)^2, \tag{3.234}$$

where n is the total number of laminations. In deriving the last formula, we simply used equation (3.232) for each lamination in the core assembly and took into account that each such lamination has volume $\frac{V}{n}$ and thickness $\frac{\Delta}{n}$.

The last formula can be written as

$$P_{ec}^{(b)} = \frac{1}{24n^2}\sigma V \omega^2 B_m^2 \Delta^2. \tag{3.235}$$

From equations (3.233) and (3.235) we find

$$P_{ec}^{(b)} = \frac{P_{ec}^{(a)}}{n^2}, \tag{3.236}$$

which suggests that for a large number of thin laminations ($n \gg 1$)

$$P_{ec}^{(b)} \ll P_{ec}^{(a)}. \tag{3.237}$$

This implies that by using the laminated design of ferromagnetic cores, eddy current losses can be substantially reduced. For this reason, the laminated design of steel cores is customarily used in many power devices. It must be remarked that the laminated design comes with a price. This design reduces the mechanical rigidity of the cores and results in noise produced by vibrations of the laminations in the core assembly. Furthermore, the laminated design may not be efficient enough for very high frequency applications which may occur in power electronics. For such applications, ferrite cores are used. Ferrites are ceramic compounds of transition metals with oxygen. They usually have relatively high permeability but very low (if any) electrical conductivity.

In the presented analysis of eddy current losses, it has been assumed that the magnetic flux density $\mathbf{B}$ is linearly polarized (see formula (3.222)), i.e., its direction does not change with time. Such situation is typical for transformers. However, in ac electrical machines, ferromagnetic cores are subject to rotating magnetic fields. For this reason, it is of interest to discuss how the polarization of magnetic flux density affects eddy current losses. To this end, consider the case when a ferromagnetic lamination is subject to circularly polarized magnetic field with magnetic flux density given by the formula

$$\mathbf{B}(t) = \mathbf{e}_x B_m \cos\omega t + \mathbf{e}_y B_m \sin\omega t. \tag{3.238}$$

It is clear that the x-component of $\mathbf{B}(t)$ will induce electric field whose lines are in planes perpendicular to $\mathbf{e}_x$ and the y-component of $\mathbf{B}(t)$ will induce electric fields whose lines are perpendicular to $\mathbf{e}_y$ (see Figure 3.47b). It is also clear that the magnetic flux due to the y-component of $\mathbf{B}(t)$ does not link L_x and, the other way around, the magnetic flux due to the x-component of $\mathbf{B}(t)$ does not link L_y. This means that $E_x(z,t)$ and $E_y(z,t)$ can be computed independently by using the same line of reasoning that has been used in the derivation of formulas (3.227) and (3.228). This implies that the following expressions are valid:

$$E_x(z,t) = -E_{mx}(z)\cos\omega t, \tag{3.239}$$

$$E_y(z,t) = E_{my}(z)\sin\omega t, \tag{3.240}$$

where

$$E_{mx}(z) = E_{my}(z) = \omega z B_m. \tag{3.241}$$

From the last three formulas we find

$$p(z,t) = \sigma \left[E_x^2(z,t) + E_y^2(z,t)\right] = \sigma\omega^2 B_m^2 z^2. \tag{3.242}$$

It is apparent from equation (3.242) that the local power loss density is constant in time. In other words, in the case of circular polarization of magnetic flux density, the eddy current energy dissipation occurs at a constant rate in time. This clearly suggests that rotational eddy current losses are higher than those for unidirectional (linearly polarized) magnetic fields. It is also clear that

$$\overline{p}(z) = p(z,t) = \sigma\omega^2 B_m^2 z^2. \tag{3.243}$$

By using the last formula in equation (3.231), we derive

$$P_{ec}^{cir} = \frac{1}{12}\sigma V \omega^2 B_m^2 \Delta^2 \tag{3.244}$$

and

$$\boxed{P_{ec}^{cir} = 2P_{ec}^{lin},} \tag{3.245}$$

where P_{ec}^{cir} is the eddy current power loss in the case of circular polarization of $\mathbf{B}(t)$, while P_{ec}^{lin} is the eddy current power loss of linear polarization of $\mathbf{B}(t)$ given by the formula (3.232).

The obtained result can be extended to the case of elliptical polarization of $\mathbf{B}(t)$:

$$\mathbf{B}(t) = \mathbf{e}_x B_{mx}\cos\omega t + \mathbf{e}_y B_{my}\sin\omega t. \tag{3.246}$$

Indeed, by using the same line of reasoning as before, it can be established that

$$P_{ec}^{el} = \frac{1}{24}\sigma V \omega^2 \left(B_{mx}^2 + B_{my}^2\right)\Delta^2, \tag{3.247}$$

where P_{ec}^{el} stands for the eddy current power loss in the case of elliptical polarization. It is apparent that the last formula can be written in the form

$$\boxed{P_{ec}^{el} = P_{ec}^{lin,x} + P_{ec}^{lin,y},} \tag{3.248}$$

where $P_{ec}^{lin,x}$ and $P_{ec}^{lin,y}$ stand for eddy current power losses associated with two unidirectional (along x- and y-axes) components of magnetic flux density. The formulas (3.245) and (3.248) are consistent with experimental results reported in [54] and [61].

It is important to stress that we derived analytical expressions for eddy current losses without invoking any assumptions concerning the magnetic properties of the laminations. The main limitation of our derivations is the tacit assumption that the magnetic flux density is uniform over a lamination cross section. This assumption ignores the magnetic field created by eddy currents that may result in nonuniform distribution of magnetic flux density. This nonuniform distribution may lead to the increase in eddy current losses, which are usually called *excess* eddy current losses. The discussion of these *excess* losses is beyond the scope of this book. For this reason, we shall just mention that this discussion can be found in [36], [37], [38], [49], where it is demonstrated that nonlinear diffusion (penetration) of electromagnetic fields occurs as an inward progress of almost rectangular fronts of magnetic flux density. As a consequence of this almost rectangular front motion, the magnetic flux density is not uniform even for relatively low frequencies, and this results in the increase of eddy current losses. Another explanation of excess eddy current losses based on the existence of domain structures within magnetic conductors was developed by G. Bertotti and it can be found in references [7], [8], [9].

Problems

(1) Give a concise summary of the basic equations of electric circuit theory (terminal relations, continuity conditions, KCL and KVL equations).

(2) Explain how to write linearly independent KCL and KVL equations.

(3) Give a concise summary of the definition of "phasor" and basic phasor relations in ac circuits.

(4) Derive formula (1.91).

(5) Consider an RC circuit shown below. The measured peak values of voltages across the resistor and capacitor are 30 V and 40 V, respectively. What is the peak value of the input voltage? Solve this problem by visualizing the appropriate phasor diagram.

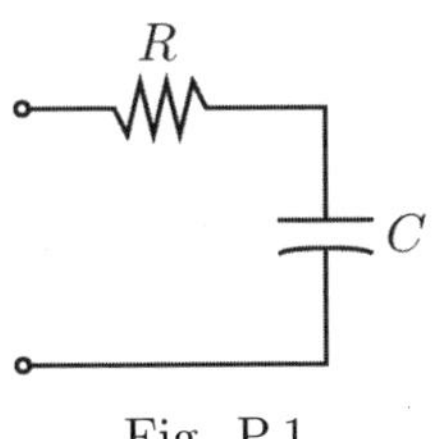

Fig. P.1

(6) Construct the phasor diagram for the electric circuit shown in Figure P.2.

(7) Construct the phasor diagram for the electric circuit shown in Figure P.3.

(8) Construct the phasor diagram for the electric circuit shown in Figure P.4.

(9) Consider an RLC circuit shown in Figure P.5 where inductive re-

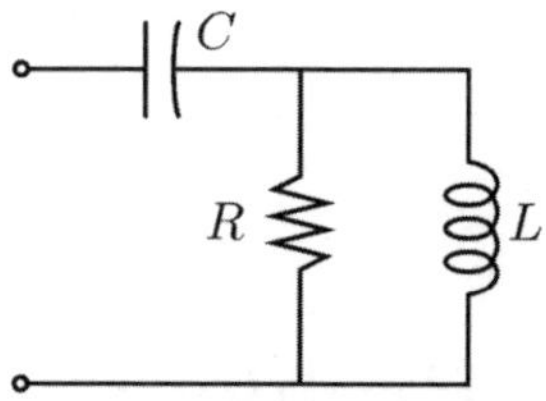

Fig. P.2

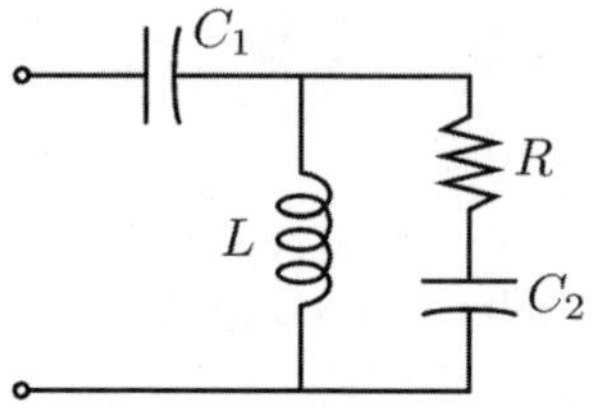

Fig. P.3

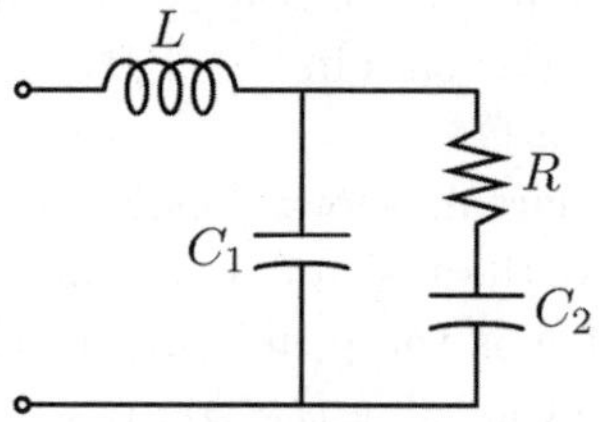

Fig. P.4

actance exceeds capacitive reactance. The measured peak values of input voltage and voltages across the resistor and capacitor are equal to 50 V, 30 V and 40 V, respectively. By using the appropriate phasor diagram, find the peak value of the voltage across the inductance.

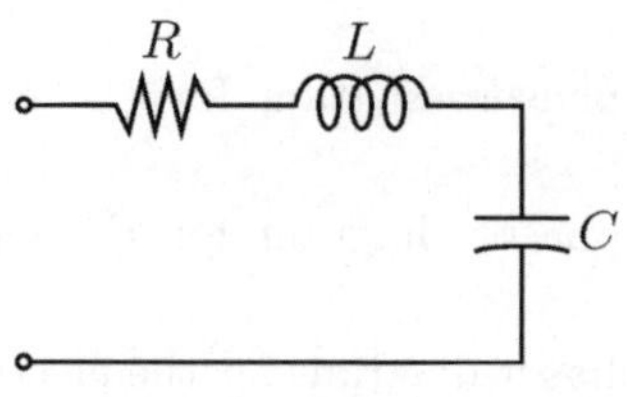

Fig. P.5

(10) Prove the orthogonality conditions (2.4)-(2.9) for trigonometric functions.

(11) By using the orthogonality conditions, derive formulas (2.10)-(2.12) for Fourier coefficients.

(12) By assuming that function $f(t)$ is differentiable, prove formulas (2.16) by integration by parts.

(13) Prove that the product of two even or two odd functions is an even function, while the product of an even and an odd function is an odd function.

(14) By using the frequency-domain technique, find $i(t)$ in the electric circuit below when $v_s(t)$ is defined by Figure 2.11.

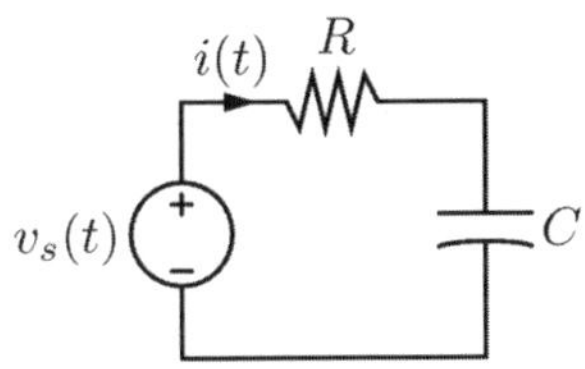

Fig. P.6

(15) By using the frequency-domain technique, find $i(t)$ in the electric circuit shown in Figure 2.16 when $v_s(t)$ is defined by Figure 2.19.

(16) By using the frequency-domain technique, find $i(t)$ in the electric circuit shown in Figure 2.14 when $v_s(t)$ is defined by Figure 2.19.

(17) By using the time-domain technique, find $i(t)$ in the electric circuit shown in Figure P.6 when $v_s(t)$ is defined by Figure 2.17.

(18) By using the time-domain technique, find $i(t)$ in the electric circuit shown in Figure 2.16 when $v_s(t)$ is defined by Figure 2.19.

(19) By using the time-domain technique, find $i(t)$ in the electric circuit shown in Figure P.6 when $v_s(t)$ is defined by Figure 2.19.

(20) By using the time-domain technique, find $i(t)$ in the electric circuit shown in Figure 2.18 when $v_s(t)$ is defined by Figure 2.17.

(21) Give a concise summary of the basic equations of magnetic circuit theory, including the main assumptions on which this theory is based.

(22) Consider the magnetic system shown in Figure P.7. The currents I_1 and I_2 as well as the number of turns N_1 and N_2 are given along with the geometry and magnetic permeabilities of each leg, ℓ_k, A_k, μ_k, $(k = 1, 2, 3)$, and the length δ of the air gap. By using the

superposition principle, derive the expression for the magnetic flux through the air gap.

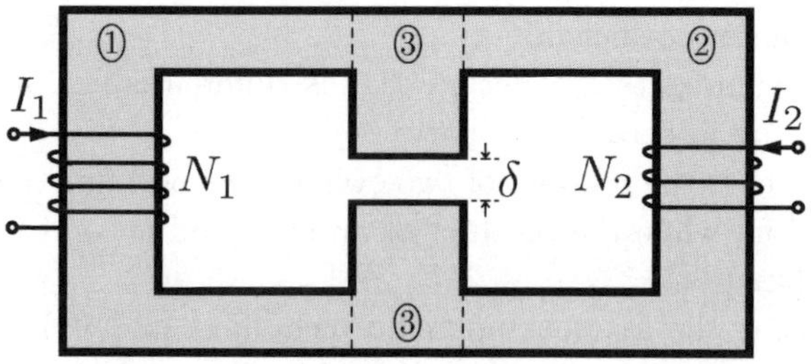

Fig. P.7

(23) State the definition of inductance. How does it depend on the number of turns and magnetic permeability? Explain how inductance can be computed by using magnetic circuit theory. What is a leakage inductance? Can it be computed by using the magnetic circuit theory?

(24) Derive the expression for the main inductance of the coil in the magnetic system shown below by assuming that the number of turns N, the geometry and magnetic permeabilities of each leg ℓ_k, A_k, μ_k, $(k = 1, 2, 3)$ and the lengths δ_2 and δ_3 of the air gaps are given.

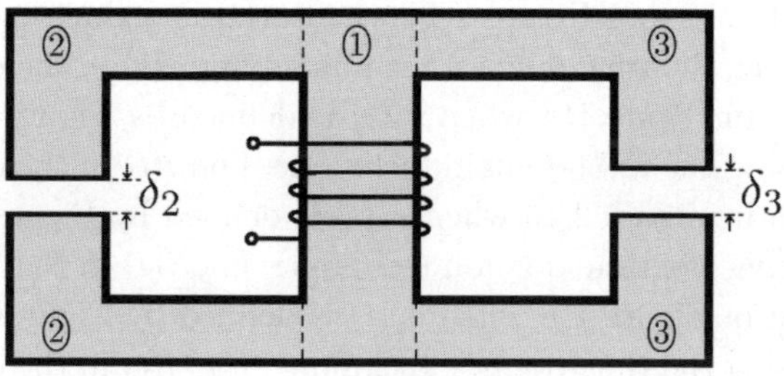

Fig. P.8

(25) State the definition of mutual inductance of two coils. Is mutual inductance always smaller than each self-inductance? Explain how the magnetic circuit theory can be used for the calculation of mutual inductances.

(26) Derive the expression for M_{21} for the two coils in the magnetic system shown in Figure 3.19 and demonstrate that M_{21}=M_{12}.

(27) How does the mutual inductance depend on the number of turns of the two coils and magnetic permeability of the ferromagnetic core? What are the main functions of ferromagnetic cores?

(28) Derive the expression for the mutual inductance between the two coils in the magnetic system shown in Figure P.7.

(29) What is a permanent magnet and what are the main figures of merit of permanent magnets? What is a demagnetizing field? What materials are used for permanent magnets?

(30) Describe concisely the two equivalent magnetic circuit models for ideal permanent magnets and provide the relevant formulas.

(31) Find the magnetic fluxes through the air gaps in the magnetic system shown in Figure P.9. Identify the quantity that must be given in order to solve this problem.

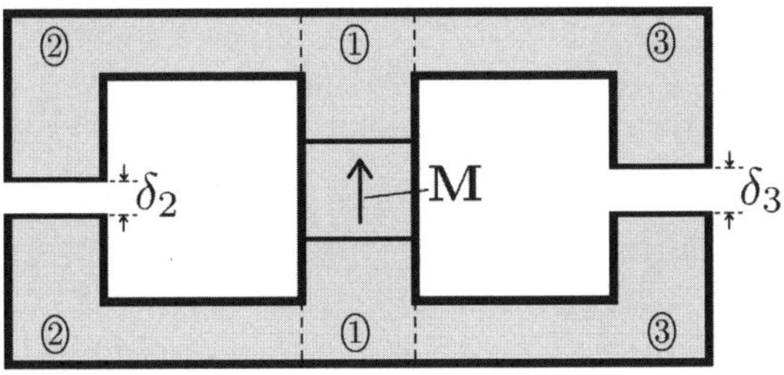

Fig. P.9

(32) How are the main equations of magnetic circuit theory modified in the case of nonlinear magnetic circuits?

(33) Explain the method of graphical analysis of nonlinear magnetic circuits and how it can be computerized.

(34) What is hysteresis of ferromagnetic materials? Give the formula for hysteresis power losses.

(35) What are eddy currents? Give the formula for eddy current losses and state the main assumptions for the validity of this formula.

(36) Explain how eddy current losses can be reduced. How does the polarization of magnetic field affect eddy current losses?

(37) Core losses are the sum of hysteresis and eddy current losses. If core losses are measured for two different frequencies, how can hysteresis and eddy current losses be separated?

(38) Suppose you measure core power losses for a particular magnetic circuit at two source excitation frequencies. For $f = 60$ Hz you

measure losses of 100 W and for $f = 75$ Hz you measure losses of 150 W, with the peak value of excitation voltage being the same in both cases. Assuming the losses are described by the classical equations for eddy current and hysteresis losses, what fractions of the total loss power are the eddy current losses and the hysteresis losses for the two cases given?

(39) For the device described in problem 38, what would the total core losses be at $f = 100$ Hz if the peak value of excitation voltage remains the same?

(40) Demonstrate the validity of the formulas (3.247) and (3.248) for eddy current losses in the case of elliptical polarization of magnetic flux density.

PART II

Power Systems

Chapter 1

Introduction to Power Systems

1.1 Brief Overview of Power System Structure

In power systems various forms of energy are converted into electric energy. This process of conversion is called electric power generation. There are the following forms of energy that may be involved in the electric power generation process:

a) chemical energy,
b) heat energy,
c) mechanical energy,
d) nuclear energy,
e) solar energy and
f) electric energy.

The questions can be asked, "What is so special about electric energy that other forms of energy are converted into it?" and "Why is electric energy the backbone of modern civilization?" The reason is that electric energy (power) has the following unique features:

(1) It can be centrally generated in huge amounts at power plants.
(2) It can be transmitted over large distances with relatively small losses by using high-voltage transmission lines.
(3) It is quite versatile and can be converted into almost any desired form of energy.
(4) It is ideally suited for encoding, transmission and processing of digital (or analog) information.
(5) It has minimally intrusive nature, being present in households, offices and industrial areas without occupying much space and without much noise or other undesirable effects.

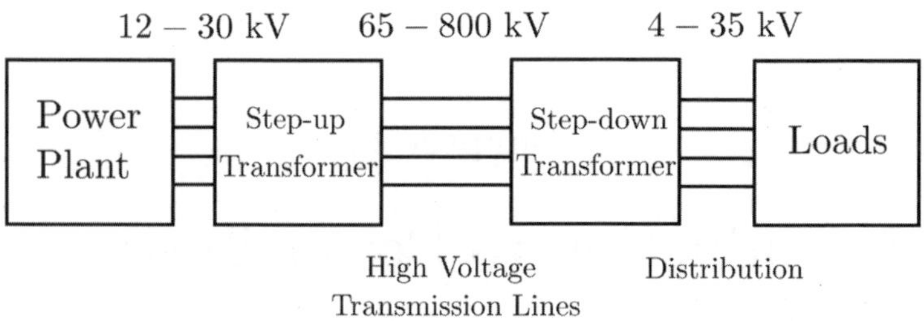

Fig. 1.1

Electric power systems are designed and constructed to utilize these unique features of electric energy. A schematic depiction of the simplest (so-called radial) model of electric power system is shown in Figure 1.1. It is apparent from this figure that electric power is generated in a power plant, where several stages of energy conversion may occur. Usually, the last stage is the conversion of mechanical energy into electric energy. This conversion is accomplished by using a synchronous generator and electric power is generated as a three-phase ac power. This power is transmitted and distributed by means of four (or three) wires. The output voltage of synchronous generators is typically between 12 kV and 30 kV. To transmit the electric power over large distances, this voltage is usually stepped up by using a transformer. Then, electric power is transmitted over large distances by using high-voltage transmission lines. The voltage of such lines may vary from one line to another and it is typically between 65 kV and 800 kV. For transmitted electric power to be consumed, its voltage must be stepped down. This is accomplished by using step-down transformers and then electric power is distributed to customers. Power distribution occurs at different voltage levels with the highest level being between 4 kV and 35 kV. Usually, there are several stages of voltage step down which are performed by transformers at power substations or by pole-mounted transformers and, finally, electric power is delivered to customers. In the US, this power is delivered to residential customers at the voltage of 120 V.

It is clear from the presented discussion that electric power systems have three major components: generation, transmission and distribution. These components are briefly discussed below. Before proceeding with this discussion, it is worthwhile to stress three important principles that have historically been adopted in the development of electric power systems.

(1) Electric power systems must be designed and operated to provide

electric power to customers at more or less constant peak (or rms) value of voltage and constant frequency despite continuously and unpredictably changing loads.

(2) Electric power systems are designed and operated to generate electric power on demand.

(3) Generation of electric power is accomplished by using high energy density conversion devices. The latter means that it is desirable to construct generators with maximum output power per unit weight.

Historically, these three principles have been instrumental in the design and operation of efficient, reliable and environmentally friendly power systems. During the past three decades, some revision of these basic principles has been under way.

Next, we briefly review the generation of electric energy which occurs in power plants. There are different types of power plants depending on the original source of energy. In a *fossil fuel* power plant the original source of energy is the chemical energy of fossil fuel (coal, natural gas or oil). There are three stages of energy conversion. The first stage of energy conversion is combustion in a furnace and the chemical energy is eventually converted into the thermal energy of steam. In the second stage, the steam drives a turbine and in this way thermal energy is converted into mechanical energy. Finally, the steam (or gas) turbine is mechanically coupled to the rotor of a synchronous generator and, in this stage, mechanical energy is converted into electric energy. According to data provided by the Energy Information Administration of the US Department of Energy, in the US in 2012 about 68% of electric power was generated by fossil fuel plants. Here, the previously dominant role of coal is being gradually reduced and replaced by gas. In 2012, about 37% of electric power was generated by coal-fired power plants, about 30% of electric power was generated by gas-fired power plants and only about 1% was generated by oil-fired power plants. Gas generation results in less pollution by carbon dioxide and, for this reason, it is more environmentally friendly. Gas generation is currently driven by the availability of relatively cheap natural gas that has been achieved due to the dramatic progress in the technology of horizontal drilling and hydraulic fracturing (commonly known as fracking).

In *nuclear* power plants, the primary source of energy is nuclear. The physical origin of nuclear energy is the strong interaction that holds (binds) neutrons and protons together in the nucleus of an atom. In nuclear reactors, a controlled nuclear fission (splitting of nucleus) process occurs which results in conversion of nuclear energy into heat and eventual production of steam. The second stage is the conversion of thermal energy of steam into mechanical energy by using steam turbines. Finally, mechanical energy is converted into electric energy by using synchronous generators. In 2012, about 19% of electric power in the US was generated by using nuclear power plants.

In *hydro*-power plants the mechanical (gravitational) energy of falling water is converted into electric energy. These power plants require the construction of dams to divert and store water, as well as to control its flow, the use of water turbines (water wheels) and salient pole synchronous generators. In 2012, about 7% of electric power in the US was generated by hydro-power plants. This percentage will drop with time because most of the available sources of hydro-power have already been developed.

Recently, a strong emphasis has been placed on using *renewable energy* in electric power generation. This emphasis has led to the increase in electric power production by using wind and solar energies as primary energy sources. Recent statistics suggest that about 5% of electric power in the US has been generated by using these two renewable sources. In the case of wind generation, airflows produced by winds drive wind turbines which are mechanically coupled with synchronous (or induction) generators that convert mechanical energy into electricity. In the case of solar generation, photovoltaic semiconductor devices are used to convert solar radiation into dc electricity. The physical mechanism of this conversion is the generation by light of electrons and holes in depletion regions of p-n junctions which are subsequently driven in opposite directions by electric fields in these depletion regions. The main advantage of wind and solar generation is that these means of electric power generation are clean, i.e., without any direct detrimental effects to the environment. However, solar and wind power plants do not generate electric power on demand but instead intermittently. Furthermore, solar and wind electric power generation is accomplished by using low energy density conversion devices. This makes renewable electric power generation less cost effective than traditional (fossil fuel or nuclear) power generation. This also raises legitimate questions concerning the overall environmental benefits of renewable sources if these benefits are measured over the entire life cycle, i.e., "from dust to dust." Finally, there are also issues of integration of renewables into existing three-phase power systems.

As has been mentioned before, electric power is generated as three-phase ac power of constant frequency. The constant frequency is achieved by maintaining constant speed of rotation of rotors of synchronous generators. In the US this frequency is 60 Hz, in Europe and some parts of Asia this frequency is 50 Hz, while in aviation ac power of 400 Hz is used. At higher frequencies, lighter and more compact power devices (generators, motors and transformers) can be used. However, higher frequencies result in higher energy losses (heat dissipation) and require more efficient cooling techniques.

Next, we shall discuss *transmission and distribution* components of electric power systems. In many respects, these components are similar. The main difference is that electric power transmission occurs at appreciably higher voltages than distribution. In general, the longer the transmission distance, the higher the voltage level. Furthermore, transmission lines are usually run from substation to substation and provide bulk power transmission, while numerous distribution lines run through populated areas to reach individual customers and deliver only some small fraction of bulk transmitted power.

There are two types of transmission and distribution lines: *overhead* lines and *underground* lines. Overhead lines are constructed by using support structures consisting of steel towers and/or wooden poles with vertical strings of suspension insulators to which multi-strand conducting wires of power lines are attached. Aluminum alloys are most frequently used for conducting wires, while the use of copper is rare. Copper has better conductivity, but aluminum is lighter and cheaper. Underground transmission and distribution lines are insulated high-voltage cables. They are predominantly used in highly populated metropolitan areas. Their construction is more expensive than the construction of overhead transmission and distribution lines, but they are much less affected by inclement weather, take less of valuable real estate (land) and have much lower visibility.

The question can be asked why high voltages are used for transmission and distribution. The first reason is that the use of high voltages reduces transmission and distribution losses. Indeed, the following formulas can be used for transmitted ($p_{tr}(t)$) and loss ($p_{loss}(t)$) powers:

$$p_{tr}(t) = v(t)i(t), \tag{1.1}$$

$$p_{loss}(t) = ri^2(t), \tag{1.2}$$

where r is the per unit length resistance of transmission (or distribution) line wire.

It is apparent from the last two formulas that if the same power is transmitted at higher voltage and lower current this will result in smaller per unit length losses. Currently, transmission and distribution losses vary mostly within 5% to 15% of transmitted power.

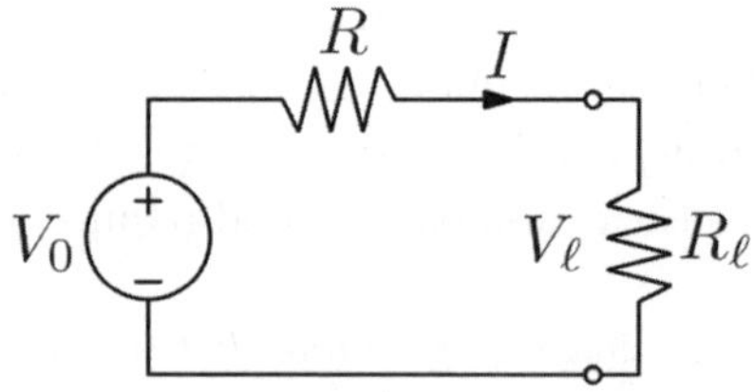

Fig. 1.2

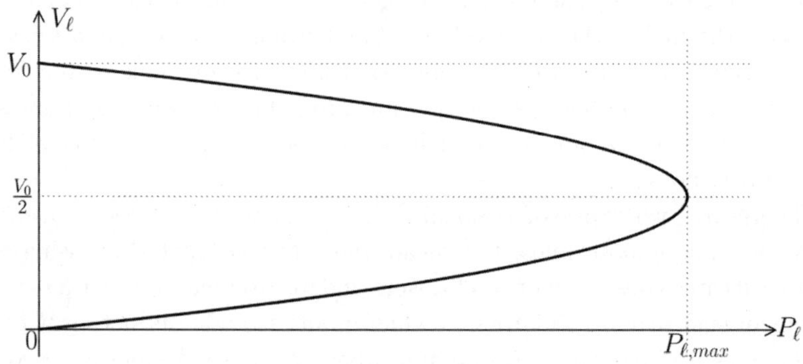

Fig. 1.3

The second reason to use high voltage transmission is to increase the transmission capacity of the transmission line. Below, we shall illustrate this by considering a very simplified model of dc transmission line shown in Figure 1.2, where R stands for the overall resistance of connecting wires and R_ℓ is the resistance of the load to which dc power P_ℓ is delivered at voltage V_ℓ. From Figure 1.2, we find that

$$P_\ell = V_\ell I \tag{1.3}$$

and

$$I = \frac{V_0 - V_\ell}{R}, \tag{1.4}$$

which results in

$$P_\ell = \frac{1}{R}\left(V_\ell V_0 - V_\ell^2\right) = \frac{V_0^2}{4R} - \frac{1}{R}\left(V_\ell - \frac{V_0}{2}\right)^2. \tag{1.5}$$

The last equation is graphically illustrated by Figure 1.3. It is clear from formula (1.5) that the maximum transmitted power (i.e., transmission capacity) is given by the formula

$$P_{\ell,max} = \frac{V_0^2}{4R}. \tag{1.6}$$

No power above $P_{\ell,max}$ can be transmitted. The last formula implies that the higher the voltage of a dc transmission line, the higher its transmission capacity. The last formula also implies that the smaller the resistance of the dc line conductors, the higher the transmission capacity. The latter fact suggests that the use of superconducting wires may dramatically increase the transmission capacity. This capacity will be limited by the value of critical current of a superconductor, i.e., by the current at which superconductivity is destroyed. This discussion clearly suggests that the progress in high-temperature superconductor physics and technology may have a dramatic effect on future structures of power systems.

Next, it is appropriate to discuss briefly the nature of *loads* in power systems. These loads can be subdivided into three distinct groups: residential, commercial and industrial. These loads vary depending on the time of day, the season of the year and from year to year. Load studies are very important and indispensable in planning the future development of a particular power system to meet future power demands. The most difficult component of load studies is the long-term prediction of future power demands associated with industrial loads. These future demands are affected by the state of the economy and manufacturing globalization.

The model of a power system shown in Figure 1.1 is quite simplistic in nature and illustrates the transmission of electric power from a single power plant in one direction toward closely localized loads. In reality, transmission lines are strongly interconnected among various power plants and utilities. These interconnected power plants and utilities form electric power grids which exist in the US on a continental scale. There are the following advantages of power grids: 1) mutual assistance in times of emergency; 2) possibilities for long-term power trade among various utilities; 3) possibilities for short-term power trades between neighboring utilities; and 4) facilitation of the creation of a global electric power market within the framework of deregulated utilities.

Finally, it is appropriate to discuss briefly the structure of utility companies which operate power systems and the concept of deregulation of the utility industry. Historically, utility companies were created and operated as *regulated and vertically integrated monopolies.* They were natural monopolies because it was recognized and accepted that it would be detrimental to the environment to have several competing utilities in the same area with different (and redundant) transmission and distribution lines. These monopolies were vertically integrated because they owned and were responsible for the operation of all three components of the power systems (generation, transmission and distribution) as well as for serving all of their customers in the area of their monopoly. These monopolies were regulated because special regulatory committees controlled how much these vertically integrated monopolies could charge their customers for the delivered electric power. This structure of electric power utilities was quite successful for many years. However, during the past two decades the new idea has been advanced that the generated electric energy as a product is fundamentally distinct from transmission and distribution as a service. On this basis, it was suggested and partially implemented that electric power generation should be open to competition rather than being the monopoly of local utilities. This resulted in formation of global energy markets where customers have opportunities to purchase electric power from different providers. It has been argued that the competition on the generation side of the utility business will eventually result in the reduction of electric energy cost. This cost reduction has not yet materialized. Deregulation raises many fundamental questions concerning the operation of interconnected power systems. However, the analysis of these issues is well beyond the scope of this brief review.

1.2 Three-Phase Circuits and Their Analysis

It is discussed in the previous section that ac electric power is generated, transmitted and distributed as three-phase power. In this section, we discuss the basic facts related to three-phase circuits and their analysis.

In three-phase circuits, there are three voltage sources $v_a(t)$, $v_b(t)$ and $v_c(t)$ which have the same peak value, the same frequency and are phase-shifted with respect to one another by $2\pi/3$:

$$v_a(t) = V_m \cos \omega t, \tag{1.7}$$

$$v_b(t) = V_m \cos\left(\omega t - \frac{2\pi}{3}\right), \tag{1.8}$$

$$v_c(t) = V_m \cos\left(\omega t - \frac{4\pi}{3}\right). \tag{1.9}$$

These voltage sources are usually connected into "star" (Y) or "delta" (Δ). We shall first discuss the star connection of voltage sources and star connection of load. This configuration of three-phase circuits is illustrated by Figure 1.4. It is clear from this figure that there are four wires that connect the star of voltage sources with the star of loads. They are lines a, b

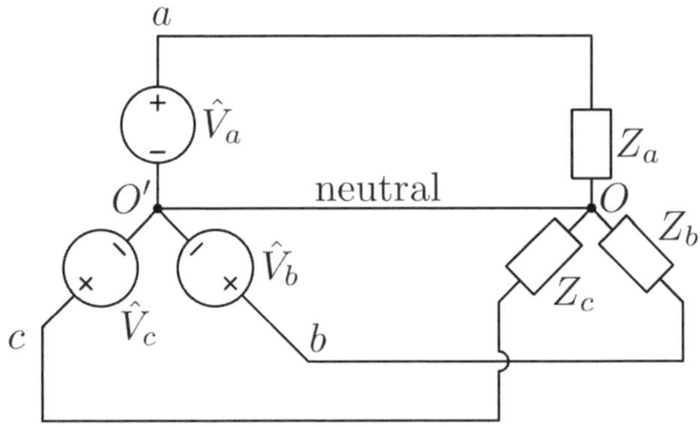

Fig. 1.4

and c as well as the neutral. Impedances Z_a, Z_b and Z_c represent different phase loads that are connected between the neutral and lines a, b and c, respectively. $\hat{V}_a$, $\hat{V}_b$ and $\hat{V}_c$ are the phasors of voltages $v_a(t)$, $v_b(t)$ and $v_c(t)$, respectively. It follows from formulas (1.7)-(1.9) that these phasors can be written as follows:

$$\hat{V}_a = V_m, \qquad \hat{V}_b = V_m e^{-j\frac{2\pi}{3}}, \qquad \hat{V}_c = V_m e^{-j\frac{4\pi}{3}}. \tag{1.10}$$

It is customary in the theory of three-phase systems to use the notation

$$\alpha = e^{-j\frac{2\pi}{3}}. \tag{1.11}$$

It is obvious that

$$\alpha^2 = e^{-j\frac{4\pi}{3}}, \qquad \alpha^3 = 1, \qquad \alpha^4 = \alpha. \tag{1.12}$$

From formulas (1.10)-(1.12) we find

$$\hat{V}_a = V_m, \qquad \hat{V}_b = \alpha\hat{V}_a, \qquad \hat{V}_c = \alpha^2\hat{V}_a. \tag{1.13}$$

It is clear from the presented discussion that multiplication of a phasor by α is equivalent to the phase shift in time by $2\pi/3$ for the corresponding sinusoidal quantity.

It can be easily proved that

$$\boxed{1 + \alpha + \alpha^2 = 0.} \tag{1.14}$$

Indeed, by treating the left-hand side of formula (1.14) as a geometric series, we find that

$$1 + \alpha + \alpha^2 = \frac{1 - \alpha^3}{1 - \alpha} = 0, \tag{1.15}$$

because $\alpha^3 = 1$ (see (1.12)).

From formulas (1.13) and (1.14), we derive

$$\boxed{\hat{V}_a + \hat{V}_b + \hat{V}_c = 0.} \tag{1.16}$$

The last equation implies that

$$\boxed{v_a(t) + v_b(t) + v_c(t) = 0.} \tag{1.17}$$

The last formula expresses a simple mathematical fact that the sum of three sinusoidal quantities of the same peak value, the same frequency and phase-shifted with respect to one another by $2\pi/3$ is equal to zero at any instant of time. This mathematical fact will be frequently used in our subsequent discussions.

It is customary in the case of three-phase circuits to speak about phase and line voltages. Phase voltages are measured between the neutral and one of the lines a, b and c. It is clear that $\hat{V}_a$, $\hat{V}_b$ and $\hat{V}_c$ are the phasors of the phase voltages and their peak values are the same:

$$\left|\hat{V}_a\right| = \left|\hat{V}_b\right| = \left|\hat{V}_c\right| = V_m = V_{ph}, \tag{1.18}$$

where V_{ph} is the notation for the common peak value of the phase voltages.

Line voltages are measured between different pairs of lines. This means that there are three line voltages

$$\hat{V}_{ab} = \hat{V}_a - \hat{V}_b, \tag{1.19}$$

$$\hat{V}_{bc} = \hat{V}_b - \hat{V}_c, \tag{1.20}$$

$$\hat{V}_{ca} = \hat{V}_c - \hat{V}_a. \tag{1.21}$$

It is clear from Figure 1.4 and the last three formulas that $\hat{V}_{ab}$ represents the phasor of the line voltage measured between lines a and b, $\hat{V}_{bc}$ is the phasor of the line voltage measured between lines b and c, while $\hat{V}_{ca}$ means the phasor of the line voltage between lines c and a. On the basis of symmetry, it is clear that the peak values of line voltages are the same:

$$\left|\hat{V}_{ab}\right| = \left|\hat{V}_{bc}\right| = \left|\hat{V}_{ca}\right| = V_\ell, \tag{1.22}$$

where V_ℓ is the common peak value of the line voltages. The last equality can be made geometrically transparent by constructing the phasor diagram shown in Figure 1.5a. In this diagram, the lengths of phasors $\hat{V}_a$, $\hat{V}_b$ and $\hat{V}_c$ and geometric angles between them are consistent with formulas (1.7), (1.8) and (1.9), while the diagram representations of phasors $\hat{V}_{ab}$, $\hat{V}_{bc}$ and $\hat{V}_{ca}$ are consistent with formulas (1.19)-(1.21). It is clear from Figure 1.5a that

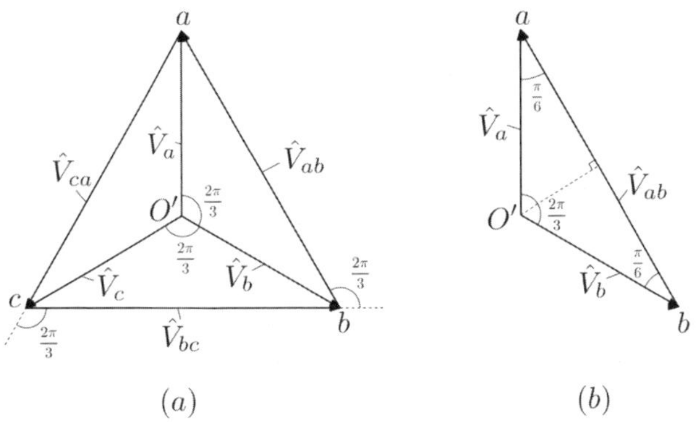

Fig. 1.5

isosceles triangles $aO'b$, $bO'c$ and $cO'a$ are identical and can be obtained from one another through rotation by $2\pi/3$. This implies the validity of formula (1.22). By using Figure 1.5b, it is easy to find the relation between the peak values of line and phase voltages. Indeed, from this figure follows that

$$V_\ell = \left|\hat{V}_{ab}\right| = 2\left|\hat{V}_a\right| \cos\left(\frac{\pi}{6}\right) = 2V_{ph}\frac{\sqrt{3}}{2}, \tag{1.23}$$

which leads to

$$\boxed{V_\ell = \sqrt{3}V_{ph}.} \tag{1.24}$$

In the US, the typical rms (root mean square) values of phase and line voltages are 120V and 208V, 208V and 360V, 360V and 620V, etc.

Furthermore, it is clear from Figure 1.5a that the line voltages are phase-shifted from one another by $2\pi/3$. Consequently,

$$\hat{V}_{bc} = \alpha\hat{V}_{ab}, \qquad \hat{V}_{ca} = \alpha^2\hat{V}_{ab}. \tag{1.25}$$

From the last formula and equation (1.14) follows that

$$\boxed{\hat{V}_{ab} + \hat{V}_{bc} + \hat{V}_{ca} = 0.} \tag{1.26}$$

The last formula implies that at any instant of time

$$v_{ab}(t) + v_{bc}(t) + v_{ca}(t) = 0. \tag{1.27}$$

It is interesting to find explicit expressions for $v_{ab}(t)$, $v_{bc}(t)$ and $v_{ca}(t)$. It is clear from Figure 1.5b that

$$v_{ab}(t) = \sqrt{3}V_m \cos\left(\omega t + \frac{\pi}{6}\right). \tag{1.28}$$

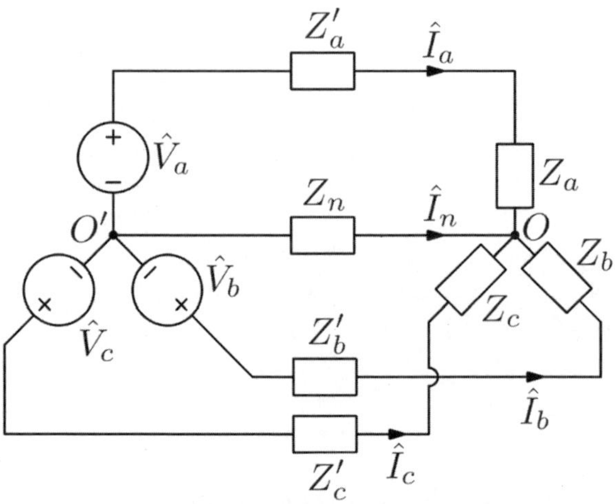

Fig. 1.6

Taking into account that line voltages are phase-shifted by $2\pi/3$, we find

$$v_{bc}(t) = \sqrt{3}V_m \cos\left(\omega t + \frac{\pi}{6} - \frac{2\pi}{3}\right), \tag{1.29}$$

$$v_{ca}(t) = \sqrt{3}V_m \cos\left(\omega t + \frac{\pi}{6} - \frac{4\pi}{3}\right). \tag{1.30}$$

Now, it is easy to write expressions for the phasors of the line voltages:

$$\hat{V}_{ab} = \sqrt{3}V_m e^{j\frac{\pi}{6}}, \qquad \hat{V}_{bc} = \sqrt{3}V_m e^{-j\frac{\pi}{2}}, \qquad \hat{V}_{ca} = \sqrt{3}V_m e^{-j\frac{7\pi}{6}}. \tag{1.31}$$

The question can be asked what function the neutral plays in the three-phase circuit shown in Figure 1.4. It is clear from this figure that as a result of the presence of the neutral the same peak (or rms) value of voltage can be maintained across the loads represented by impedances Z_a, Z_b and Z_c despite their possible variations in time due to changing loads. This conclusion is achieved by ignoring in the three-phase circuit shown in Figure 1.4 the impedances of connecting lines a, b and c and the neutral. To analyze the effect of these impedances on the ability to maintain more or less constant peak value of voltages across the loads, we shall analyze the electric circuit shown in Figure 1.6. The intrinsic simplicity of the circuit shown in this figure is revealed by the observation that there are only two nodes (O' and O) in this circuit. To take advantage of this simplicity, we choose node O' as a reference node with zero potential,

$$\hat{V}_{O'} = 0. \tag{1.32}$$

Then, the analysis of the circuit can be performed by using the following two steps.

Step 1. It is apparent from Figure 1.6 and equation (1.32) that

$$\hat{I}_a = \frac{\hat{V}_a - \hat{V}_O}{Z_a + Z'_a}, \quad \hat{I}_b = \frac{\hat{V}_b - \hat{V}_O}{Z_b + Z'_b}, \quad \hat{I}_c = \frac{\hat{V}_c - \hat{V}_O}{Z_c + Z'_c}, \quad \hat{I}_n = -\frac{\hat{V}_O}{Z_n}. \tag{1.33}$$

Next, we introduce admittances

$$Y_a = \frac{1}{Z_a + Z'_a}, \quad Y_b = \frac{1}{Z_b + Z'_b}, \quad Y_c = \frac{1}{Z_c + Z'_c}, \quad Y_n = \frac{1}{Z_n} \tag{1.34}$$

and rewrite the equations in (1.33) as follows:

$$\hat{I}_a = Y_a\left(\hat{V}_a - \hat{V}_O\right), \quad \hat{I}_b = Y_b\left(\hat{V}_b - \hat{V}_O\right), \quad \hat{I}_c = Y_c\left(\hat{V}_c - \hat{V}_O\right),$$
$$\hat{I}_n = -Y_n\hat{V}_O. \tag{1.35}$$

It is apparent that if $\hat{V}_O$ is found then all currents can be found by using the last formulas in (1.35) or the formulas in (1.33).

Step 2. From the Kirchhoff Current Law for node O, we find

$$\hat{I}_a + \hat{I}_b + \hat{I}_c + \hat{I}_n = 0. \tag{1.36}$$

By substituting the expressions from (1.35) into the last equation, we find

$$Y_a\left(\hat{V}_a - \hat{V}_O\right) + Y_b\left(\hat{V}_b - \hat{V}_O\right) + Y_c\left(\hat{V}_c - \hat{V}_O\right) - Y_n\hat{V}_O = 0, \tag{1.37}$$

which can be further transformed as

$$Y_a\hat{V}_a + Y_b\hat{V}_b + Y_c\hat{V}_c = (Y_a + Y_b + Y_c + Y_n)\hat{V}_O \tag{1.38}$$

and leads to

$$\boxed{\hat{V}_O = \frac{Y_a\hat{V}_a + Y_b\hat{V}_b + Y_c\hat{V}_c}{Y_a + Y_b + Y_c + Y_n}.} \tag{1.39}$$

This completes the analysis of the electric circuit shown in Figure 1.6. Indeed, by computing $\hat{V}_O$ from formula (1.39) and then using the computed value of $\hat{V}_O$ in formulas from (1.33), we can find all the currents.

Formula (1.39) is of special interest because the deviations of $\hat{V}_O$ from zero reflect deviations of voltages across the phase loads. Indeed, if $|Z_n|$ is very small and, consequently, $|Y_n|$ is very large, then according to (1.39)

$$\hat{V}_O \approx 0. \tag{1.40}$$

Furthermore, if line impedances Z'_a, Z'_b and Z'_c are small in comparison with load impedances Z_a, Z_b and Z_c, then the peak values of the voltages across the load impedances can be maintained approximately the same.

Next, we consider the important case of balanced load when

$$Z_a + Z_a' = Z_b + Z_b' = Z_c + Z_c'. \tag{1.41}$$

Then, according to (1.34), we have

$$Y_a = Y_b = Y_c = Y. \tag{1.42}$$

By using the last formula as well as formula (1.16) in equation (1.39), we derive

$$\hat{V}_O = \frac{Y\left(\hat{V}_a + \hat{V}_b + \hat{V}_c\right)}{3Y + Y_n} = 0. \tag{1.43}$$

From (1.33) and (1.43), we conclude that

$$\hat{I}_a = \frac{\hat{V}_a}{Z_a + Z_a'}, \quad \hat{I}_b = \frac{\hat{V}_b}{Z_b + Z_b'}, \quad \hat{I}_c = \frac{\hat{V}_c}{Z_c + Z_c'}, \quad \hat{I}_n = 0. \tag{1.44}$$

Now, taking into account formulas (1.13) and (1.41), we find

$$\hat{I}_b = \alpha \hat{I}_a, \qquad \hat{I}_c = \alpha^2 \hat{I}_a. \tag{1.45}$$

Thus, in the case of balanced load, the current through the neutral is equal to zero, while $\hat{I}_a$, $\hat{I}_b$ and $\hat{I}_c$ have the same peak values and are phase-shifted with respect to one another by $2\pi/3$. This is a very important fact because, as discussed in Chapter 4, such currents through stationary (but distributed) windings may create uniformly rotating magnetic fields. Such fields are at the very foundation of the principles of operation of ac electric machines (i.e., generators and motors). The ability to create uniformly rotating magnetic fields was historically one of the main reasons why three-phase circuits (and three-phase power systems) were introduced. It is also discussed in Chapter 4 that unbalanced loads are very detrimental to the operation of synchronous generators because they may result in induction of appreciable eddy currents in the solid rotors of these generators, leading to their heating. Thus, the balanced load is the preferable mode of operation of power systems, and utility companies usually take special measures to achieve it.

In the case of balanced load, the analysis of three-phase circuits can be substantially simplified by using the concept of "per-phase analysis." Indeed, consider instead of the three-phase circuit shown in Figure 1.6 a very simple single-phase circuit shown in Figure 1.7. According to this figure, we have

$$\hat{I}_a = \frac{\hat{V}_a}{Z_a + Z_a'}, \tag{1.46}$$

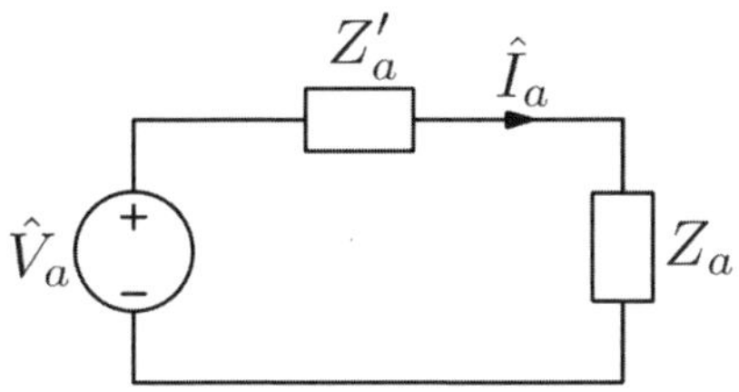

Fig. 1.7

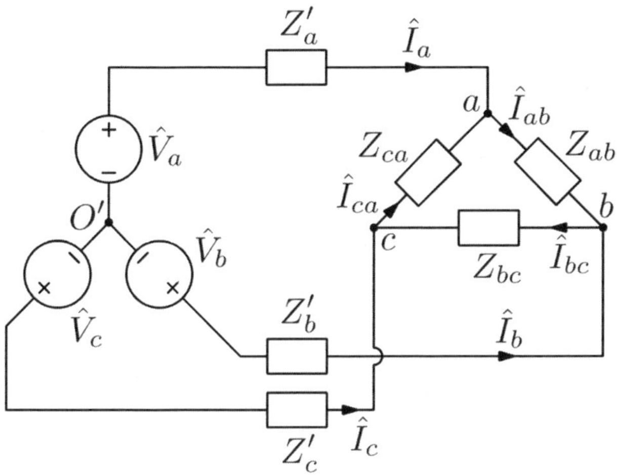

Fig. 1.8

which is identical to the first formula in (1.44). As soon as $\hat{I}_a$ is found, formulas in (1.45) can be used to find $\hat{I}_b$ and $\hat{I}_c$. Thus, in the case of balanced load, per-phase analysis leads to the same result as the analysis of the three-phase circuit shown in Figure 1.6.

We conclude this section by considering the analysis of a more complicated three-phase circuit with a delta connection of loads shown in Figure 1.8. In this circuit, three-phase voltages as well as all impedances are given, and the purpose of analysis is to find all marked currents. It is interesting to note that in the case when line impedances Z'_a, Z'_b and Z'_c are very small and negligible, then load impedances Z_{ab}, Z_{bc} and Z_{ca} are subject to line voltages and all currents can be easily found. Furthermore, in this case peak (or rms) values of load voltages are maintained the same regardless of variations of load impedances. The analysis of the circuit shown in Figure

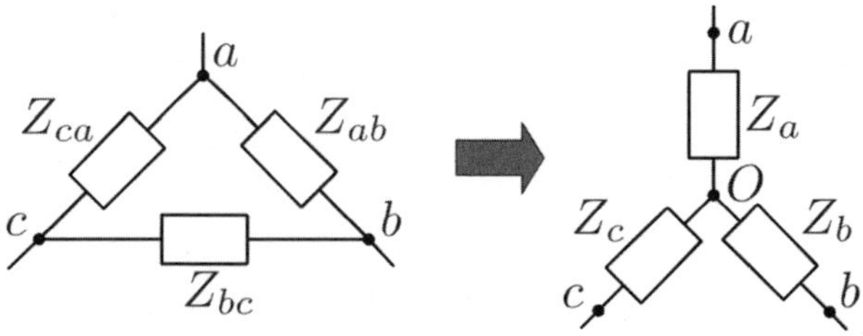

Fig. 1.9

1.8 may be of interest in order to understand how line impedances Z'_a, Z'_b and Z'_c may affect the variation of load voltages. This analysis can be accomplished by using the following three steps.

Step 1. is to replace the delta connection by the equivalent star connection (see Figure 1.9). This means that we have to find Z_a, Z_b and Z_c in terms of Z_{ab}, Z_{bc} and Z_{ca}. It can be proved (the proof is based on a superposition argument and it is omitted here) that "star" and "delta" connections are equivalent if impedances between any two identical nodes of these connections are the same. By computing impedances between nodes a and b, b and c, and c and a for star and delta connections and equating them, we arrive at the following linear simultaneous equations with respect to Z_a, Z_b and Z_c:

$$Z_a + Z_b = \frac{Z_{ab}(Z_{bc} + Z_{ca})}{Z_{ab} + Z_{bc} + Z_{ca}}, \tag{1.47}$$

$$Z_b + Z_c = \frac{Z_{bc}(Z_{ab} + Z_{ca})}{Z_{ab} + Z_{bc} + Z_{ca}}, \tag{1.48}$$

$$Z_c + Z_a = \frac{Z_{ca}(Z_{ab} + Z_{bc})}{Z_{ab} + Z_{bc} + Z_{ca}}. \tag{1.49}$$

It is easy to check that the solution of these equations is given by the formulas

$$Z_a = \frac{Z_{ab}Z_{ca}}{Z_{ab} + Z_{bc} + Z_{ca}}, \tag{1.50}$$

$$Z_b = \frac{Z_{bc}Z_{ab}}{Z_{ab} + Z_{bc} + Z_{ca}}, \tag{1.51}$$

$$Z_c = \frac{Z_{ca}Z_{bc}}{Z_{ab} + Z_{bc} + Z_{ca}}. \tag{1.52}$$

Step 2. is to analyze the three-phase circuit shown in Figure 1.10, which is obtained as a result of equivalent transformation of delta into star. Since

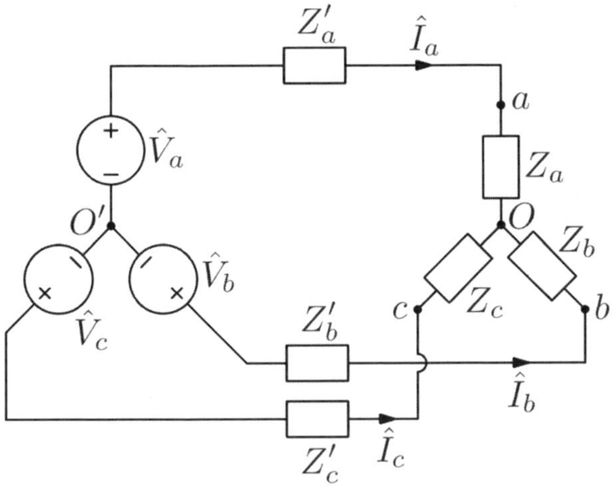

Fig. 1.10

this circuit is equivalent to the circuit shown in Figure 1.8, currents $\hat{I}_a$, $\hat{I}_b$ and $\hat{I}_c$ are the same in both circuits. These currents can be found in the same way as for the circuit shown in Figure 1.6 with only one simplification that for the circuit in Figure 1.10 $Z_n = \infty$ and, consequently, $Y_n = 0$. The currents $\hat{I}_a$, $\hat{I}_b$ and $\hat{I}_c$ can be computed by using sequentially the following formulas:

$$Y_a = \frac{1}{Z_a + Z'_a}, \quad Y_b = \frac{1}{Z_b + Z'_b}, \quad Y_c = \frac{1}{Z_c + Z'_c}, \tag{1.53}$$

$$\hat{V}_O = \frac{Y_a\hat{V}_a + Y_b\hat{V}_b + Y_c\hat{V}_c}{Y_a + Y_b + Y_c}, \tag{1.54}$$

$$\hat{I}_a = Y_a\left(\hat{V}_a - \hat{V}_O\right), \quad \hat{I}_b = Y_b\left(\hat{V}_b - \hat{V}_O\right), \quad \hat{I}_c = Y_c\left(\hat{V}_c - \hat{V}_O\right). \tag{1.55}$$

Step 3. Now, by taking into account that the voltages between a and b, b and c, and c and a for the circuits shown in Figures 1.8 and 1.10 are the same, we derive

$$\hat{I}_{ab} = \frac{\hat{V}_{ab}}{Z_{ab}} = \frac{\hat{I}_a Z_a - \hat{I}_b Z_b}{Z_{ab}}, \tag{1.56}$$

$$\hat{I}_{bc} = \frac{\hat{V}_{bc}}{Z_{bc}} = \frac{\hat{I}_b Z_b - \hat{I}_c Z_c}{Z_{bc}}, \tag{1.57}$$

$$\hat{I}_{ca} = \frac{\hat{V}_{ca}}{Z_{ca}} = \frac{\hat{I}_c Z_c - \hat{I}_a Z_a}{Z_{ca}}. \tag{1.58}$$

This completes the analysis of the three-phase circuit in Figure 1.8.

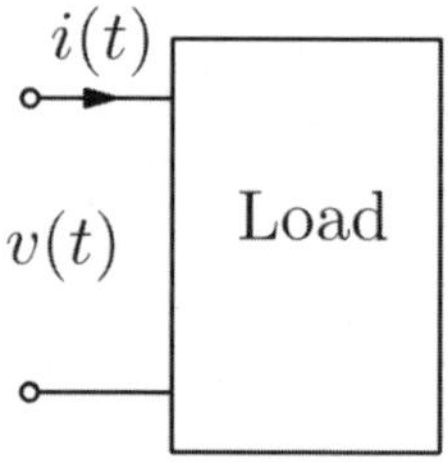

Fig. 1.11

1.3 AC Power and Power Factor

In this section, we shall discuss issues related to ac power. First, we consider the case of single-phase load shown in Figure 1.11. We assume that input voltage and current are sinusoidal,

$$v(t) = V_m \cos(\omega t + \varphi_V), \tag{1.59}$$

$$i(t) = I_m \cos(\omega t + \varphi_I). \tag{1.60}$$

Then, for instantaneous power $p(t)$ delivered to the load, we have

$$p(t) = V_m I_m \cos(\omega t + \varphi_V) \cos(\omega t + \varphi_I). \tag{1.61}$$

By using a simple trigonometric identity, the last formula can be transformed as

$$p(t) = \frac{V_m I_m}{2}[\cos(\varphi_V - \varphi_I) + \cos(2\omega t + \varphi_V + \varphi_I)]. \tag{1.62}$$

By taking into account that

$$\varphi_V - \varphi_I = \varphi, \tag{1.63}$$

we find

$$p(t) = \frac{V_m I_m}{2} \cos\varphi + \frac{V_m I_m}{2} \cos(2\omega t + \varphi_V + \varphi_I). \tag{1.64}$$

Thus, the instantaneous power has two distinct components: one that does not change with time and another that oscillates with double frequency. Power consumption is usually characterized by average power P defined as

$$P = \frac{1}{T} \int_0^T p(t)dt, \tag{1.65}$$

where $T = \frac{2\pi}{\omega}$ is the period of the ac voltage and current.

From formulas (1.64) and (1.65) we derive

$$\boxed{P = \frac{V_m I_m}{2} \cos\varphi.} \tag{1.66}$$

Indeed, the integration of the second term in the right-hand side of formula (1.64) yields zero because the integration is performed over two periods of the cosine function, while the integration of the first term in the right-hand side leads to equation (1.66).

The last formula is often written in terms of rms values of voltage and current,

$$P = VI\cos\varphi, \tag{1.67}$$

where $V = V_m/\sqrt{2}$ and $I = I_m/\sqrt{2}$.

We shall next introduce the notion of complex power, which allows to compute P by using phasors of voltage $\hat{V}$ and current $\hat{I}$. The complex power $\hat{S}$ is defined as

$$\boxed{\hat{S} = \frac{1}{2}\hat{V}\hat{I}^*,} \tag{1.68}$$

where $\hat{I}^*$ stands for the complex conjugate of $\hat{I}$.

The last formula admits the following transformation:

$$\hat{S} = \frac{1}{2}V_m e^{j\varphi_V} I_m e^{-j\varphi_I} = \frac{V_m I_m}{2} e^{j(\varphi_V - \varphi_I)} = \frac{V_m I_m}{2} e^{j\varphi}. \tag{1.69}$$

This means that

$$\hat{S} = \frac{V_m I_m}{2}\cos\varphi + j\frac{V_m I_m}{2}\sin\varphi. \tag{1.70}$$

Recalling equation (1.66), the last formula can be written as

$$\boxed{\hat{S} = P + jQ,} \tag{1.71}$$

where

$$\boxed{Q = \frac{V_m I_m}{2}\sin\varphi} \tag{1.72}$$

is called reactive power. This power is generated but not consumed. This power is provided to increase the energy stored in electromagnetic fields of power loads. It is clear that the origin of this power is due to the presence of energy storage elements (capacitors and inductors) in power loads which are responsible for time phase shift between input voltages and currents. However, at resonance conditions when input voltages and currents are in phase ($\sin\varphi = 0$), no reactive power is generated and provided to the loads. At these resonance conditions, exchange of reactive power occurs between capacitive and inductive components of the power loads.

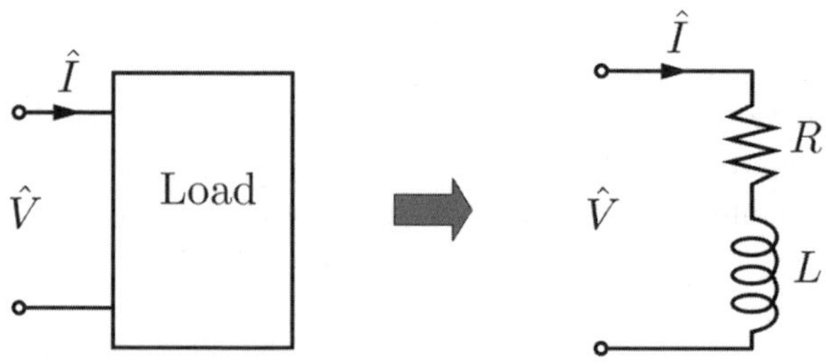

Fig. 1.12

From formulas (1.68) and (1.71), we obtain the following expressions for active (P) and reactive (Q) powers in terms of phasors:

$$P = \mathrm{Re}\left(\hat{S}\right) = \mathrm{Re}\left(\frac{1}{2}\hat{V}\hat{I}^*\right), \tag{1.73}$$

$$Q = \mathrm{Im}\left(\hat{S}\right) = \mathrm{Im}\left(\frac{1}{2}\hat{V}\hat{I}^*\right). \tag{1.74}$$

Next, we shall return to formula (1.66) and discuss the very important issue of power factor, which is defined as $\cos\varphi$ in that formula. It has already been discussed that one of the principles adopted in operation of power systems is to provide electric power to a customer at more or less constant peak value V_m of voltage. Thus, if a certain (fixed) amount of power has to be delivered to a load at constant V_m, this means that the product $I_m \cos\varphi$ is fixed as well. As a result, there are two options of delivering the same power: at higher peak value of current and lower value of power factor, or lower peak value of current and higher value of power factor. It is obvious that the second option is preferable. This is because if the same power is delivered at smaller peak value of current, then this power delivery results in smaller power losses in the connecting wires of distribution and transmission lines. These power losses are unproductive and they result in heating of the environment as well as in the overall increase of the cost of consumed power. Thus, the problem of adjustment of power factor presents itself. For inductive-type loads where input voltages lead in time input currents, the adjustment of power factor can be accomplished by placing capacitors across loads and choosing proper values of their capacitances. We shall discuss this matter in detail below. The first step in choosing the proper capacitance is the representation of inductive load by the equivalent RL circuit shown in Figure 1.12. The resistance and inductance of this equivalent circuit can be determined (identified) by applying to the load an ac voltage of known peak value V_m and then measuring the

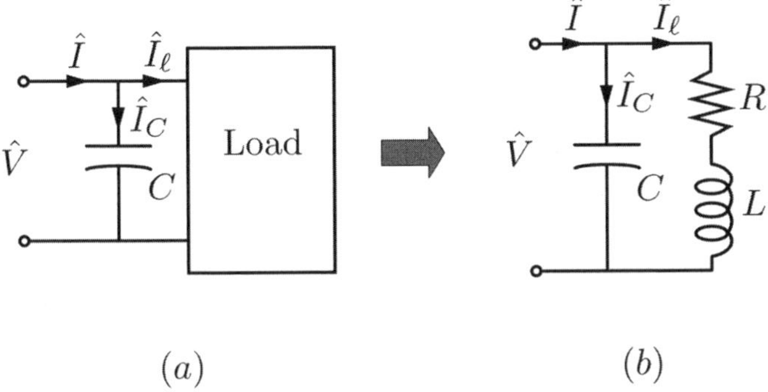

Fig. 1.13

peak value I_m of the input current and active power P. The parameters of the RL circuit can be chosen to result in the same I_m and P for the same V_m. Indeed, the impedance Z of the RL circuit can be written as

$$Z = |Z|e^{j\varphi} = |Z|\cos\varphi + j|Z|\sin\varphi = R + j\omega L, \tag{1.75}$$

and, consequently,

$$R = |Z|\cos\varphi, \qquad \omega L = |Z|\sin\varphi. \tag{1.76}$$

From the measured data we find

$$|Z| = \frac{V_m}{I_m}, \qquad \cos\varphi = \frac{2P}{V_m I_m}. \tag{1.77}$$

By using the last two expressions in (1.76), we can find the resistance and inductance of the equivalent RL circuit.

Now, we can proceed to the calculation of capacitance C of the capacitor placed across the terminals of the inductive-type load (see Figure 1.13a) that adjusts the power factor to one. The calculation is based on the equivalent circuit shown in Figure 1.13b. According to KCL, we have

$$\hat{I} = \hat{I}_C + \hat{I}_\ell. \tag{1.78}$$

On the other hand, the following expressions can be written for $\hat{I}_C$ and $\hat{I}_\ell$:

$$\hat{I}_C = j\omega C\hat{V}, \tag{1.79}$$

$$\hat{I}_\ell = \frac{\hat{V}}{R + j\omega L} = \hat{V}\frac{R - j\omega L}{R^2 + \omega^2 L^2} = \hat{V}\left(\frac{R}{R^2 + \omega^2 L^2} - j\frac{\omega L}{R^2 + \omega^2 L^2}\right). \tag{1.80}$$

By substituting the last two formulas into equation (1.78), after simple transformations we find

$$\hat{I} = \frac{R}{R^2 + \omega^2 L^2}\hat{V} + j\left(\omega C - \frac{\omega L}{R^2 + \omega^2 L^2}\right)\hat{V}. \tag{1.81}$$

We choose capacitance C in such a way that

$$\omega C - \frac{\omega L}{R^2 + \omega^2 L^2} = 0, \tag{1.82}$$

which is the case if

$$\boxed{C = \frac{L}{R^2 + \omega^2 L^2}.} \tag{1.83}$$

If the capacitance is chosen in accordance with formula (1.83), then the equation (1.81) is reduced to

$$\hat{I} = \frac{R}{R^2 + \omega^2 L^2}\hat{V}, \tag{1.84}$$

or

$$I_m e^{j\varphi_I} = \frac{RV_m}{R^2 + \omega^2 L^2} e^{j\varphi_V}. \tag{1.85}$$

It is apparent from the last formula that

$$\varphi_V = \varphi_I, \qquad \varphi = \varphi_V - \varphi_I = 0, \qquad \cos\varphi = 1, \tag{1.86}$$

and the adjustment of power factor is accomplished.

It is apparent that the choice of capacitance in accordance with formula (1.83) results in a resonance condition where no reactive power is supplied to the load from generators and instead there exists an exchange of reactive power between the inductor of the load and the capacitor.

Practically, adjustment of power factor exactly to one does not occur because loads (and the inductances and resistances representing them) vary with time. There is another, more fundamental reason not to adjust the power factor exactly to one. This is because a leading power factor ($\varphi < 0$) may have a beneficial effect on maintaining constant peak value voltage across the loads as they vary in time. We consider this problem in detail because it is quite interesting in its own right. To start the discussion, consider a simple (per-phase) circuit model of ac transmission line shown in Figure 1.14. Here, the effect of the transmission line is modeled by reactance X, while the load (with parallel capacitance for power factor adjustment) is represented by impedance Z_ℓ. Resistance of the transmission line is neglected because it is usually much smaller than reactance X. We

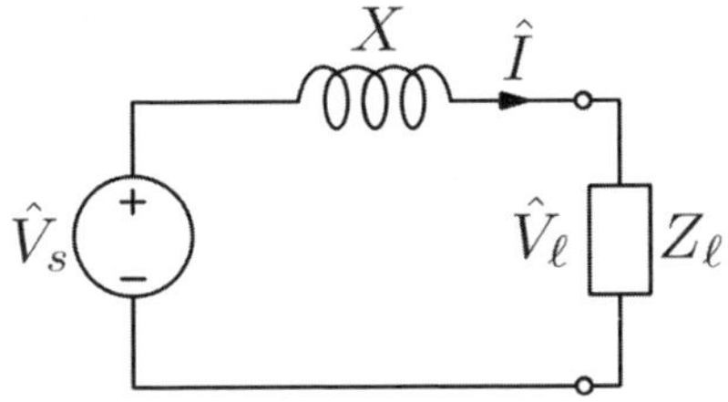

Fig. 1.14

want to study how the peak value $V_{m\ell}$ of the voltage across the load varies with consumed power P as well as what the maximum power P_{max} that can be delivered to the load is. It is apparent that this problem is similar to the problem discussed in the first section of this chapter (see Figure 1.2) for the case of dc transmission. It is apparent that there exists some phase shift θ in time between the source voltage $v_s(t)$ and the load voltage $v_\ell(t)$:

$$v_s(t) = V_{ms} \cos \omega t, \tag{1.87}$$

$$v_\ell(t) = V_{m\ell} \cos(\omega t - \theta). \tag{1.88}$$

This means that

$$\hat{V}_s = V_{ms}, \tag{1.89}$$

$$\hat{V}_\ell = V_{m\ell} e^{-j\theta}. \tag{1.90}$$

From the circuit shown in Figure 1.14, we find

$$\hat{I} = \frac{\hat{V}_s - \hat{V}_\ell}{jX}, \tag{1.91}$$

and

$$\hat{I}^* = j\frac{\hat{V}_s^* - \hat{V}_\ell^*}{X} = \frac{\hat{V}_s^* - \hat{V}_\ell^*}{X} e^{j\frac{\pi}{2}}. \tag{1.92}$$

Now by using formula (1.68), we find

$$\hat{S} = \frac{\hat{V}_\ell \hat{I}^*}{2} = \frac{\hat{V}_\ell \left(\hat{V}_s^* - \hat{V}_\ell^*\right)}{2X} e^{j\frac{\pi}{2}}, \tag{1.93}$$

or

$$\hat{S} = \frac{\hat{V}_\ell \hat{V}_s^*}{2X} e^{j\frac{\pi}{2}} - j\frac{V_{m\ell}^2}{2X}. \tag{1.94}$$

Next, by using formulas (1.89), (1.90) and (1.94), we obtain

$$\hat{S} = \frac{V_{m\ell} V_{ms}}{2X} e^{j\left(\frac{\pi}{2} - \theta\right)} - j\frac{V_{m\ell}^2}{2X}, \tag{1.95}$$

which can be further transformed as follows:

$$\hat{S} = \frac{V_{m\ell}V_{ms}}{2X}\sin\theta + j\left(\frac{V_{m\ell}V_{ms}}{2X}\cos\theta - \frac{V_{m\ell}^2}{2X}\right). \tag{1.96}$$

According to formulas (1.66), (1.71) and (1.72), we can write

$$\hat{S} = P(1 + j\alpha), \tag{1.97}$$

where

$$\alpha = \tan\varphi. \tag{1.98}$$

By comparing formulas (1.96) and (1.97), we conclude that

$$P = \frac{V_{ms}V_{m\ell}}{2X}\sin\theta, \tag{1.99}$$

$$\alpha P = \frac{V_{ms}V_{m\ell}}{2X}\cos\theta - \frac{V_{m\ell}^2}{2X}. \tag{1.100}$$

The last equation can be written in the form

$$\alpha P + \frac{V_{m\ell}^2}{2X} = \frac{V_{ms}V_{m\ell}}{2X}\cos\theta. \tag{1.101}$$

From formulas (1.99) and (1.101), we derive

$$\boxed{P^2 + \left(\alpha P + \frac{V_{m\ell}^2}{2X}\right)^2 = \frac{V_{ms}^2V_{m\ell}^2}{4X^2}.} \tag{1.102}$$

This is the sought equation that relates the peak value $V_{m\ell}$ of the voltage across the load to the consumed power P. It is clear that this is a quadratic equation for $V_{m\ell}^2$. If its discriminant D is positive for a given P, then it has two real solutions. It has one real solution when the discriminant is equal to zero and no real solution when the discriminant is negative. Thus, there are two distinct branches of the relation $V_{m\ell}$ vs. P.

Consider first the case of $P = 0$. Then, according to equation (1.102), we find

$$\frac{V_{m\ell}^4}{4X^2} = \frac{V_{ms}^2V_{m\ell}^2}{4X^2}. \tag{1.103}$$

It is obvious that the last equation has two solutions,

$$V_{m\ell} = 0 \qquad \text{and} \qquad V_{m\ell} = V_{ms}. \tag{1.104}$$

For the sake of notational simplicity, we introduce new variable

$$\lambda = V_{m\ell}^2 \tag{1.105}$$

and rewrite the equation (1.102) as

$$P^2 + \left(\alpha P + \frac{\lambda}{2X}\right)^2 = \lambda \frac{V_{ms}^2}{4X^2}, \tag{1.106}$$

which can be further transformed to result in

$$\lambda^2 - \lambda\left(V_{ms}^2 - 4\alpha XP\right) + 4\left(1+\alpha^2\right)X^2P^2 = 0. \tag{1.107}$$

For the discriminant of this equation, we find

$$D = \frac{1}{2}\left[\left(V_{ms}^2 - 4\alpha XP\right)^2 - 16\left(1+\alpha^2\right)X^2P^2\right]^{\frac{1}{2}}. \tag{1.108}$$

It is clear from the last formula that $D > 0$ for $P = 0$ and that D is decreased as P is increased. Thus, the value P_{max} at which

$$D = 0 \tag{1.109}$$

satisfies the equation

$$V_{ms}^2 - 4\alpha XP_{max} = 4\sqrt{1+\alpha^2}XP_{max}, \tag{1.110}$$

and, consequently,

$$\boxed{P_{max} = \frac{V_{ms}^2}{4X\left(\sqrt{1+\alpha^2}+\alpha\right)}.} \tag{1.111}$$

It is clear that P_{max} has the physical meaning of maximum power that can be delivered to the load. Indeed, a power larger than P_{max} is not consistent with peak value $V_{m\ell}$ of the load voltage being a real number. In the particular case when the power factor is adjusted to one (i.e., $\varphi = 0$), from formulas (1.98) and (1.111) we find

$$P_{max} = \frac{V_{ms}^2}{4X}, \tag{1.112}$$

which is similar to formula (1.6). It is also easy to see from formula (1.111) that P_{max} is monotonically increased if α is negative and decreased. This suggests that a leading power factor ($\varphi < 0$) may be beneficial in increasing P_{max}.

Furthermore, from equation (1.107) we find that under the condition (1.109),

$$\lambda = \frac{V_{ms}^2 - 4\alpha XP_{max}}{2}. \tag{1.113}$$

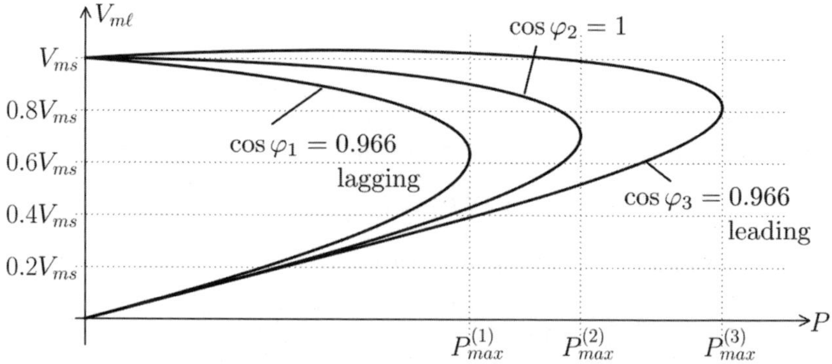

Fig. 1.15

According to formula (1.105), this means that the maximum power P_{max} can be delivered at load voltage peak value $V_{m\ell}^{max}$ which is given by the equation

$$\boxed{V_{m\ell}^{max} = \left(\frac{V_{ms}^2 - 4\alpha X P_{max}}{2}\right)^{\frac{1}{2}}.} \tag{1.114}$$

This is the value of $V_{m\ell}$ at which the lower and upper branches of $V_{m\ell}$ vs. P are connected. The last formula suggests that $V_{m\ell}^{max}$ is increased and approaches V_{ms} when α is negative. This means that a leading power factor ($\varphi < 0$) may be beneficial in maintaining $V_{m\ell}$ for varying loads.

The presented discussion is illustrated by Figure 1.15 where relations $V_{m\ell}$ vs. P are plotted for different values of power factor. It is interesting to mention that the existence of lower branches of relations $V_{m\ell}$ vs. P is sometimes used for the explanation of the "voltage collapse" phenomenon that has been observed in power systems.

We conclude this section by the discussion of ac power in three-phase circuits with balanced load (Figure 1.16). In the case of balanced load

$$Z_a = Z_b = Z_c, \tag{1.115}$$

and, for this reason, the phase shifts in time between load voltage and load current are the same for all three phases:

$$\varphi_a = \varphi_b = \varphi_c = \varphi. \tag{1.116}$$

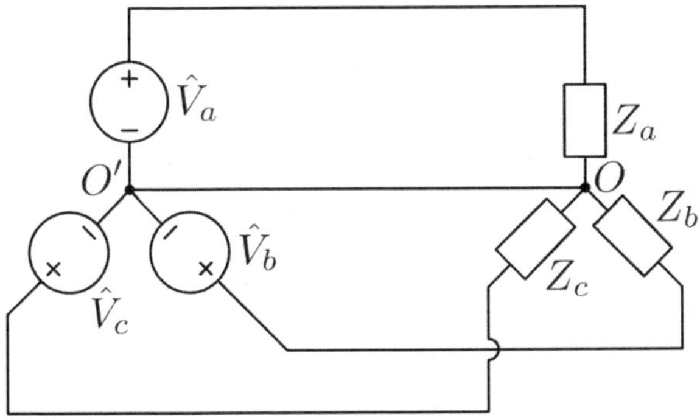

Fig. 1.16

By using this fact, we find

$$\begin{aligned} p_a(t) &= v_a(t)i_a(t) = [V_m \cos\omega t][I_m \cos(\omega t - \varphi)] \\ &= \frac{V_m I_m}{2}\cos\varphi + \frac{V_m I_m}{2}\cos(2\omega t - \varphi), \end{aligned} \tag{1.117}$$

$$\begin{aligned} p_b(t) &= v_b(t)i_b(t) = \left[V_m \cos\left(\omega t - \frac{2\pi}{3}\right)\right]\left[I_m \cos\left(\omega t - \frac{2\pi}{3} - \varphi\right)\right] \\ &= \frac{V_m I_m}{2}\cos\varphi + \frac{V_m I_m}{2}\cos\left(2\omega t - \varphi - \frac{4\pi}{3}\right), \end{aligned} \tag{1.118}$$

$$\begin{aligned} p_c(t) &= v_c(t)i_c(t) = \left[V_m \cos\left(\omega t - \frac{4\pi}{3}\right)\right]\left[I_m \cos\left(\omega t - \frac{4\pi}{3} - \varphi\right)\right] \\ &= \frac{V_m I_m}{2}\cos\varphi + \frac{V_m I_m}{2}\cos\left(2\omega t - \varphi - \frac{2\pi}{3}\right). \end{aligned} \tag{1.119}$$

The total instantaneous power is given by the equation

$$p(t) = p_a(t) + p_b(t) + p_c(t). \tag{1.120}$$

By using formulas (1.117), (1.118), (1.119) and (1.120), we get

$$\begin{aligned} p(t) = \frac{3V_m I_m}{2}\cos\varphi + \frac{V_m I_m}{2}\Bigg[&\cos(2\omega t - \varphi) + \cos\left(2\omega t - \varphi - \frac{2\pi}{3}\right) \\ &+ \cos\left(2\omega t - \varphi - \frac{4\pi}{3}\right)\Bigg]. \end{aligned} \tag{1.121}$$

It is clear that the expression in brackets in the last formula is equal to zero, because the sum of three sinusoidal functions of the same frequency and phase-shifted by $\frac{2\pi}{3}$ with respect to one another is equal to zero at any instant of time. Thus,

$$p(t) = \frac{3}{2}V_m I_m \cos\varphi = const, \tag{1.122}$$

which means that *in three-phase circuits with balanced loads electric energy is consumed at a constant rate in time.* The last equation also implies that

$$P = p(t) = \frac{3}{2}V_m I_m \cos\varphi. \tag{1.123}$$

In the last equation V_m and I_m are peak values of phase voltage and current. The last formula can be written in the equivalent form

$$P = \sqrt{3}V_\ell I_{ph} \cos\varphi, \tag{1.124}$$

where V_ℓ and I_{ph} are rms values of line voltage and phase current. It is left to the reader as a simple exercise to prove the last formula.

Chapter 2

Fault Analysis

2.1 Fault Analysis by Using the Thevenin Theorem.

It has been discussed previously that it is very desirable to operate power systems under balanced load conditions and usually special efforts are made to achieve this mode of operation. However, these balanced load conditions may be disrupted by faults in power systems. Power line faults are the most common because the power lines are exposed to inclement weather conditions such as lightning strikes, icing, high winds, trees falling, etc. Line faults may result in very large currents that may damage very expensive power equipment. For this reason, proper relay protection systems are used to detect faults and minimize their destructive effects. The design of the relay protection systems is based on accurate predictions of fault currents in power systems. Such predictions can be obtained through accurate analysis of line faults.

The most typical line faults are the single line-to-ground (SLG) faults, line-to-line (LL) faults and double line-to-ground (DLG) faults. The analysis of these faults is discussed in this chapter, and two techniques are developed to carry out this analysis. The first technique, which is presented in this section, is based on the Thevenin theorem, while the second technique is based on the very important concept of symmetrical components and extensively discussed in subsequent sections of this chapter.

We begin with a basic review of the Thevenin theorem. Consider a branch with impedance Z connected to an active linear circuit (see Figure 2.1a). The term "active" implies that this circuit contains voltage (and/or current) sources. The Thevenin theorem states that as far as the current $\hat{I}$ through this branch is concerned, the linear active circuit can be replaced by an equivalent nonideal voltage source shown in Figure 2.1b. If the pa-

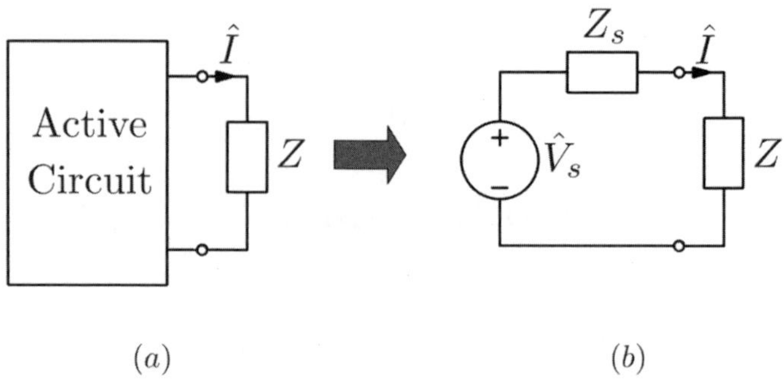

Fig. 2.1

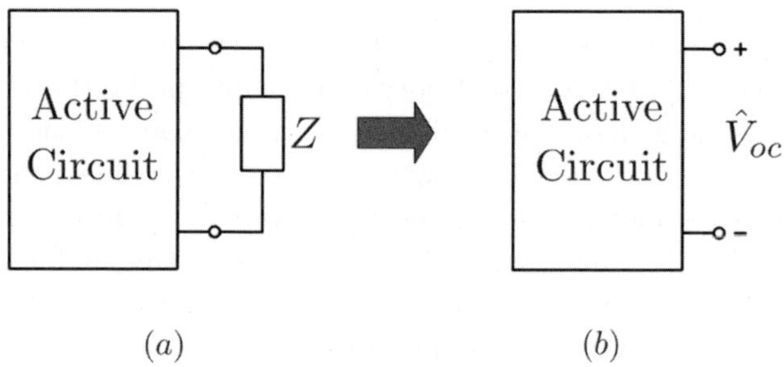

Fig. 2.2

rameters $\hat{V}_s$ and Z_s of the nonideal voltage source are computed, then the current $\hat{I}$ can be easily determined:

$$\boxed{\hat{I} = \frac{\hat{V}_s}{Z_s + Z}.} \tag{2.1}$$

The proof of the Thevenin theorem is based on the linearity of the active circuit and it is usually given in textbooks on electric circuit theory (see, for instance, [35]). This proof also reveals the physical meaning of $\hat{V}_s$ and Z_s, and this allows to formulate the following three-step technique for the analysis of electric circuits based on the Thevenin theorem.

Step 1. The branch with impedance Z is removed (see Figure 2.2) and the open-circuit voltage $\hat{V}_{oc}$ across the open terminals is computed. Usually, the removal of impedance Z greatly simplifies the circuit and makes its

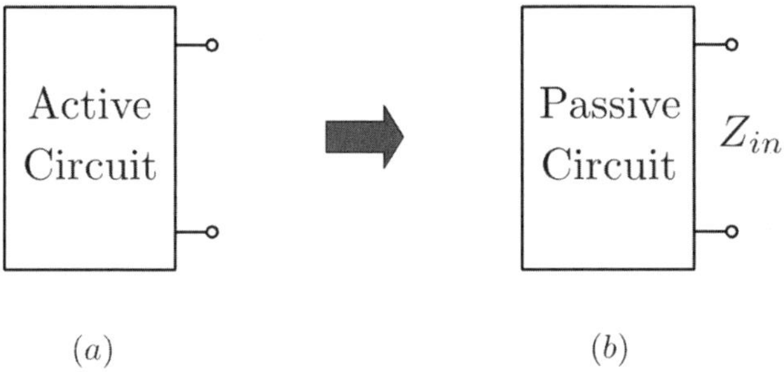

Fig. 2.3

analysis relatively easy. It turns out that (see [35])

$$\boxed{\hat{V}_s = \hat{V}_{oc}.} \tag{2.2}$$

Step 2. In this step, the active circuit shown in Figure 2.3a is transformed into a passive circuit by replacing voltage sources by short-circuit branches and current sources by open-circuit branches (see Figure 2.3). Then, the input impedance Z_{in} of this passive circuit with respect to the open terminals is computed (see Figure 2.3b). It turns out that (see [35])

$$\boxed{Z_s = Z_{in}.} \tag{2.3}$$

Step 3. Once the open-circuit voltage and input impedance are computed, then the finding of current $\hat{I}$ is accomplished by using formula (2.1).

Now, we proceed with the application of the Thevenin theorem to the analysis of SLG fault shown in Figure 2.4a. The location of SLG fault is marked by letters a, b and c on lines connected to $\hat{V}_a$, $\hat{V}_b$ and $\hat{V}_c$, respectively. Impedances of these lines before the location of the fault are the same and equal to Z'. Impedances of the lines after the fault location are connected in series with the load impedances and their lumped (equivalent) impedances are equal to Z. It is apparent that before the fault the current $\hat{I}_n$ through the grounded neutral is equal to zero. It is also clear from the circuit shown in Figure 2.4a that immediately after the SLG fault this current is equal to the fault current $\hat{I}_f$. Thus, the measured current $\hat{I}_n$ can be used in principle for the detection of fault occurrence as well as its location. The latter may be possible because the analysis performed below results in explicit formulas for $\hat{I}_f$ in terms of Z', which depends on the fault location.

To find the fault current $\hat{I}_f$ (and subsequently all currents), we shall use the Thevenin equivalent circuit shown in Figure 2.4b. The analysis consists of the following four steps.

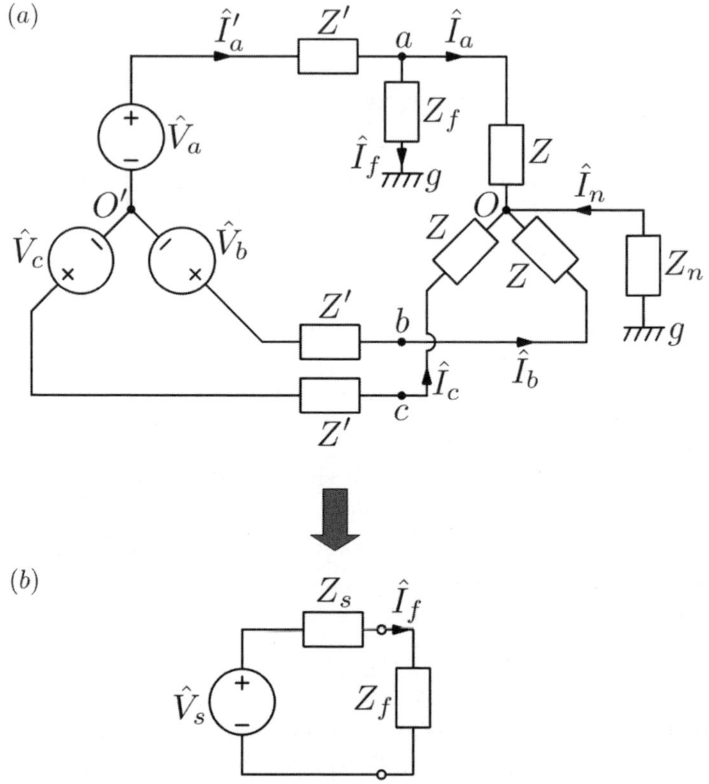

Fig. 2.4

Step 1. The branch with impedance Z_f is removed and voltage $\hat{V}_{oc}$ across the open terminals is computed. The removal of Z_f results in the circuit shown in Figure 2.5a. This is a three-phase circuit under balanced load conditions. For this reason, the potentials of nodes O' and O are the same and equal to zero ($\hat{V}_{O'} = \hat{V}_O = 0$) and the current $\hat{I}_n$ is equal to zero as well. To find $\hat{V}_{oc}$, the per-phase analysis (see Figure 2.5b) can be used, which easily leads to the following formula:

$$\hat{V}_s = \hat{V}_{oc} = \hat{V}_a \frac{Z}{Z+Z'}. \tag{2.4}$$

Step 2. The active circuit shown in Figure 2.5a is transformed into the passive circuit (see Figure 2.6a) by replacing voltage sources by short-circuit branches. Now, the input impedance of this passive circuit with respect to the open terminals a and g must be computed. The computation is

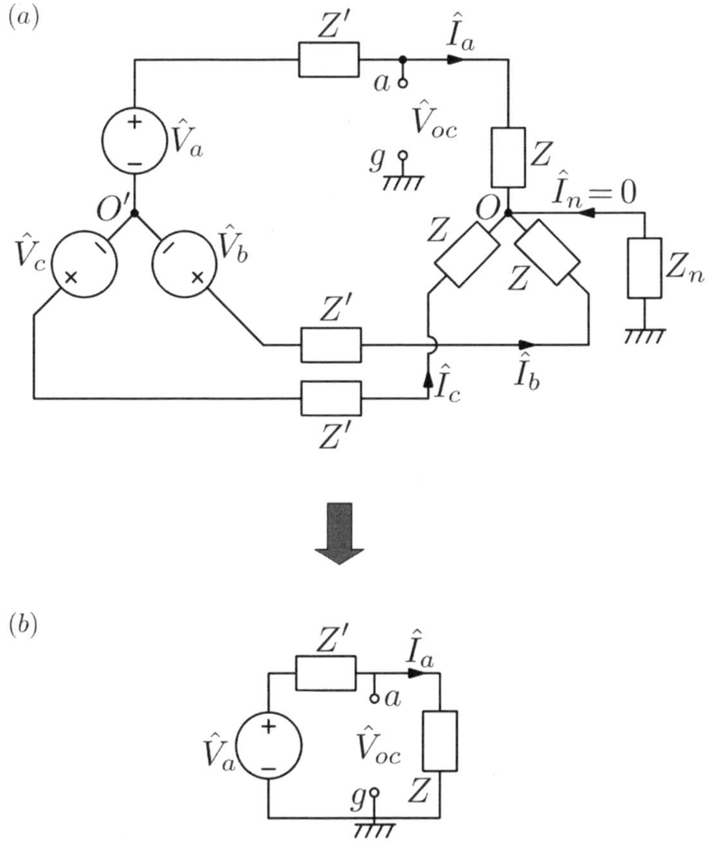

Fig. 2.5

appreciably simplified by redrawing this circuit in the form shown in Figure 2.6b. Indeed, as a result of this redrawing, series and parallel connections of different impedances become easily recognizable, and the input impedance can be computed as follows:

$$Z_{in} = \frac{\left(\frac{Z'+Z}{2} + Z'\right) Z}{\frac{Z'+Z}{2} + Z' + Z} + Z_n. \tag{2.5}$$

After simple algebraic transformation, the last formula can be written as

$$Z_s = Z_{in} = \frac{(3Z' + Z)\, Z + 3Z_n\,(Z' + Z)}{3\,(Z' + Z)}. \tag{2.6}$$

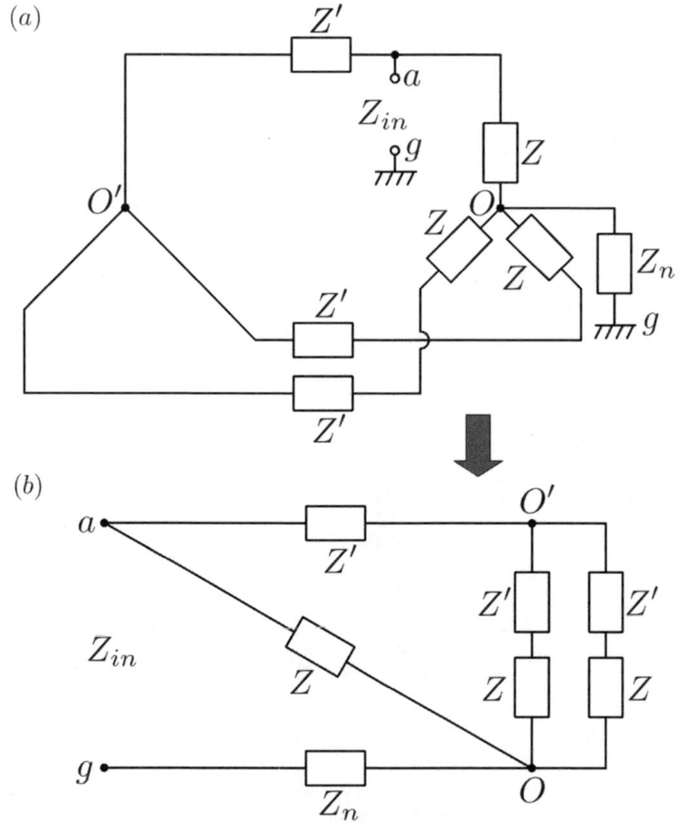

Fig. 2.6

Step 3. Now, we can compute the fault current $\hat{I}_f$ by using the formula

$$\hat{I}_f = \frac{\hat{V}_s}{Z_s + Z_f}. \tag{2.7}$$

According to formula (2.6), we find

$$Z_s + Z_f = \frac{(3Z' + Z)\,Z + 3(Z_n + Z_f)\,(Z' + Z)}{3\,(Z' + Z)}. \tag{2.8}$$

By substituting formulas (2.4) and (2.8) into equation (2.7), we obtain

$$\boxed{\hat{I}_f = \hat{V}_a \frac{3Z}{(3Z' + Z)\,Z + 3(Z_n + Z_f)\,(Z' + Z)}.} \tag{2.9}$$

Step 4. Finally, we can find all currents in the three-phase circuit shown in Figure 2.4a. Indeed, it is clear from this figure that Z_f and Z_n are connected in series. Consequently, we have

$$\hat{I}_n = \hat{I}_f. \tag{2.10}$$

Then, by using KVL for the loop traced from a to g, from g to O and from O to a, we find

$$\hat{I}_f(Z_f + Z_n) - \hat{I}_a Z = 0 \tag{2.11}$$

and

$$\hat{I}_a = \hat{I}_f \frac{Z_f + Z_n}{Z}. \tag{2.12}$$

Next, by applying KCL to the node a, we conclude that

$$\hat{I}'_a = \hat{I}_a + \hat{I}_f. \tag{2.13}$$

Now, by applying KVL to the loop traced from O to a, from a to O', from O' to b and from b to O, we obtain

$$\hat{I}'_a Z' + \hat{I}_a Z - \hat{I}_b \left(Z' + Z\right) = \hat{V}_a - \hat{V}_b, \tag{2.14}$$

which leads to

$$\hat{I}_b = \frac{\hat{V}_b - \hat{V}_a + \hat{I}'_a Z' + \hat{I}_a Z}{Z' + Z}. \tag{2.15}$$

Finally, by using KCL for the node O', we have

$$\hat{I}_c = -\left(\hat{I}'_a + \hat{I}_b\right). \tag{2.16}$$

Thus, by computing the fault current according to equation (2.9) and then by using formulas (2.10), (2.12), (2.13), (2.15) and (2.16), we can compute all currents in the three-phase circuit shown in Figure 2.4a. This concludes the analysis of a single line-to-ground (SLG) fault.

Next, by using the Thevenin Theorem, we consider analysis of line-to-line (LL) fault shown in Figure 2.7a. As before, the central idea of our analysis is to reduce the three-phase circuit shown in Figure 2.7a to the Thevenin equivalent circuit shown in Figure 2.7b. The analysis consists of the following four steps.

Step 1. The impedance Z_f is removed and voltage $\hat{V}_{oc}$ across the open terminals is computed. The removal of Z_f results in the circuit shown in Figure 2.8. This is a three-phase circuit with balanced load. Consequently,

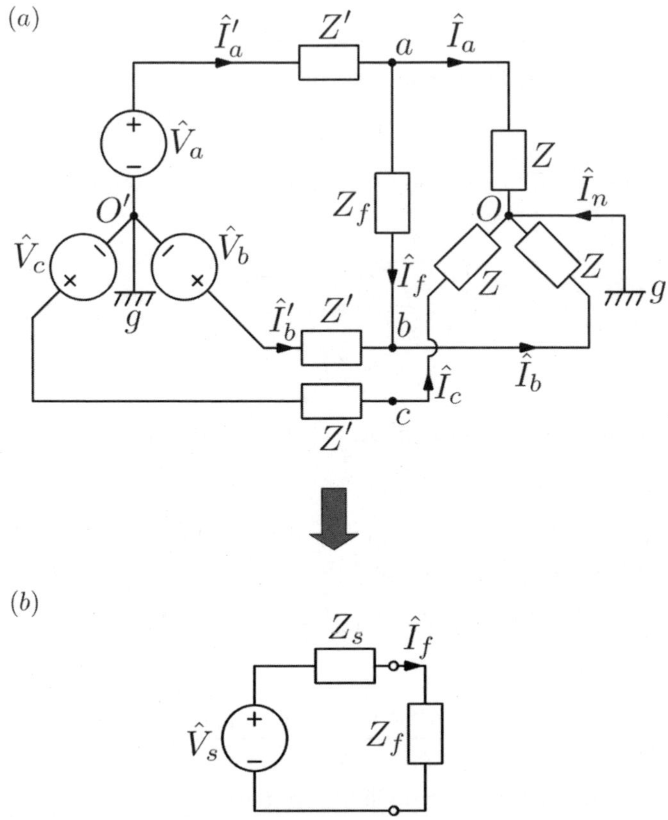

Fig. 2.7

$$\hat{I}_a = \frac{\hat{V}_a}{Z' + Z}, \qquad \hat{I}_b = \alpha \hat{I}_a, \qquad \hat{I}_c = \alpha^2 \hat{I}_a. \tag{2.17}$$

Now, by using KVL for the loop traced from a to O, from O to b and from b to a, we find

$$\hat{I}_a Z - \hat{I}_b Z - \hat{V}_{oc} = 0, \tag{2.18}$$

which leads to

$$\hat{V}_{oc} = \left(\hat{I}_a - \hat{I}_b\right) Z. \tag{2.19}$$

According to the first two formulas from (2.17), the last equation can be transformed as follows:

$$\hat{V}_s = \hat{V}_{oc} = \frac{(1-\alpha)Z}{Z' + Z} \hat{V}_a. \tag{2.20}$$

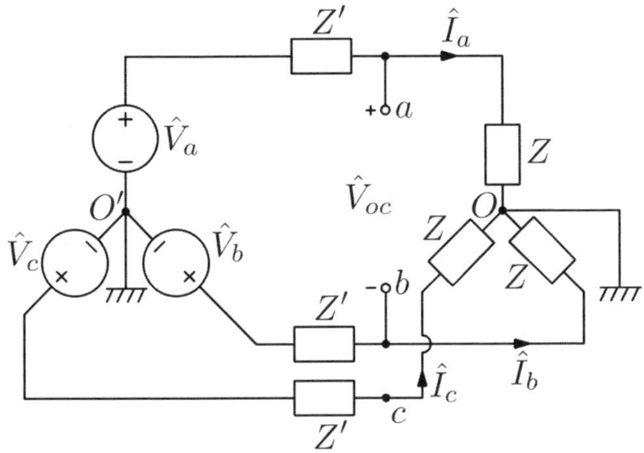

Fig. 2.8

Step 2. The active circuit shown in Figure 2.8 is transformed to the passive circuit shown in Figure 2.9a. This is done by replacing the voltage sources by short-circuit branches. Now, the input impedance of this passive circuit with respect to the open terminals a and b must be computed. The computation is greatly simplified by redrawing this circuit as shown in Figure 2.9b. (Please note that the line c branch consisting of Z' and Z is omitted because it is short-circuited.) From Figure 2.9b, we find

$$Z_s = Z_{in} = 2\frac{Z'Z}{Z' + Z}. \tag{2.21}$$

Step 3. Now, we can compute the fault current $\hat{I}_f$ by using the formula

$$\hat{I}_f = \frac{\hat{V}_s}{Z_s + Z_f}. \tag{2.22}$$

According to equation (2.21), we have

$$Z_s + Z_f = \frac{2Z'Z + Z_f\,(Z' + Z)}{Z' + Z}. \tag{2.23}$$

By substituting formulas (2.20) and (2.23) into equation (2.22), we arrive at

$$\boxed{\hat{I}_f = \hat{V}_a\frac{(1-\alpha)Z}{2Z'Z + Z_f\,(Z' + Z)}}. \tag{2.24}$$

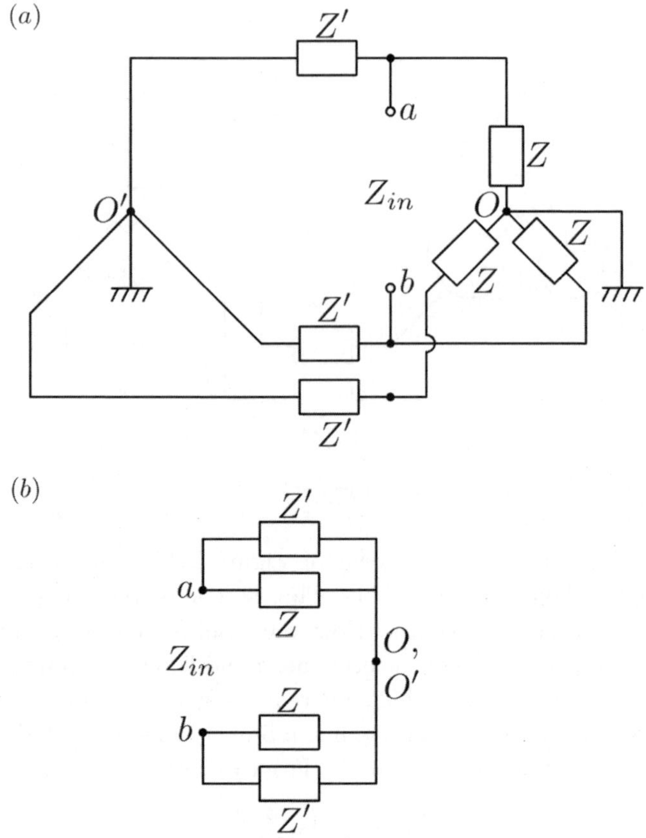

Fig. 2.9

Step 4. Finally, we can find all currents in the three-phase circuit shown in Figure 2.7a. Indeed, by using KVL for the loop traced from O' to a, from a to O and from O to O', we find

$$\left(\hat{I}_a + \hat{I}_f\right) Z' + \hat{I}_a Z = \hat{V}_a, \tag{2.25}$$

which leads to

$$\hat{I}_a = \frac{\hat{V}_a - \hat{I}_f Z'}{Z' + Z}. \tag{2.26}$$

Then,

$$\hat{I}'_a = \hat{I}_a + \hat{I}_f. \tag{2.27}$$

Similarly, by using KVL for the loop traced from O' to b, from b to O and from O to O', we obtain

$$\left(\hat{I}_b - \hat{I}_f\right) Z' + \hat{I}_b Z = \hat{V}_b, \tag{2.28}$$

which leads to

$$\hat{I}_b = \frac{\hat{V}_b + \hat{I}_f Z'}{Z' + Z}, \tag{2.29}$$

and then

$$\hat{I}'_b = \hat{I}_b - \hat{I}_f. \tag{2.30}$$

Current $\hat{I}_c$ is found by applying KVL to the loop traced from O' to c, from c to O and from O to O':

$$\hat{I}_c = \frac{\hat{V}_c}{Z' + Z}. \tag{2.31}$$

By using KCL for node O, we find

$$\hat{I}_n = -\left(\hat{I}_a + \hat{I}_b + \hat{I}_c\right). \tag{2.32}$$

Thus, by computing the fault current $\hat{I}_f$ using equation (2.24) and then by using formulas (2.26), (2.27), (2.29), (2.30), (2.31) and (2.32), we can compute all currents in the three-phase circuit shown in Figure 2.7a. This concludes the analysis of LL fault.

In the conclusion of this section, we shall sketch the analysis of double line-to-ground (DLG) fault by using the Thevenin theorem. This fault is presented in Figure 2.10a. The central idea of our analysis is to reduce the three-phase circuit shown in Figure 2.10a to the Thevenin equivalent circuit shown in Figure 2.10b. As before, the analysis consists of the following four steps.

Step 1. The impedance Z_f^b is removed and voltage $\hat{V}_{oc}$ across the open terminals is computed. The removal of Z_f^b results in the circuit shown in Figure 2.11. This circuit is identical to the three-phase circuit of SLG fault analyzed at the beginning of this section. By using this analysis $\hat{V}_s = \hat{V}_{oc}$ can be found.

Step 2. The active circuit shown in Figure 2.11 is replaced by the passive circuit shown in Figure 2.12a. This is done by replacing voltage sources by short-circuit branches. Then, this passive circuit is redrawn as shown

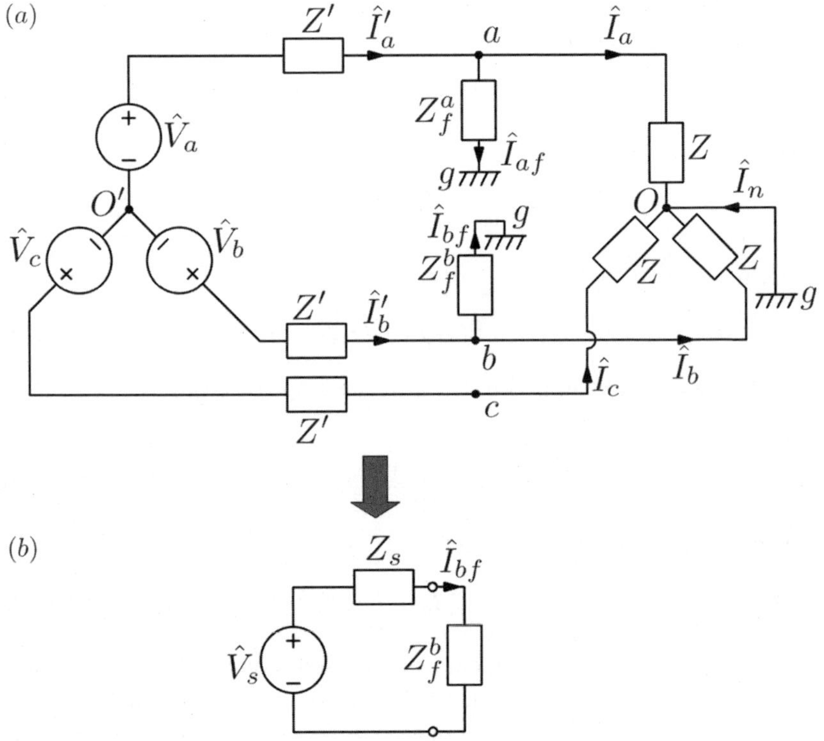

Fig. 2.10

in Figure 2.12b to make series and parallel connections apparent. The redrawn circuit is used to compute $Z_s = Z_{in}$.

Step 3. Now, the fault current $\hat{I}_{bf}$ can be computed by using the Thevenin equivalent circuit as well as $\hat{V}_s$ and Z_s found in the first and second steps, respectively.

Step 4. By using the found value of $\hat{I}_{bf}$, all currents in the three-phase circuit shown in Figure 2.10a can be computed. This can be done by computing these currents in the following order: $\hat{I}_b$, $\hat{I}'_b$, $\hat{I}_c$, $\hat{I}'_a$, $\hat{I}_a$ and $\hat{I}_{af}$. It is left to the reader as an exercise to perform all the computations and derivations sketched in our discussion.

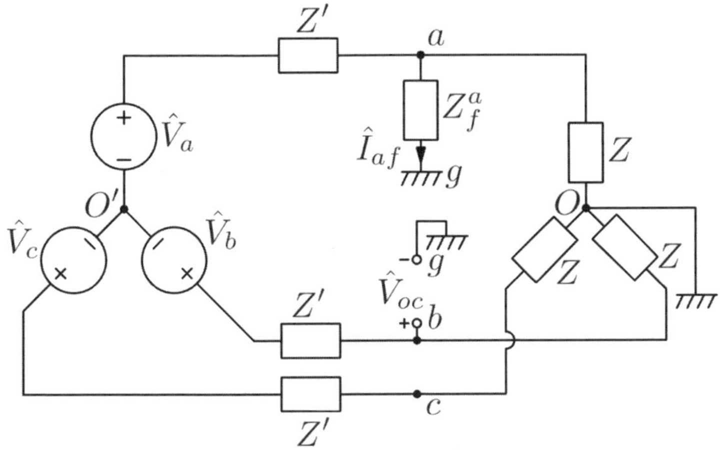

Fig. 2.11

2.2 Symmetrical Components

The method of symmetrical components was developed in electric power engineering for the analysis of unbalanced regimes and faults in power systems. In this section, we present the definition and the basic mathematical facts related to the symmetrical components. In the subsequent two sections, these facts are extensively used for the analysis of faults in power systems.

Symmetrical components can be introduced for three-phase currents and three-phase voltages or any other three-phase quantities. We first consider three-phase currents. There are three distinct sets of symmetrical components: the positive-sequence set, the negative-sequence set and the zero-sequence set. These sets are defined below.

In the case of the positive-sequence set, one deals with three phasors

$$\hat{I}_a^+, \quad \hat{I}_b^+, \quad \hat{I}_c^+ \tag{2.33}$$

with the properties

$$\left|\hat{I}_a^+\right| = \left|\hat{I}_b^+\right| = \left|\hat{I}_c^+\right|, \tag{2.34}$$

$$\hat{I}_b^+ = \alpha \hat{I}_a^+, \quad \hat{I}_c^+ = \alpha^2 \hat{I}_a^+, \tag{2.35}$$

where, as before, $\alpha = e^{-j\frac{2\pi}{3}}$. It is worthwhile to point out that the property (2.34) follows from the property (2.35). These three phasors can be

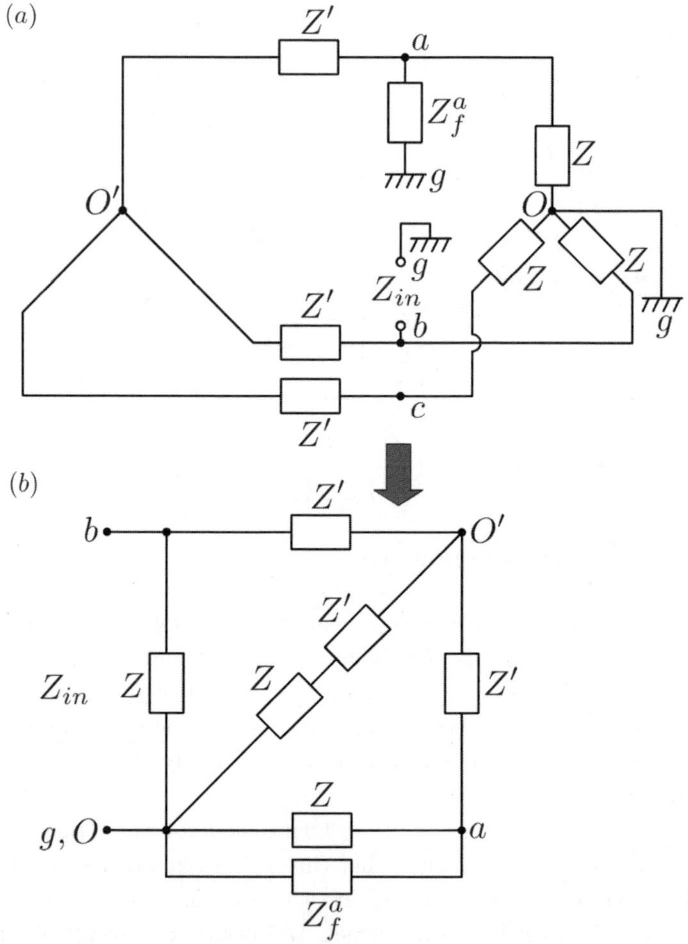

Fig. 2.12

considered as the phasors of three electric currents of the same frequency and the same peak values but phase-shifted with respect to one another by $\frac{2\pi}{3}$. A phasor diagram for these currents is shown in Figure 2.13a.

In the case of the negative-sequence set, one deals with three phasors

$$\hat{I}_a^-, \quad \hat{I}_b^-, \quad \hat{I}_c^- \tag{2.36}$$

with the properties

$$\left|\hat{I}_a^-\right| = \left|\hat{I}_b^-\right| = \left|\hat{I}_c^-\right|, \tag{2.37}$$

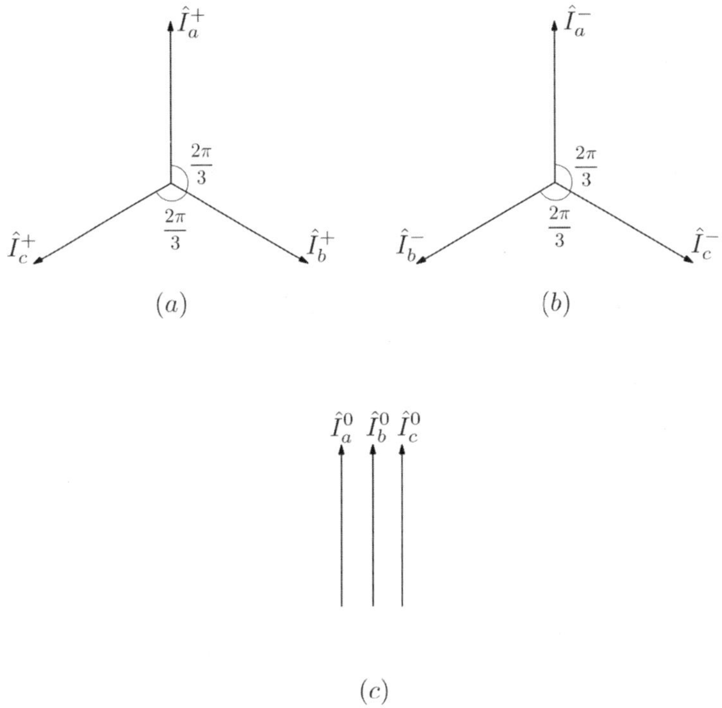

Fig. 2.13

$$\hat{I}_b^- = \alpha^2 \hat{I}_a^-, \quad \hat{I}_c^- = \alpha \hat{I}_a^-. \tag{2.38}$$

Again, it is apparent that relation (2.37) follows from (2.38). These three phasors can be considered as the phasors of three electric currents of the same frequency and the same peak values but phase-shifted by $\frac{4\pi}{3}$ with respect to one another. A phasor diagram for these currents is shown in Figure 2.13b.

Finally, in the case of the zero-sequence set, one deals with three phasors

$$\hat{I}_a^0, \quad \hat{I}_b^0, \quad \hat{I}_c^0 \tag{2.39}$$

with the properties

$$\hat{I}_a^0 = \hat{I}_b^0 = \hat{I}_c^0. \tag{2.40}$$

These three phasors can be considered as the phasors of three electric currents of the same frequency and the same peak values and which are in phase with one another. A phasor diagram for these currents is shown in Figure 2.13c.

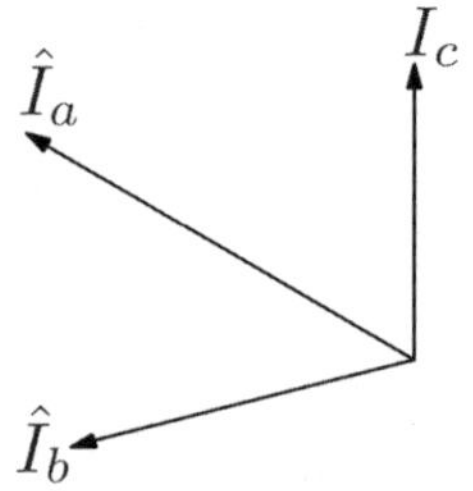

Fig. 2.14

It turns out that an arbitrary set of three currents with phasors $\hat{I}_a$, $\hat{I}_b$ and $\hat{I}_c$ (see Figure 2.14) can be decomposed into symmetrical components. Mathematically, this means that

$$\hat{I}_a = \hat{I}_a^0 + \hat{I}_a^+ + \hat{I}_a^-, \tag{2.41}$$

$$\hat{I}_b = \hat{I}_b^0 + \hat{I}_b^+ + \hat{I}_b^-, \tag{2.42}$$

$$\hat{I}_c = \hat{I}_c^0 + \hat{I}_c^+ + \hat{I}_c^-. \tag{2.43}$$

By using the properties of symmetrical components expressed by formulas (2.34), (2.35), (2.37), (2.38) and (2.40), the last three equations can be written as follows:

$$\boxed{\begin{aligned}\hat{I}_a &= \hat{I}_a^0 + \hat{I}_a^+ + \hat{I}_a^-, & (2.44)\\ \hat{I}_b &= \hat{I}_a^0 + \alpha\hat{I}_a^+ + \alpha^2\hat{I}_a^-, & (2.45)\\ \hat{I}_c &= \hat{I}_a^0 + \alpha^2\hat{I}_a^+ + \alpha\hat{I}_a^-. & (2.46)\end{aligned}}$$

The last three equations can be construed as linear simultaneous equations with respect to $\hat{I}_a^0$, $\hat{I}_a^+$ and $\hat{I}_a^-$. It is shown below that the solution of the simultaneous equations is given by the following formulas:

$$\boxed{\begin{aligned}\hat{I}_a^0 &= \frac{1}{3}\left(\hat{I}_a + \hat{I}_b + \hat{I}_c\right), & (2.47)\\ \hat{I}_a^+ &= \frac{1}{3}\left(\hat{I}_a + \alpha^2\hat{I}_b + \alpha\hat{I}_c\right), & (2.48)\\ \hat{I}_a^- &= \frac{1}{3}\left(\hat{I}_a + \alpha\hat{I}_b + \alpha^2\hat{I}_c\right). & (2.49)\end{aligned}}$$

The proof of formulas (2.47), (2.48) and (2.49) proceeds as follows. To derive formula (2.47), we add up all three equations (2.44), (2.45) and (2.46). This yields

$$\hat{I}_a + \hat{I}_b + \hat{I}_c = 3\hat{I}_a^0 + \left(1 + \alpha + \alpha^2\right)\hat{I}_a^+ + \left(1 + \alpha^2 + \alpha\right)\hat{I}_a^-. \tag{2.50}$$

Now, by taking into account that $1 + \alpha + \alpha^2 = 0$, we arrive at formula (2.47). To derive formula (2.48), we multiply both sides of equation (2.45) by α^2, both sides of equation (2.46) by α and add them all with equation (2.44). This yields

$$\hat{I}_a + \alpha^2 \hat{I}_b + \alpha \hat{I}_c = \left(1 + \alpha^2 + \alpha\right) \hat{I}_a^0 + \left(1 + \alpha^3 + \alpha^3\right) \hat{I}_a^+ + \left(1 + \alpha^4 + \alpha^2\right) \hat{I}_a^- . \tag{2.51}$$

By taking into account that $\alpha^3 = 1$ and $\alpha^4 = \alpha$, from the last expression we easily obtain formula (2.48). By using the same line of reasoning, formula (2.49) can be derived. Its derivation is left to the reader as a simple exercise.

So far, we have discussed symmetrical components for three-phase currents. Symmetrical components for three-phase voltages can be introduced in the identical way. Namely, we consider the positive-sequence set, negative-sequence set and zero-sequence set of voltage phasors

$$\hat{V}_a^+, \hat{V}_b^+, \hat{V}_c^+; \quad \hat{V}_a^-, \hat{V}_b^-, \hat{V}_c^-; \quad \hat{V}_a^0, \hat{V}_b^0, \hat{V}_c^0, \tag{2.52}$$

which are related to one another by formulas similar to (2.34)-(2.35), (2.37)-(2.38) and (2.40), respectively. Then, for an arbitrary set of three voltages with phasors $\hat{V}_a$, $\hat{V}_b$ and $\hat{V}_c$ we have relations mathematically identical to formulas (2.44)-(2.46) and (2.47)-(2.49):

$$\hat{V}_a = \hat{V}_a^0 + \hat{V}_a^+ + \hat{V}_a^-, \tag{2.53}$$

$$\hat{V}_b = \hat{V}_a^0 + \alpha \hat{V}_a^+ + \alpha^2 \hat{V}_a^-, \tag{2.54}$$

$$\hat{V}_c = \hat{V}_a^0 + \alpha^2 \hat{V}_a^+ + \alpha \hat{V}_a^- \tag{2.55}$$

and

$$\hat{V}_a^0 = \frac{1}{3}\left(\hat{V}_a + \hat{V}_b + \hat{V}_c\right), \tag{2.56}$$

$$\hat{V}_a^+ = \frac{1}{3}\left(\hat{V}_a + \alpha^2 \hat{V}_b + \alpha \hat{V}_c\right), \tag{2.57}$$

$$\hat{V}_a^- = \frac{1}{3}\left(\hat{V}_a + \alpha \hat{V}_b + \alpha^2 \hat{V}_c\right). \tag{2.58}$$

It is worthwhile to point out that by using formulas (2.47)-(2.49) and (2.56)-(2.58) we can compute $\hat{I}_a^0$, $\hat{I}_a^+$, $\hat{I}_a^-$ and $\hat{V}_a^0$, $\hat{V}_a^+$, $\hat{V}_a^-$, i.e., symmetrical components associated with phase a. Then, symmetrical components for phases b and c can be determined by using formulas (2.35), (2.38) and (2.40) for currents and similar formulas for voltages.

It is apparent from the above discussion that the transformations from three-phase quantities to their symmetrical components are linear transformations. For this reason, any linear combination of three-phase quantities can be represented as the same linear combination of their symmetrical components.

Now, we consider some examples that will be used in our future discussions.

Example 1. Given

$$\hat{I}_a \neq 0, \quad \hat{I}_b = \hat{I}_c = 0, \tag{2.59}$$

it is required to find symmetrical components $\hat{I}_a^0$, $\hat{I}_a^+$ and $\hat{I}_a^-$.

By substituting formulas (2.59) into equations (2.47), (2.48) and (2.49), we find

$$\hat{I}_a^0 = \hat{I}_a^+ = \hat{I}_a^- = \frac{1}{3}\hat{I}_a. \tag{2.60}$$

Example 2. Given

$$\hat{V}_a = \hat{V}_b = \hat{V}_c, \tag{2.61}$$

it is required to find symmetrical components $\hat{V}_a^0$, $\hat{V}_a^+$ and $\hat{V}_a^-$.

By using the relation (2.61) in formulas (2.56), (2.57) and (2.58) and taking into account that $1 + \alpha + \alpha^2 = 0$, we find

$$\hat{V}_a^0 = \hat{V}_a, \qquad \text{while } \hat{V}_a^+ = \hat{V}_a^- = 0. \tag{2.62}$$

Example 3. Given a set of three-phase voltages

$$\hat{V}_a, \quad \hat{V}_b = \alpha\hat{V}_a, \quad \hat{V}_c = \alpha^2\hat{V}_a, \tag{2.63}$$

it is required to find $\hat{V}_a^0$, $\hat{V}_a^+$ and $\hat{V}_a^-$.

By using relations (2.63) in formulas (2.56), (2.57) and (2.58), we easily establish that

$$\hat{V}_a^+ = \hat{V}_a, \qquad \text{while } \hat{V}_a^0 = \hat{V}_a^- = 0. \tag{2.64}$$

Example 4. Consider three currents $\hat{I}_a$, $\hat{I}_b$ and $\hat{I}_c$ in three branches connected into star (without a neutral) as shown in Figure 2.15. It is required to prove that

$$\hat{I}_a^0 = 0. \tag{2.65}$$

According to KCL, we have

$$\hat{I}_a + \hat{I}_b + \hat{I}_c = 0. \tag{2.66}$$

Then, by using formula (2.47), we conclude that

$$\hat{I}_a^0 = \frac{1}{3}\left(\hat{I}_a + \hat{I}_b + \hat{I}_c\right) = 0. \tag{2.67}$$

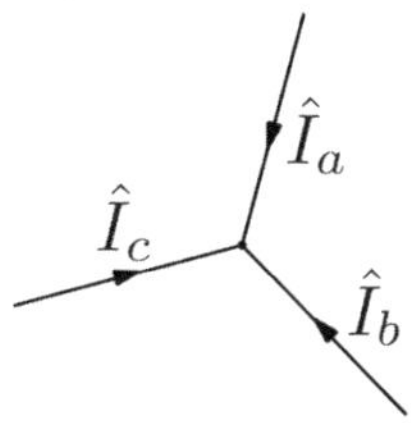

Fig. 2.15

Thus, it can be concluded that a star connection (without neutral) can be used as a filter of zero-sequence components of three-phase currents. It is also easy to prove that a delta connection may serve as a filter of zero-sequence components of three-phase voltages.

Next, we introduce some matrices associated with symmetrical components. To this end, we interpret phasors $\hat{V}_a$, $\hat{V}_b$ and $\hat{V}_c$ as well as $\hat{V}_a^0$, $\hat{V}_a^+$ and $\hat{V}_a^-$ as components of three-dimensional vectors

$$\begin{pmatrix} \hat{V}_a \\ \hat{V}_b \\ \hat{V}_c \end{pmatrix} \text{ and } \begin{pmatrix} \hat{V}_a^0 \\ \hat{V}_a^+ \\ \hat{V}_a^- \end{pmatrix}. \tag{2.68}$$

Then, equations (2.53)-(2.55) can be written in the matrix form as

$$\begin{pmatrix} \hat{V}_a \\ \hat{V}_b \\ \hat{V}_c \end{pmatrix} = \mathbf{A} \begin{pmatrix} \hat{V}_a^0 \\ \hat{V}_a^+ \\ \hat{V}_a^- \end{pmatrix}, \tag{2.69}$$

where

$$\mathbf{A} = \begin{pmatrix} 1 & 1 & 1 \\ 1 & \alpha & \alpha^2 \\ 1 & \alpha^2 & \alpha \end{pmatrix}. \tag{2.70}$$

Similarly, equations (2.56)-(2.58) can be written in the form

$$\begin{pmatrix} \hat{V}_a^0 \\ \hat{V}_a^+ \\ \hat{V}_a^- \end{pmatrix} = \mathbf{B} \begin{pmatrix} \hat{V}_a \\ \hat{V}_b \\ \hat{V}_c \end{pmatrix}, \tag{2.71}$$

where

$$\mathbf{B} = \frac{1}{3} \begin{pmatrix} 1 & 1 & 1 \\ 1 & \alpha^2 & \alpha \\ 1 & \alpha & \alpha^2 \end{pmatrix}. \tag{2.72}$$

It is apparent from formulas (2.69) and (2.71) that

$$\mathbf{B} = \mathbf{A}^{-1} \text{ and } \mathbf{A} = \mathbf{B}^{-1}, \tag{2.73}$$

i.e., matrices **B** and **A** are inverses of one another. It is apparent that equations (2.44)-(2.46) and (2.47)-(2.49) can also be written in terms of matrices **A** and **B**.

It is also useful to write equations (2.53)-(2.55) in the following vector form:

$$\begin{pmatrix} \hat{V}_a \\ \hat{V}_b \\ \hat{V}_c \end{pmatrix} = \hat{V}_a^0 \begin{pmatrix} 1 \\ 1 \\ 1 \end{pmatrix} + \hat{V}_a^+ \begin{pmatrix} 1 \\ \alpha \\ \alpha^2 \end{pmatrix} + \hat{V}_a^- \begin{pmatrix} 1 \\ \alpha^2 \\ \alpha \end{pmatrix}. \tag{2.74}$$

It is clear from the last formula that vectors

$$\mathbf{e}^0 = \begin{pmatrix} 1 \\ 1 \\ 1 \end{pmatrix}, \quad \mathbf{e}^+ = \begin{pmatrix} 1 \\ \alpha \\ \alpha^2 \end{pmatrix}, \quad \mathbf{e}^- = \begin{pmatrix} 1 \\ \alpha^2 \\ \alpha \end{pmatrix} \tag{2.75}$$

serve as basis vectors in the symmetrical component decomposition of arbitrary vector $(\hat{V}_a, \hat{V}_b, \hat{V}_c)^T$. It is easy and interesting to demonstrate that vectors $\mathbf{e}^0$, $\mathbf{e}^+$ and $\mathbf{e}^-$ are orthogonal with respect to inner product

$$\langle \mathbf{x}, \mathbf{y} \rangle = \sum_{k=1}^{3} x_k y_k^*. \tag{2.76}$$

Indeed, we have

$$\left\langle \mathbf{e}^+, \mathbf{e}^0 \right\rangle = 1 + \alpha + \alpha^2 = 0. \tag{2.77}$$

Similarly,

$$\left\langle \mathbf{e}^-, \mathbf{e}^0 \right\rangle = 1 + \alpha^2 + \alpha = 0. \tag{2.78}$$

Finally,

$$\begin{aligned} \left\langle \mathbf{e}^+, \mathbf{e}^- \right\rangle &= 1 + \alpha \left(\alpha^2\right)^* + \alpha^2 \alpha^* = 1 + \alpha\alpha^* \left(\alpha^* + \alpha\right) \\ &= 1 + |\alpha|^2 \left(e^{j\frac{2\pi}{3}} + e^{-j\frac{2\pi}{3}}\right) = 1 + 2\cos\frac{2\pi}{3} = 0. \end{aligned} \tag{2.79}$$

(Alternatively, the previous result can be shown by noting that $\alpha^* = \alpha^2$.) Thus, vectors $\mathbf{e}^0$, $\mathbf{e}^+$ and $\mathbf{e}^-$ form an orthogonal basis in the decomposition (2.74). It is easy to see that the components of these vectors represent the zero-sequence set, the positive-sequence set and the negative-sequence set of phasors with magnitudes normalized (equal) to one. This explains the use of superscripts "0", "+" and "−" for these vectors. Furthermore, vectors $\mathbf{e}^0$, $\mathbf{e}^+$ and $\mathbf{e}^-$ are eigenvectors of specific matrices frequently encountered in power systems with balanced loads. In generic form, these matrices can be written as follows:

$$\mathbf{T} = \begin{pmatrix} d+c & c & c \\ c & d+c & c \\ c & c & d+c \end{pmatrix}, \tag{2.80}$$

where d and c are arbitrary complex numbers. These matrices have the same diagonal elements and the same off-diagonal elements. We shall next prove that

$$\mathbf{T}\mathbf{e}^0 = (d+3c)\mathbf{e}^0, \tag{2.81}$$

$$\mathbf{T}\mathbf{e}^+ = d\mathbf{e}^+ \tag{2.82}$$

and

$$\mathbf{T}\mathbf{e}^- = d\mathbf{e}^-. \tag{2.83}$$

First, we prove formula (2.81) by using the following calculations:

$$\begin{aligned} \mathbf{T}\mathbf{e}^0 &= \begin{pmatrix} d+c & c & c \\ c & d+c & c \\ c & c & d+c \end{pmatrix} \begin{pmatrix} 1 \\ 1 \\ 1 \end{pmatrix} \\ &= \begin{pmatrix} d+3c \\ d+3c \\ d+3c \end{pmatrix} = (d+3c) \begin{pmatrix} 1 \\ 1 \\ 1 \end{pmatrix} = (d+3c)\mathbf{e}^0. \end{aligned} \tag{2.84}$$

Next, we prove formula (2.82):

$$\begin{aligned} \mathbf{T}\mathbf{e}^+ &= \begin{pmatrix} d+c & c & c \\ c & d+c & c \\ c & c & d+c \end{pmatrix} \begin{pmatrix} 1 \\ \alpha \\ \alpha^2 \end{pmatrix} \\ &= \begin{pmatrix} d + c\left(1+\alpha+\alpha^2\right) \\ \alpha d + c\left(1+\alpha+\alpha^2\right) \\ \alpha^2 d + c\left(1+\alpha+\alpha^2\right) \end{pmatrix} = d \begin{pmatrix} 1 \\ \alpha \\ \alpha^2 \end{pmatrix} = d\mathbf{e}^+, \end{aligned} \tag{2.85}$$

where again the fact that $1+\alpha+\alpha^2 = 0$ has been used.

The proof of formula (2.83) is similar to the proof of formula (2.82) and is left to the reader as a simple exercise.

Thus, it has been established that $\mathbf{e}^0$, $\mathbf{e}^+$ and $\mathbf{e}^-$ are eigenvectors of $\mathbf{T}$-type matrices with eigenvalues $d+3c$, d and d, respectively. This means that the symmetrical component decomposition (2.74) can be construed as an eigenvector decomposition. It is also interesting to observe that the columns and rows of matrix $\mathbf{A}$ coincide with the vectors $\mathbf{e}^0$, $\mathbf{e}^+$ and $\mathbf{e}^-$,

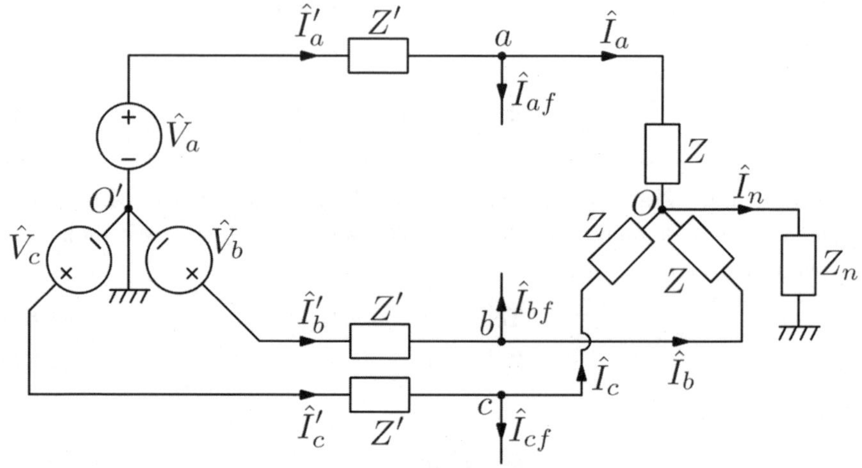

Fig. 2.16

i.e., with the eigenvectors of **T** matrices. For this reason, **T** matrices can be diagonalized by the following similarity transformation:

$$\mathbf{A}^{-1}\mathbf{T}\mathbf{A} = \begin{pmatrix} d+3c & 0 & 0 \\ 0 & d & 0 \\ 0 & 0 & d \end{pmatrix}. \tag{2.86}$$

This fact is well known from linear algebra and it will be instrumental in the derivation of the sequence networks discussed in the next section.

2.3 Sequence Networks

In this section, we derive the sequence networks for symmetrical components applicable to a general fault case of three-phase circuits shown in Figure 2.16. In the next section, these sequence networks will be used for the analysis of particular (SLG, LL and DLG) faults. The derivation consists of four distinct steps.

Step 1. In this step, three coupled equations for $\hat{I}_a$, $\hat{I}_b$, $\hat{I}_c$ and $\hat{I}_{af}$, $\hat{I}_{bf}$, $\hat{I}_{cf}$ are constructed. This is done by writing first the following four KCL equations for nodes a, b, c and O:

$$\hat{I}'_a = \hat{I}_a + \hat{I}_{af}, \tag{2.87}$$

$$\hat{I}'_b = \hat{I}_b + \hat{I}_{bf}, \tag{2.88}$$

$$\hat{I}'_c = \hat{I}_c + \hat{I}_{cf}, \tag{2.89}$$

$$\hat{I}_n = \hat{I}_a + \hat{I}_b + \hat{I}_c. \tag{2.90}$$

Then, we use these formulas in writing the following three KVL equations for the loops that can be easily recognized from the equation structures:

$$\left(\hat{I}_a + \hat{I}_{af}\right) Z' + \hat{I}_a Z + \left(\hat{I}_a + \hat{I}_b + \hat{I}_c\right) Z_n = \hat{V}_a, \tag{2.91}$$

$$\left(\hat{I}_b + \hat{I}_{bf}\right) Z' + \hat{I}_b Z + \left(\hat{I}_a + \hat{I}_b + \hat{I}_c\right) Z_n = \hat{V}_b, \tag{2.92}$$

$$\left(\hat{I}_c + \hat{I}_{cf}\right) Z' + \hat{I}_c Z + \left(\hat{I}_a + \hat{I}_b + \hat{I}_c\right) Z_n = \hat{V}_c. \tag{2.93}$$

Step 2. In this step, we transform these coupled equations to the form that can be represented in terms of a **T**-type matrix. In doing so, we combine terms with the same currents $\hat{I}_a$, $\hat{I}_b$ and $\hat{I}_c$ and move terms with the fault currents to the right-hand sides. These transformations yield

$$\hat{I}_a \left(Z' + Z + Z_n\right) + \hat{I}_b Z_n + \hat{I}_c Z_n = \hat{V}_a - \hat{I}_{af} Z', \tag{2.94}$$

$$\hat{I}_a Z_n + \hat{I}_b \left(Z' + Z + Z_n\right) + \hat{I}_c Z_n = \hat{V}_b - \hat{I}_{bf} Z', \tag{2.95}$$

$$\hat{I}_a Z_n + \hat{I}_b Z_n + \hat{I}_c \left(Z' + Z + Z_n\right) = \hat{V}_c - \hat{I}_{cf} Z'. \tag{2.96}$$

Now, we introduce the matrix

$$\mathbf{T} = \begin{pmatrix} Z' + Z + Z_n & Z_n & Z_n \\ Z_n & Z' + Z + Z_n & Z_n \\ Z_n & Z_n & Z' + Z + Z_n \end{pmatrix} \tag{2.97}$$

and write the coupled equations in the matrix form

$$\mathbf{T} \begin{pmatrix} \hat{I}_a \\ \hat{I}_b \\ \hat{I}_c \end{pmatrix} = \begin{pmatrix} \hat{V}_a - \hat{I}_{af} Z' \\ \hat{V}_b - \hat{I}_{bf} Z' \\ \hat{V}_c - \hat{I}_{cf} Z' \end{pmatrix}. \tag{2.98}$$

Step 3. In this step, we demonstrate that the last coupled equations can be completely decoupled when they are written in terms of symmetrical components. To this end, we introduce the following change of variables:

$$\begin{pmatrix} \hat{I}_a \\ \hat{I}_b \\ \hat{I}_c \end{pmatrix} = \mathbf{A} \begin{pmatrix} \hat{I}_a^0 \\ \hat{I}_a^+ \\ \hat{I}_a^- \end{pmatrix}, \tag{2.99}$$

where $\hat{I}_a^0$, $\hat{I}_a^+$ and $\hat{I}_a^-$ are symmetrical components of $\hat{I}_a$, $\hat{I}_b$ and $\hat{I}_c$.

Similarly, we represent the vector in the right-hand side of equation (2.98) in terms of symmetrical components,

$$\begin{pmatrix} \hat{V}_a - \hat{I}_{af}Z' \\ \hat{V}_b - \hat{I}_{bf}Z' \\ \hat{V}_c - \hat{I}_{cf}Z' \end{pmatrix} = \mathbf{A} \begin{pmatrix} \hat{V}_a^0 - \hat{I}_{af}^0 Z' \\ \hat{V}_a^+ - \hat{I}_{af}^+ Z' \\ \hat{V}_a^- - \hat{I}_{af}^- Z' \end{pmatrix}, \tag{2.100}$$

where $\hat{V}_a^0$, $\hat{V}_a^+$ and $\hat{V}_a^-$ are symmetrical components of $\hat{V}_a$, $\hat{V}_b$ and $\hat{V}_c$, while $\hat{I}_{af}^0$, $\hat{I}_{af}^+$ and $\hat{I}_{af}^-$ are symmetrical components of $\hat{I}_{af}$, $\hat{I}_{bf}$ and $\hat{I}_{cf}$. In writing formula (2.100), we also tacitly used the linearity of the transformation between three physical quantities and their symmetrical components.

It has been shown in Example 3 of the previous section that for three-phase voltages $\hat{V}_a$, $\hat{V}_b = \alpha\hat{V}_a$ and $\hat{V}_c = \alpha^2\hat{V}_a$, we have

$$\hat{V}_a^+ = \hat{V}_a, \quad \text{while} \quad \hat{V}_a^0 = \hat{V}_a^- = 0. \tag{2.101}$$

By using the relations in (2.101), we can rewrite formula (2.100) as follows:

$$\begin{pmatrix} \hat{V}_a - \hat{I}_{af}Z' \\ \hat{V}_b - \hat{I}_{bf}Z' \\ \hat{V}_c - \hat{I}_{cf}Z' \end{pmatrix} = \mathbf{A} \begin{pmatrix} -\hat{I}_{af}^0 Z' \\ \hat{V}_a - \hat{I}_{af}^+ Z' \\ -\hat{I}_{af}^- Z' \end{pmatrix}. \tag{2.102}$$

By substituting formulas (2.99) and (2.102) into equation (2.98), we obtain

$$\mathbf{TA} \begin{pmatrix} \hat{I}_a^0 \\ \hat{I}_a^+ \\ \hat{I}_a^- \end{pmatrix} = \mathbf{A} \begin{pmatrix} -\hat{I}_{af}^0 Z' \\ \hat{V}_a - \hat{I}_{af}^+ Z' \\ -\hat{I}_{af}^- Z' \end{pmatrix}. \tag{2.103}$$

Next, we multiply both sides of equation (2.103) by $\mathbf{A}^{-1}$ to get

$$\mathbf{A}^{-1}\mathbf{TA} \begin{pmatrix} \hat{I}_a^0 \\ \hat{I}_a^+ \\ \hat{I}_a^- \end{pmatrix} = \begin{pmatrix} -\hat{I}_{af}^0 Z' \\ \hat{V}_a - \hat{I}_{af}^+ Z' \\ -\hat{I}_{af}^- Z' \end{pmatrix}. \tag{2.104}$$

Now, by recalling formula (2.86) and taking into account that according to (2.97), $d = Z' + Z$ and $c = Z_n$, we find

$$\mathbf{A}^{-1}\mathbf{TA} = \begin{pmatrix} Z' + Z + 3Z_n & 0 & 0 \\ 0 & Z' + Z & 0 \\ 0 & 0 & Z' + Z \end{pmatrix}. \tag{2.105}$$

This means that equation (2.104) can be written in the form

$$\begin{pmatrix} Z' + Z + 3Z_n & 0 & 0 \\ 0 & Z' + Z & 0 \\ 0 & 0 & Z' + Z \end{pmatrix} \begin{pmatrix} \hat{I}_a^0 \\ \hat{I}_a^+ \\ \hat{I}_a^- \end{pmatrix} = \begin{pmatrix} -\hat{I}_{af}^0 Z' \\ \hat{V}_a - \hat{I}_{af}^+ Z' \\ -\hat{I}_{af}^- Z' \end{pmatrix}. \tag{2.106}$$

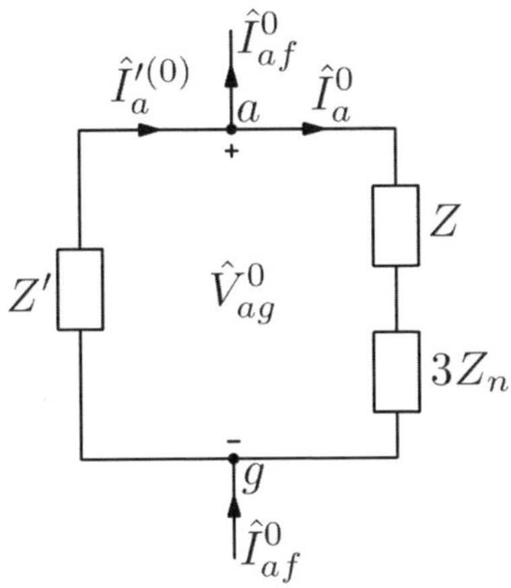

Fig. 2.17

Thus, we have arrived at the set of three decoupled equations for symmetrical components $\hat{I}^0_a$, $\hat{I}^+_a$ and $\hat{I}^-_a$.

Step 4. In this step, each of the decoupled equations (2.106) is interpreted as a KVL equation for a specific circuit called a sequence network. We start with the first equation,

$$\hat{I}^0_a\left(Z' + Z + 3Z_n\right) = -\hat{I}^0_{af}Z', \tag{2.107}$$

and represent it in the form

$$\left(\hat{I}^0_a + \hat{I}^0_{af}\right)Z' + \hat{I}^0_a\left(Z + 3Z_n\right) = 0. \tag{2.108}$$

Next, we can write the three equations (2.87), (2.88) and (2.89) in terms of symmetrical components as follows:

$$\hat{I}'^{(0)}_a = \hat{I}^0_a + \hat{I}^0_{af}, \tag{2.109}$$

$$\hat{I}'^{(+)}_a = \hat{I}^+_a + \hat{I}^+_{af}, \tag{2.110}$$

$$\hat{I}'^{(-)}_a = \hat{I}^-_a + \hat{I}^-_{af}, \tag{2.111}$$

where $\hat{I}'^{(0)}_a$, $\hat{I}'^{(+)}_a$ and $\hat{I}'^{(-)}_a$ are symmetrical components of $\hat{I}'_a$, $\hat{I}'_b$ and $\hat{I}'_c$.

By using formula (2.109), equation (2.108) is transformed as follows:

$$\hat{I}'^{(0)}_a Z' + \hat{I}^0_a(Z + 3Z_n) = 0. \tag{2.112}$$

Now, it is clear that the last equation can be interpreted as the KVL equation for the zero-sequence network shown in Figure 2.17. It is also clear

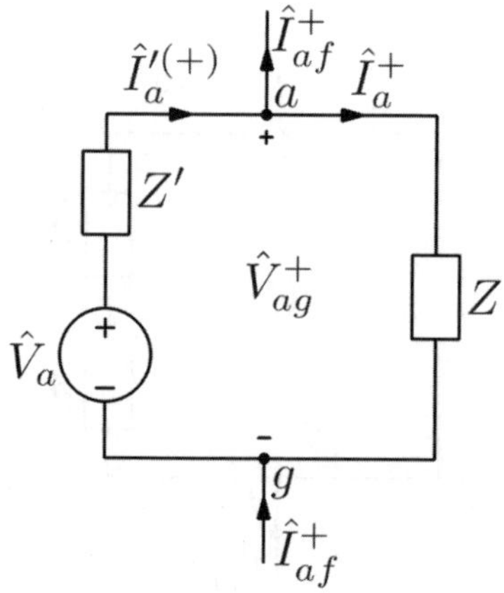

Fig. 2.18

that the KCL equations at nodes a and g are identical to equation (2.109). The reason why these nodes are marked as a and g and the voltage across these nodes is marked as $\hat{V}_{ag}^0$ will be explained later.

The second equation in (2.106) is

$$\hat{I}_a^+ (Z' + Z) = \hat{V}_a - \hat{I}_{af}^+ Z'. \tag{2.113}$$

This equation can be written in the form

$$\left(\hat{I}_a^+ + \hat{I}_{af}^+\right) Z' + \hat{I}_a^+ Z = \hat{V}_a. \tag{2.114}$$

By taking into account equation (2.110), we find

$$\hat{I}_a'^{(+)} Z' + \hat{I}_a^+ Z = \hat{V}_a. \tag{2.115}$$

Now, it is clear that the last equation can be interpreted as the KVL equation for the positive-sequence network shown in Figure 2.18. Furthermore, the KCL equations at nodes a and g are equivalent to equation (2.110).

Finally, the last equation in (2.106) is

$$\hat{I}_a^- (Z' + Z) = -\hat{I}_{af}^- Z', \tag{2.116}$$

which is equivalent to

$$\left(\hat{I}_a^- + \hat{I}_{af}^-\right) Z' + \hat{I}_a^- Z = 0. \tag{2.117}$$

By using equation (2.111), we find

$$\hat{I}_a'^{(-)} Z' + \hat{I}_a^- Z = 0. \tag{2.118}$$

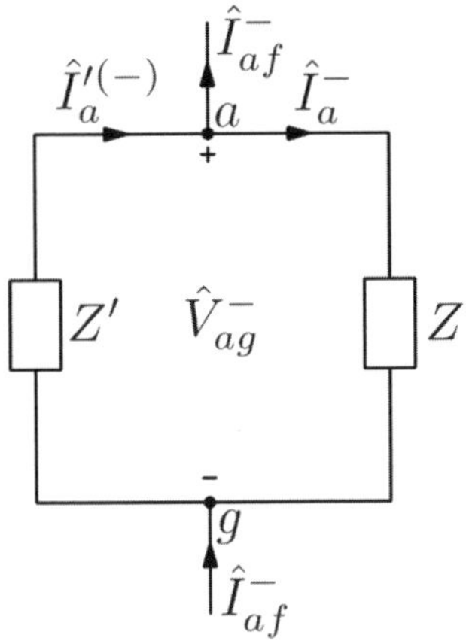

Fig. 2.19

The last equation can be interpreted as the KVL equation for the negative-sequence network shown in Figure 2.19. It is also clear that the KCL equations for nodes a and g are identical to equation (2.111).

To conclude the derivation of the sequence networks shown in Figures 2.17, 2.18 and 2.19, it is necessary to explain the rationale behind the markings of the nodes and the voltages across these nodes in these networks. To this end, we shall use the original three-phase circuit shown in Figure 2.16 and write the following KVL equations for the loops that can be easily identified from the equation structures:

$$\hat{V}_a = \hat{I}'_a Z' + \hat{V}_{ag}, \tag{2.119}$$

$$\hat{V}_b = \hat{I}'_b Z' + \hat{V}_{bg}, \tag{2.120}$$

$$\hat{V}_c = \hat{I}'_c Z' + \hat{V}_{cg}. \tag{2.121}$$

These three equations can be written in terms of symmetrical components as follows:

$$0 = \hat{I}'^{(0)}_a Z' + \hat{V}^0_{ag}, \tag{2.122}$$

$$\hat{V}_a = \hat{I}'^{(+)}_a Z' + \hat{V}^+_{ag}, \tag{2.123}$$

$$0 = \hat{I}'^{(-)}_a Z' + \hat{V}^-_{ag}, \tag{2.124}$$

where $\hat{V}^0_{ag}$, $\hat{V}^+_{ag}$ and $\hat{V}^-_{ag}$ are symmetrical components of $\hat{V}_{ag}$, $\hat{V}_{bg}$ and $\hat{V}_{cg}$, while the symmetrical components of $\hat{V}_a$, $\hat{V}_b$ and $\hat{V}_c$ are given by formula (2.101).

Now, it is clear that equations (2.122), (2.123) and (2.124) coincide with KVL equations written for the zero-sequence, positive-sequence

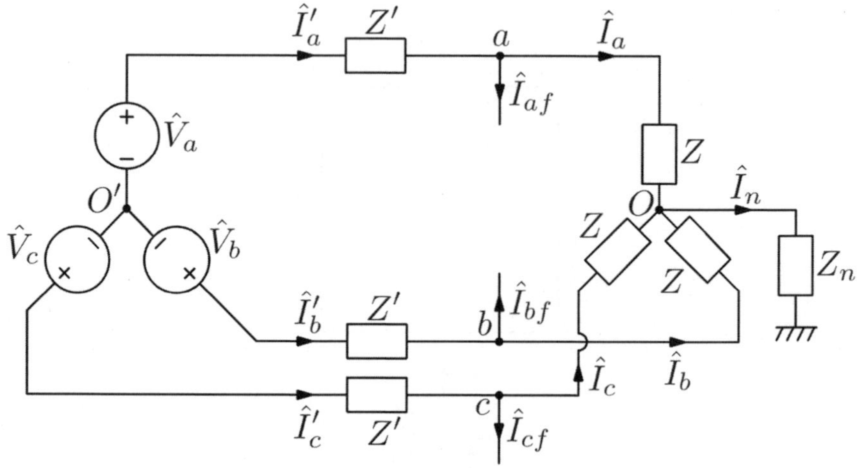

Fig. 2.20

and negative-sequence networks, respectively. This justifies the notations adopted for the nodes and voltages across these nodes in those networks.

The previous derivation has been performed for the three-phase circuit (see Figure 2.16) with the node O' being grounded. Some modification in the derivation of the sequence networks is needed in the case when this node is not grounded (see Figure 2.20). It is clear that in this case, we have

$$\hat{I}'_a + \hat{I}'_b + \hat{I}'_c = 0. \tag{2.125}$$

Furthermore, the following three KVL equations can be written:

$$\hat{I}'_a Z' + \hat{I}_a Z + \hat{I}_n Z_n + \hat{V}_{gO'} = \hat{V}_a, \tag{2.126}$$

$$\hat{I}'_b Z' + \hat{I}_b Z + \hat{I}_n Z_n + \hat{V}_{gO'} = \hat{V}_b, \tag{2.127}$$

$$\hat{I}'_c Z' + \hat{I}_c Z + \hat{I}_n Z_n + \hat{V}_{gO'} = \hat{V}_c. \tag{2.128}$$

By summing up the last three equations and by taking into account formula (2.125) and the fact that for three-phase voltage sources we have

$$\hat{V}_a + \hat{V}_b + \hat{V}_c = 0, \tag{2.129}$$

we derive

$$\left(\hat{I}_a + \hat{I}_b + \hat{I}_c\right) Z + 3\left(\hat{I}_n Z_n + \hat{V}_{gO'}\right) = 0. \tag{2.130}$$

This means that

$$\hat{I}_n Z_n + \hat{V}_{gO'} = -\left(\hat{I}_a + \hat{I}_b + \hat{I}_c\right) \frac{Z}{3}. \tag{2.131}$$

By substituting formula (2.131) into equations (2.126), (2.127) and (2.128) as well as taking into account formulas (2.87), (2.88) and (2.89), we obtain

$$\hat{I}_a\left(Z' + \frac{2Z}{3}\right) - \hat{I}_b\frac{Z}{3} - \hat{I}_c\frac{Z}{3} = \hat{V}_a - \hat{I}_{af}Z', \tag{2.132}$$

$$-\hat{I}_a\frac{Z}{3} + \hat{I}_b\left(Z' + \frac{2Z}{3}\right) - \hat{I}_c\frac{Z}{3} = \hat{V}_b - \hat{I}_{bf}Z', \tag{2.133}$$

$$-\hat{I}_a\frac{Z}{3} - \hat{I}_b\frac{Z}{3} + \hat{I}_c\left(Z' + \frac{2Z}{3}\right) = \hat{V}_c - \hat{I}_{cf}Z'. \tag{2.134}$$

It is clear that the last three equations can be written in the form (2.98) with matrix $\mathbf{T}$ being

$$\mathbf{T} = \begin{pmatrix} Z' + \frac{2Z}{3} & -\frac{Z}{3} & -\frac{Z}{3} \\ -\frac{Z}{3} & Z' + \frac{2Z}{3} & -\frac{Z}{3} \\ -\frac{Z}{3} & -\frac{Z}{3} & Z' + \frac{2Z}{3} \end{pmatrix}. \tag{2.135}$$

Now, by literally repeating the same reasoning which led from equation (2.98) to equation (2.106), we shall arrive at the following decoupled equations:

$$\begin{pmatrix} Z' & 0 & 0 \\ 0 & Z' + Z & 0 \\ 0 & 0 & Z' + Z \end{pmatrix}\begin{pmatrix} \hat{I}_a^0 \\ \hat{I}_a^+ \\ \hat{I}_a^- \end{pmatrix} = \begin{pmatrix} -\hat{I}_{af}^0 Z' \\ \hat{V}_a - \hat{I}_{af}^+ Z' \\ -\hat{I}_{af}^- Z' \end{pmatrix}. \tag{2.136}$$

It is apparent that the second and third equations in (2.136) are identical to the second and third equations in (2.106). This implies that the positive-sequence and negative-sequence networks for the three-phase circuit shown in Figure 2.20 are the same as for the circuit shown in Figure 2.16.

The first equation in (2.136) is

$$\hat{I}_a^0 = -\hat{I}_{af}^0, \tag{2.137}$$

which is consistent with equation (2.109) because according to formula (2.125) (see also Example 4 from the previous section) we have $\hat{I}_a'^{(0)} = 0$. This equation is not sufficient to derive the zero-sequence network. To accomplish the latter, we shall use the following three KVL equations:

$$\hat{I}_a Z + \hat{I}_n Z_n = \hat{V}_{ag}, \tag{2.138}$$

$$\hat{I}_b Z + \hat{I}_n Z_n = \hat{V}_{bg}, \tag{2.139}$$

$$\hat{I}_c Z + \hat{I}_n Z_n = \hat{V}_{cg}. \tag{2.140}$$

By summing up the last three equations and taking into account formula (2.90), we find

$$\left(\hat{I}_a + \hat{I}_b + \hat{I}_c\right)(Z + 3Z_n) = \hat{V}_{ag} + \hat{V}_{bg} + \hat{V}_{cg}. \tag{2.141}$$

Now, by recalling formulas (2.47) and (2.56), the last equation can be written as

$$\hat{I}_a^0(Z + 3Z_n) = \hat{V}_{ag}^0. \tag{2.142}$$

This suggests that the zero-sequence network has the form shown in Figure 2.21a. It is clear that this circuit is consistent with equations (2.137) and (2.142). For the sake of completeness, we present in Figure 2.21 all three sequence networks for the three-phase circuit shown in Figure 2.20.

It is worthwhile to stress in the conclusion of this section that the derived sequence networks are general in nature and valid for any particular (SLG, LL and DLG) fault. They are also remarkably simple in comparison with the three-phase circuits shown in Figures 2.16 and 2.20. They represent relations between symmetrical components of physical quantities related to only one phase a. In this sense, they can be construed as the far-reaching generalization of per-phase analysis to unbalanced (fault) conditions.

2.4 Analysis of Faults by Using Sequence Networks

This analysis is performed by using the following steps:

a) from the nature of the fault, find the relation between symmetrical components $\hat{I}_{af}^0$, $\hat{I}_{af}^+$ and $\hat{I}_{af}^-$;
b) from the nature of the fault, find the relation between symmetrical components $\hat{V}_{ag}^0$, $\hat{V}_{ag}^+$ and $\hat{V}_{ag}^-$;
c) interconnect the sequence networks in accordance with these relations;
d) carry out the analysis of the circuit obtained as a result of interconnection of the sequence networks.

We shall illustrate this outlined approach by the three examples of analysis of SLG, DLG and LL faults.

Example 1. *SLG Fault*

Consider the circuit shown in Figure 2.22.

Step 1. According to the nature of the fault shown in this figure, we find that

$$\hat{I}_{af} = \hat{I}_f \neq 0, \tag{2.143}$$

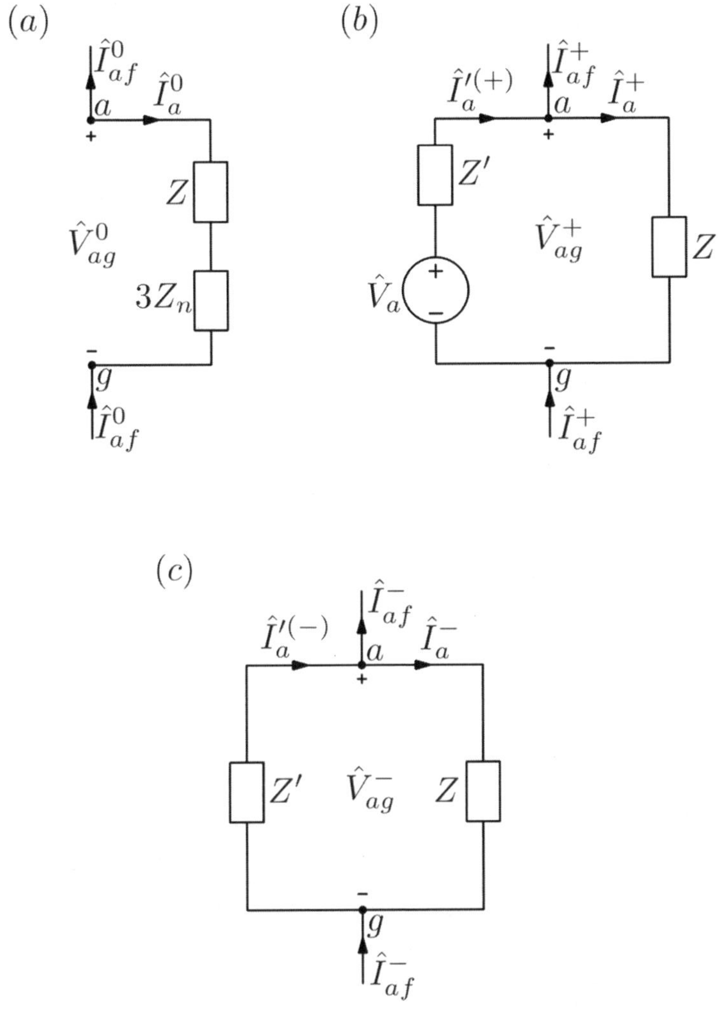

Fig. 2.21

while

$$\hat{I}_{bf} = \hat{I}_{cf} = 0. \tag{2.144}$$

This implies (see Example 1 from section 2 of this chapter) that

$$\boxed{\hat{I}^0_{af} = \hat{I}^+_{af} = \hat{I}^-_{af} = \frac{1}{3}\hat{I}_f.} \tag{2.145}$$

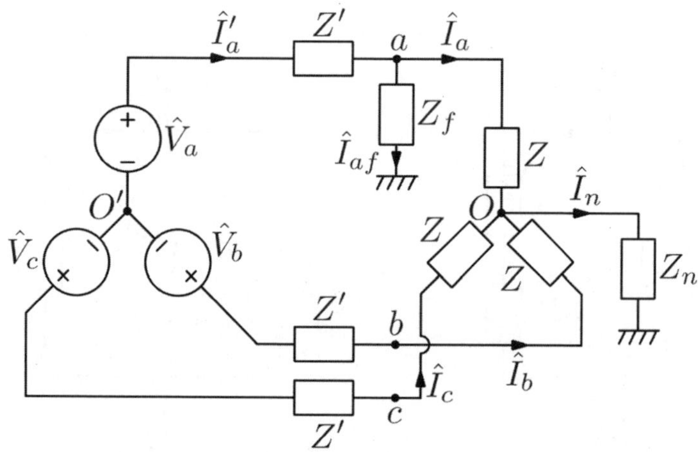

Fig. 2.22

Step 2. From the nature of the fault shown in Figure 2.22, we observe that

$$\hat{V}_{ag} = \hat{I}_f Z_f = \frac{1}{3}\hat{I}_f(3Z_f). \tag{2.146}$$

Next, we use the relation

$$\hat{V}_{ag} = \hat{V}^0_{ag} + \hat{V}^+_{ag} + \hat{V}^-_{ag}. \tag{2.147}$$

From the last two formulas we derive

$$\boxed{\hat{V}^0_{ag} + \hat{V}^+_{ag} + \hat{V}^-_{ag} = \frac{1}{3}\hat{I}_f(3Z_f).} \tag{2.148}$$

Step 3. From relations (2.145) and (2.148) we conclude that the three sequence networks shown in Figure 2.21 (as well as the impedance $3Z_f$) must be connected in series to form the loop as shown in Figure 2.23a. It is clear from this figure that the relations (2.145) and (2.148) between the symmetrical components are satisfied for the electric circuit constructed in this figure.

Step 4. Finally, we shall carry out the analysis of this circuit to find $\hat{I}_f$ and all other currents. It is easy to see that the circuit in Figure 2.23a can be equivalently transformed into the circuit in Figure 2.23b with

$$\tilde{Z} = \frac{Z'Z}{Z'+Z} + Z + 3(Z_n + Z_f) = \frac{(2Z'+Z)Z + 3(Z_n+Z_f)(Z'+Z)}{Z'+Z}. \tag{2.149}$$

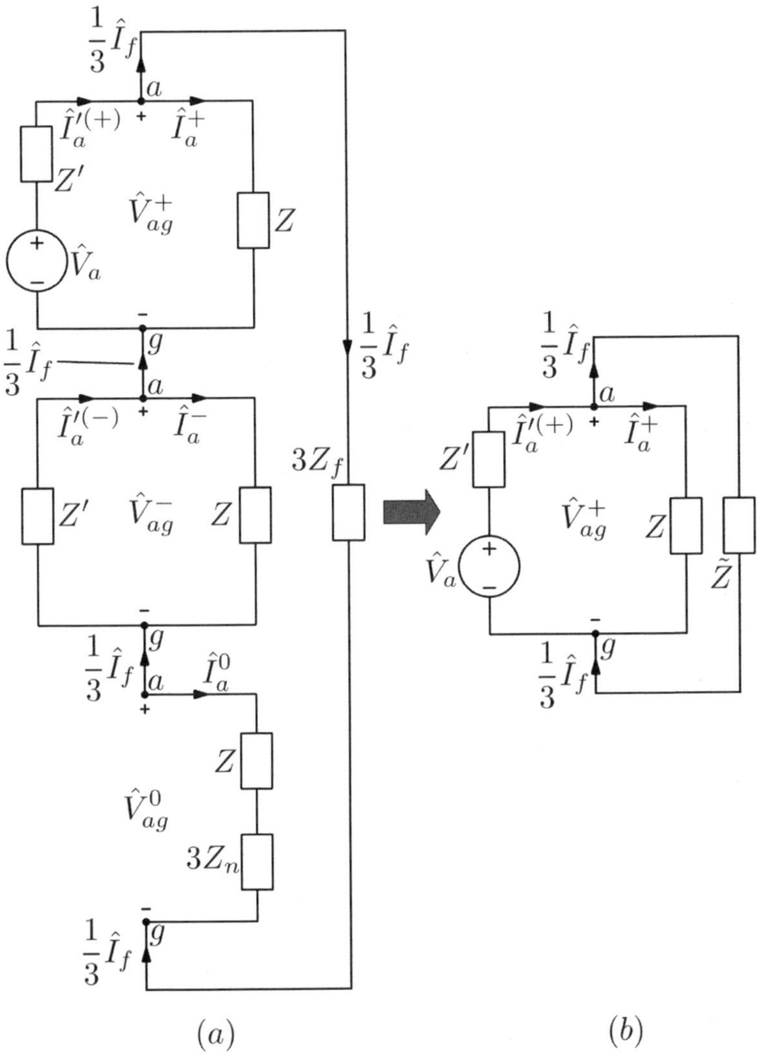

Fig. 2.23

Then, according to the voltage divider rule, we find

$$\hat{V}_{ag}^{+} = \hat{V}_a \frac{\dfrac{Z\tilde{Z}}{Z+\tilde{Z}}}{Z' + \dfrac{Z\tilde{Z}}{Z+\tilde{Z}}} = \hat{V}_a \frac{Z\tilde{Z}}{Z'Z + \tilde{Z}\left(Z'+Z\right)}. \tag{2.150}$$

Furthermore,

$$\frac{1}{3}\hat{I}_f = \frac{\hat{V}_{ag}^+}{\tilde{Z}} = \hat{V}_a \frac{Z}{Z'Z + \tilde{Z}\,(Z' + Z)}. \tag{2.151}$$

Finally, by using formula (2.149), we obtain

$$\boxed{\hat{I}_f = \hat{V}_a \frac{3Z}{(3Z' + Z)\,Z + 3(Z_n + Z_f)\,(Z' + Z)},} \tag{2.152}$$

which is identical to formula (2.9) derived by using the Thevenin theorem.

Now, all other currents can be determined by using the same reasoning as presented after formula (2.9) or by using the circuit shown in Figure 2.23a. Indeed, from formula (2.150), we find

$$\hat{I}_a^+ = \frac{\hat{V}_{ag}^+}{Z} = \hat{V}_a \frac{\tilde{Z}}{Z'Z + \tilde{Z}\,(Z' + Z)}. \tag{2.153}$$

Furthermore, we have

$$\hat{I}_a^- = -\frac{1}{3}\hat{I}_f \frac{Z'}{Z' + Z}, \tag{2.154}$$

and

$$\hat{I}_a^0 = -\frac{1}{3}\hat{I}_f. \tag{2.155}$$

By using symmetrical components $\hat{I}_a^+$, $\hat{I}_a^-$ and $\hat{I}_a^0$, currents $\hat{I}_a$, $\hat{I}_b$ and $\hat{I}_c$ are found according to formulas

$$\hat{I}_a = \hat{I}_a^0 + \hat{I}_a^+ + \hat{I}_a^-, \tag{2.156}$$

$$\hat{I}_b = \hat{I}_a^0 + \alpha\hat{I}_a^+ + \alpha^2\hat{I}_a^-, \tag{2.157}$$

$$\hat{I}_c = \hat{I}_a^0 + \alpha^2\hat{I}_a^+ + \alpha\hat{I}_a^-. \tag{2.158}$$

Finally,

$$\hat{I}_n = -\hat{I}_f, \qquad \hat{I}'_a = \hat{I}_a + \hat{I}_f. \tag{2.159}$$

This concludes the analysis.

Example 2. *DLG Fault*

Consider the circuit shown in Figure 2.24.

Step 1. According to the nature of the fault shown in this figure, we find that

$$\hat{I}_{af} = 0. \tag{2.160}$$

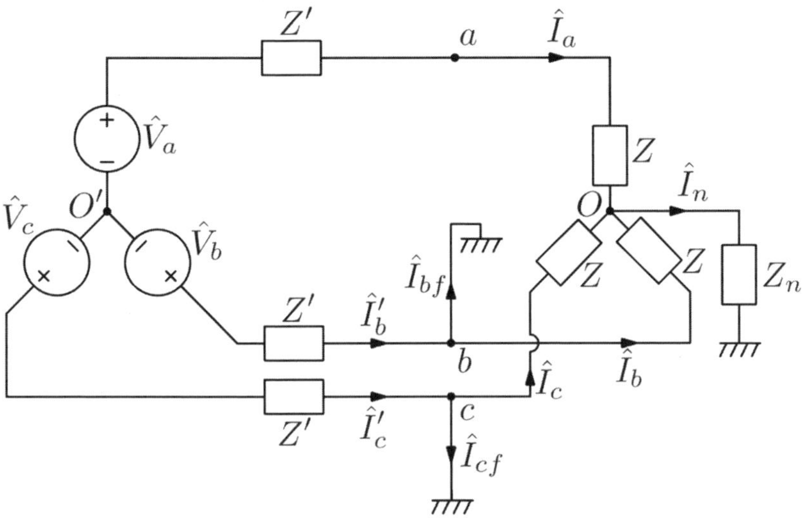

Fig. 2.24

On the other hand,

$$\hat{I}_{af} = \hat{I}^0_{af} + \hat{I}^+_{af} + \hat{I}^-_{af}. \tag{2.161}$$

Consequently,

$$\boxed{\hat{I}^0_{af} + \hat{I}^+_{af} + \hat{I}^-_{af} = 0.} \tag{2.162}$$

Step 2. From the nature of the fault shown in Figure 2.24, we observe that

$$\hat{V}_{ag} \neq 0, \qquad \hat{V}_{bg} = \hat{V}_{cg} = 0. \tag{2.163}$$

This means (see Example 1 from section 2 of this chapter) that

$$\boxed{\hat{V}^0_{ag} = \hat{V}^+_{ag} = \hat{V}^-_{ag} = \frac{1}{3}\hat{V}_{ag}.} \tag{2.164}$$

Step 3. From relations (2.162) and (2.164), we conclude that the three sequence networks shown in Figure 2.21 must be connected in parallel as shown in Figure 2.25a. Indeed, it is clear from this figure that the relations (2.162) and (2.164) are satisfied by the electric circuit shown in this figure.

Step 4. Finally, we shall carry out the analysis of the electric circuit shown in Figure 2.25a. It is easy to see that this circuit can be equivalently transformed into the circuit shown in Figure 2.25b, where

$$\tilde{Z} = \frac{1}{\tilde{Y}}, \qquad \tilde{Y} = \frac{1}{Z'} + \frac{1}{Z} + \frac{1}{Z + 3Z_n}. \tag{2.165}$$

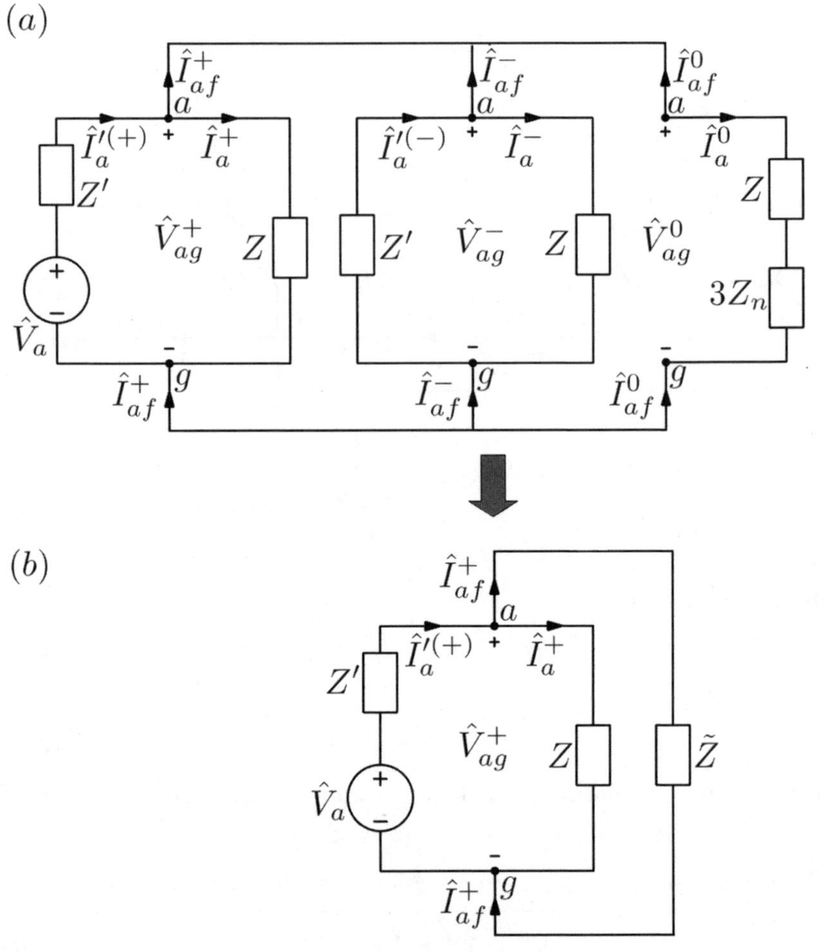

Fig. 2.25

Then, we find

$$\hat{V}^{+}_{ag} = \hat{V}_a \frac{\dfrac{Z\tilde{Z}}{Z+\tilde{Z}}}{Z' + \dfrac{Z\tilde{Z}}{Z+\tilde{Z}}} = \hat{V}_a \frac{Z\tilde{Z}}{Z'Z + \tilde{Z}\,(Z'+Z)}. \tag{2.166}$$

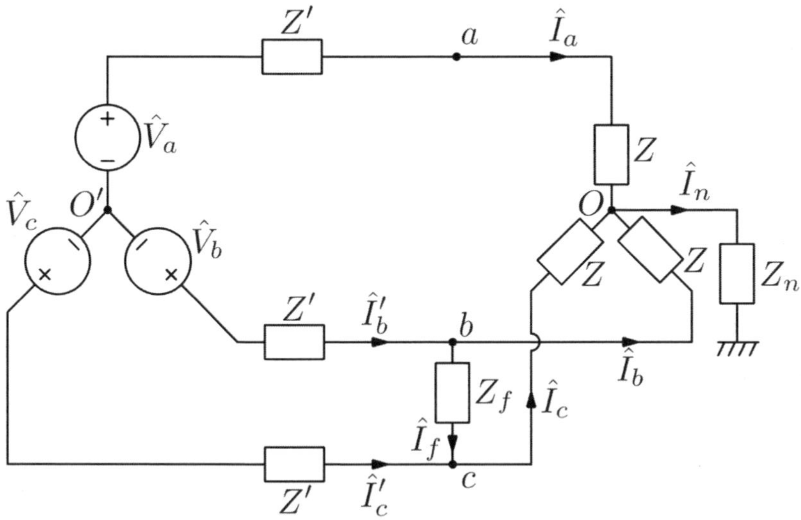

Fig. 2.26

Furthermore,

$$\hat{I}_{af}^{+} = \frac{\hat{V}_{ag}^{+}}{\tilde{Z}} = \hat{V}_a \frac{Z}{Z'Z + \tilde{Z}\,(Z' + Z)}, \tag{2.167}$$

$$\hat{I}_{af}^{-} = -\frac{\hat{V}_{ag}^{+}\,(Z' + Z)}{Z'Z}, \tag{2.168}$$

$$\hat{I}_{af}^{0} = -\frac{\hat{V}_{ag}^{+}}{Z + 3Z_n}. \tag{2.169}$$

As soon as symmetrical components $\hat{I}_{af}^{+}$, $\hat{I}_{af}^{-}$ and $\hat{I}_{af}^{0}$ are found, the fault currents $\hat{I}_{bf}$ and $\hat{I}_{cf}$ can be computed by using the formulas

$$\hat{I}_{bf} = \hat{I}_{af}^{0} + \alpha \hat{I}_{af}^{+} + \alpha^2 \hat{I}_{af}^{-}, \tag{2.170}$$

$$\hat{I}_{cf} = \hat{I}_{af}^{0} + \alpha^2 \hat{I}_{af}^{+} + \alpha \hat{I}_{af}^{-}. \tag{2.171}$$

It is left to the reader as a simple exercise to find all other currents.

Example 3. *LL Fault*

Consider the circuit shown in Figure 2.26.

Step 1. According to the nature of the fault shown in the above figure, we have

$$\hat{I}_{af} = 0, \tag{2.172}$$

while

$$\hat{I}_{bf} = -\hat{I}_{cf} = \hat{I}_f. \tag{2.173}$$

By recalling the relation

$$\hat{I}^0_{af} = \frac{1}{3}\left(\hat{I}_{af} + \hat{I}_{bf} + \hat{I}_{cf}\right), \tag{2.174}$$

from the last three formulas we find

$$\boxed{\hat{I}^0_{af} = 0.} \tag{2.175}$$

Furthermore, we derive

$$\hat{I}^+_{af} = \frac{1}{3}\left(\hat{I}_{af} + \alpha^2\hat{I}_{bf} + \alpha\hat{I}_{cf}\right) = \frac{\alpha^2 - \alpha}{3}\hat{I}_f, \tag{2.176}$$

and

$$\hat{I}^-_{af} = \frac{1}{3}\left(\hat{I}_{af} + \alpha\hat{I}_{bf} + \alpha^2\hat{I}_{cf}\right) = -\frac{\alpha^2 - \alpha}{3}\hat{I}_f. \tag{2.177}$$

The last two equations imply that

$$\boxed{\hat{I}^+_{af} + \hat{I}^-_{af} = 0.} \tag{2.178}$$

Step 2. From the nature of the fault shown in Figure 2.26, we observe that

$$\hat{V}_{cg} = \hat{V}_{bg} - \hat{I}_f Z_f. \tag{2.179}$$

Next, we shall use the last formula to derive

$$\hat{V}^+_{ag} = \frac{1}{3}\left(\hat{V}_{ag} + \alpha^2\hat{V}_{bg} + \alpha\hat{V}_{cg}\right) = \frac{1}{3}\left(\hat{V}_{ag} + \alpha^2\hat{V}_{bg} + \alpha\hat{V}_{bg}\right) - \frac{\alpha}{3}\hat{I}_f Z_f, \tag{2.180}$$

as well as

$$\hat{V}^-_{ag} = \frac{1}{3}\left(\hat{V}_{ag} + \alpha\hat{V}_{bg} + \alpha^2\hat{V}_{cg}\right) = \frac{1}{3}\left(\hat{V}_{ag} + \alpha\hat{V}_{bg} + \alpha^2\hat{V}_{bg}\right) - \frac{\alpha^2}{3}\hat{I}_f Z_f. \tag{2.181}$$

By subtracting formula (2.181) from formula (2.180), we find

$$\hat{V}^+_{ag} - \hat{V}^-_{ag} = \frac{\alpha^2 - \alpha}{3}\hat{I}_f Z_f. \tag{2.182}$$

By taking into account relation (2.176) in the last equation, we find

$$\boxed{\hat{V}^+_{ag} - \hat{V}^-_{ag} = \hat{I}^+_{af} Z_f.} \tag{2.183}$$

Step 3. It follows from formulas (2.178) and (2.183) that the positive-sequence and negative-sequence networks shown in Figure 2.21 must be

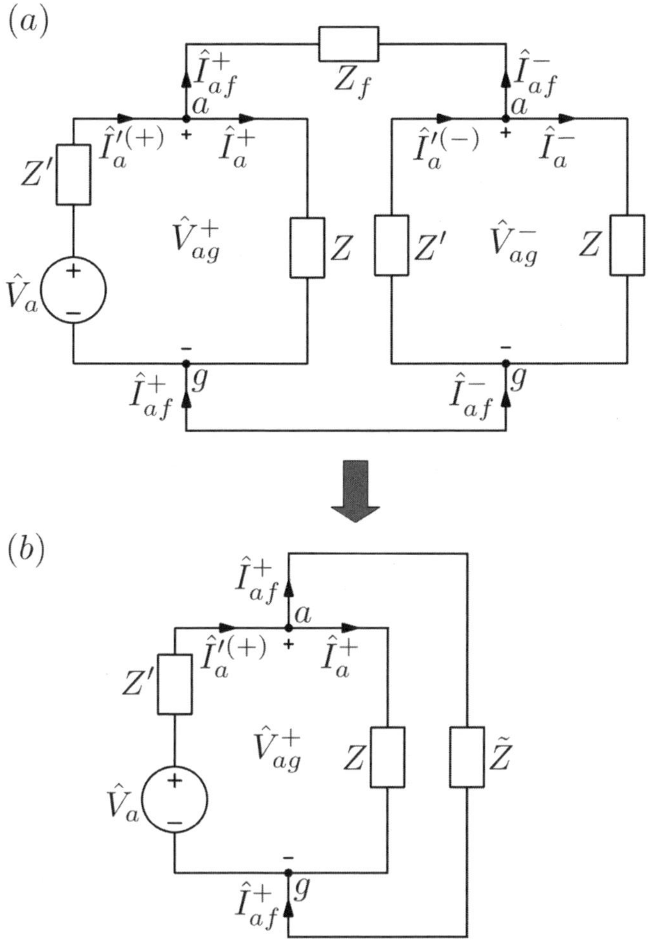

Fig. 2.27

connected in the manner illustrated by Figure 2.27a. Indeed, it is clear from this figure that the relations (2.178) and (2.183) are satisfied by the electric circuit in this figure.

Step 4. Next, we shall carry out the analysis of the electric circuit presented in Figure 2.27a. To this end, the above circuit is equivalently transformed into the circuit shown in Figure 2.27b. This transformation implies that

$$\tilde{Z} = Z_f + \frac{Z'Z}{Z' + Z} = \frac{Z'Z + Z_f\,(Z' + Z)}{Z' + Z}. \tag{2.184}$$

Then, we find that

$$\hat{V}_{ag}^{+} = \hat{V}_a \frac{\dfrac{\tilde{Z}Z}{\tilde{Z}+Z}}{Z' + \dfrac{\tilde{Z}Z}{\tilde{Z}+Z}} = \hat{V}_a \frac{Z\tilde{Z}}{Z'Z + \tilde{Z}\left(Z'+Z\right)}. \tag{2.185}$$

Furthermore,

$$\hat{I}_{af}^{+} = \frac{\hat{V}_{ag}^{+}}{\tilde{Z}} = \hat{V}_a \frac{Z}{Z'Z + \tilde{Z}\left(Z'+Z\right)}. \tag{2.186}$$

Now, by substituting formula (2.184) into the last equation, we derive

$$\hat{I}_{af}^{+} = \hat{V}_a \frac{Z}{2Z'Z + Z_f\left(Z'+Z\right)}. \tag{2.187}$$

Next, we recall that

$$\hat{I}_f = \hat{I}_{bf} = \hat{I}_{af}^{0} + \alpha \hat{I}_{af}^{+} + \alpha^2 \hat{I}_{af}^{-}, \tag{2.188}$$

which, according to formulas (2.175) and (2.178), leads to

$$\hat{I}_f = \left(\alpha - \alpha^2\right) \hat{I}_{af}^{+}. \tag{2.189}$$

By substituting formula (2.187) into the last equation, we find

$$\hat{I}_f = \alpha \hat{V}_a \frac{(1-\alpha)Z}{2Z'Z + Z_f\left(Z'+Z\right)}. \tag{2.190}$$

Since $\hat{V}_b = \alpha \hat{V}_a$, we obtain

$$\boxed{\hat{I}_f = \hat{V}_b \frac{(1-\alpha)Z}{2Z'Z + Z_f\left(Z'+Z\right)},} \tag{2.191}$$

which is consistent with formula (2.24) derived by using the Thevenin theorem. Actually, these two formulas are identical after the proper change of phase markings. It is left to the reader to derive the expressions for all other currents.

Chapter 3

Transformers

3.1 Design and Principle of Operation of the Transformer; The Ideal Transformer

A transformer is a device that finds numerous applications which range from power systems to power electronics and further to computer-communication networks. In power systems, transformers are used to step up and step down ac voltages. For this reason, they are essential for the transmission, distribution and utilization of electric power. Transformers can also be used to electrically isolate sources from loads and for impedance matching purposes. This explains why transformers are vital components in many low-power and high-frequency applications.

A power transformer is a static device in which two (or more) coils (usually called windings) are *strongly* electromagnetically coupled. One of the windings, known as the primary, receives power at a certain voltage and frequency from the source, while the other winding, known as the secondary, delivers power to the load at a different voltage but the same frequency. To enhance electromagnetic coupling between the primary and secondary windings, they are placed around the same leg of iron (ferromagnetic) core. This iron core is subject to a time-varying magnetic flux which links the primary and secondary windings. Since the iron core usually has a finite (nonzero) conductivity, this time-varying magnetic flux induces eddy currents in the iron core. These eddy currents may produce substantial power losses called eddy current losses (see section 3.5 of Part I). To reduce eddy current losses, the iron core is laminated. This means that the iron core is assembled of a very large number of very thin steel laminations which are electrically isolated from one another by very thin oxidation or varnish layers. The steel used for transformer laminations is customarily called transformer steel. It is usually deliberately doped with silicon to reduce its intrinsic conductivity without appreciably affecting its high magnetic permeability. This reduction in steel conductivity further diminishes eddy current losses. Transformer steel is a soft magnetic material with a narrow hysteresis loop, and this leads to the reduction of hysteresis losses. Eddy current and hysteresis losses are called core losses. There are also winding (joule) losses due to winding resistances as well as stray losses due to eddy currents induced by stray (leakage) magnetic fields in conductive materials of the transformer's support structures. High-power and high-voltage transformers have elaborate cooling and insulating systems. For instance, such transformers can be placed in metallic tanks filled with transformer (highly refined mineral) oil that both cools and insulates the windings.

The basic design of small high-frequency transformers used in power electronics and communication networks is quite different from the design of power transformers. To illustrate this point, consider briefly the design of Ethernet transformers widely used for interfacing computers with communication networks. The main function of these transformers is not to step up or step down ac voltages but rather to suppress common-mode (noise) signals and transmit with minimal distortions the differential-mode (information carrier) signals in the wide frequency range of 0.1 MHz-100 MHz. In these wideband Ethernet transformers, toroidal ferrite cores are used, and their primary and secondary windings usually have the same number of turns and they are wound together in bifilar manner. The midpoints of primary and secondary windings are grounded, and this midpoint grounding results in low impedances of primary and secondary windings to common-mode signals and their effective filtering out.

To stress better the main principle of operation of transformers, we shall first consider a two-winding ideal transformer whose schematic depiction is shown in Figure 3.1. In the case of the ideal transformer, we neglect the small resistances R_1 and R_2 of the primary and secondary windings, respectively. We also neglect leakage flux linkages ψ_ℓ, which are due to the small number of magnetic field lines that partially (or completely) go through air and link only one of the two coils. In other words, it is assumed that all magnetic field lines are entirely confined to the ferromagnetic core and, consequently, all these magnetic field lines link both coils. We also assume that the magnetic permeability μ_c of the ferromagnetic core is infinite, while the conductivity of the same core σ_c is equal to zero. All the mentioned

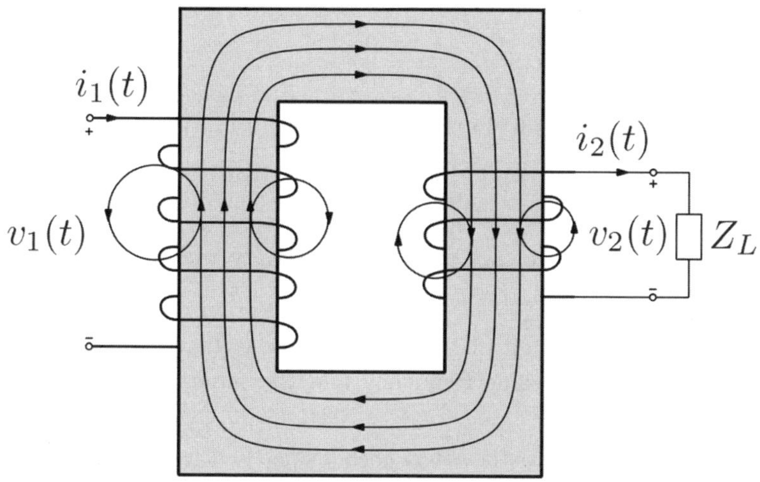

Fig. 3.1

assumptions are summarized below:

$$R_1 = R_2 = 0, \quad \psi_\ell = 0, \quad \mu_c = \infty, \quad \sigma_c = 0. \tag{3.1}$$

As mentioned above, by neglecting the leakage flux, we assume that all turns of the primary and secondary windings are linked with the same flux $\Phi(t)$ which is formed by the magnetic field lines entirely confined to the core. This means that the flux linkages $\psi_1(t)$ and $\psi_2(t)$ of the primary and secondary windings are given by the following formulas, respectively:

$$\psi_1(t) = N_1\Phi(t), \tag{3.2}$$

$$\psi_2(t) = N_2\Phi(t), \tag{3.3}$$

where N_1 and N_2 are the numbers of turns of the primary and secondary windings. Since we neglect the resistances of the primary and secondary windings, the primary and secondary voltages are equal to the voltages induced due to the time variations of $\psi_1(t)$ and $\psi_2(t)$:

$$v_1(t) = \frac{d\psi_1(t)}{dt} = N_1\frac{d\Phi(t)}{dt}, \tag{3.4}$$

$$v_2(t) = \frac{d\psi_2(t)}{dt} = N_2\frac{d\Phi(t)}{dt}. \tag{3.5}$$

From the last two equations we find

$$\frac{v_1(t)}{v_2(t)} = \frac{N_1}{N_2}. \tag{3.6}$$

It is customary to introduce the turns ratio

$$\boxed{a = \frac{N_1}{N_2}} \tag{3.7}$$

and to write equation (3.6) in the form

$$\boxed{\frac{v_1(t)}{v_2(t)} = a.} \tag{3.8}$$

If the applied (source) voltage $v_1(t)$ of the primary winding is sinusoidal,

$$v_1(t) = V_{m1}\cos(\omega t + \varphi_V), \tag{3.9}$$

then, according to formula (3.8), the secondary voltage $v_2(t)$ is sinusoidal as well and it has the same frequency and the same initial phase. This means that

$$v_2(t) = V_{m2}\cos(\omega t + \varphi_V). \tag{3.10}$$

It is also clear from formula (3.8) that

$$\boxed{\frac{V_{m1}}{V_{m2}} = a.} \tag{3.11}$$

If we introduce the phasors $\hat{V}_1$ and $\hat{V}_2$ of the primary and secondary voltages, then from formulas (3.9), (3.10) and (3.11) we find

$$\boxed{\frac{\hat{V}_1}{\hat{V}_2} = a.} \tag{3.12}$$

Expressions (3.8) and (3.11) clearly reveal the principle of operation of the power transformer. They suggest that by manipulating the turns ratio a, the desired peak value of the secondary voltage can be achieved.

In the case of Ethernet (signal) transformers when $a = 1$, formula (3.8) suggests that the secondary voltage $v_2(t)$ replicates the primary voltage $v_1(t)$ without any distortion. This implies that it is very desirable that the performance of signal transformers closely imitates the performance of an ideal transformer.

Next, we shall derive the expression for the ratio of primary and secondary currents. To this end, we shall use the assumptions that $R_1 = R_2 = 0$ and $\sigma_c = 0$. These assumptions imply that there are no power losses in the ideal transformer. Consequently, the instantaneous primary power $p_1(t)$ delivered to the terminals of the primary winding must be equal to the instantaneous secondary power $p_2(t)$ delivered to the load,

$$p_1(t) = p_2(t). \tag{3.13}$$

By taking into account that

$$p_1(t) = v_1(t)i_1(t) \tag{3.14}$$

and

$$p_2(t) = v_2(t)i_2(t), \tag{3.15}$$

formula (3.13) can be written as

$$v_1(t)i_1(t) = v_2(t)i_2(t). \tag{3.16}$$

The last equation implies that

$$\frac{i_1(t)}{i_2(t)} = \frac{v_2(t)}{v_1(t)}. \tag{3.17}$$

Now, by recalling formula (3.8), we find

$$\boxed{\frac{i_1(t)}{i_2(t)} = \frac{1}{a}.} \tag{3.18}$$

In the case when the primary current is sinusoidal,

$$i_1(t) = I_{m1}\cos(\omega t + \varphi_I), \tag{3.19}$$

the formula (3.18) implies that the secondary current is sinusoidal as well and it has the same frequency and the same initial phase:

$$i_2(t) = I_{m2}\cos(\omega t + \varphi_I). \tag{3.20}$$

Moreover, it is also clear from formula (3.18) that

$$\frac{I_{m1}}{I_{m2}} = \frac{1}{a}. \tag{3.21}$$

If we introduce the phasors $\hat{I}_1$ and $\hat{I}_2$ of the primary and secondary currents, then from formulas (3.19), (3.20) and (3.21) we find

$$\boxed{\frac{\hat{I}_1}{\hat{I}_2} = \frac{1}{a}.} \tag{3.22}$$

It is convenient to write equations (3.12) and (3.22) in the form

$$\hat{V}_1 = a\hat{V}_2, \tag{3.23}$$

$$\hat{I}_1 = \frac{1}{a}\hat{I}_2, \tag{3.24}$$

which provides complete terminal characterization of the ideal transformer. These terminal relations should be combined with KVL and KCL equations

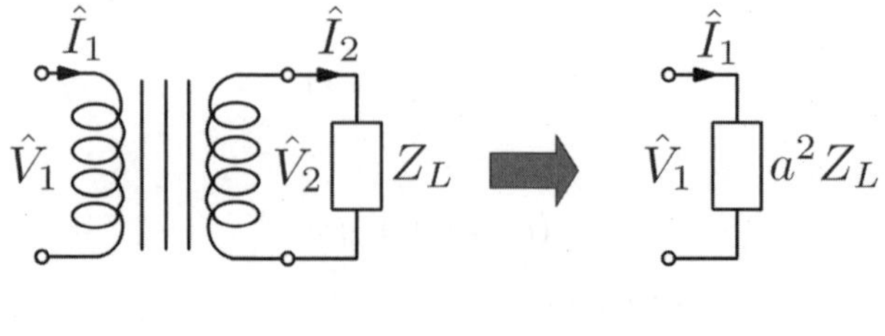

Fig. 3.2

for electric circuits connected to transformer terminals to completely analyze electric circuits with ideal transformers.

Now suppose that some load impedance Z_L is connected to the terminals of the secondary winding as shown in Figure 3.2a. This figure also presents the concise circuit notation for a two-winding transformer. This notation is used in drawings of electric circuits with transformers instead of schematics shown in Figure 3.1. We want to find the input impedance of a loaded ideal transformer. This input impedance is defined as

$$Z_{in} = \frac{\hat{V}_1}{\hat{I}_1}. \tag{3.25}$$

By substituting terminal relations (3.23) and (3.24) in the last formula, we find

$$Z_{in} = a^2 \frac{\hat{V}_2}{\hat{I}_2}. \tag{3.26}$$

However, the ratio of $\hat{V}_2$ to $\hat{I}_2$ is the load impedance

$$Z_L = \frac{\hat{V}_2}{\hat{I}_2}. \tag{3.27}$$

By combining the last two formulas, we derive

$$\boxed{Z_{in} = a^2 Z_L.} \tag{3.28}$$

It is clear from the last expression that, with respect to the primary terminals, the ideal transformer can be represented by the equivalent circuit shown in Figure 3.2b. The equivalence here is understood in the sense that, as far as the relationship between the primary voltage $\hat{V}_1$ and primary current $\hat{I}_1$ is concerned, the ideal transformer and the circuit shown in Figure 3.2b are indistinguishable. It is a very powerful idea to replace an actual complicated device by a simple equivalent electric circuit which replicates

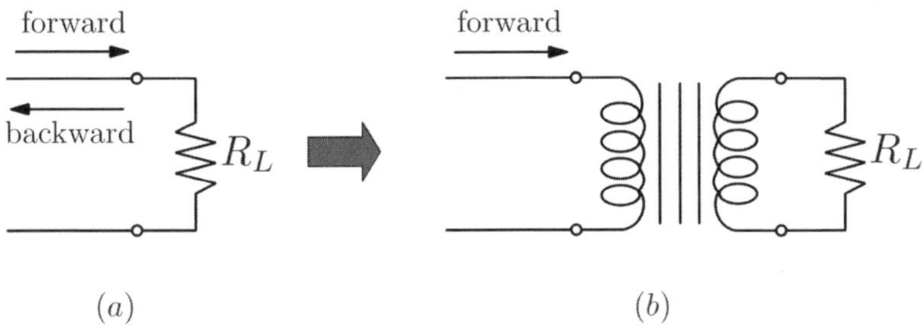

Fig. 3.3

the terminal relations between the voltages and currents of the actual device. This idea permeates many different areas of electrical engineering.

Formula (3.28) implies that the load impedance Z_L viewed from the primary terminals of the ideal transformer is equal to $a^2 Z_L$. This fact suggests that ideal (and real) transformers can be used for impedance matching purposes. We shall illustrate this point by the following example. Consider a lossless transmission line with some load resistance R_L (see Figure 3.3a). If this resistance is not equal to the characteristic impedance Z_0,

$$R_L \neq Z_0 = \sqrt{\frac{L}{C}}, \tag{3.29}$$

then there will be forward and backward propagating waves. If the load resistance R_L is connected to the transmission line through an ideal transformer with turns ratio a such that

$$a^2 R_L = Z_0, \tag{3.30}$$

then there will be no reflection at the end of the transmission line. This is because load resistance R_L acts as impedance Z_0 with respect to the terminals of this line.

3.2 Coupled Circuit Equations and Equivalent Circuit for the Transformer

In this section, we proceed to the discussion of the transformer theory by removing the first three assumptions in (3.1). The only assumption that still will be in place is that σ_c is equal to zero, which is tantamount to neglecting eddy current losses. These losses will be taken into account at the very end of our discussion, albeit in a somewhat *ad hoc* manner.

Our discussion will be based on the coupled circuit equations for the primary and secondary windings. These equations can be written as follows:

$$\begin{cases} v_1(t) = R_1 i_1(t) + L_1 \dfrac{di_1(t)}{dt} - M\dfrac{di_2(t)}{dt}, & (3.31) \\ v_2(t) = R_2 i_2(t) + L_2 \dfrac{di_2(t)}{dt} - M\dfrac{di_1(t)}{dt}. & (3.32) \end{cases}$$

Each right-hand side of the last two equations contains three terms which have distinct physical meanings. The first terms represent drops of voltages due to the finite resistances of the windings, the second terms represent voltages induced due to the time variations of self-flux linkages of the windings, while the third terms represent voltages induced as a result of time variations of mutual flux linkages of the windings. We remark that, in general, there may be some ambiguity concerning the signs of the third terms in the right-hand sides of the last two equations. This ambiguity may exist because mutual flux linkages may add to or subtract from self-flux linkages. Which of these two cases occurs depends on the relative winding directions of the two windings as well as the relative reference directions of their currents. This ambiguity can be removed by introducing the dot convention. However, in the case of transformers, the form of coupled circuit equations with negative signs is somewhat preferable from the physical point of view. This is the case because the second winding does not have an independent source of excitation and it is excited due to the electromagnetic coupling with the first winding. In accordance with Lenz's law, the current $i_2(t)$ in the second winding is always induced in such a way as to counteract the cause of induction. The cause of induction is the primary voltage $v_1(t)$ and the minus sign in equation (3.31) is the reflection of counteraction.

In the case of ac steady state, the coupled circuit equations can be written in the phasor form as follows:

$$\begin{cases} \hat{V}_1 = R_1\hat{I}_1 + jX_{11}\hat{I}_1 - jX_{12}\hat{I}_2, & (3.33) \\ \hat{V}_2 = R_2\hat{I}_2 + jX_{22}\hat{I}_2 - jX_{12}\hat{I}_1, & (3.34) \end{cases}$$

where X_{11} and X_{22} are self-reactances of the primary and secondary windings, respectively, while X_{12} is the mutual reactance. These reactances are given by the formulas

$$X_{11} = \omega L_1, \qquad X_{22} = \omega L_2, \tag{3.35}$$

$$X_{12} = \omega M. \tag{3.36}$$

Next, we shall use the coupled circuit equations to derive the expression for the secondary (load) voltage in terms of the load current $\hat{I}_2$. To this end, we find from the first equation (3.33) that

$$\hat{I}_1 = \frac{\hat{V}_1}{R_1 + jX_{11}} + \frac{jX_{12}}{R_1 + jX_{11}}\hat{I}_2. \tag{3.37}$$

By substituting the last formula into the second equation (3.34), we obtain

$$\hat{V}_2 = -\frac{jX_{12}}{R_1 + jX_{11}}\hat{V}_1 + (R_2 + jX_{22})\left[1 - \frac{(jX_{12})^2}{(R_1 + jX_{11})(R_2 + jX_{22})}\right]\hat{I}_2. \tag{3.38}$$

It is apparent that the expression for $\hat{V}_2$ has two distinct terms. The first term

$$\hat{V}_2^{ind} = -\frac{jX_{12}}{R_1 + jX_{11}}\hat{V}_1 \tag{3.39}$$

has the meaning of the secondary voltage in the case when $\hat{I}_2 = 0$. This means that this voltage can be physically interpreted as the voltage induced in the secondary winding by the magnetic flux created by the current in the primary winding. This explains the use of the superscript "*ind*" for $\hat{V}_2$ in the last formula.

Now, we shall discuss the second term in the right-hand side of equation (3.38):

$$\hat{V}_2^{drop} = (R_2 + jX_{22})\left[1 - \frac{(jX_{12})^2}{(R_1 + jX_{11})(R_2 + jX_{22})}\right]\hat{I}_2. \tag{3.40}$$

It is clear that this term has the physical meaning of the voltage drop due to the load current $\hat{I}_2$. It is very desirable to have this term as small as possible in order to maintain the secondary voltage (voltage across the load terminals) as constant as possible in the face of continuously changing load and load current $\hat{I}_2$. It is understandable that the smaller $\hat{V}_2^{drop}$, the better the quality of the transformer.

Next, we consider another important quantity, namely, secondary short-circuit current $\hat{I}_2^{sc}$. This current occurs when the secondary winding is accidentally short-circuited, that is, when

$$\hat{V}_2 = 0. \tag{3.41}$$

From formulas (3.38) and (3.41) we find

$$\hat{I}_2^{sc} = \frac{jX_{12}\hat{V}_1}{(R_1 + jX_{11})(R_2 + jX_{22})\left[1 - \frac{(jX_{12})^2}{(R_1 + jX_{11})(R_2 + jX_{22})}\right]}. \tag{3.42}$$

It is apparent that the same quantity

$$D = 1 - \frac{(jX_{12})^2}{(R_1 + jX_{11})(R_2 + jX_{22})} \tag{3.43}$$

appears in formulas (3.40) and (3.42) for $\hat{V}_2^{drop}$ and $\hat{I}_2^{sc}$, respectively. The smaller D, the better the quality of the transformer with respect to its ability to maintain more or less constant voltage across the load terminals in the face of changing load current $\hat{I}_2$. On the other hand, small D may result in large short-circuit current $\hat{I}_2^{sc}$, which is not desirable. This implies that transformers with small D are more vulnerable to fault occurrence and must be properly protected against such faults.

The previous discussion reveals the importance of D. This suggests the careful analysis of this quantity which is presented below. We start with the remark that usually the resistances are much smaller than the reactances,

$$R_1 \ll X_{11}, \qquad R_2 \ll X_{22}. \tag{3.44}$$

By using this fact, formula (3.43) for D can be simplified as follows:

$$D \approx 1 - \frac{X_{12}^2}{X_{11}X_{22}}. \tag{3.45}$$

By using equations (3.35) and (3.36) in (3.45), we find

$$D \approx 1 - \frac{M^2}{L_1 L_2}. \tag{3.46}$$

As discussed in section 3.2 of Part I, inductance L_1 can be split into two distinct components,

$$L_1 = L_1^m + L_1^\ell, \tag{3.47}$$

where L_1^m is the main inductance which is due to the flux formed by the magnetic field lines that are entirely confined to the ferromagnetic core and link all turns of the first winding, while L_1^ℓ is the leakage inductance which is due to the magnetic field lines that "leak" out of the ferromagnetic core and may not link all the turns of the winding.

Next, we shall establish the connection between mutual inductance M and the main inductance L_1^m. First, we recall that

$$M = \frac{\psi_{21}}{i_1}. \tag{3.48}$$

It is also clear that

$$\psi_{21} = N_2 \Phi_c^{(1)}, \tag{3.49}$$

where, as before, $\Phi_c^{(1)}$ is the magnetic flux through the core created by i_1. This flux links all N_2 turns of the secondary winding. By substituting formula (3.49) into equation (3.48), we find

$$M = \frac{N_2 \Phi_c^{(1)}}{i_1} = \frac{N_2}{N_1} \frac{N_1 \Phi_c^{(1)}}{i_1}. \tag{3.50}$$

If the two windings are placed around the same leg of the core (which is usually the case in order to achieve strong electromagnetic coupling), then

$$L_1^m = \frac{N_1 \Phi_c^{(1)}}{i_1}, \tag{3.51}$$

and, according to (3.50), we have

$$M = \frac{N_2}{N_1} L_1^m. \tag{3.52}$$

By literally repeating the same line of reasoning as has been used in the derivation of formula (3.52), we find

$$L_2 = L_2^m + L_2^\ell, \tag{3.53}$$

$$M = \frac{N_1}{N_2} L_2^m. \tag{3.54}$$

From equations (3.47) and (3.52) as well as equations (3.53) and (3.54), we obtain

$$\frac{N_1}{N_2} M = L_1 - L_1^\ell, \tag{3.55}$$

$$\frac{N_2}{N_1} M = L_2 - L_2^\ell. \tag{3.56}$$

By multiplying the last two equations, we arrive at

$$M^2 = L_1 L_2 - L_1 L_2^\ell - L_2 L_1^\ell + L_1^\ell L_2^\ell, \tag{3.57}$$

which can be further transformed as follows:

$$\frac{M^2}{L_1 L_2} = 1 - \frac{L_1^\ell}{L_1} - \frac{L_2^\ell}{L_2} + \frac{L_1^\ell L_2^\ell}{L_1 L_2}. \tag{3.58}$$

It is clear that the last term in formula (3.58) is quite small in comparison with the preceding two terms. Consequently,

$$\frac{M^2}{L_1 L_2} \approx 1 - \frac{L_1^\ell}{L_1} - \frac{L_2^\ell}{L_2}. \tag{3.59}$$

By substituting the last formula into equation (3.46), we find

$$D \approx \frac{L_1^\ell}{L_1} + \frac{L_2^\ell}{L_2}. \tag{3.60}$$

The last expression clearly reveals the importance of leakage inductances. These inductances determine the value of D, which, in turn, controls the ability of the transformer to maintain more or less constant voltage across the load terminals as well as the vulnerability of the transformer to accidental shorts of the load terminals.

The importance of leakage inductances can also be elucidated from the purely mathematical point of view. Consider the coupled circuit equations (3.33)-(3.34) as a set of two linear simultaneous equations with respect to $\hat{I}_1$ and $\hat{I}_2$. It is clear that the determinant Δ of these equations is equal to

$$\Delta = (R_1 + jX_{11})(R_2 + jX_{22})D. \tag{3.61}$$

It is evident from formulas (3.60) and (3.61) that this determinant is quite small and it is equal to zero when the leakage inductances (along with small resistances R_1 and R_2) are neglected. This means that the set of coupled circuit equations (3.33)-(3.34) becomes degenerate (singular) if the leakage inductances are neglected. In mathematics, the problems that become degenerate if some small parameters are neglected are called *singularly perturbed* problems. Thus, in the case of *strong* electromagnetic coupling between the windings, the coupled circuit equations are singularly perturbed. It has been understood in mathematics that small parameters in singularly perturbed problems are very important because these small parameters make the singularly perturbed problems well defined (nonsingular). This discussion brings mathematical evidence for the importance of the leakage inductances and also suggests that these inductances are important for any strongly electromagnetically coupled systems.

The coupled circuit equations (3.33)-(3.34) do not contain the leakage inductances (and corresponding reactances) explicitly. These leakage reactances are absorbed by and hidden within the total reactances X_{11} and X_{22} and, as a result, their significance is masked and not immediately apparent. For this reason, it is highly desirable to modify the coupled circuit equations (3.33)-(3.34) in such a way that the leakage inductances will be exposed and *explicitly accounted for*. This will also lead to the equivalent electric circuit of the transformer.

The modification of coupled circuit equations and the derivation of the equivalent circuit consists of the following four steps.

Step 1. As is typical in singularly perturbed problems, the small parameters, i.e., leakage inductances, can be exposed as a result of the appropriate *scaling*. The essence of this scaling is the introduction of scaled (primed) secondary voltage and secondary current

$$\hat{V}_2' = a\hat{V}_2, \tag{3.62}$$

$$\hat{I}_2' = \frac{1}{a}\hat{I}_2, \tag{3.63}$$

where, as before, a is the turns ratio. Then, the coupled circuit equations (3.33)-(3.34) can be written in terms of $\hat{V}_2'$ and $\hat{I}_2'$ as follows:

$$\begin{cases} \hat{V}_1 = R_1\hat{I}_1 + jX_{11}\hat{I}_1 - jaX_{12}\hat{I}_2', & (3.64) \\ \hat{V}_2' = a^2R_2\hat{I}_2' + ja^2X_{22}\hat{I}_2' - jaX_{12}\hat{I}_1. & (3.65) \end{cases}$$

Now, we introduce the scaled secondary resistance and secondary reactance

$$R_2' = a^2R_2, \tag{3.66}$$

$$X_{22}' = a^2X_{22}, \tag{3.67}$$

and rewrite equations (3.64) and (3.65) in the form

$$\begin{cases} \hat{V}_1 = R_1\hat{I}_1 + jX_{11}\hat{I}_1 - jaX_{12}\hat{I}_2', & (3.68) \\ \hat{V}_2' = R_2'\hat{I}_2' + jX_{22}'\hat{I}_2' - jaX_{12}\hat{I}_1. & (3.69) \end{cases}$$

Step 2. Next, we perform the following mathematical transformation of the last two equations:

$$\begin{cases} \hat{V}_1 = R_1\hat{I}_1 + j(X_{11} - aX_{12})\hat{I}_1 + jaX_{12}\left(\hat{I}_1 - \hat{I}_2'\right), & (3.70) \\ \hat{V}_2' = R_2'\hat{I}_2' + j\left(X_{22}' - aX_{12}\right)\hat{I}_2' + jaX_{12}\left(\hat{I}_2' - \hat{I}_1\right). & (3.71) \end{cases}$$

Step 3. Now, we consider the physical meaning of the coefficients in the above coupled equations. By using formulas (3.35), (3.36) and (3.52), we derive

$$aX_{12} = \frac{N_1}{N_2}\omega M = \frac{N_1}{N_2}\omega\frac{N_2}{N_1}L_1^m = \omega L_1^m = X_{11}^m, \tag{3.72}$$

where X_{11}^m has the physical meaning of the main reactance of the primary winding.

Next, by using formulas (3.35), (3.47) and (3.72), we find

$$X_{11} - aX_{12} = X_{11} - X_{11}^m = \omega L_1 - \omega L_1^m = \omega L_1^\ell = X_1^\ell, \tag{3.73}$$

where X_1^ℓ stands for the leakage reactance of the primary winding.

Finally, by using formula (3.67), we find

$$X_{22}' - aX_{12} = a^2\left(X_{22} - \frac{1}{a}X_{12}\right), \tag{3.74}$$

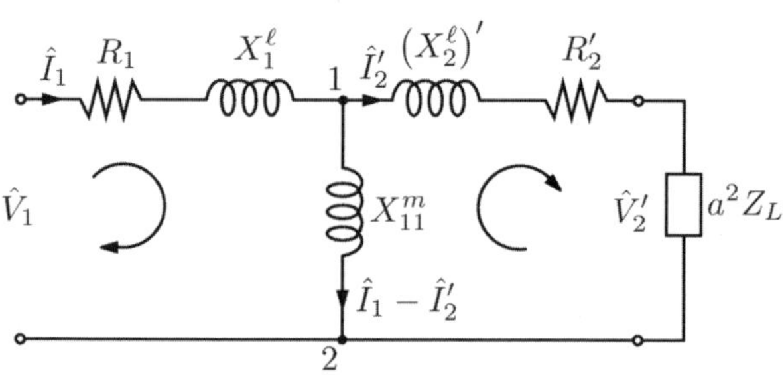

Fig. 3.4

which, according to equations (3.35), (3.36) and (3.54), can be further transformed as follows:

$$X_{22}' - aX_{12} = a^2\left(X_{22} - \frac{1}{a}X_{12}\right) = a^2\left(\omega L_2 - \frac{N_2}{N_1}\omega M\right) = a^2\omega L_2^\ell$$
$$= \left(X_2^\ell\right)', \tag{3.75}$$

where $X_2^\ell = \omega L_2^\ell$ is the leakage reactance of the secondary winding, while $\left(X_2^\ell\right)'$ is the scaled value of this reactance. By substituting formulas (3.72), (3.73) and (3.75) into coupled equations (3.70) and (3.71) we end up with

$$\begin{cases} \hat{V}_1 = R_1\hat{I}_1 + jX_1^\ell\hat{I}_1 + jX_{11}^m\left(\hat{I}_1 - \hat{I}_2'\right), & (3.76) \\ \hat{V}_2' = R_2'\hat{I}_2' + j\left(X_2^\ell\right)'\hat{I}_2' + jX_{11}^m\left(\hat{I}_2' - \hat{I}_1\right). & (3.77) \end{cases}$$

Step 4. Thus, by using equivalent mathematical transformations, we have reduced the original coupled circuit equations (3.33)-(3.34) to the coupled equations (3.76) and (3.77) in which the leakage reactances are exposed and explicitly accounted for. The useful by-product of these transformations is the fact that equations (3.76) and (3.77) coincide with KVL equations for the electric circuit shown in Figure 3.4. This circuit can be considered as an equivalent circuit for the transformer. This is because this circuit and the transformer are described by mathematically identical sets of equations. For this reason, the transformer and the circuit shown in Figure 3.4 are indistinguishable as far as the relationship between the terminal voltages and terminal currents is concerned. In other words, if the circuit in Figure 3.4 were connected to a network instead of the transformer, the currents and voltages in the network would not be changed because in both cases the network is described by identical sets of equations.

The load impedance of the transformer is modeled in the equivalent circuit by its scaled value $a^2 Z_L$ (see Figure 3.4). Indeed, from formulas

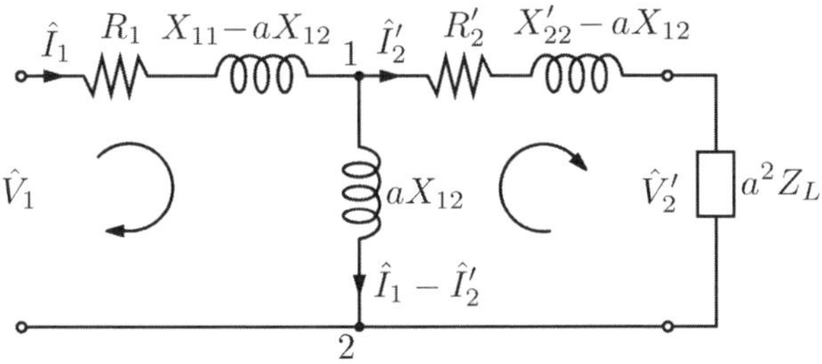

Fig. 3.5

(3.62) and (3.63) we find

$$\frac{\hat{V}'_2}{\hat{I}'_2} = \frac{a\hat{V}_2}{\frac{1}{a}\hat{I}_2} = a^2\frac{\hat{V}_2}{\hat{I}_2} = a^2 Z_L. \tag{3.78}$$

It is interesting to point out that the equivalent circuit for the transformer is not unique. Indeed, coupled equations (3.70) and (3.71) have been derived from the original coupled circuit equations (3.33) and (3.34) by using scaling (3.62) and (3.63). This derivation is valid for any value of the scaling parameter a, not only when a is the turns ratio. The coupled equations (3.70) and (3.71) coincide with KVL equations for the circuit shown in Figure 3.5. Consequently, this circuit can be considered as an equivalent circuit for the transformer as well. The choice of a as the turns ratio leads to the exposure of the leakage reactances. This choice is preferable in the case when the primary and secondary windings are strongly electromagnetically coupled because in this case the leakage parameters are very important. However, for not strongly coupled windings, another choice of a and another equivalent circuit may be preferable.

Now, we shall return to the discussion of the equivalent circuit shown in Figure 3.4. In deriving this equivalent circuit, we neglected eddy current losses in the transformer core by assuming that $\sigma_c = 0$. These losses can be accounted for in the equivalent circuit in the following *ad hoc* manner. It has been shown in section 3.5 of Part I that the eddy current losses are proportional to B_m^2, where B_m is the peak value of magnetic flux density in the core:

$$P_{ec} \sim B_m^2. \tag{3.79}$$

However, B_m is proportional to the peak value of magnetic flux Φ_{cm} through the core which, in turn, is proportional to the peak value of the voltage induced by the core flux. In the equivalent circuit shown in Figure

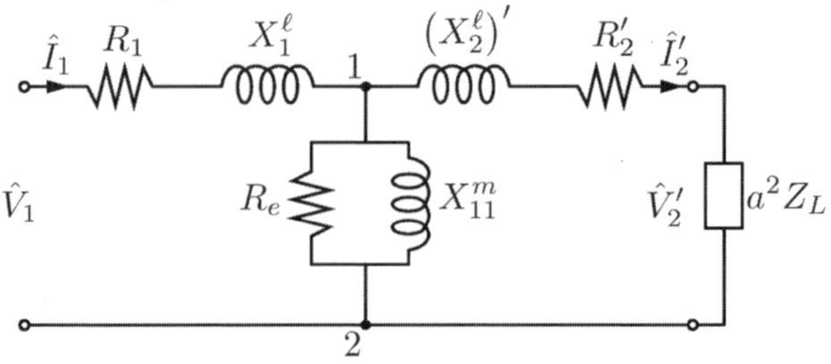

Fig. 3.6

3.4, this voltage can be identified as voltage V_{m12} across the terminals 1 and 2. Thus, we conclude that

$$P_{ec} \sim V_{m12}^2. \tag{3.80}$$

The last formula suggests the idea of modeling the eddy current losses in the transformer iron core by the ohmic losses P_{R_e} in an equivalent resistor connected across the terminals 1 and 2, i.e., in parallel with X_{11}^m (see Figure 3.6). The rationale behind this idea is the fact that the ohmic losses in R_e are also proportional to V_{m12}^2:

$$P_{R_e} = \frac{V_{m12}^2}{2R_e}. \tag{3.81}$$

The resistor R_e can be chosen from the condition

$$P_{ec} = P_{R_e}, \tag{3.82}$$

which leads to

$$R_e = \frac{V_{m12}^2}{2P_{ec}}. \tag{3.83}$$

The electric circuit shown in Figure 3.6 is the complete equivalent circuit for a power transformer. It is worthwhile to stress again that the important feature of this equivalent circuit is that leakage reactances are explicitly accounted for. In power system applications, this equivalent circuit is often simplified by neglecting small (in comparison with X_1^ℓ and $\left(X_2^\ell\right)'$) resistances R_1 and R_2' and by assuming that R_e and X_{11}^m are very large so that the 1-2 branch can be regarded as open. This leads to the equivalent circuit shown in Figure 3.7, which clearly reveals the unique importance of leakage reactances in the performance of power transformers. It is also clear that the last equivalent circuit is reduced to the equivalent circuit of the ideal transformer (see Figure 3.2b) if the leakage reactances are neglected.

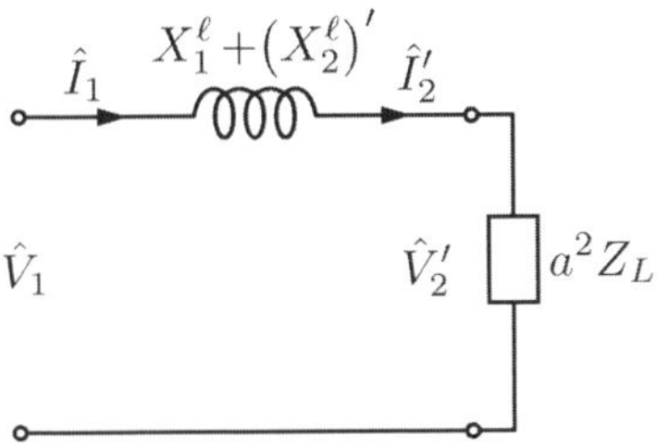

Fig. 3.7

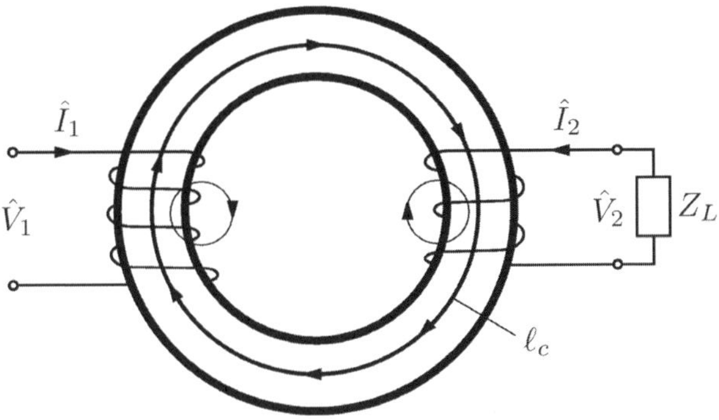

Fig. 3.8

In our derivation of the equivalent circuit shown in Figure 3.6, resistance R_e takes into account the core losses which are due to the macroscopic eddy currents induced in laminated ferromagnetic cores. In power electronics (as well as in other areas of electrical engineering) toroidal transformers (see Figure 3.8) with ferrite cores are used for high-frequency applications. For ferrite cores, losses can be modeled by using complex magnetic permeability

$$\mu = \mu' - j\mu''. \tag{3.84}$$

The imaginary part μ'' of such permeability accounts for losses, and μ'' (as well as μ') is frequency dependent. The equivalent circuit for such transformers can be derived by using the somewhat different reasoning presented below.

As before, we shall make the distinction between the magnetic field lines that are entirely confined to the ferrite core and field lines that partially leak out (see Figure 3.8). This means that the coupled circuit equations can be written in the form

$$\begin{cases} \hat{V}_1 = R_1\hat{I}_1 + jX_1^\ell \hat{I}_1 + \hat{V}_{1c}, & (3.85) \\ \hat{V}_2 = R_2\hat{I}_2 + jX_2^\ell \hat{I}_2 + \hat{V}_{2c}, & (3.86) \end{cases}$$

where $\hat{V}_{1c}$ and $\hat{V}_{2c}$ are the voltages induced in the primary and secondary windings, respectively, by the core magnetic flux formed by the field lines confined to the core, while leakage reactances X_1^ℓ and X_2^ℓ account for voltages induced by leakage fluxes.

To compute voltages $\hat{V}_{1c}$ and $\hat{V}_{2c}$, consider a magnetic field line ℓ_c. By applying Ampere's Law, we find

$$\oint_{\ell_c} \hat{\mathbf{H}}_c \cdot d\boldsymbol{\ell} = N_1\hat{I}_1 + N_2\hat{I}_2. \tag{3.87}$$

The right-hand side in the last formula can be modified as follows:

$$\oint_{\ell_c} \hat{\mathbf{H}}_c \cdot d\boldsymbol{\ell} = N_1\left(\hat{I}_1 + \hat{I}_2'\right), \tag{3.88}$$

where $\hat{I}_2' = \frac{N_2}{N_1}\hat{I}_2 = \frac{1}{a}\hat{I}_2$.

By assuming that the magnetic field in the ferrite core is uniform, from equation (3.88) we derive

$$\hat{H}_c = \frac{N_1\left(\hat{I}_1 + \hat{I}_2'\right)}{\ell_c}, \tag{3.89}$$

where ℓ_c in (3.89) can be construed as some average length of the toroidal core. From the last formula we find

$$\hat{B}_c = \mu\hat{H}_c = \frac{\mu N_1\left(\hat{I}_1 + \hat{I}_2'\right)}{\ell_c}, \tag{3.90}$$

and

$$\hat{\psi}_c^{(1)} = N_1 A_c \hat{B}_c = \frac{\mu A_c N_1^2}{\ell_c}\left(\hat{I}_1 + \hat{I}_2'\right), \tag{3.91}$$

where $\hat{\psi}_c^{(1)}$ is the phasor of the flux linkages of the primary winding formed by magnetic field lines confined to the core, while A_c is the cross-sectional area of the core.

Now, $\hat{V}_{1c}$ can be computed as

$$\hat{V}_{1c} = j\omega\psi_c^{(1)} = Z_c\left(\hat{I}_1 + \hat{I}_2'\right), \tag{3.92}$$

where

$$Z_c = \frac{j\omega\mu A_c N_1^2}{\ell_c} = \frac{\omega\left(\mu'' + j\mu'\right)A_c N_1^2}{\ell_c}. \tag{3.93}$$

It is clear that Z_c can be presented as

$$Z_c = R_c + jX_c, \tag{3.94}$$

where

$$\boxed{R_c = \frac{\omega\mu'' A_c N_1^2}{\ell_c}} \tag{3.95}$$

and

$$\boxed{X_c = \frac{\omega\mu' A_c N_1^2}{\ell_c}.} \tag{3.96}$$

Similarly,

$$\hat{\psi}_c^{(2)} = N_2 A_c \hat{B}_c = \frac{N_2}{N_1} N_1 A_c \hat{B}_c = \frac{1}{a}\hat{\psi}_c^{(1)}, \tag{3.97}$$

where $\hat{\psi}_c^{(2)}$ is the phasor of the flux linkages of the secondary winding due to the magnetic field lines confined to the toroidal core.

It is clear from the last formula that

$$\hat{V}_{2c} = j\omega\psi_c^{(2)} = \frac{1}{a}\hat{V}_{1c} = \frac{1}{a}Z_c\left(\hat{I}_1 + \hat{I}_2'\right). \tag{3.98}$$

Now, by substituting formulas (3.92) and (3.98) into equations (3.85) and (3.86), respectively, and then multiplying both sides of equation (3.86) by a and taking into account that $\hat{I}_2 = a\hat{I}_2'$ and $\hat{V}_2' = a\hat{V}_2$, we derive

$$\begin{cases} \hat{V}_1 = R_1\hat{I}_1 + jX_1^\ell \hat{I}_1 + Z_c\left(\hat{I}_1 + \hat{I}_2'\right), & (3.99) \\ \hat{V}_2' = R_2'\hat{I}_2' + j\left(X_2^\ell\right)'\hat{I}_2' + Z_c\left(\hat{I}_1 + \hat{I}_2'\right), & (3.100) \end{cases}$$

where, as before, $R_2' = a^2 R_2$ and $\left(X_2^\ell\right)' = a^2 X_2^\ell$.

It is clear that equations (3.99) and (3.100) coincide with KVL equations for the electric circuit shown in Figure 3.9. In this sense, this circuit can be construed as the equivalent circuit for the toroidal transformer with ferrite core. It is worthwhile to stress that R_c and X_c in this equivalent circuit are explicitly expressed by formulas (3.95) and (3.96) in terms of geometry of the toroidal core and the real and imaginary parts of its magnetic permeability.

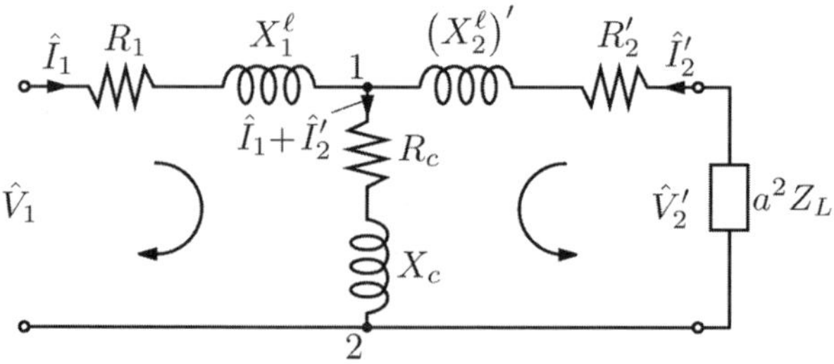

Fig. 3.9

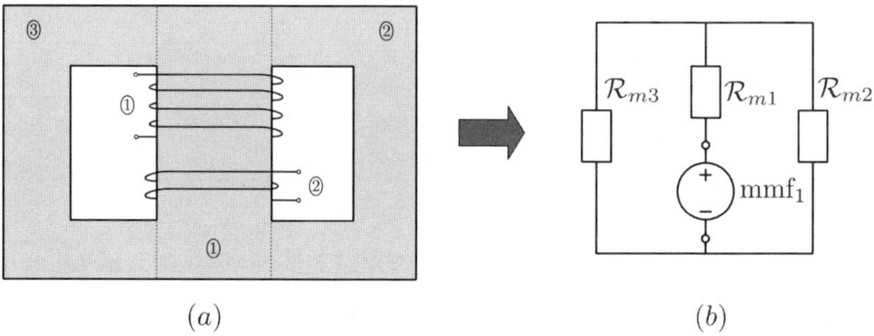

Fig. 3.10

3.3 Determination of Parameters of Equivalent Circuits; Three-Phase Transformers

We start this section with the discussion of how some parameters of the transformer equivalent circuit shown in Figure 3.6 can be computed. Such computations are possible for X_{11}^m and R_e by using the magnetic circuit theory presented in Chapter 3 of Part I. We shall illustrate these computations for a shell-type transformer shown in Figure 3.10a. According to formula (3.72), we have

$$X_{11}^m = \omega L_1^m. \tag{3.101}$$

On the other hand, according to formula (3.83) from Part I, we find

$$L_1^m = \frac{N_1^2}{\mathcal{R}_{me}}, \tag{3.102}$$

where $\mathcal{R}_{me}$ is the equivalent magnetic reluctance of the magnetic circuit shown in Figure 3.10b with respect to the terminals of mmf_1 representing the primary winding.

It is clear that

$$\mathcal{R}_{me} = \mathcal{R}_{m1} + \frac{\mathcal{R}_{m2}\mathcal{R}_{m3}}{\mathcal{R}_{m2} + \mathcal{R}_{m3}}, \tag{3.103}$$

where

$$\mathcal{R}_{mk} = \frac{\ell_k}{\mu_c A_k}, \quad (k = 1, 2, 3), \tag{3.104}$$

and the meaning of the notations is the same as in Chapter 3 of Part I. Typically, legs 2 and 3 have the same geometry and permeability. Consequently,

$$\mathcal{R}_{m2} = \mathcal{R}_{m3} \tag{3.105}$$

and formula (3.103) can be written as

$$\mathcal{R}_{me} = \mathcal{R}_{m1} + \frac{\mathcal{R}_{m2}}{2}. \tag{3.106}$$

Furthermore, shell cores are often designed in such a way that

$$A_2 = A_3 = \frac{A_1}{2}. \tag{3.107}$$

In this case, from formulas (3.104) and (3.106), we obtain

$$\mathcal{R}_{me} = \frac{\ell_1 + \ell_2}{\mu_c A_1}. \tag{3.108}$$

Finally, by combining formulas (3.101), (3.102) and (3.108), we find

$$\boxed{X_{11}^m = \frac{\omega \mu_c A_1 N_1^2}{\ell_1 + \ell_2}.} \tag{3.109}$$

Next, we consider the computation of R_e by using formula (3.83) of Part II. It is clear that total eddy current losses P_{ec} in the ferromagnetic core are equal to the sum of losses in legs 1, 2 and 3:

$$P_{ec} = P_{ec1} + P_{ec2} + P_{ec3}. \tag{3.110}$$

According to formula (3.235) from Part I, we have

$$P_{ec1} = \frac{1}{24}\sigma_c V_1 \omega^2 B_{m1}^2 \tau^2, \tag{3.111}$$

$$P_{ec2} = \frac{1}{24}\sigma_c V_2 \omega^2 B_{m2}^2 \tau^2, \tag{3.112}$$

$$P_{ec3} = \frac{1}{24}\sigma_c V_3 \omega^2 B_{m3}^2 \tau^2, \tag{3.113}$$

where V_1, V_2 and V_3 are the volumes of legs 1, 2 and 3, respectively, B_{m1}, B_{m2} and B_{m3} are peak values of magnetic flux density in those legs, and $\tau = \frac{\Delta}{n}$ is the thickness of core laminations.

It is apparent that

$$B_{m1} = \frac{\Phi_{m1}}{A_1}, \tag{3.114}$$

$$B_{m2} = \frac{\Phi_{m2}}{A_2} = \frac{1}{A_2}\frac{\mathcal{R}_{m3}}{\mathcal{R}_{m2} + \mathcal{R}_{m3}}\Phi_{m1}, \tag{3.115}$$

$$B_{m3} = \frac{\Phi_{m3}}{A_3} = \frac{1}{A_3}\frac{\mathcal{R}_{m2}}{\mathcal{R}_{m2} + \mathcal{R}_{m3}}\Phi_{m1}. \tag{3.116}$$

Furthermore,

$$V_{m12} = \omega N_1 \Phi_{m1} \tag{3.117}$$

and

$$\Phi_{m1} = \frac{V_{m12}}{\omega N_1}. \tag{3.118}$$

Substituting the last expression into formulas (3.114)-(3.116) and then by inserting these formulas into formulas (3.111)-(3.113), using the last formulas in equation (3.110), we derive

$$P_{ec} = V_{m12}^2 \frac{\sigma_c \tau^2}{24 N_1^2}\left[\frac{V_1}{A_1^2} + \frac{V_2}{A_2^2}\left(\frac{\mathcal{R}_{m3}}{\mathcal{R}_{m2} + \mathcal{R}_{m3}}\right)^2 + \frac{V_3}{A_3^2}\left(\frac{\mathcal{R}_{m2}}{\mathcal{R}_{m2} + \mathcal{R}_{m3}}\right)^2\right]. \tag{3.119}$$

By using the last formula in equation (3.83), we obtain

$$R_e = \frac{12 N_1^2}{\sigma_c \tau^2\left[\frac{V_1}{A_1^2} + \frac{V_2}{A_2^2}\left(\frac{\mathcal{R}_{m3}}{\mathcal{R}_{m2} + \mathcal{R}_{m3}}\right)^2 + \frac{V_3}{A_3^2}\left(\frac{\mathcal{R}_{m2}}{\mathcal{R}_{m2} + \mathcal{R}_{m3}}\right)^2\right]}. \tag{3.120}$$

This is a general formula for R_e which can be simplified by taking into account relations (3.105) and (3.107). This leads to

$$\boxed{R_e = \frac{12 A_1^2 N_1^2}{\sigma_c \tau^2 (V_1 + 2V_2)}.} \tag{3.121}$$

Resistances R_1 and R_2 of the primary and secondary windings in the transformer equivalent circuit can be computed by using standard formulas provided that the skin and proximity effects are negligible. The most challenging are the computations of leakage inductances L_1^ℓ and L_2^ℓ and the reactances X_1^ℓ and X_2^ℓ corresponding to them. Such computations cannot be performed by using the magnetic circuit theory because this theory neglects leakage phenomena. Such computations are performed through solving magnetic field equations by using existing numerical techniques such as finite elements or integral equations, for instance. The discussion of this matter is beyond the scope of this text.

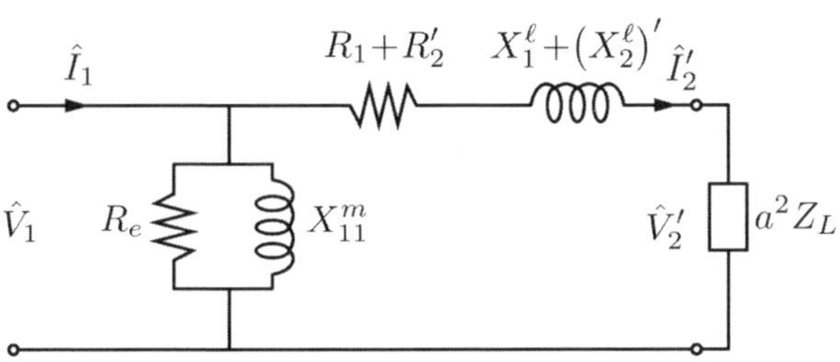

Fig. 3.11

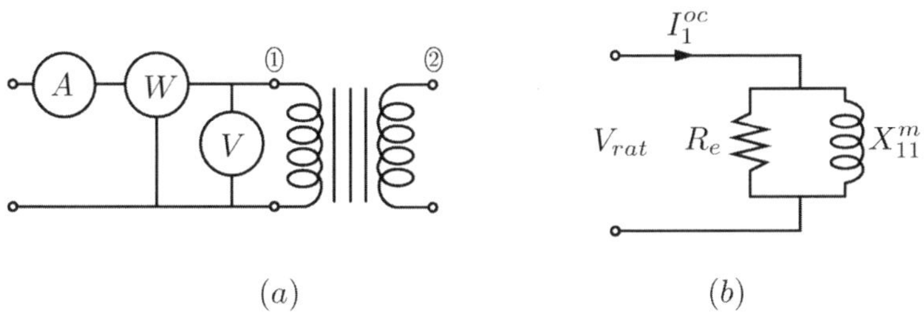

(*a*) (*b*)

Fig. 3.12

Next, we turn to the discussion of how parameters of the transformer equivalent circuit can be determined experimentally. This is usually done by performing open-circuit (OC) and short-circuit (SC) tests. These tests are actually used for the determination of parameters of the approximate equivalent circuit shown in Figure 3.11. This approximate equivalent circuit is obtained from the equivalent circuit shown in Figure 3.6 by moving the parallel R_e-X_{11}^m connection directly across the primary terminals. This transformation is usually justified on the grounds that R_e and X_{11}^m are fairly large and, consequently, current $\hat{I}_1 - \hat{I}_2'$ (see Figure 3.6) is small in magnitude. For this reason, current $\hat{I}_1$ through R_1 and X_1^ℓ is almost equal to the current $\hat{I}_2'$. Moreover, since R_1 and X_1^ℓ are small, the voltage across the terminals 1-2 is almost equal to $\hat{V}_1$. The above approximations are consistent with the equivalent circuit shown in Figure 3.11.

The open-circuit test is illustrated by Figure 3.12a. In this test, the secondary winding is open and, consequently,

$$\hat{I}_2 = a\hat{I}_2' = 0. \tag{3.122}$$

Furthermore, the rms value of the applied primary voltage V_1 in this test is usually equal to the rated voltage V_{rat} of the transformer,

$$V_1 = V_{rat}. \tag{3.123}$$

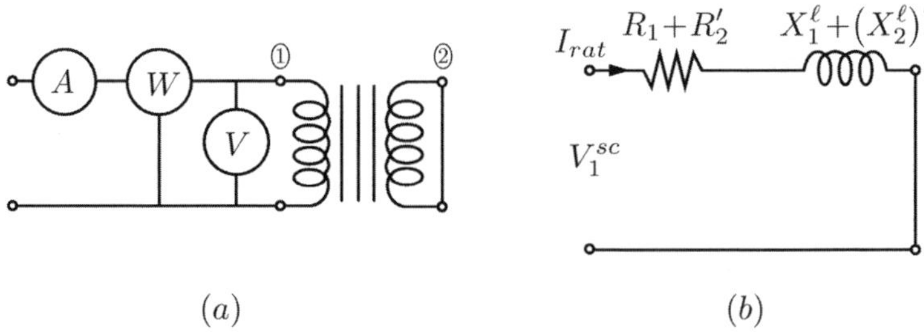

Fig. 3.13

The measurements produce

$$I_1^{oc} \quad \text{and} \quad P^{oc}, \tag{3.124}$$

which are called the primary open-circuit current and the open-circuit power, respectively.

Since $\hat{I}_2' = 0$, the approximate equivalent circuit shown in Figure 3.11 is reduced to the circuit shown in Figure 3.12b. From this circuit, we find

$$P^{oc} = \frac{V_{rat}^2}{R_e}, \tag{3.125}$$

and

$$\boxed{R_e = \frac{V_{rat}^2}{P^{oc}}.} \tag{3.126}$$

Next, from Figure 3.12b we obtain

$$|Y| = \frac{I_1^{oc}}{V_{rat}} = \sqrt{\left(\frac{1}{R_e}\right)^2 + \left(\frac{1}{X_{11}^m}\right)^2}, \tag{3.127}$$

which leads to

$$\boxed{X_{11}^m = \frac{1}{\sqrt{\left(\frac{I_1^{oc}}{V_{rat}}\right)^2 - \left(\frac{1}{R_e}\right)^2}}.} \tag{3.128}$$

Thus, by using formulas (3.126) and (3.128), parameters R_e and X_{11}^m can be identified from the measurements obtained through the open-circuit test.

The short-circuit test is illustrated by Figure 3.13a. In this test, the secondary winding is short-circuited and, consequently,

$$\hat{V}_2' = a\hat{V}_2 = 0. \tag{3.129}$$

Next, the rms value of the primary voltage is gradually increased until the rms value of the primary current reaches the rated value I_{rat},

$$V_1 = V_1^{sc} \rightarrow I_1 = I_{rat}, \tag{3.130}$$

where V_1^{sc} is called short-circuit voltage. This voltage is usually quite small; it is mostly below 6% of the rated voltage.

As soon as $V_1 = V_1^{sc}$, the short-circuit power P^{sc} is measured.

When the secondary winding is short-circuited, the approximate equivalent circuit in Figure 3.11 is reduced to the circuit shown in Figure 3.13b. This is the case because $R_1 + R_2'$ and $X_1^\ell + \left(X_2^\ell\right)'$ are much smaller than R_e and X_{11}^m. From the circuit in Figure 3.13b, we find

$$P^{sc} = I_{rat}^2 \left(R_1 + R_2'\right), \tag{3.131}$$

and

$$\boxed{R_1 + R_2' = \frac{P^{sc}}{I_{rat}^2}.} \tag{3.132}$$

Next, from Figure 3.13b we obtain

$$|Z| = \frac{V_1^{sc}}{I_{rat}} = \sqrt{\left(R_1 + R_2'\right)^2 + \left[X_1^\ell + \left(X_2^\ell\right)'\right]^2}, \tag{3.133}$$

which leads to

$$\boxed{X_1^\ell + \left(X_2^\ell\right)' = \sqrt{\left(\frac{V_1^{sc}}{I_{rat}}\right)^2 - \left(R_1 + R_2'\right)^2}.} \tag{3.134}$$

Thus, by using formulas (3.132) and (3.134), parameters $R_1 + R_2'$ and $X_1^\ell + \left(X_2^\ell\right)'$ can be identified from the measurements obtained through the short-circuit test. Now, the conclusion can be drawn that by using the open-circuit and short-circuit tests the parameters of the approximate equivalent circuit in Figure 3.11 can be completely identified. After this identification is performed, the above equivalent circuit can be used for the analysis of the transformer at any loading conditions.

Next, we turn to the discussion of three-phase transformers. These transformers have three primary windings and three secondary windings. The three-phase transformers can be designed in the way that three sets of phase windings share the same ferromagnetic core, or three single-phase transformers with separate ferromagnetic cores can be combined to form a three-phase transformer. These different core designs of three-phase transformers are illustrated in Figures 3.14a, b and c. In these figures, the terminals of primary phase windings are marked by capital letters, while

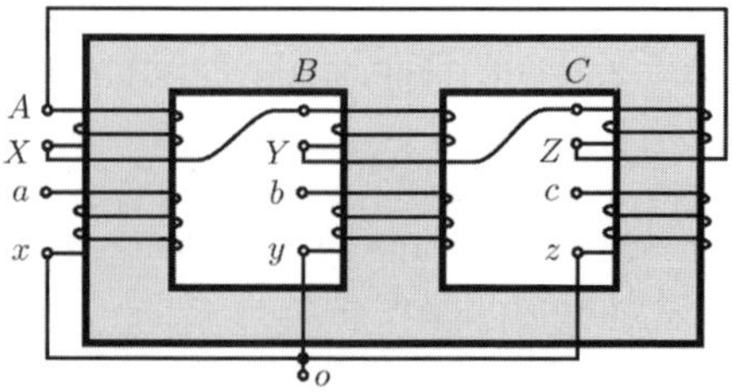

(a) core type (Δ-Y connection)

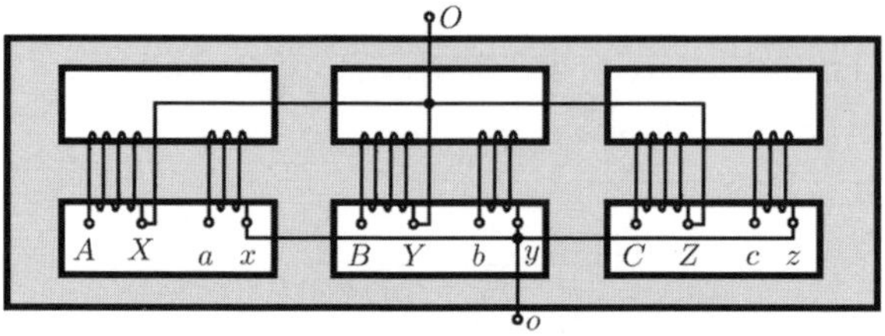

(b) shell type (Y-Y connection)

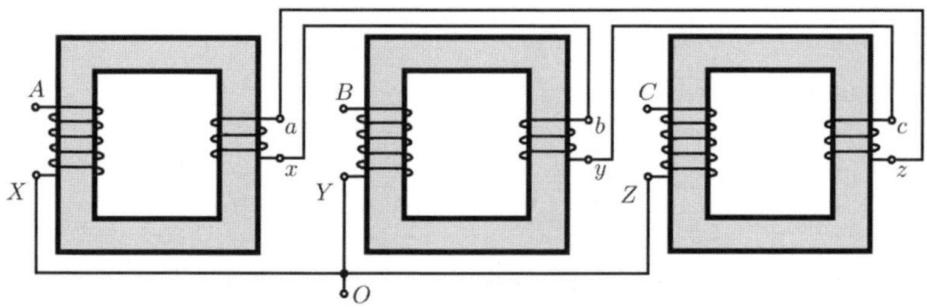

(c) three-phase connection of single-phase units (Y-Δ connection)

Fig. 3.14

the secondary terminals are marked by lowercase letters. The main advantage of the designs shown in Figures 3.14a and 3.14b is that there are some savings in core materials which make these designs cheaper and reduce overall core losses. However, a fault in any phase may damage the entire three-phase transformer, while a similar fault in the design shown in Figure 3.14c may damage only one single-phase unit. Furthermore, single-phase units can be separately shipped and installed, which may facilitate on-site construction.

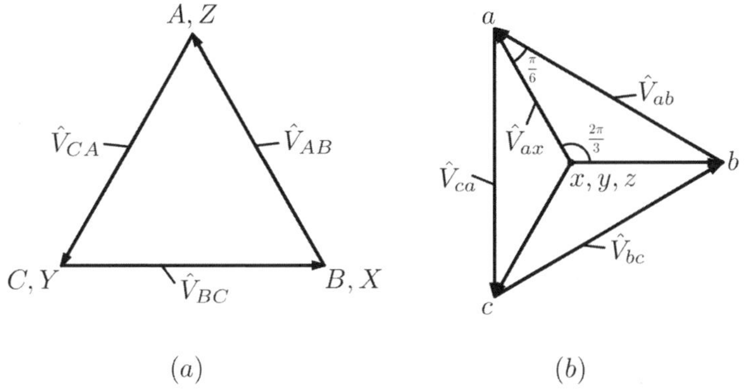

Fig. 3.15

As far as connectivity of the primary and secondary phase windings is concerned, there are four distinct possibilities listed below:

$$\text{Y-Y}, \quad \Delta\text{-}\Delta, \quad \text{Y-}\Delta \quad \text{and} \quad \Delta\text{-Y}. \tag{3.135}$$

Some of these connections are illustrated in Figures 3.14a, b and c. It is worthwhile to mention that Δ-Y connection (with Y on the high-voltage side) is often favored for the following reasons. There is line voltage gain of $\sqrt{3}$ achieved beyond the voltage gain due to the turns ratio and there is a possibility of having neutral on the high-voltage side. The Y-Δ connection is often used at the receiving ends of power transmission lines because it provides step-down line voltage ratios larger by $\sqrt{3}$. In both (Δ-Y and Y-Δ) connections there is a $\frac{\pi}{6}$ phase shift between the primary and secondary line voltages. To illustrate these facts, consider a Δ-Y connection shown in Figure 3.14a. It is clear from this figure that the primary and secondary windings corresponding to the same phase are linked by the same core flux. Consequently, in the framework of ideal transformer assumptions, the primary line voltage $\hat{V}_{AX} = \hat{V}_{AB}$ and secondary phase voltage $\hat{V}_{ax}$ have the same phase and

$$\frac{1}{a}\hat{V}_{AB} = \hat{V}_{ax}, \tag{3.136}$$

provided that the primary and secondary windings have the same winding directions (i.e., A and a are the "dotted" terminals). Formula (3.136) is illustrated by the phasor diagrams (a) and (b) shown in Figure 3.15. In these diagrams, vectors representing phasors $\hat{V}_{AB}$ and $\hat{V}_{ax}$ are parallel as

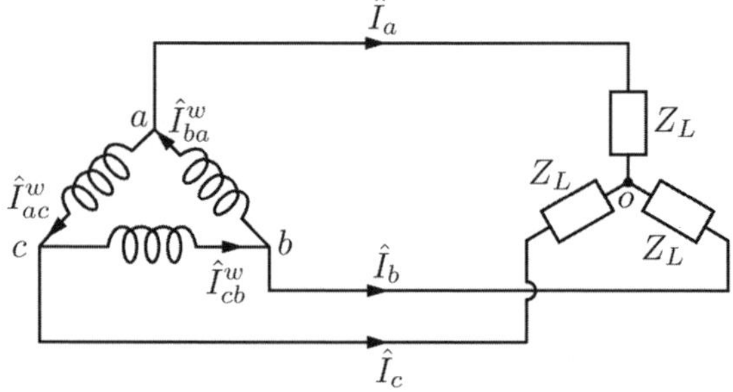

Fig. 3.16

required by formula (3.136). From Figure 3.15b, we easily conclude that

$$\hat{V}_{ab} = \sqrt{3}\hat{V}_{ax}e^{j\frac{\pi}{6}}. \tag{3.137}$$

From the last two formulas we find

$$\boxed{\hat{V}_{ab} = \frac{\sqrt{3}}{a}\hat{V}_{AB}e^{j\frac{\pi}{6}},} \tag{3.138}$$

which clearly reveals the $\frac{\pi}{6}$ phase shift between line voltages as well as $\sqrt{3}$ line voltage gain due to the Δ-Y connectivity.

The theory of the single-phase transformer discussed in the previous section can be extended to three-phase transformers. Indeed, for each pair of primary and secondary windings corresponding to the same phase we can write coupled circuit equations of the type (3.33)-(3.34) and mathematically transform them to obtain equations of the type (3.76)-(3.77). On the secondary side, these equations can be coupled to the circuit equations for the three-phase loads written in terms of primed (scaled) voltages and currents. In this way, complete sets of equations can be obtained. In the case of balanced loads, substantial simplifications can be achieved by using per-phase analysis. We shall illustrate this point by considering the case of Δ configuration of secondary windings connected to Y configuration of balanced load (see Figure 3.16). In order to construct the per-phase equivalent circuit of the three-phase transformer with the depicted secondary winding and load connectivities, we have to find the equivalent load impedance which is equal to the ratio

$$\frac{\hat{V}_2'}{\hat{I}_2'} = \frac{\left(\hat{V}^w_{ab}\right)'}{\left(\hat{I}^w_{ba}\right)'} = a^2\frac{\hat{V}^w_{ab}}{\hat{I}^w_{ba}}. \tag{3.139}$$

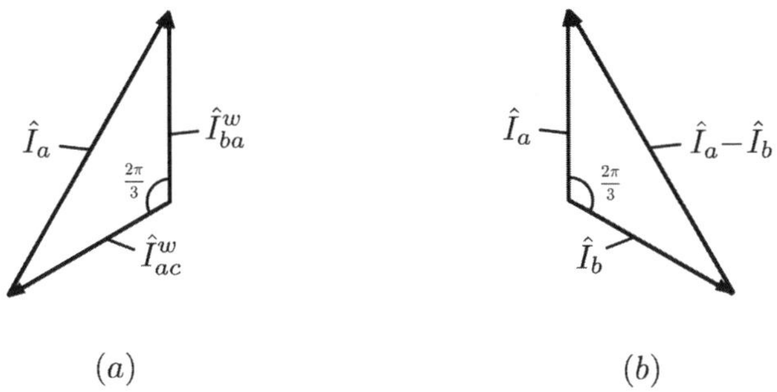

Fig. 3.17

To this end, we first remark that according to KCL

$$\hat{I}_a = \hat{I}^w_{ba} - \hat{I}^w_{ac}. \tag{3.140}$$

By using the phasor diagram shown in Figure 3.17a, we derive

$$\hat{I}_a = \sqrt{3}\hat{I}^w_{ba} e^{-j\frac{\pi}{6}}. \tag{3.141}$$

Then, from Figure 3.16, it follows that

$$\hat{V}^w_{ab} = \left(\hat{I}_a - \hat{I}_b\right) Z_L. \tag{3.142}$$

By using the phasor diagram shown in Figure 3.17b, we obtain

$$\hat{I}_a - \hat{I}_b = \sqrt{3}\hat{I}_a e^{j\frac{\pi}{6}}, \tag{3.143}$$

and, according to formula (3.142), we have

$$\hat{V}^w_{ab} = \sqrt{3}\hat{I}_a e^{j\frac{\pi}{6}} Z_L. \tag{3.144}$$

By substituting formula (3.141) into the last equation, we find

$$\hat{V}^w_{ab} = 3\hat{I}^w_{ba} Z_L. \tag{3.145}$$

This implies in accordance with formula (3.139) that the equivalent load impedance is

$$\frac{\hat{V}_2'}{\hat{I}_2'} = a^2 \frac{\hat{V}^w_{ab}}{\hat{I}^w_{ba}} = 3a^2 Z_L. \tag{3.146}$$

This leads to the per-phase equivalent circuit shown in Figure 3.18.

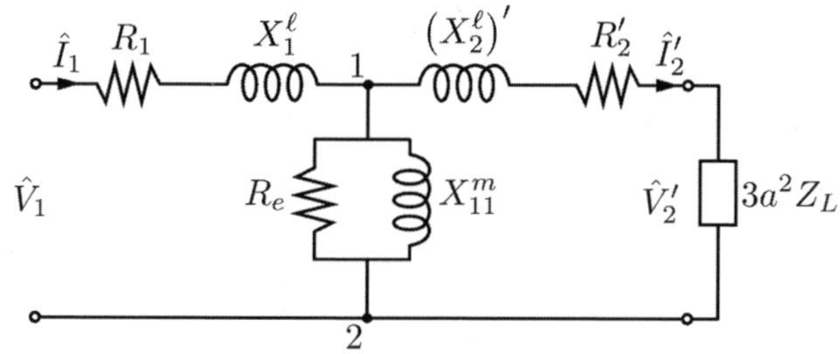

Fig. 3.18

Chapter 4

Synchronous Generators

4.1 Design and Principle of Operation of Synchronous Generators

Synchronous generators are electrical machines that convert mechanical energy of prime movers (turbines, water wheels, etc.) into electric energy supplied to power systems. This energy conversion is often called electric power generation. Most of the electric power in conventional (utility) power systems is generated by using synchronous generators. In this sense, these generators are indispensable components of power systems.

Synchronous generators have two major parts (see the schematic cross section in Figure 4.1): *stator* and *rotor* separated by an air gap. The stator is stationary as implied by its name and it is also referred to as the *armature*. The stator has a laminated structure; this means that it is usually assembled of a very large number of very thin varnished (or oxidized) silicon steel laminations. This is done to reduce eddy current losses. The stator has slots uniformly distributed over its interior surface. A three-phase winding is embedded in these slots. This is a *distributed* winding. The latter means that each phase of the stator winding consists of several coils connected in series and embedded in different (but adjacent) slots. Furthermore, these phase windings are shifted with respect to one another along the interior circumference of the stator by $120°$ in the case of two-pole machines (or by $240°/p$ in the case of p-pole machines). As discussed later in this chapter, these three-phase stationary windings create uniformly rotating magnetic fields when energized (excited) by three-phase electric currents of the same frequency and peak value but phase-shifted (in time) by $\frac{2\pi}{3}$ (or $120°$). This creation of uniformly rotating magnetic fields is the main reason for the special design of stator windings outlined

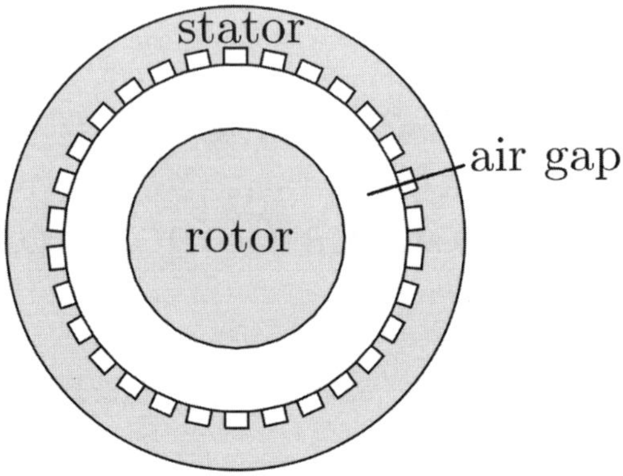

Fig. 4.1

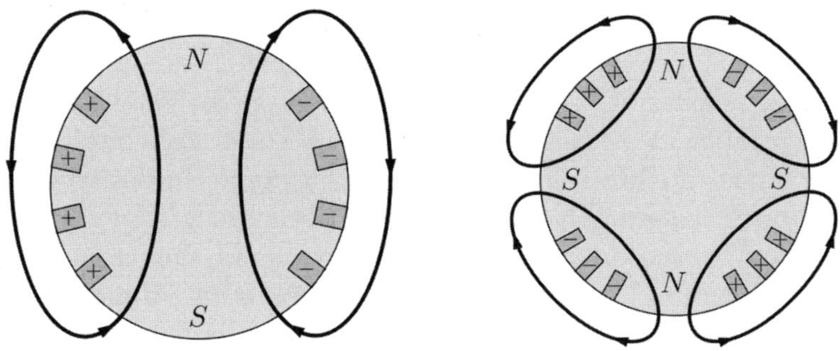

(a) (b)

Fig. 4.2

above. The magnetic fields created by the stator windings are usually called *armature reaction magnetic fields.*

Rotors are rotating parts of synchronous generators which are mechanically driven by prime movers connected to the rotor shafts. There are two distinct designs of rotors of synchronous generators: *cylindrical* rotors and *salient pole*-type rotors. The basic design of cylindrical rotors is illustrated by Figure 4.2, which schematically depicts rotor cross sections in the case of two- and four-pole machines. Two-pole machines are typical for fossil fuel power plants, while four (or six)-pole machines are typical for nuclear power plants. These rotors are usually driven by steam or gas turbines. For this reason, synchronous generators with cylindrical rotors are called *turbo-*

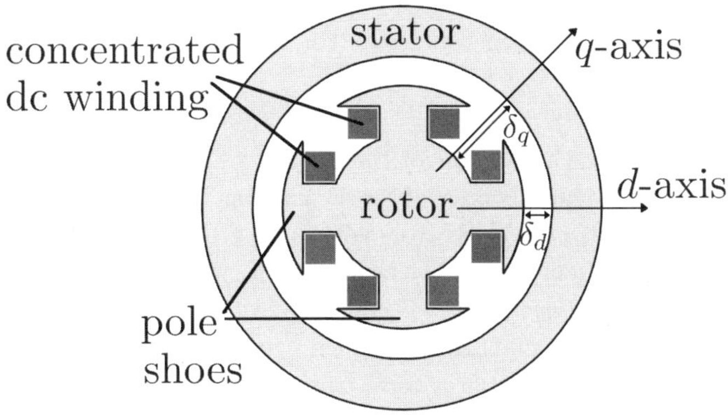

Fig. 4.3

generators. Cylindrical rotors are not laminated but forged solid pieces of conducting ferromagnetic (steel) material. Cylindrical rotors have slots and the conductors of distributed rotor windings (also called field windings) are embedded in these slots. Rotor windings are excited by dc currents which create static (with respect to the rotor) magnetic fields. The lines of such magnetic fields are illustrated in Figure 4.2. It is clear from this figure that the large (wide) teeth of cylindrical rotors serve as north (N) or south (S) poles where the magnetic field lines emanate from the rotor and enter the rotor, respectively. Recently, permanent magnet synchronous generators have been developed and used in certain applications such as, for instance, wind power generation. In such generators, the rotor field windings are replaced by permanent magnets. The main advantage of such generators is that they do not need dc power supplies for excitation of rotor windings. However, large permanent magnets are costly and the magnetic field strength of permanent magnets is limited and not controllable.

The design of *salient pole* rotors is illustrated by Figure 4.3 for the case of four-pole rotor machines. In most applications, salient pole rotor synchronous generators have a large number of poles (72 poles, for instance). These rotors are usually driven by water wheels and, for this reason, synchronous generators with salient pole rotors are called hydro-generators. As seen from Figure 4.3, salient poles have concentrated (not distributed) windings which are wrapped around each protruding pole. These windings are excited by dc currents which create static (with respect to the rotor) magnetic field. The poles have pole shoes whose geometry is chosen to create sinusoidal magnetic field distribution in the air gap. This air gap is strongly nonuniform and has two symmetry axes called the direct axis (d-axis) and quadrature axis (q-axis). It is apparent (see Figure 4.3) that

$$\delta_d < \delta_q, \tag{4.1}$$

where δ_d and δ_q are the lengths of the air gap along the d-axis and q-axis, respectively. The nonuniformity of the air gap has important implications in the theory of synchronous generators with salient pole rotors (see section 4.4 of this chapter). Salient poles have laminated structure and their pole shoes are provided with *damper windings* consisting of embedded conducting bars interconnected at their ends by conducting rings. These windings are used to damp electromechanical rotor oscillations caused by disturbances. There is no need for such damper windings in the case of cylindrical solid rotors because of their intrinsic conductivity.

Each synchronous generator is equipped with an exciter which provides dc current for the rotor (field) winding. The structure of the exciter is usually quite complex and it has evolved over the years due to the progress in power electronics and in permanent magnet technology. The current tendency is to avoid sliding contacts (i.e., slip rings and brushes) and generate needed dc currents in the frames of rotating rotors. This is done, for instance, by using rectifiers mounted on rotor shafts. As a result, dc current produced by these rectifiers can be supplied to the rotor winding by a direct connection. The ac inputs to rectifiers may be obtained from ac armature windings of an auxiliary (small) synchronous generator. These small auxiliary generators have an inverted structure in the sense that their ac armatures are mounted on the rotor shafts of the main synchronous generators, while their "rotors" are stationary and produce static (dc) magnetic fields by using, for instance, permanent magnets. In this way, the synchronous generator may operate without depending on external sources of electricity. There are many modifications of the described excitation system (using, for instance, a pilot exciter), and a particular choice of excitation system depends on the unit power of a synchronous generator.

Synchronous generators are usually designed to maximize their power per unit weight. This is achieved by using high currents in the rotor and stator windings. The latter necessitates efficient cooling of these windings. Synchronous generators are equipped with sophisticated cooling systems, which typically employ direct water cooling of stator windings and hydrogen cooling of rotor windings. Detailed discussion of synchronous generator cooling systems is beyond the scope of this book.

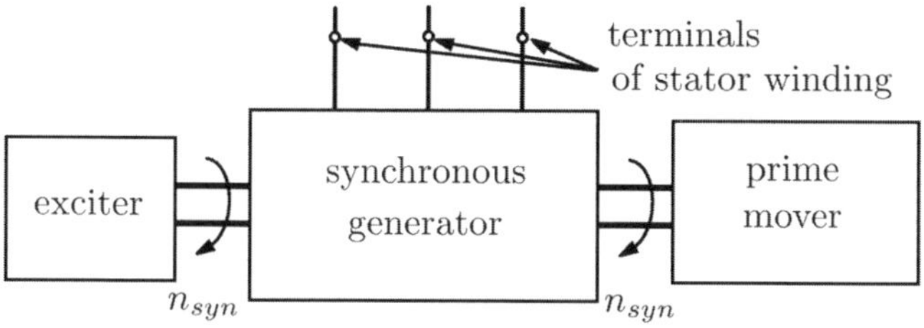

Fig. 4.4

Now, we shall proceed to the discussion of the *principle of operation* of the synchronous generator. During synchronous generator operation, its rotor is driven by a prime mover (see Figure 4.4) with a certain speed n_{syn}, called synchronous speed and measured in terms of revolutions per minute (rpm). This speed depends on the number of poles p of the synchronous generator and it is chosen (as discussed below) to guarantee a desired frequency (60 Hz in US utility power systems) of generated ac electric power. As the rotor of the synchronous generator is driven by the prime mover, an exciter provides dc current excitation to the rotor winding. This excitation results in a magnetic field static (time-invariant) with respect to the rotor but rotating with speed n_{syn} with respect to the stator. This rotor magnetic field induces emfs (internal voltages) in the three phases of the stator windings. Due to the special design of the stator windings (i.e., their spatial shift by 120°), these emfs have the same frequency and peak values but are *phase-shifted (in time) by* $\frac{2\pi}{3}$. When the stator winding is connected to a *balanced load*, the currents of the same frequency and peak values but phase-shifted (in time) by $\frac{2\pi}{3}$ will flow through the three phases of the stator winding. These currents create the magnetic field which is called the armature reaction field. It turns out (and this is demonstrated in the next section) that the armature reaction magnetic field is uniformly rotating and the speed $\tilde{n}_{syn}$ of rotation of this field is equal to the mechanical speed of the rotor, namely,

$$\boxed{n_{syn} = \tilde{n}_{syn}.} \tag{4.2}$$

In other words, the rotor and the armature reaction magnetic field rotate in *synchronism*. This is the reason for the name "synchronous generator." As a result of interaction between the armature reaction field and dc current in the conductors of the rotor winding, the electromagnetic torque appears which tends to slow down the speed of the rotor. The latter is true because, according to Lenz's law, the stator currents are always induced in such a way to counteract the cause of induction, which is in our case the rotating rotor and its magnetic field. The mechanical power of the prime mover is needed to maintain the constant synchronous speed n_{syn} of the rotor in the presence of the "slowing down" action of electromagnetic torque. An increase in consumption of electric power from the generator terminals results in an increase in currents in the stator three-phase windings because the voltage across the generator terminals is usually maintained more or less constant. The increase in stator currents results in the increase in the armature reaction magnetic fields which leads to the increase in electromagnetic torque which tends to reduce the rotor speed. Consequently, more mechanical power of the prime mover needs to be supplied to maintain the synchronous speed n_{syn} of the rotor constant. On the other hand, a decrease in consumption of electrical power from the generator terminals results in the decrease in stator currents which leads to the reduction of braking electromagnetic torque caused by interaction between the armature reaction magnetic field and currents in the rotor windings. This means that the prime mover mechanical power must be reduced to maintain the synchronous speed n_{syn} of the rotor constant. Thus, *it is clear that preserving the speed of the rotor n_{syn} constant in the face of continuously varying interaction between armature reaction magnetic field and rotor currents is the physical mechanism for conversion of mechanical energy into electric energy.* It is also clear from the above discussion that through maintaining constant synchronous speed of the rotor the electric power is generated *on demand.* Furthermore, it is worthwhile to stress here again that preserving constant synchronous speed n_{syn} of the rotor is needed for maintaining the constant frequency of generated ac electric power. In addition, the loss of synchronism is very detrimental to the operation of the generator because it may result in induction of appreciable eddy currents in the solid conducting rotors of turbo-generators and large losses. The same detrimental effect of eddy current induction in solid rotors occurs in the case of *unbalanced loads* of synchronous generators. Indeed, in the case of such loads, currents in the stator three-phase windings can be decomposed into positive, negative and zero-sequence symmetrical components. Currents of positive sequence will create armature reaction magnetic fields rotating in synchronism with the rotor. However, as can be shown, currents of negative sequence will create armature fields rotating with speed $\tilde{n}_{syn}$ in the direction opposite to the rotation of the rotor. As a result, eddy currents of double frequency (120 Hz) will be induced by these fields. Zero-sequence currents will also cause induction of eddy currents. However, these zero-sequence currents can be eliminated by using star without neutral connection of the stator winding. Thus, it is clear that synchronous generators *are vulnerable to unbalanced loads* and every effort must be made to maintain more or less balanced load across the generator terminals.

It is evident from the presented discussion that the frequency of ac voltage across the synchronous generator terminals is maintained by preserving the mechanical speed n_{syn} of the rotor, while the peak value of terminal voltage can be controlled through proper adjustment of dc excitation of the rotor winding. However, there is no way to control the initial phase of the terminal voltage of the synchronous generator. This phase varies with changes in loading conditions. For this reason, a synchronous generator cannot be construed as an ac voltage source; it is rather a (P, V)-source. The latter means that the real power P supplied to the power network and the peak value V_m of the terminal voltage can be controlled by controlling the mechanical power of the prime mover and dc excitation of the rotor winding, respectively. This representation of the synchronous generator as a (P, V)-source is very instrumental in the analysis of power flow in power networks (see the next chapter of this part of the book).

Next, we shall derive the important formula for the mechanical synchronous speed n_{syn} of the rotor that is required to generate ac electric power of specific (desired) frequency f. The starting point of our derivation is the observation that one cycle (i.e., one positive half-cycle and one negative half-cycle) is induced in the stationary stator winding when adjacent north and south poles of the rotor pass by this winding. This observation implies that one revolution of the rotor results in the induction of $p/2$ cycles where, as before, p is the number of rotor poles. This, in turn, suggests that $n_{syn}p/2$ cycles are induced per one minute because n_{syn} has the meaning of number of revolutions per minute. Since the frequency f of induced emf is measured in number of cycles per second, we find

$$f = \frac{n_{syn}p}{120}, \tag{4.3}$$

which leads to

$$\boxed{n_{syn} = \frac{120f}{p}.} \tag{4.4}$$

The table below presents the typical values of n_{syn} for synchronous generators with different numbers of poles.

Table

f	p	n_{syn}
60 Hz	2	3600 rpm
60 Hz	4	1800 rpm
60 Hz	6	1200 rpm
60 Hz	72	100 rpm
400 Hz	2	24000 rpm

As mentioned before, the case $p = 2$ is typical for turbo-generators in fossil fuel power plants, the case $p = 4$ is usual for turbo-generators in nuclear power plants, the case $p = 72$ is representative for hydro-generators and the case $f = 400$ Hz is realized in aviation. It is clear that the smaller p, the faster the synchronous generators and the smaller their geometric dimensions for the same output electric power. The latter is true because in faster synchronous generators the same input mechanical power can be achieved for smaller rotating masses. Fast synchronous generators also have large air gaps which may approach 15 cm. One may say that "an air gap of a synchronous generator is so large that birds can fly through it." As will be seen in subsequent discussions in this chapter, large air gaps decrease the reactances of the stator windings and this is beneficial to the overall quality of the synchronous generators.

Synchronous machines can also operate as motors. Nowadays, synchronous motors with permanent magnets on rotors are widely used in various applications. One example is spindle motors of hard disk drives in magnetic data storage. The mechanical speed of synchronous motors is given by formula (4.4). It is clear from this formula that this speed can be controlled by varying frequency f of the ac voltage applied to the stator windings of the synchronous motors. This frequency control of speed can be realized by using ac-to-ac converters or dc-to-ac inverters which are discussed in Part III of this book, which deals with power electronics.

4.2 Ideal Cylindrical Rotor Synchronous Generators and Their Armature Reaction Magnetic Fields

In this section, ideal cylindrical rotor generators are discussed and their armature reaction magnetic fields are analytically studied. In the case of ideal cylindrical rotor machines, the analysis of electromagnetic fields is performed under the following assumptions.

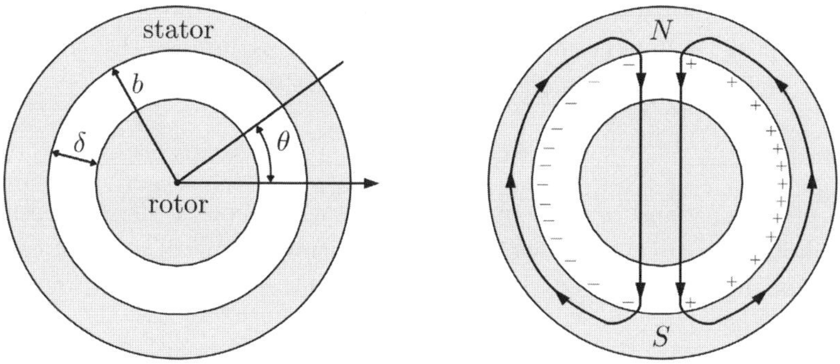

(a) (b)

Fig. 4.5

(1) Rotor and stator slots are neglected and the air gap is assumed to be *uniform.*
(2) Magnetic saturation of the rotor and stator steel is neglected and the magnetic permeabilities of the stator and rotor are assumed to be *infinite.*
(3) Electromagnetic fields in the air gap are assumed to be two-dimensional.
(4) *Each phase* of the stator winding is modeled as surface current continuously distributed along the stator interior boundary.

For phase a, the surface density of these currents is given by the formula

$$\mathbf{i}_a(\theta, t) = \mathbf{e}_z i_m \cos \nu\theta \sin \omega t, \tag{4.5}$$

where $\mathbf{e}_z$ is the unit vector along the rotational z-axis normal to the cross-sectional plane of the generator, while θ is the polar angle in this plane (see Figure 4.5a). It is apparent from formula (4.5) that the surface density is periodic with respect to θ and ν is the number of periods per one revolution (i.e., for $0 \leq \theta \leq 2\pi$). It turns out that each period of surface current density can be associated with two (north-N and south-S) poles. This is illustrated in Figure 4.5b in the particular case when $\nu = 1$ and

$$\mathbf{i}_a(\theta, t) = \mathbf{e}_z i_m \cos \theta \sin \omega t. \tag{4.6}$$

In this figure the signs of $i_a(\theta, t)$ are shown for time intervals when $\sin \omega t > 0$, and it is clear that magnetic field lines emanate from the area of the stator marked by N and enter the stator in the area marked by S. It is also clear that for time intervals when $\sin \omega t < 0$ the signs of $i_a(\theta, t)$ are reversed along with directions of magnetic field lines, however the two-pole structure of the magnetic field is preserved. This implies that $p = 2$ for $\nu = 1$. It is easy to conclude that in the general case

$$\boxed{\nu = \frac{p}{2}.} \tag{4.7}$$

It is customary and convenient to deal with continuously distributed magnetomotive force (mmf) associated with continuous distribution of surface currents. In the theory of electric machines this mmf is denoted by F and is defined as

$$F_a(\theta, t) = \int_0^\theta i_a(\theta', t) d\ell_{\theta'} = b \int_0^\theta i_a(\theta', t) d\theta', \tag{4.8}$$

where b is the interior radius of the stator.

From formulas (4.5) and (4.8), we find

$$F_a(\theta, t) = F_m \sin \nu\theta \sin \omega t, \tag{4.9}$$

where

$$F_m = \frac{i_m b}{\nu}. \tag{4.10}$$

The windings for phases b and c are also modeled by continuously distributed surface currents with the same i_m and ν, but phase-shifted in time with respect to one another by $\frac{2\pi}{3}$ and shifted along θ by $\frac{2\pi}{3\nu}$. Consequently, the mmfs of these phase windings are described, respectively, by the following formulas:

$$F_b(\theta, t) = F_m \sin\left(\nu\theta - \frac{2\pi}{3}\right) \sin\left(\omega t - \frac{2\pi}{3}\right), \tag{4.11}$$

$$F_c(\theta, t) = F_m \sin\left(\nu\theta - \frac{4\pi}{3}\right) \sin\left(\omega t - \frac{4\pi}{3}\right). \tag{4.12}$$

For the sake of conciseness, we shall refer in our subsequent discussion to this three-phase winding as being shifted in space by $\frac{2\pi}{3}$. In other words, it will be tacitly understood that the polar angle shift is scaled by $\frac{1}{\nu}$.

Now, the total magnetomotive force created by the ideal three-phase winding can be computed as

$$\begin{aligned} F(\theta, t) &= F_a(\theta, t) + F_b(\theta, t) + F_c(\theta, t) \\ &= F_m \bigg[\sin \nu\theta \sin \omega t + \sin\left(\nu\theta - \frac{2\pi}{3}\right) \sin\left(\omega t - \frac{2\pi}{3}\right) \\ &\qquad + \sin\left(\nu\theta - \frac{4\pi}{3}\right) \sin\left(\omega t - \frac{4\pi}{3}\right) \bigg]. \end{aligned} \tag{4.13}$$

By using the trigonometric identity

$$\sin\alpha\sin\beta = \frac{1}{2}\left[\cos(\alpha-\beta)-\cos(\alpha+\beta)\right], \tag{4.14}$$

formula (4.13) can be transformed as follows:

$$\begin{aligned} F(\theta,t) &= \frac{3F_m}{2}\cos(\omega t-\nu\theta) \\ &- \frac{F_m}{2}\left[\cos(\omega t+\nu\theta)+\cos\left(\omega t+\nu\theta-\frac{2\pi}{3}\right)+\cos\left(\omega t+\nu\theta-\frac{4\pi}{3}\right)\right]. \end{aligned} \tag{4.15}$$

By recalling the fact (see the remark after formula (1.17) in this part of the book) that the sum of sinusoidal quantities of the same peak value and frequency but phase-shifted (in time) with respect to one another by $\frac{2\pi}{3}$ is equal to zero, we find from the last equation that

$$\boxed{F(\theta,t) = \frac{3F_m}{2}\cos(\omega t-\nu\theta).} \tag{4.16}$$

Next, we shall demonstrate that formula (4.16) represents a uniformly rotating mmf. To do this, consider an "observer" that moves around the interior surface of the stator and whose polar angle position at any instant of time is defined by function $\tilde{\theta}(t)$. This "observer" will "see" at any time t the magnetomotive force

$$F\left[\tilde{\theta}(t),t\right] = \frac{3F_m}{2}\cos\left[\omega t-\nu\tilde{\theta}(t)\right]. \tag{4.17}$$

Now, the question can be asked how the "observer" should move in order to see at any instant of time t the same value of F. It is clear that this will be the case if

$$\omega t-\nu\tilde{\theta}(t) = const. \tag{4.18}$$

By differentiating the last equation with respect to t, we find

$$\omega-\nu\frac{d\tilde{\theta}}{dt} = 0, \tag{4.19}$$

and

$$\frac{d\tilde{\theta}}{dt} = \frac{\omega}{\nu} = \frac{4\pi f}{p}\ \text{(rad/s)}. \tag{4.20}$$

Thus, if the "observer" moves around the stator with constant angular speed

$$\tilde{\Omega}_{syn} = \frac{d\tilde{\theta}}{dt} = \frac{4\pi f}{p}\ \text{(rad/s)}, \tag{4.21}$$

then the same value of F is observed. However, this is only possible if magnetomotive force F uniformly rotates with angular speed $\tilde{\Omega}_{syn}$. This speed is measured in terms of radians per second. To represent this speed in terms of revolutions per minute (rpm), we use the relation

$$\tilde{n}_{syn} = \frac{\tilde{\Omega}_{syn}}{2\pi} \cdot 60 \text{ (rpm)}. \tag{4.22}$$

By substituting formula (4.21) into the last equation, we find

$$\boxed{\tilde{n}_{syn} = \frac{120f}{p}.} \tag{4.23}$$

If the number of poles of the stator windings is the same as the number of poles of the rotor and frequency f is the same as the frequency of the internal voltage induced in the stator windings by the rotating magnetic field of the rotor, then by comparing formulas (4.4) and (4.23), we find

$$\boxed{n_{syn} = \tilde{n}_{syn}.} \tag{4.24}$$

In other words, the magnetomotive force F rotates in synchronism with the rotor. As we shall see below, this uniformly rotating mmf of the stator winding creates uniformly rotating magnetic field which moves with the same speed as the mmf. This implies that formula (4.24) has the same meaning as formula (4.2), which means that the stator armature reaction magnetic field and the rotor rotate in *synchronism.*

Now, we shall proceed to the analysis of magnetic field created by the ideal three-phase stator winding with magnetomotive force specified by formula (4.16). It is clear from formulas (4.8) and (4.13) that the *total* surface current density

$$i(\theta,t) = i_a(\theta,t) + i_b(\theta,t) + i_c(\theta,t) \tag{4.25}$$

of the ideal three-phase stator winding is related to the *total* mmf, $F(\theta,t)$, by the equation

$$i(\theta,t) = \frac{1}{b}\frac{dF(\theta,t)}{d\theta}, \tag{4.26}$$

which, according to formula (4.16), leads to

$$\boxed{i(\theta,t) = \frac{3\nu}{2b}F_m \sin(\omega t - \nu\theta).} \tag{4.27}$$

It is clear that the phasor $\hat{i}(\theta)$ of this current density can be written as

$$\hat{i}(\theta) = -j\frac{3\nu}{2b}F_m e^{-j\nu\theta}. \tag{4.28}$$

The phasor of the magnetic field created by the ideal stator winding inside the air gap satisfies the following homogeneous equations:

$$\text{curl } \hat{\mathbf{H}} = 0, \tag{4.29}$$

$$\text{div } \hat{\mathbf{B}} = 0, \tag{4.30}$$

$$\hat{\mathbf{B}} = \mu_0 \hat{\mathbf{H}} \tag{4.31}$$

and the boundary conditions

$$\hat{H}_\theta^{st} - \hat{H}_\theta^{g} = \hat{i} \quad \text{for} \quad r = b, \tag{4.32}$$

$$\hat{H}_\theta^{rot} - \hat{H}_\theta^{g} = 0 \quad \text{for} \quad r = a, \tag{4.33}$$

where $\hat{H}_\theta^{st}$, $\hat{H}_\theta^{rot}$ and $\hat{H}_\theta^{g}$ are tangential θ-components of magnetic field from the stator, rotor and gap sides, respectively, while a is the rotor radius.

Since the phasor of magnetic flux density $\hat{\mathbf{B}}$ is always finite, by using the second assumption of the ideal machine we find that

$$\hat{H}_\theta^{st} = \frac{\hat{B}_\theta^{st}}{\mu} \to 0 \quad \text{as} \quad \mu \to \infty. \tag{4.34}$$

Similarly, we conclude that

$$\hat{H}_\theta^{rot} = 0. \tag{4.35}$$

By using the last two formulas in boundary conditions (4.32) and (4.33), we obtain

$$\hat{H}_\theta^{g} = -\hat{i} \quad \text{for} \quad r = b, \tag{4.36}$$

$$\hat{H}_\theta^{g} = 0 \quad \text{for} \quad r = a. \tag{4.37}$$

Now, we introduce the vector magnetic potential $\hat{\mathbf{A}}(r, \theta)$ by formulas

$$\hat{\mathbf{B}} = \text{curl } \hat{\mathbf{A}}, \tag{4.38}$$

$$\text{div } \hat{\mathbf{A}} = 0. \tag{4.39}$$

Since the magnetic field in the air gap is assumed to be two-dimensional, the vector potential has only one (z) component,

$$\hat{\mathbf{A}}(r, \theta) = \mathbf{e}_z \hat{A}(r, \theta). \tag{4.40}$$

From equations (4.29)-(4.31) and (4.38)-(4.39) follows that $\hat{A}(r, \theta)$ satisfies the Laplace equation, which in the polar coordinates (r, θ) can be written as follows:

$$\frac{1}{r}\frac{\partial}{\partial r}\left(r\frac{\partial \hat{A}}{\partial r}\right) + \frac{1}{r^2}\frac{\partial^2 \hat{A}}{\partial \theta^2} = 0 \quad \text{for} \quad a < r < b. \tag{4.41}$$

Furthermore, this vector potential satisfies the following boundary conditions:

$$\frac{\partial \hat{A}}{\partial r}(b,\theta) = -j\frac{3\mu_0\nu}{2b}F_m e^{-j\nu\theta}, \tag{4.42}$$

$$\frac{\partial \hat{A}}{\partial r}(a,\theta) = 0. \tag{4.43}$$

These boundary conditions follow from the relation

$$\hat{H}_\theta^g = -\frac{1}{\mu_0}\frac{\partial \hat{A}}{\partial r}, \tag{4.44}$$

boundary conditions (4.36) and (4.37), respectively, and formula (4.28).

Thus, the analysis of magnetic field in the air gap of the ideal machine is reduced to the solution of boundary value problem (4.41)-(4.43). To find the solution to this problem, we shall use the method of separation of variables and represent $\hat{A}(r,\theta)$ in the form

$$\hat{A}(r,\theta) = T(r)\Phi(\theta). \tag{4.45}$$

By substituting formula (4.45) into the boundary condition (4.42), we find

$$T'(b)\Phi(\theta) = -j\frac{3\mu_0\nu}{2b}F_m e^{-j\nu\theta}. \tag{4.46}$$

From the last equation follows that

$$\Phi(\theta) = e^{-j\nu\theta} \tag{4.47}$$

and, consequently,

$$\hat{A}(r,\theta) = T(r)e^{-j\nu\theta}. \tag{4.48}$$

Next, by writing the Laplace equation (4.41) in the form

$$r^2\frac{\partial^2 \hat{A}}{\partial r^2} + r\frac{\partial \hat{A}}{\partial r} + \frac{\partial^2 \hat{A}}{\partial \theta^2} = 0 \quad \text{for} \quad a < r < b \tag{4.49}$$

and by substituting formula (4.48) into the last equation as well as in boundary conditions (4.42) and (4.43), we derive that function $T(r)$ is the solution of the following boundary value problem:

$$r^2T''(r) + rT'(r) - \nu^2T(r) = 0 \quad \text{for} \quad a < r < b, \tag{4.50}$$

$$T'(b) = -j\frac{3\mu_0\nu}{2b}F_m, \tag{4.51}$$

$$T'(a) = 0. \tag{4.52}$$

Equation (4.50) is the Euler equation whose solution can be sought in the form

$$T(r) = Cr^{\alpha}, \tag{4.53}$$

where C and α are some constants.

By substituting the last formula into equation (4.50), we find

$$\alpha(\alpha - 1) + \alpha - \nu^2 = 0, \tag{4.54}$$

and, consequently, function (4.53) satisfies equation (4.50) for the following two values of α:

$$\alpha_1 = \nu, \qquad \alpha_2 = -\nu. \tag{4.55}$$

This implies that a general solution of equation (4.50) has the form

$$T(r) = C_1 r^{\nu} + C_2 r^{-\nu}, \tag{4.56}$$

where C_1 and C_2 are some constants. These constants can be found from the boundary conditions (4.51) and (4.52), which lead to the following equations for C_1 and C_2:

$$C_1 b^{\nu-1} - C_2 b^{-\nu-1} = -j\frac{3\mu_0}{2b}F_m, \tag{4.57}$$

$$C_1 a^{\nu-1} - C_2 a^{-\nu-1} = 0. \tag{4.58}$$

These two equations can be easily solved and the following formulas for C_1 and C_2 can be derived:

$$C_1 = -j\frac{3\mu_0 F_m}{2b\left(b^{\nu-1} - a^{2\nu}b^{-\nu-1}\right)}, \tag{4.59}$$

$$C_2 = -j\frac{3\mu_0 a^{2\nu} F_m}{2b\left(b^{\nu-1} - a^{2\nu}b^{-\nu-1}\right)}. \tag{4.60}$$

Finally, by using formulas (4.48), (4.56), (4.59) and (4.60) as well as simple algebra, the following expression for $\hat{A}(r,\theta)$ can be derived:

$$\boxed{\hat{A}(r,\theta) = -j\frac{3}{2}\mu_0 F_m a^{\nu} b^{\nu}\frac{\left(\frac{r}{a}\right)^{\nu} + \left(\frac{r}{a}\right)^{-\nu}}{b^{2\nu} - a^{2\nu}}e^{-j\nu\theta}.} \tag{4.61}$$

The last equation can be simplified by using the relation

$$a = b - \delta \tag{4.62}$$

where

$$\delta \ll b. \tag{4.63}$$

Indeed, from the last two formulas we derive that

$$b^{2\nu} - a^{2\nu} \approx 2\nu b^{2\nu-1}\delta, \tag{4.64}$$

$$a^{\nu} b^{\nu} \approx b^{2\nu}. \tag{4.65}$$

By using the last two formulas in equation (4.61) we arrive at

$$\hat{A}(r,\theta) \approx -j\frac{3\mu_0 b}{4\nu\delta}F_m\left[\left(\frac{r}{a}\right)^{\nu} + \left(\frac{r}{a}\right)^{-\nu}\right]e^{-j\nu\theta}. \tag{4.66}$$

Further simplification is also possible because $a < r < b$ and according to formulas (4.62)-(4.63)

$$\left(\frac{r}{a}\right)^{\nu} + \left(\frac{r}{a}\right)^{-\nu} \approx 2. \tag{4.67}$$

This leads to

$$\boxed{\hat{A}(r,\theta) \approx -j\frac{3\mu_0 b}{2\nu\delta}F_m e^{-j\nu\theta}.} \tag{4.68}$$

The last formula implies that

$$\boxed{A(r,\theta,t) \approx \frac{3\mu_0 b}{2\nu\delta}F_m \sin(\omega t - \nu\theta).} \tag{4.69}$$

It is evident from formulas (4.16) and (4.69) that the mmf and A have spatial and temporal dependence on the same argument $(\omega t - \nu\theta)$. This suggests that the vector magnetic potential is uniformly rotating with synchronous speed $\tilde{n}_{syn}$. The latter can be proved by literally repeating the same reasoning which led to the derivation of formula (4.23) from formula (4.16). Since it is known that in 2D the level lines of the vector potential coincide with magnetic flux density lines, it is naturally concluded that the magnetic field in the gap represented by this vector potential is uniformly rotating with synchronous speed $\tilde{n}_{syn}$. A similar conclusion can be achieved by deriving the expression for the radial component of the magnetic flux density by using formula (4.69).

The previous discussion is based on formula (4.69) derived for cylindrical rotor machines with uniform air gaps. However, the conclusion that the surface current density (4.27) creates uniformly rotating magnetic field is valid for salient pole machines (with nonuniform air gaps) as well under the condition of synchronism $n_{syn} = \tilde{n}_{syn}$. Indeed, under the condition of synchronism, the surface current density does not depend on time in the rotor frame of reference and creates static magnetic field in this reference frame. Consequently, this field is uniformly rotating in the stator frame of reference.

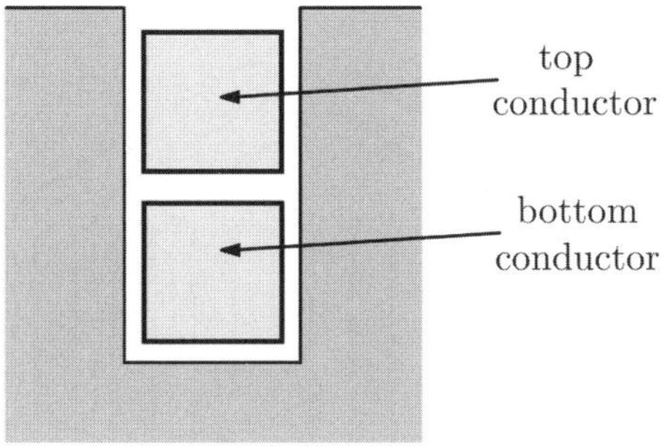

Fig. 4.6

4.3 Design of Stator Windings and Their Reactances

In the previous section, the ideal three-phase stator windings have been discussed. Actual stator windings are designed as *discrete approximations* of ideal windings. This can be accomplished in many ways and, for this reason, there are many different designs of stator windings. We shall not attempt the detailed discussion of these designs, but rather we shall illustrate the central ideas and stress the basic facts involved in the design of stator windings.

It is quite often that two-layer windings are used. In such windings, there are "top" and "bottom" conductors embedded in each slot (see Figure 4.6). Furthermore, each phase winding consists of several coils connected usually in series and embedded in several adjacent slots. The parts of coils which are embedded in slots are called active parts, while the parts of coils outside the slots are called end parts (see Figure 4.7). Usually, one active (direct) part of a coil is embedded in a slot as a top conductor, while another active (reverse) part of the same coil is embedded in a different slot as a bottom conductor. An example of the described design of stator phase windings is illustrated (in developed view form) by Figure 4.7 for the case of one phase (phase a) consisting of four coils for a two-pole machine with 12 slots in the stator. It is clear from this figure that the active parts of coil 1 are embedded in slots 1 and 7 as top and bottom conductors, respectively; those of coil 2 are in slots 2 and 8; those of coil 3 are in slots 3 and 9; and those of coil 4 are in slots 4 and 10. It is apparent that these coils are connected in series and that the difference between polar angles θ_k and θ'_k of the centers of the slots in which top and bottom parts of the same coil

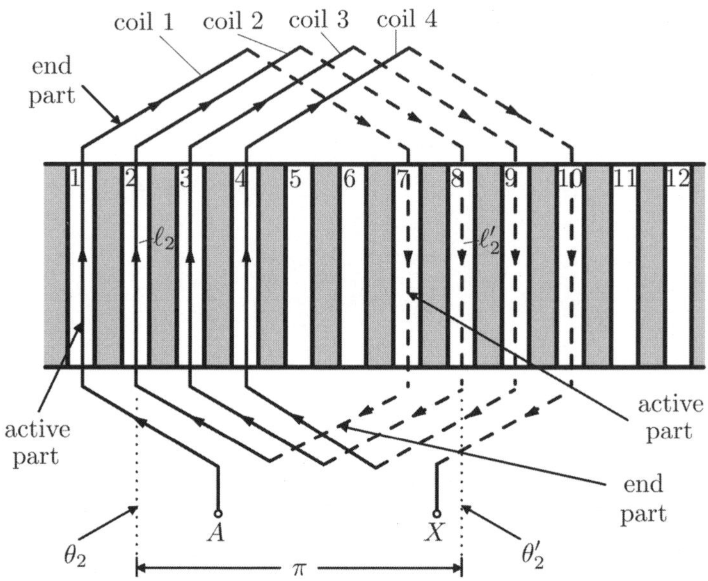

Fig. 4.7

Occupancy Table 1

slot #	1	2	3	4	5	6	7	8	9	10	11	12
top	A_1^+	A_2^+	A_3^+	A_4^+	B_1^+	B_2^+	B_3^+	B_4^+	C_1^+	C_2^+	C_3^+	C_4^+
bottom	B_3^-	B_4^-	C_1^-	C_2^-	C_3^-	C_4^-	A_1^-	A_2^-	A_3^-	A_4^-	B_1^-	B_2^-

number k are embedded is equal to π, i.e.,

$$\theta_k - \theta'_k = \pi. \tag{4.70}$$

This is the so-called *full-pitch* winding. Arrows in Figure 4.7 mark the reference directions of coil currents and they reveal why the top and bottom parts of the coils can be called direct and reverse, respectively. The stator windings for phases b and c are designed in the same way as for phase a, but shifted in space with respect to one another by $120°$, i.e., by four slots. It is somewhat cumbersome to show all three phase windings in the same Figure 4.7. However, the described design of the stator three-phase winding can be illustrated by the Occupancy Table 1. In this table, $A_k^{\pm}$, $B_k^{\pm}$ and $C_k^{\pm}$ (with $k = 1, 2, 3, 4$) correspond to active parts of coil number k of the stator windings for phases a, b and c, respectively. It is clear from this table that each phase has four full-pitch coils embedded in adjacent slots

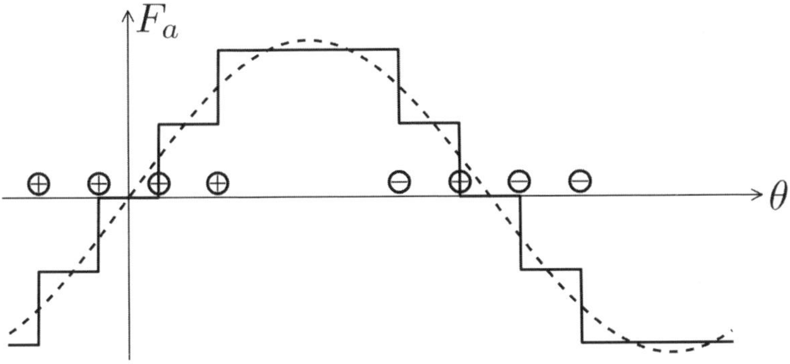

Fig. 4.8

and that the three windings are shifted with respect to one another by 120° along the stator interior circumference.

Next, we consider the mmf of each phase winding. To simplify the reasoning, we represent each coil of the phase winding by two (direct and reverse) *filamentary* conductors located on the stator surface at polar angles θ_k and θ'_k, respectively (see Figure 4.8). By using the definition of mmf given by formula (4.8) and taking into account that each current-carrying filamentary conductor can be construed as a delta-type function of surface current density, we find after integration that the mmf F_a of phase a is a piecewise step function represented by the solid line in Figure 4.8. It is clear that this multi-step function is a discrete approximation of sinusoidal mmf shown in Figure 4.8 by a dashed line. Actually, this sinusoidal mmf is the first (fundamental) harmonic of the Fourier series expansion of the multi-step mmf of the actual phase winding. Since the stator windings for phases b and c are identical to the stator phase a winding except that they are shifted in θ by 120° and 240°, respectively, then the fundamental harmonics of F_b and F_c have the same peak values as the fundamental harmonic for F_a but are shifted in θ by 120° and 240°, respectively. When the phase windings are excited by currents of the same peak values and frequency but phase-shifted in time by $\frac{2\pi}{3}$ and $\frac{4\pi}{3}$, respectively, then the fundamental harmonics of their mmfs will have the same mathematical forms as given by formulas (4.9), (4.11) and (4.12). This implies that the total fundamental harmonic of the mmf of the three-phase winding is given by formula (4.16). In other words, the fundamental harmonic of mmf of the actual three-phase stator winding replicates the mmf of the ideal machine and, consequently, creates magnetic field that rotates in synchronism with the rotor. However,

Occupancy Table 2

slot #	1	2	3	4	5	6	7	8	9	10	11	12
top	A_1^+	A_2^+	A_3^+	A_4^+	B_1^+	B_2^+	B_3^+	B_4^+	C_1^+	C_2^+	C_3^+	C_4^+
bottom	B_4^-	C_1^-	C_2^-	C_3^-	C_4^-	A_1^-	A_2^-	A_3^-	A_4^-	B_1^-	B_2^-	B_3^-

Occupancy Table 3

slot #	1	2	3	4	5	6	7	8	9	10	11	12
top	A_1^+	A_3^+	B_1^+	B_3^+	C_1^+	C_3^+	A_2^-	A_4^-	B_2^-	B_4^-	C_2^-	C_4^-
bottom	A_2^+	A_4^+	B_2^+	B_4^+	C_2^+	C_4^+	A_1^-	A_3^-	B_1^-	B_3^-	C_1^-	C_3^-

the multi-step nature of mmfs of actual stator windings results in higher-order spatial harmonics of mmf which correspond to higher-order terms in Fourier series expansions of these mmfs. These higher-order harmonics create magnetic fields which are not in synchronism with the rotor. As a result, these fields generate eddy currents in the solid conducting rotor which are detrimental to the overall performance of synchronous generators. For this reason, it is desirable to suppress (or "filter out") these higher-order spatial harmonics. This can be accomplished by using stator windings whose coils have *fractional pitch*. The latter means that the relation (4.70) is replaced by the following:

$$\theta_k - \theta'_k = \beta\pi, \qquad (0 < \beta < 1). \tag{4.71}$$

An example of the design of a fractional-pitch winding is illustrated by Occupancy Table 2. It is clear that $\beta = 5/6$ for the winding design presented in this table. However, this design of fractional-pitch winding is not practically useful. The reason is that the mmf of each coil of such winding does not have half-wave symmetry, and this leads to the appearance of even harmonics in this mmf and, consequently, in the total mmf of the winding. This difficulty can be circumvented by fully exploiting the two-layer structure of the winding. This is first accomplished by designing the full-pitch winding as illustrated by Occupancy Table 3. Then, the fractional-pitch winding with $\beta = 5/6$ is produced by shifting the bottom layer by one slot as shown in Occupancy Table 4. It is clear from this occupancy table that even coils (2 and 4) and odd coils (1 and 3) have the same pitch $\beta = 5/6$ when the inner circumference of the stator is traversed in *opposite* directions. This preserves the half-wave symmetry of the total mmf of each phase winding. Another way to illustrate this fact is to observe that a two-layer fractional-

Occupancy Table 4

slot #	1	2	3	4	5	6	7	8	9	10	11	12
top	A_1^+	A_3^+	B_1^+	B_3^+	C_1^+	C_3^+	A_2^-	A_4^-	B_2^-	B_4^-	C_2^-	C_4^-
bottom	C_3^-	A_2^+	A_4^+	B_2^+	B_4^+	C_2^+	C_4^+	A_1^-	A_3^-	B_1^-	B_3^-	C_1^-

pitch phase winding represented in Table 4 can be viewed as two single-layer full-pitch windings shifted with respect to one another by the angle $(1-\beta)\pi$ and connected in series. Indeed, for phase a, these two full-pitch windings are formed by conductors A_1^+-A_2^-, A_3^+-A_4^- in the top layer and by conductors A_2^+-A_1^-, A_4^+-A_3^- in the bottom layer. This equivalent representation of the two-layer fractional-pitch winding as two (top and bottom) full-pitch windings shifted by angle $(1-\beta)\pi$ is legitimate because all winding coils are connected in series. It is also apparent that half-wave symmetry holds for any full-pitch winding and, consequently, for the fractional-pitch winding designed as shown in Occupancy Table 4 there is no generation of even harmonics in the mmf, while some odd mmf harmonics can be suppressed by using the fractional pitch (as discussed somewhat later). It is worthwhile to note that the fractional-pitch design of stator windings is also beneficial for suppression of higher-order time harmonics in induced internal voltage (emf). These time harmonics appear because the distribution of rotor magnetic field deviates from the sinusoidal case. Thus, it can be concluded that three-phase stator windings are designed as filters of higher-order spatial and temporal harmonics.

The design of stator windings has been discussed above for the case of two-pole machines. When the number of poles is larger than two, the pattern of the winding is periodically repeated along the stator circumference. This results in ν identical groups of coils which can be connected in series or parallel.

Next, we proceed to the calculation of reactances of stator windings. For this purpose, we shall recall that the magnetic flux which links a filamentary conductor L (see Figure 4.9) can be computed by using Stokes' theorem and magnetic vector potential. Namely,

$$\Phi = \iint_S \mathbf{B} \cdot \mathbf{ds} = \iint_S \text{curl } \mathbf{A} \cdot \mathbf{ds} = \oint_L \mathbf{A} \cdot d\boldsymbol{\ell}. \tag{4.72}$$

Thus, the flux Φ_k which links one turn of the coil number k can be computed

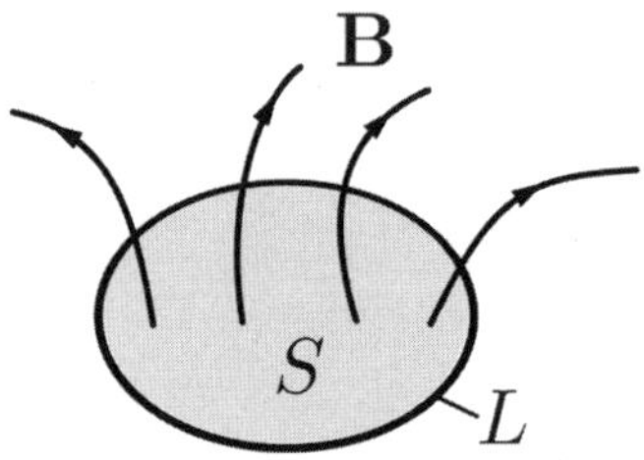

Fig. 4.9

as follows:

$$\Phi_k = \oint_{L_k} \mathbf{A} \cdot d\boldsymbol{\ell} = \int_{\ell_k} \mathbf{A} \cdot d\boldsymbol{\ell} + \int_{\ell'_k} \mathbf{A} \cdot d\boldsymbol{\ell} + \int_{\substack{\text{end}\\ \text{parts}}} \mathbf{A} \cdot d\boldsymbol{\ell}, \tag{4.73}$$

where ℓ_k and ℓ'_k are direct and reverse active parts of the coil (see Figure 4.7). In our calculations below, we shall neglect the last integral (over "end parts") in formula (4.73), while the integrals over ℓ_k and ℓ'_k can be simplified by assuming that $\mathbf{A}$ has only z-components and, consequently, is parallel and anti-parallel to $d\boldsymbol{\ell}$ for ℓ_k and ℓ'_k, respectively. Furthermore, it will be assumed that ℓ_k and ℓ'_k are active parts of full-pitch top (or bottom) coils in the equivalent representation of double-layer fractional-pitch winding discussed above. This yields the expression

$$\Phi_k^{(\text{top})} \approx \ell \left[A(\theta_k, t) - A\left(\theta'_k, t\right)\right], \tag{4.74}$$

where ℓ is the length of the active parts, which is the same as the stator length, while superscript "(top)" denotes that the flux is computed for a top full-pitch coil.

Next, we shall use the expression (4.69) for A that was derived for the ideal machine. It is understandable and in agreement with the previous discussion that F_m in this expression can be construed as the peak value of the fundamental harmonic of mmf of the actual ("discrete") stator phase winding. For the sake of simplicity of derivations, we consider the case of the two-pole machine when $\nu = 1$. Now, by substituting formula (4.69) into equation (4.74), we obtain

$$\Phi_k^{(\text{top})}(t) = \frac{3\mu_0 b\ell}{2\delta} F_m \left[\sin(\omega t - \theta_k) - \sin\left(\omega t - \theta'_k\right)\right]. \tag{4.75}$$

In the case of full-pitch coils when formula (4.70) is valid, the last formula can be simplified as follows:

$$\Phi_k^{(\text{top})}(t) = \frac{3\mu_0 b\ell}{\delta} F_m \sin(\omega t - \theta_k). \tag{4.76}$$

It is apparent from the last formula that fluxes linking different coils have *the same peak value but different initial phases.* The latter is the manifestation of the distributed nature of the winding. Similarly, for the bottom full-pitch coils, shifted with respect to the top full-pitch coils by the angle $(1-\beta)\pi$, we can write

$$\Phi_k^{(\mathrm{bot})}(t) = \frac{3\mu_0 b\ell}{\delta} F_m \sin\left[\omega t - \theta_k + (1-\beta)\pi\right]. \tag{4.77}$$

The total flux linkages $\psi(t)$ of the phase winding can be computed as follows:

$$\psi(t) = N_c \sum_{k=1}^{q} \left[\Phi_k^{(\mathrm{top})}(t) + \Phi_k^{(\mathrm{bot})}(t)\right], \tag{4.78}$$

where N_c is the number of turns of each coil, while q is the number of full-pitch coils in the top (or bottom) layer.

By substituting formulas (4.75) and (4.77) into equation (4.78), we find

$$\psi(t) = \frac{3\mu_0 b\ell N}{2\delta q} F_m \sum_{k=1}^{q} \left\{\sin(\omega t - \theta_k) + \sin\left[\omega t - \theta_k + (1-\beta)\pi\right]\right\}, \tag{4.79}$$

where

$$N = 2qN_c \tag{4.80}$$

is the total number of turns in the phase winding.

Formula (4.79) can be written in the phasor form as

$$\hat{\psi} = -j\frac{3\mu_0 b\ell N}{2\delta q} F_m \sum_{k=1}^{q} \left(e^{-j\theta_k} + e^{-j\theta_k} e^{j(1-\beta)\pi}\right), \tag{4.81}$$

which can be further transformed as follows:

$$\hat{\psi} = -j\frac{3\mu_0 b\ell N}{2\delta q} F_m \left(1 - e^{-j\beta\pi}\right) \sum_{k=1}^{q} e^{-j\theta_k}. \tag{4.82}$$

Next, we can write that

$$\theta_k = \theta_1 + (k-1)\alpha, \tag{4.83}$$

where

$$\alpha = \theta_k - \theta_{k-1} \tag{4.84}$$

is the polar angle increment between two adjacent slots.

By using formula (4.83), we find

$$\sum_{k=1}^{q} e^{-j\theta_k} = e^{-j\theta_1} \sum_{k=1}^{q} e^{-j(k-1)\alpha}. \tag{4.85}$$

The sum in the right-hand side of the last formula can be recognized as the finite geometric series with the ratio $e^{-j\alpha}$. Consequently, the last formula can be further transformed as follows:

$$\sum_{k=1}^{q} e^{-j\theta_k} = e^{-j\theta_1}\frac{1-e^{-jq\alpha}}{1-e^{-j\alpha}}. \tag{4.86}$$

By substituting formula (4.86) into formula (4.82), we find

$$\hat{\psi} = -j\frac{3\mu_0 b\ell N}{2\delta q}F_m e^{-j\theta_1}\left(1-e^{-j\beta\pi}\right)\frac{1-e^{-jq\alpha}}{1-e^{-j\alpha}}. \tag{4.87}$$

Consequently,

$$\psi_m = \frac{3\mu_0 b\ell N}{2\delta q}F_m\left|1-e^{-j\beta\pi}\right|\frac{\left|1-e^{-jq\alpha}\right|}{\left|1-e^{-j\alpha}\right|}. \tag{4.88}$$

Next, we shall use the formula

$$\left|1-e^{-jx}\right| = 2\sin\frac{x}{2}, \tag{4.89}$$

assuming that $0 < x < \pi$. The latter formula is derived as follows:

$$1-e^{-jx} = 1-\cos x + j\sin x = 2\sin^2\frac{x}{2} + j2\sin\frac{x}{2}\cos\frac{x}{2}. \tag{4.90}$$

This implies that

$$1-e^{-jx} = 2\sin\frac{x}{2}\left[\sin\frac{x}{2} + j\cos\frac{x}{2}\right], \tag{4.91}$$

which leads to formula (4.89).

By using equality (4.89), formula (4.88) can be written as follows:

$$\psi_m = \frac{3\mu_0 b\ell N}{\delta}F_m\sin\frac{\beta\pi}{2}\frac{\sin\frac{q\alpha}{2}}{q\sin\frac{\alpha}{2}}. \tag{4.92}$$

Now, we shall introduce two very important coefficients (factors): the pitch coefficient

$$\boxed{k_p = \sin\frac{\beta\pi}{2},} \tag{4.93}$$

and the distribution (or breadth) coefficient

$$\boxed{k_d = \frac{\sin\frac{q\alpha}{2}}{q\sin\frac{\alpha}{2}}.} \tag{4.94}$$

These two coefficients are usually combined into one winding coefficient

$$\boxed{k_w = k_p k_d.} \tag{4.95}$$

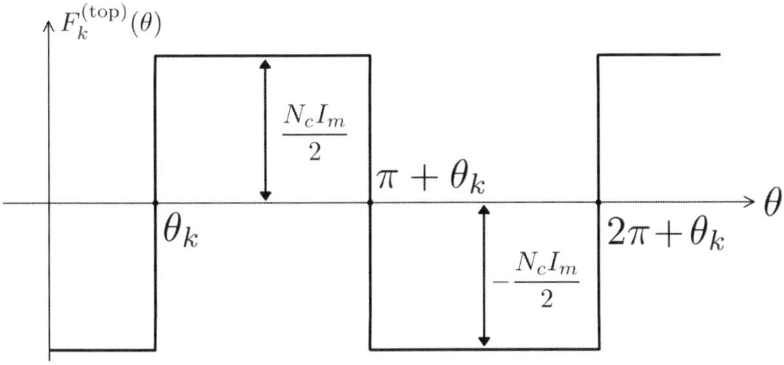

Fig. 4.10

By using the last three formulas, equation (4.92) can be written in the form

$$\psi_m = \frac{3\mu_0 b\ell N k_w}{\delta} F_m. \tag{4.96}$$

This implies that the peak value of the voltage induced by flux linkages $\psi(t)$ can be computed as

$$\boxed{V_m = \omega\psi_m = \frac{3\mu_0 b\ell N k_w}{\delta}\omega F_m.} \tag{4.97}$$

Next, we shall compute F_m in terms of peak value I_m of the phase winding current. To this end, we shall first consider the mmf $F_k^{(\text{top})}(\theta)$ of the filamentary full-pitch coil number k in the top layer at the moment of time when the current in the phase winding achieves its maximum value. This mmf is represented in Figure 4.10. The Fourier coefficients for the first-order terms of the Fourier expansion of $F_k^{(\text{top})}(\theta)$ can be computed by using the formulas

$$a_1^{(k)} = \frac{1}{\pi}\int_{\theta_k}^{2\pi+\theta_k} F_k^{(\text{top})}(\theta)\cos\theta d\theta, \tag{4.98}$$

$$b_1^{(k)} = \frac{1}{\pi}\int_{\theta_k}^{2\pi+\theta_k} F_k^{(\text{top})}(\theta)\sin\theta d\theta. \tag{4.99}$$

After integration, we find

$$a_1^{(k)} = -\frac{2N_c I_m}{\pi}\sin\theta_k, \tag{4.100}$$

$$b_1^{(k)} = \frac{2N_c I_m}{\pi}\cos\theta_k, \tag{4.101}$$

where, as before, N_c is the number of turns of the coil. Thus, the fundamental harmonic $F_{k1}^{(\text{top})}$ of the coil mmf can be found as follows:

$$F_{k1}^{(\text{top})}(\theta) = a_1^{(k)} \cos\theta + b_1^{(k)} \sin\theta$$
$$= \frac{2N_c I_m}{\pi} \left[-\cos\theta \sin\theta_k + \sin\theta \cos\theta_k\right]. \tag{4.102}$$

After simple trigonometric transformations, we derive

$$F_{k1}^{(\text{top})}(\theta) = \frac{2N_c I_m}{\pi} \sin(\theta - \theta_k). \tag{4.103}$$

Since the full-pitch coils of the bottom layer are shifted by the angle $(1-\beta)\pi$, we find

$$F_{k1}^{(\text{bot})}(\theta) = \frac{2N_c I_m}{\pi} \sin\left[\theta - \theta_k + (1-\beta)\pi\right]. \tag{4.104}$$

Now, the fundamental harmonic of the phase winding mmf can be computed as follows:

$$F_1(\theta) = \sum_{k=1}^{q} \left[F_{k1}^{(\text{top})}(\theta) + F_{k1}^{(\text{bot})}(\theta)\right]. \tag{4.105}$$

From formulas (4.103), (4.104) and (4.105), we find

$$F_1(\theta) = \frac{NI_m}{\pi q} \sum_{k=1}^{q} \left\{\sin(\theta - \theta_k) + \sin[\theta - \theta_k + (1-\beta)\pi]\right\}, \tag{4.106}$$

where N is given by formula (4.80).

It can be observed that the mathematical form of the sum in formula (4.106) is similar to the mathematical form of the sum in formula (4.79); they become formally identical if ωt is replaced by θ. This means that by literally repeating the same line of reasoning used in the derivation of formula (4.96) for ψ_m, we can derive the following expression for the peak value F_m of $F_1(\theta)$:

$$\boxed{F_m = \frac{2Nk_w}{\pi} I_m.} \tag{4.107}$$

By substituting the last formula into equation (4.97), we obtain

$$V_m = \frac{12\mu_0 b\ell}{\delta} f N^2 k_w^2 I_m. \tag{4.108}$$

The last formula implies the validity of the following expression for the main reactance $X_s^{(m)}$ of the stator phase winding:

$$\boxed{X_s^{(m)} = \frac{12\mu_0 b\ell}{\delta} f N^2 k_w^2.} \tag{4.109}$$

This reactance is termed "main" reactance (and marked by the superscript "(m)") in order to emphasize that this is not the total reactance of the stator phase winding but rather its main part. Indeed, in the derivation of expression (4.109), the formula (4.69) was used. The latter formula was derived for the ideal machine by neglecting end parts of the stator winding as well as slots on the stator and rotor and by assuming that the armature reaction magnetic field has only the fundamental spatial harmonic. For this reason, the actual stator winding reactance is somewhat different from $X_s^{(m)}$ and can be represented as follows:

$$X_s = X_s^{(m)} + X_s^{\ell}, \tag{4.110}$$

where X_s^{ℓ} is the so-called leakage reactance that accounts for end-part leakage, slot leakage and differential leakage associated with higher-order spatial harmonics of magnetic field.

Another very important point is that formula (4.109) has been derived by using the expression for vector potential A of the rotating magnetic field, i.e., magnetic field created by three currents in three stator phase windings. In this sense, reactance X_s accounts for self and mutual reactances of the three stator phase windings. It is interesting to find the relation between X_s and self and mutual reactances. To do this, consider the KVL equation for phase a:

$$\hat{V}_a = jX_{aa}\hat{I}_a + jX_{ab}\hat{I}_b + jX_{ac}\hat{I}_c, \tag{4.111}$$

where X_{aa} is the self-reactance, while X_{ab} and X_{ac} are mutual reactances. It is apparent that due to the symmetry of the three-phase stator winding under rotation by 120°, we have

$$X_{ab} = X_{ac}. \tag{4.112}$$

Furthermore, in the case of a balanced load,

$$\hat{I}_b = \hat{I}_a e^{-j\frac{2\pi}{3}}, \qquad \hat{I}_c = \hat{I}_a e^{-j\frac{4\pi}{3}}. \tag{4.113}$$

By using formulas (4.112) and (4.113), we easily derive that

$$\hat{V}_a = j(X_{aa} - X_{ab})\hat{I}_a. \tag{4.114}$$

The latter implies that

$$V_m = (X_{aa} - X_{ab})I_m, \tag{4.115}$$

which means that

$$X_s = X_{aa} - X_{ab}. \tag{4.116}$$

Thus, in the case of balanced load, self and mutual reactances can be accounted for by one reactance X_s, which is quite often called synchronous reactance.

Now, we shall briefly return to the discussion of higher-order spatial harmonics of the stator winding. As was mentioned before, even spatial harmonics are absent due to the half-wave symmetry of the stator winding mmf. Next, we illustrate that all "multiple of three" higher-order harmonics are equal to zero as well. The illustration is based on the fact that for a full-pitch coil number k of phase a the η-th harmonic of its mmf at time t is given by the formula

$$F_{k\eta}^{a}(\theta, t) = \frac{2N_c i_a(t)}{\pi\eta} \sin(\eta\theta - \eta\theta_k). \tag{4.117}$$

The derivation of the last relation is similar to the derivation of formula (4.103) and it takes into account that η is an odd number. Since the coils number k of phases a, b and c are shifted with respect to one another by $\frac{2\pi}{3}$, we find

$$F_{k\eta}^{b}(\theta, t) = \frac{2N_c i_b(t)}{\pi\eta} \sin\left[\eta\theta - \eta\left(\theta_k - \frac{2\pi}{3}\right)\right], \tag{4.118}$$

$$F_{k\eta}^{c}(\theta, t) = \frac{2N_c i_c(t)}{\pi\eta} \sin\left[\eta\theta - \eta\left(\theta_k - \frac{4\pi}{3}\right)\right]. \tag{4.119}$$

Now, by taking into account that η is a multiple of three and that for a balanced load

$$i_a(t) + i_b(t) + i_c(t) = 0, \tag{4.120}$$

we conclude that

$$F_{k\eta}^{a}(\theta, t) + F_{k\eta}^{b}(\theta, t) + F_{k\eta}^{c}(\theta, t) = 0. \tag{4.121}$$

The latter implies that there are no multiple of three spatial harmonics in the mmf of the three-phase winding in the case of balanced load. The fifth and seventh harmonics can be suppressed by using fractional-pitch winding design. Indeed, it can be shown that for η-th spatial harmonics equation (4.107) has the form

$$F_{\eta m} = \frac{2Nk_{w\eta}}{\pi\eta} I_m, \tag{4.122}$$

where

$$k_{w\eta} = k_{p\eta} k_{d\eta} \tag{4.123}$$

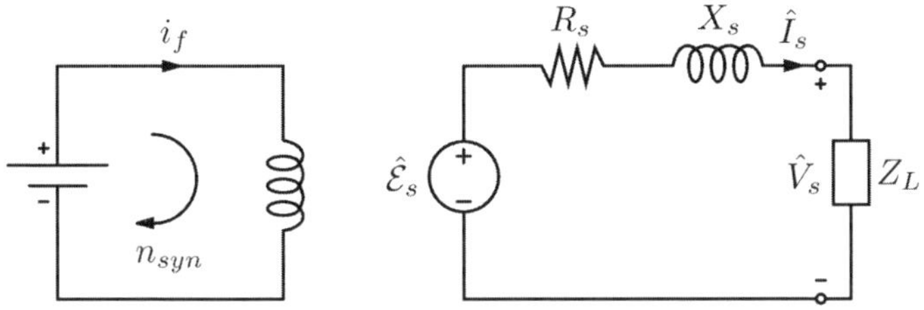

Fig. 4.11

and

$$k_{p\eta} = \sin\frac{\eta\beta\pi}{2}. \tag{4.124}$$

For the fractional pitch $\beta = 5/6$, we find

$$k_{p5} = \sin\frac{25\pi}{12} = \sin\frac{\pi}{12} \approx 0.26, \tag{4.125}$$

$$k_{p7} = \sin\frac{35\pi}{12} = \sin\frac{\pi}{12} \approx 0.26, \tag{4.126}$$

while $k_p = k_{p1} = \sin\frac{5\pi}{12} \approx 0.97$. Thus, the fifth and seventh spatial harmonics are equally suppressed by the small values of k_{p5} and k_{p7} in comparison with k_{p1}. Additional suppression occurs due to the appearance of η in the denominator of formula (4.122).

Next, we shall discuss the equivalent circuit of the synchronous generator in the case of balanced load. By using per-phase analysis, this equivalent circuit can be represented as shown in Figure 4.11, where R_s stands for the phase resistance of the stator winding. This equivalent circuit is convenient for the analysis of terminal voltage of a synchronous generator supplying an isolated system characterized by load impedance Z_L.

From the equivalent circuit shown in Figure 4.11, we find

$$\hat{\mathcal{E}}_s = R_s\hat{I}_s + jX_s\hat{I}_s + \hat{V}_s, \tag{4.127}$$

which leads to

$$\hat{V}_s = \hat{\mathcal{E}}_s - R_s\hat{I}_s - jX_s\hat{I}_s. \tag{4.128}$$

Usually,

$$R_s \ll X_s \tag{4.129}$$

and

$$X_s \simeq X_s^{(m)}. \tag{4.130}$$

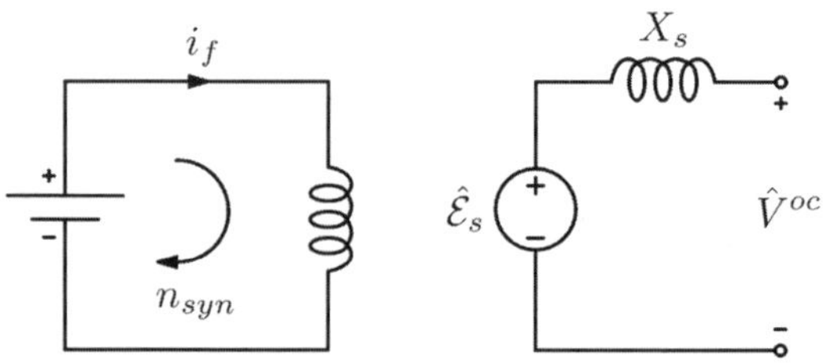

Fig. 4.12

Consequently,

$$\hat{V}_s \approx \hat{\mathcal{E}}_s - jX_s^{(m)}\hat{I}_s. \tag{4.131}$$

It follows from the last formula that the smaller $X_s^{(m)}$, the less the terminal voltage $\hat{V}_s$ is sensitive to the variation of load current $\hat{I}_s$. The smallness of $X_s^{(m)}$ can be achieved according to formula (4.109) by increasing the length δ of the air gap. The case of a synchronous generator supplying an isolated load is not typical for utility power systems where usually a large number of synchronous generators are connected to the same power network. In the latter case, the terminal voltage of a single generator and its frequency are by and large fixed by the presence of other generators in the network. This situation is usually described as a synchronous generator connected to an *infinite bus*, and it is discussed in the next section. Nevertheless, the smallness of $X_s^{(m)}$ that can be achieved for large δ is still very important from the point of view of static stability of the synchronous generator. This issue is also discussed in the next section.

We shall conclude this section by discussing how the synchronous reactance X_s can be determined experimentally. This is usually done by performing two tests, the open-circuit test and short-circuit test. Remarkably, these tests can be performed without connecting a synchronous generator to an outside power network but rather only using the available dc excitation of its rotor winding. The schematics of the open-circuit test are presented in Figure 4.12. It is apparent from this figure that by measuring the peak value of open-circuit voltage V_m^{oc} we find the peak value $\mathcal{E}_{sm}$ of the internal voltage (emf) for a fixed level of rotor winding excitation:

$$\mathcal{E}_{sm} = V_m^{oc}. \tag{4.132}$$

The schematics of the short-circuit test are presented in Figure 4.13. By measuring the peak value of the short-circuit current I_m^{sc} for the same level of rotor winding excitation, according to formula (4.132) and Figure 4.13,

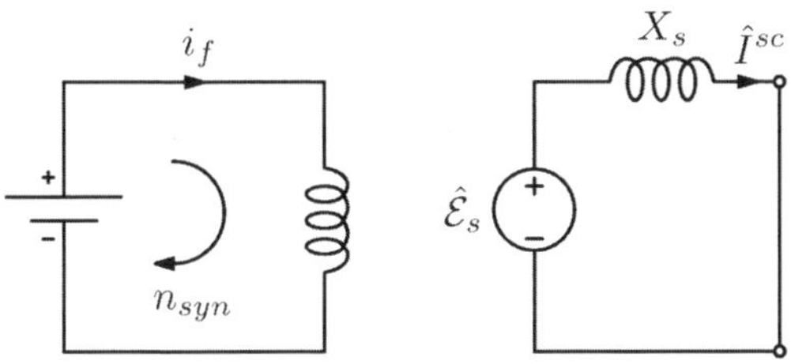

Fig. 4.13

the synchronous reactance X_s can be found by using the formula

$$X_s = \frac{V_m^{oc}}{I_m^{sc}}. \tag{4.133}$$

It is evident that by using the last formula we neglect the phase resistance R_s of the stator winding, which is usually very small in comparison with X_s.

4.4 Two-Reactance Theory for Salient Pole Synchronous Generators; Power of Synchronous Generators

In the previous section, we have discussed the design of stator windings and derived the formula (4.109) for their reactance. This derivation has been carried out for cylindrical rotor machines whose air gaps are mostly uniform. In the case of salient pole machines, the air gaps are strongly nonuniform; they assume the smallest values δ_d along the direct axes and the largest values δ_q along the quadrature axes. For this reason, it is not clear what value of δ can be used in the formula (4.109) or even if this formula is still applicable to salient pole synchronous machines. It is clear that the value of δ depends on the spatial orientation of armature reaction magnetic field with respect to the salient pole rotor. In other words, it is clear that different values of δ must be used when magnetic field lines of armature reaction field are mostly aligned along the direct axis than when these field lines are aligned along the quadrature axis. Here, an interesting physical phenomenon occurs. *It turns out that the spatial orientation of armature reaction magnetic field with respect to a salient pole rotor is controlled by the phase shift in time between internal voltage (induced emf) $\hat{\mathcal{E}}_s$ and stator current $\hat{I}_s$.* To illustrate this, consider two particular cases when a) $\hat{\mathcal{E}}_s$ and $\hat{I}_s$ are in phase and b) $\hat{I}_s$ lags behind $\hat{\mathcal{E}}_s$ by $\frac{\pi}{2}$. For the sake of simplicity of our reasoning, we consider a two-pole machine and represent the stator

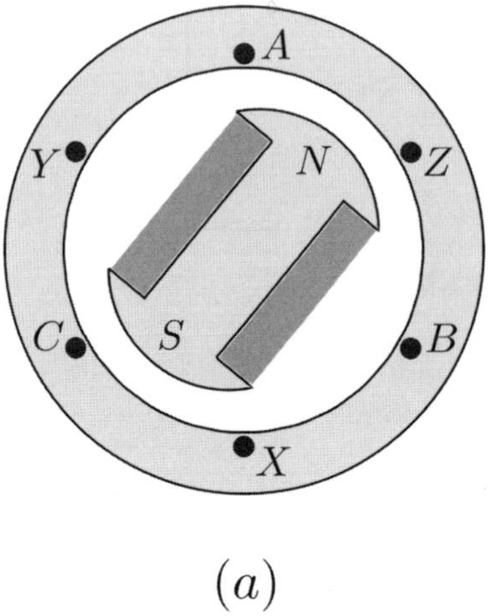

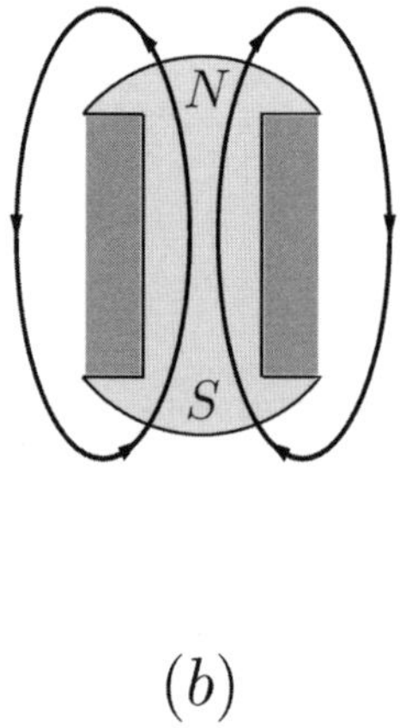

Fig. 4.14

phase windings by single coils AX, BY and CZ shifted in space by 120° (see Figure 4.14a). Magnetic field of the rotor is shown in Figure 4.14b.

We start with the case a) which is illustrated by the phasor diagram in Figure 4.15a. We shall also consider a special instant of time when the position of the rotor is such that its direct axis is in the plane of the AX coil (see Figure 4.15b). We shall use this instant of time as the initial time instant $t = 0$. It is clear from the structure of the magnetic field lines of the rotor shown in Figure 4.14b that at this time instant the flux linkage of the coil AX is equal to zero. Since flux linkage is sinusoidal in time, its derivative achieves its maximum at the time instant when the flux linkage is equal to zero. Consequently, the internal voltage induced in the coil AX assumes its maximum value at $t = 0$. Since the current in this coil is in phase with the induced internal voltage (see the phasor diagram in Figure 4.15a), this current also assumes its maximum positive value at $t = 0$. This means that this current can be written as

$$i_a(t) = I_m \cos \omega t. \tag{4.134}$$

Since we deal with the case of balanced load, currents in coils BY and CZ

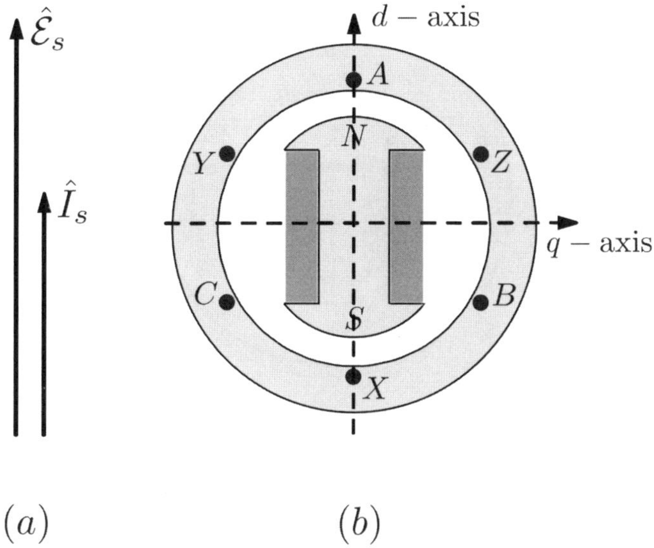

Fig. 4.15

are given by the formulas

$$i_b(t) = I_m \cos\left(\omega t - \frac{2\pi}{3}\right), \tag{4.135}$$

$$i_c(t) = I_m \cos\left(\omega t - \frac{4\pi}{3}\right). \tag{4.136}$$

At time instant $t = 0$, from the last three formulas we find

$$i_a(0) = I_m > 0, \qquad i_b(0) = i_c(0) = -\frac{I_m}{2} < 0. \tag{4.137}$$

These current signs are marked in Figure 4.16, and it is clear that due to the symmetry of the current distribution the lines of magnetic field created by these currents are directed along the quadrature axis. This fact is established for the time instant $t = 0$, that is, for a very specific position of the rotor. *However, since the rotor and the armature reaction magnetic field rotate in synchronism (i.e., with the same speed), it can be concluded that the lines of the armature reaction magnetic field are directed along the quadrature axis at any instant of time, i.e., for any position of the rotor.* This suggests that δ_q can be used instead of δ in formula (4.109) and this leads to the reactance

$$X_{sq}^{(m)} = \frac{12\mu_0 b\ell}{\delta_q} f N^2 k_w^2, \tag{4.138}$$

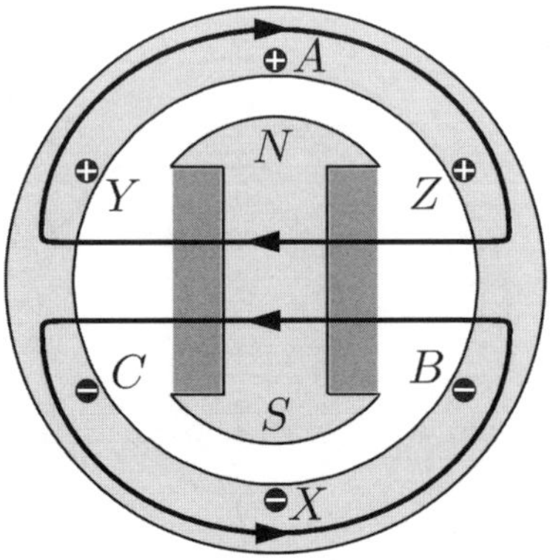

Fig. 4.16

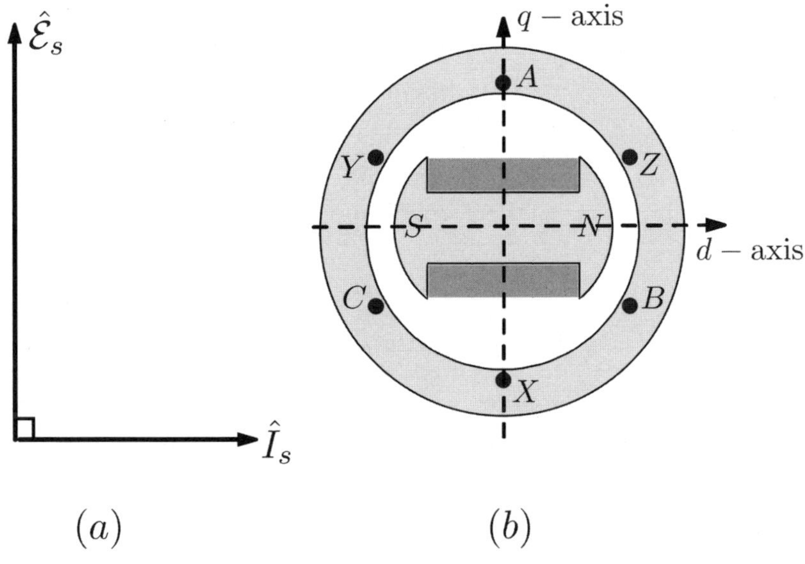

Fig. 4.17

which can be called quadrature axis main reactance.

Now, we turn to the discussion of the case b) when the stator current in each phase winding lags behind induced internal voltage by $\frac{\pi}{2}$ as illustrated by the phasor diagram shown in Figure 4.17a. We shall also consider a special instant of time when the position of the rotor is such that its direct axis is normal to the plane of the AX coil (see Figure 4.17b). We shall

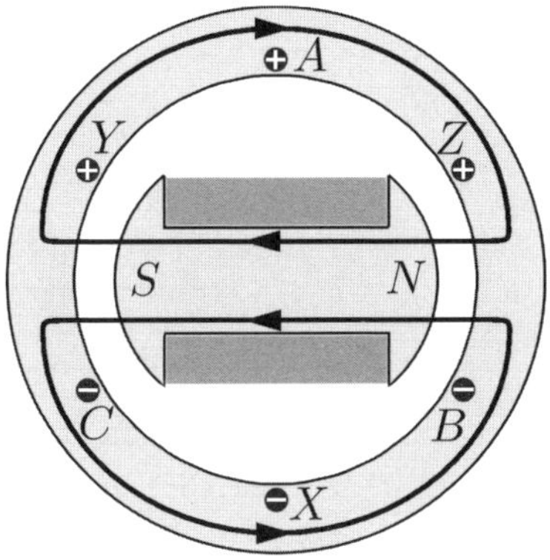

Fig. 4.18

use this instant of time as the time instant $t = 0$. It is clear from the structure of magnetic field lines of the rotor (see Figure 4.14b) that at this time instant the flux linkage of the coil AX achieves its maximum value. This implies that the time derivative of this flux linkage is equal to zero. The latter indicates that the internal voltage induced by the rotor magnetic field in the coil AX is equal to zero at $t = 0$. Since the current in this coil lags behind the induced internal voltage by $\frac{\pi}{2}$ (see the phasor diagram in Figure 4.17a), we conclude that this current achieves its maximum positive value at $t = 0$. This means that this current can be written as

$$i_a(t) = I_m \cos \omega t. \tag{4.139}$$

Since we deal with the case of balanced load, currents in coils BY and CZ are described by the formulas

$$i_b(t) = I_m \cos\left(\omega t - \frac{2\pi}{3}\right), \tag{4.140}$$

$$i_c(t) = I_m \cos\left(\omega t - \frac{4\pi}{3}\right). \tag{4.141}$$

From the last three formulas we find that at time instant $t = 0$ we have

$$i_a(0) = I_m > 0, \qquad i_b(0) = i_c(0) = -\frac{I_m}{2} < 0. \tag{4.142}$$

These current signs are marked in Figure 4.18, and it is clear that due to the symmetry of the current distribution the lines of magnetic field created by these currents are along the direct axis of the rotor. This fact is established for the time instant $t = 0$, that is, for a very specific position of the rotor.

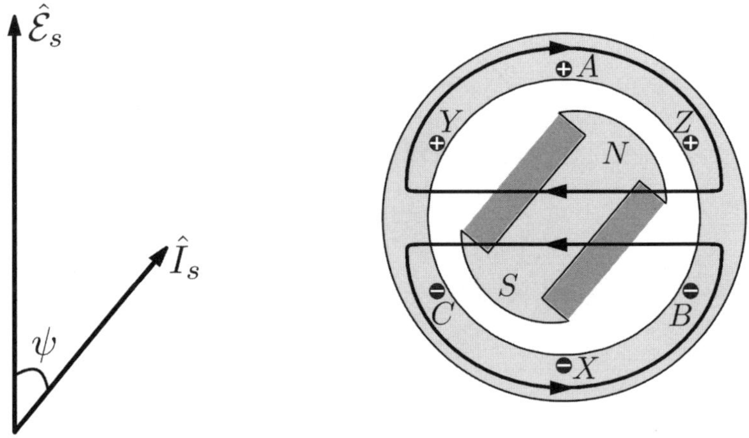

(a) (b)

Fig. 4.19

However, since the rotor and the armature reaction magnetic field rotate in synchronism (i.e., with the same speed), it can be concluded that the lines of the armature reaction magnetic field are along the direct axis at any instant of time, i.e., for any position of the rotor. This suggests that δ_d can be used instead of δ in formula (4.109) and this leads to the reactance

$$X_{sd}^{(m)} = \frac{12\mu_0 b\ell}{\delta_d} f N^2 k_w^2, \tag{4.143}$$

which can be termed the direct axis main reactance.

Now, consider a general case when the phase shift in time between the induced internal voltage and stator current is equal to some angle $0 \leq \psi \leq \frac{\pi}{2}$ (see Figure 4.19a). It is clear that in this general case the positive maximum value of current $i_a(t)$ will be achieved at some position of the rotor (see Figure 4.19b) which is intermediate between the rotor positions shown in Figures 4.16 and 4.18. It is also clear that this rotor position depends on the phase shift ψ between $\hat{\mathcal{E}}_s$ and $\hat{I}_s$. Thus, the conclusion can be reached that the orientation of field lines of the armature reaction magnetic field with respect to the salient pole rotor is controlled by ψ, and different synchronous reactances $X_{s\psi}^{(m)}$ should be used for different phase shifts ψ. The phase shift ψ does not remain constant but changes with load variations. This clearly suggests that for salient pole machines the stator windings

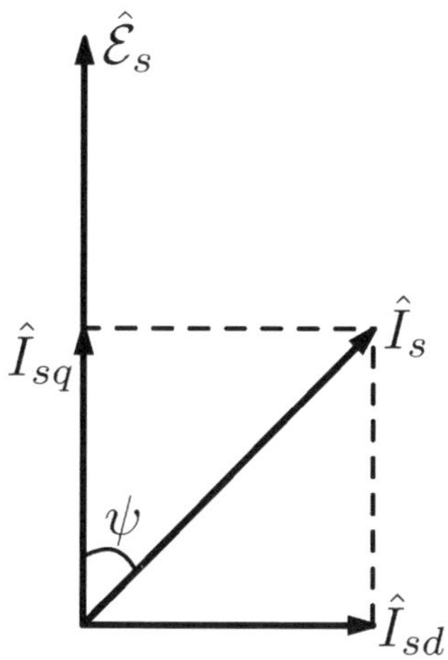

Fig. 4.20

cannot be characterized by one reactance as in the case of cylindrical rotor machines. Luckily, the stator windings of salient pole machines can be characterized by two reactances: direct axis and quadrature axis reactances. This characterization is based on the superposition principle and the following decomposition of stator current $\hat{I}_s$:

$$\hat{I}_s = \hat{I}_{sq} + \hat{I}_{sd}, \tag{4.144}$$

which is illustrated by the phasor diagram shown in Figure 4.20. It is clear from the previous discussion that the current $\hat{I}_{sq}$ will produce the magnetic field whose lines are directed along the quadrature axis, while the current $\hat{I}_{sd}$ will produce the magnetic field whose lines are along the direct axis. The actual magnetic field is the superposition of these two magnetic fields. This implies that the flux linkage of the stator windings $\hat{\psi}_s$ can be represented as the sum of two flux linkages

$$\hat{\psi}_s = \hat{\psi}_{sq} + \hat{\psi}_{sd}, \tag{4.145}$$

which are due to the magnetic fields created by currents $\hat{I}_{sq}$ and $\hat{I}_{sd}$, respectively.

From the last formula we find that the KVL equation for a stator phase winding can be written in the form

$$\hat{\mathcal{E}}_s = \hat{V}_{sq} + \hat{V}_{sd} + \hat{V}_s, \tag{4.146}$$

where $\hat{V}_{sq}$ is the voltage induced due to the time variations of flux linkage created by the component of the armature reaction magnetic field whose

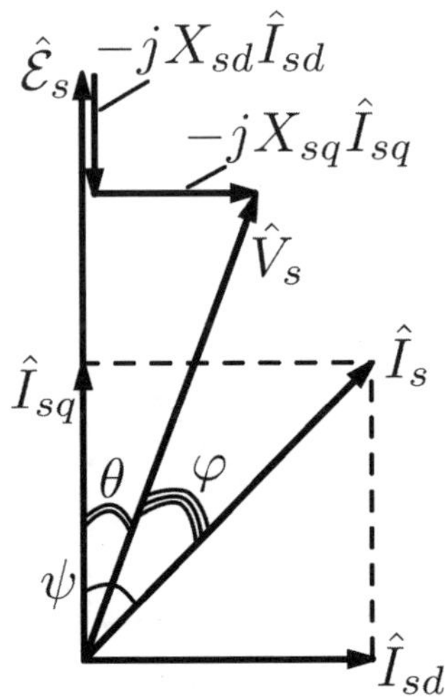

Fig. 4.21

lines are directed along the quadrature axis, while $\hat{V}_{sd}$ is the voltage induced due to the time variations of flux linkage of the stator winding created by the component of the armature reaction magnetic field whose lines are along the direct axis. Consequently, these two voltages can be written as

$$\hat{V}_{sq} = jX_{sq}\hat{I}_{sq}, \tag{4.147}$$

$$\hat{V}_{sd} = jX_{sd}\hat{I}_{sd}, \tag{4.148}$$

where reactances X_{sq} and X_{sd} include not only the main reactances (see formulas (4.138) and (4.143)) but leakage reactances as well.

By substituting formulas (4.147) and (4.148) into equation (4.146), we end up with the following expression for the terminal voltage $\hat{V}_s$:

$$\boxed{\hat{V}_s = \hat{\mathcal{E}}_s - jX_{sd}\hat{I}_{sd} - jX_{sq}\hat{I}_{sq}.} \tag{4.149}$$

The last formula is illustrated by the phasor diagram shown in Figure 4.21. This phasor diagram as well as equation (4.149) is the essence of the two-reactance theory that was first developed by A. Blondel.

It must be remarked that the two-reactance theory is only applicable under steady-state conditions and balanced loads. In the case of transients, this theory is not applicable and one is resigned to use coupled circuit equations which account for coupling between stator phase windings and rotor dc and damper windings. These are quite complicated equations with variable-in-time coefficients. The latter is the consequence of the fact that inductances and mutual inductances of windings depend on the temporal position of the salient pole rotor. It turns out that by using special Park transformations associated with rotating direct and quadrature axes of the rotor these coupled equations with variable coefficients can be reduced to differential equations with constant coefficients. The detailed discussion of the Park theory is beyond the scope of this book.

Now, we shall proceed to the discussion of electric power generated by the synchronous generator. In the case of balanced load, this power can be written as

$$P = \frac{3}{2} V_{sm} I_{sm} \cos\varphi, \tag{4.150}$$

and if we use rms values V_s and I_s instead of peak values V_{sm} and I_{sm}, then

$$P = 3V_s I_s \cos\varphi. \tag{4.151}$$

It is clear from Figure 4.21 that

$$\cos\varphi = \cos(\psi - \theta) = \cos\psi\cos\theta + \sin\psi\sin\theta. \tag{4.152}$$

By substituting the last formula into equation (4.151), we arrive at

$$P = 3V_s I_s \cos\psi\cos\theta + 3V_s I_s \sin\psi\sin\theta. \tag{4.153}$$

It is apparent from Figure 4.21 that

$$I_s \cos\psi = I_{sq}, \tag{4.154}$$

$$I_s \sin\psi = I_{sd}. \tag{4.155}$$

By substituting the last two equations into formula (4.153), we find

$$P = 3V_s I_{sq} \cos\theta + 3V_s I_{sd} \sin\theta. \tag{4.156}$$

From the phasor diagram in Figure 4.21 we derive

$$\mathcal{E}_s = X_{sd} I_{sd} + V_s \cos\theta, \tag{4.157}$$

which leads to

$$I_{sd} = \frac{\mathcal{E}_s - V_s \cos\theta}{X_{sd}}. \tag{4.158}$$

Similarly, from the same phasor diagram we find

$$X_{sq} I_{sq} = V_s \sin\theta, \tag{4.159}$$

which results in

$$I_{sq} = \frac{V_s \sin\theta}{X_{sq}}. \tag{4.160}$$

By inserting formulas (4.158) and (4.160) into equation (4.156), we obtain

$$P = \frac{3V_s^2}{X_{sq}} \sin\theta\cos\theta + \frac{3V_s(\mathcal{E}_s - V_s\cos\theta)}{X_{sd}} \sin\theta, \tag{4.161}$$

which leads to

$$P = \frac{3\mathcal{E}_s V_s}{X_{sd}} \sin\theta + 3V_s^2 \left(\frac{1}{X_{sq}} - \frac{1}{X_{sd}}\right) \sin\theta\cos\theta. \tag{4.162}$$

Finally, taking into account that

$$\sin\theta\cos\theta = \frac{1}{2}\sin 2\theta, \tag{4.163}$$

we arrive at

$$\boxed{P = \frac{3\mathcal{E}_s V_s}{X_{sd}} \sin\theta + \frac{3V_s^2}{2} \left(\frac{1}{X_{sq}} - \frac{1}{X_{sd}}\right) \sin 2\theta.} \tag{4.164}$$

The last formula has two distinct terms. The first term contains the internal voltage $\mathcal{E}_s$ induced by the rotating magnetic field of the rotor. The second term does not contain $\mathcal{E}_s$ and appears due to the saliency of the rotor, that is, due to the difference in magnetic reluctance along the direct and quadrature axes. For this reason, this second term is sometimes called *reluctance power.* Usually, this reluctance power has a value of about 20% to 25% of the total (rated) power of the synchronous machine. This second ("reluctance") term in the expression of the power is quite interesting from the physical point of view because it reveals that the electric power can be generated without dc excitation of the rotor but solely due to its saliency. However, due to the relative smallness of this "reluctance power," this possibility of power generation without rotor excitation is practically not utilized in conventional power systems.

In the case of cylindrical rotor turbine generators we have

$$X_{sd} = X_{sq} = X_s \tag{4.165}$$

and formula (4.164) is reduced to

$$\boxed{P = \frac{3\mathcal{E}_s V_s}{X_s} \sin\theta.} \tag{4.166}$$

The last formula can also be written as

$$\boxed{P(\theta) = P_m \sin\theta,} \tag{4.167}$$

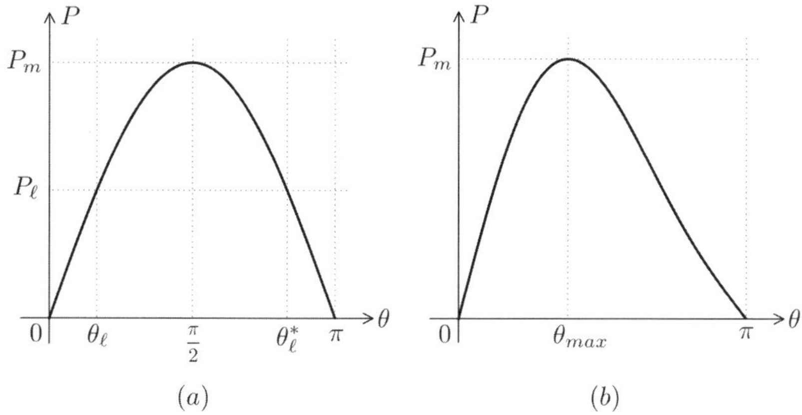

Fig. 4.22

where

$$\boxed{P_m = \frac{3\mathcal{E}_s V_s}{X_s}}. \tag{4.168}$$

A plot of $P(\theta)$ defined by formula (4.167) is presented in Figure 4.22a. It is apparent from this figure that a supplied load power P_ℓ is delivered at some angle θ_ℓ which is called "load" or "power" angle. It is also clear from the same figure that $P(\theta)$ is equal to P_ℓ at the angle θ_ℓ^*. However, the performance of the synchronous generator is *unstable* at this angle. This is discussed in the last section of the next chapter, where it is demonstrated that θ_ℓ^* corresponds to the *saddle point* of synchronous machine dynamics.

It is interesting to discuss the geometric (or mechanical) meaning of load angle θ. According to the phasor diagram shown in Figure 4.21, θ is the phase shift in time between $\hat{\mathcal{E}}_s$ and $\hat{V}_s$. $\hat{\mathcal{E}}_s$ is induced by the rotating magnetic field of the rotor, while $\hat{V}_s$ is induced by the total rotating field in the gap of the synchronous generator (as before, we neglect here the small resistance of the stator winding). This total field is the superposition of the rotating magnetic field of the rotor and the rotating armature reaction magnetic field. As a result of this superposition, the axes of these two fields do not coincide. In other words, these two magnetic fields move in synchronism, however the total magnetic field lags behind the rotor magnetic field by some angle. It is clear that this angle is equal to θ, that is, to the time phase shift between $\hat{\mathcal{E}}_s$ and $\hat{V}_s$ induced by these fields. The axis of the rotor magnetic field coincides with the rotor axis. For this reason, it can be stated that the axis of the total rotating magnetic field lags behind the rotor axis by angle θ.

Formula (4.167) has been derived under the tacit assumption of synchronous steady-state balanced load operation of the synchronous generator. For this reason, P_m can be interpreted as the maximum of power that can be generated by the synchronous generator without losing synchronism. In other words, if a synchronous generator is loaded above P_m, the synchronism between the mechanical speed of the rotor and the armature reaction magnetic field will be broken. As a result, large eddy currents may be induced in the conducting solid rotor, which may result in large heat dissipation and eventual damage of the synchronous generator. Thus, P_m can be considered as a limit of static stability. The larger this limit, the better the quality of the synchronous machine. It is clear from formula (4.168) that for fixed $\mathcal{E}_s$ and V_s, larger P_m can be achieved by making X_s smaller. Since $X_s \approx X_s^{(m)}$ and $X_s^{(m)}$ according to formula (4.109) can be reduced by increasing the air gap length δ, it can be concluded that the larger the air gap, the better the quality of the synchronous generator. This explains why synchronous generators, especially two-pole turbine generators, have large air gaps which may approach 15 cm.

A plot of P given by formula (4.164) is shown in Figure 4.22b. It is apparent that this plot is somewhat similar in the qualitative sense to the plot shown in Figure 4.22a for a cylindrical rotor machine. The main difference is that due to the second (saliency) term in equation (4.164) the maximum power P_m is achieved at the load angle θ_{max} which is less than $\frac{\pi}{2}$.

In the conclusion of this chapter, we shall briefly discuss the performance of a synchronous generator connected to *an infinite bus*. The latter means that a synchronous generator is connected to a power network with a large number of other synchronous generators whose overall delivered power is much above the power of the single generator. Under these conditions, the terminal voltage of a single generator and its frequency are by and large fixed by the presence of other generators in the power network:

$$V_s = const, \quad f = const. \tag{4.169}$$

We shall also consider the situation when the active power delivered by a generator is also fixed,

$$P = const, \tag{4.170}$$

and we are interested what can be achieved under these circumstances by varying the rotor excitation current i_f of the generator. We shall limit our discussion to the case of cylindrical rotor machines, although it can be extended to the case of salient pole rotor machines as well.

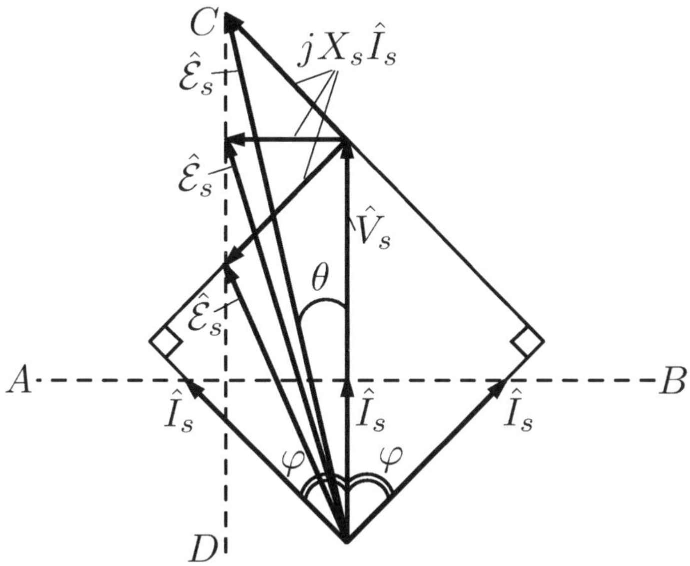

Fig. 4.23

From constraints (4.169) and (4.170) and formulas (4.151) and (4.166) immediately follows that during the variations of rotor excitation current i_f the following two quantities are conserved:

$$I_s \cos\phi = const, \tag{4.171}$$

$$\mathcal{E}_s \sin\theta = const. \tag{4.172}$$

These two conservation equations can be geometrically interpreted as follows: the end of the phasor $\hat{I}_s$ moves along the line AB, while the end of the phasor $\hat{\mathcal{E}}_s$ moves along the line CD as the excitation current i_f is varied (see Figure 4.23). In this figure, the phasor diagrams of the equation

$$\hat{\mathcal{E}}_s = \hat{V}_s + jX_s\hat{I}_s \tag{4.173}$$

are presented for three distinct cases: a) when $\varphi = 0$, b) $\varphi > 0$, i.e., $\hat{I}_s$ lags behind $\hat{V}_s$ and c) $\varphi < 0$, i.e., $\hat{I}_s$ leads $\hat{V}_s$. These phasor diagrams reveal that I_s achieves its minimum value for $\varphi = 0$ and $\cos\varphi = 1$. I_s is increased as $\varphi > 0$ is increased and this results in increase of $\mathcal{E}_s$, which can only be achieved by increasing i_f. Furthermore, I_s is increased as $\varphi < 0$ and the magnitude of φ is increased. This results in decrease of $\mathcal{E}_s$, which can

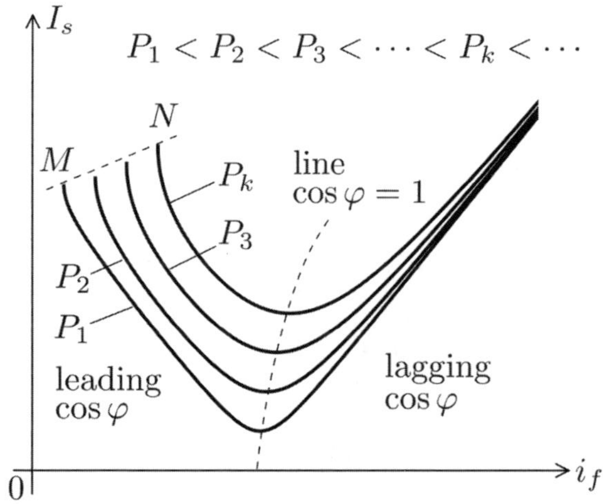

Fig. 4.24

only be achieved by decreasing i_f. This discussion implies that the relation between I_s and i_f is represented by a V-type curve k shown in Figure 4.24 for some fixed value of active power P_k. As the value of the power is changed, this results in the change of the positions of lines AB and CD in Figure 4.23; this yields different V-type curves. The line that goes through the minima of the V-curves is the line of $\cos\varphi = 1$. This line divides the I_s-i_f plane into two parts with leading power factors and lagging power factors, respectively. A decrease in i_f results in a decrease of $\mathcal{E}_s$ and P_m (see formula (4.168)), which eventually results in loss of synchronism when $P_m < P_k$. This occurs along the line MN.

It is immediately evident from the V-curves shown in Figure 4.24 that by varying the excitation current i_f the power factor of the synchronous generator and its reactive power can be controlled. The achieving of leading power factor is especially intriguing. However, this possibility is limited only to very small angles φ because it is associated with reduction of $\mathcal{E}_s$ and P_m, which is detrimental to the stability of the synchronous generator. Nevertheless, this possibility to generate reactive power with leading power factor can be realized if a synchronous machine is used as a motor with practically no mechanical load. It can be shown that for a synchronous motor the regions of lagging $\cos\varphi$ and leading $\cos\varphi$ are switched. For this reason, an appreciable reactive power with leading power factor can be generated by increasing excitation current i_f. In this case, a synchronous machine is operated as a *synchronous condenser* which can be used instead of capacitor banks for adjustment of power factor in power systems. The main advantage of such condensers over capacitor banks is that their reactive power can be continuously controlled through the variation of excitation current i_f, that is, without any switching as required in the case of capacitor banks.

Chapter 5

Power Flow Analysis and Stability of Power Systems

5.1 Power Flow Analysis

In this section, the power flow analysis in power systems is discussed. This analysis provides *voltage profiles* in power systems. Namely, it leads to the determination of voltage at load terminals at various operational conditions. The knowledge of voltage profiles is very important because the delivering of electric power at voltages with more or less constant peak (or rms) values despite continuously changing loads is one of the main challenges and obligations in operating power systems. The power flow analysis is also very important for *contingency studies* of cases when, for instance, one or more generators may go off-line due to some accidents and this may affect voltage profiles and also may result in overloading of the generating units remaining in operation. Finally, the power flow analysis is very useful for the *planning of future developments and extensions* of existing power systems. For the reasons presented above, the power flow analysis is one of the most common computer calculations routinely performed in operation of power systems.

As it is usually done, the power flow analysis is discussed below on the transmission level when it is assumed that different generators and bulk loads are interconnected by transmission lines. However, the technique of power flow analysis presented here is applicable to different levels (layers) of power systems as well. In particular, it can be used for the analysis of "subtransmission" systems used for distribution and further dispersal of electric power to actual loads.

Conceptually, power flow analysis is the nodal analysis of electric networks formulated in terms of nodal electric powers. For this reason, the power flow equations are *strongly nonlinear* even if the underlying nodal potential equations are linear. According to the terminology used in power systems, nodes of a power network are called "buses." We shall be concerned with three-phase balanced load operation of power systems when per-phase analysis can be used. In per-phase analysis, three-phase loads as well as three-phase generators can be represented by single buses. Each bus can be fully characterized by four quantities, i.e., active power, reactive power, peak value of the voltage and its initial phase:

$$P_k, Q_k, V_{mk}, \varphi_k, \quad (k = 1, 2, ..., N), \tag{5.1}$$

where N is the total number of buses. It is tacitly assumed that voltages V_{mk} are specified with respect to the common neutral which is regarded as the *reference node* (bus).

It turns out that all buses can be divided into two groups:

(1) Generator buses where

$$P_k \text{ and } V_{mk}, \quad (k = 1, 2, ..., G), \tag{5.2}$$

are specified. This specification is consistent with the interpretation of synchronous generators as (P, V) sources as discussed in the previous chapter (see section 4.1). In a typical power system, generator buses are about 15% of the total number of buses.

(2) Load buses where

$$P_k \text{ and } Q_k, \quad (k = G+1, G+2, ..., N), \tag{5.3}$$

are specified.

Besides generator and load buses, it is very convenient in practice to introduce the so-called "*slack bus.*" This is a special generator bus used to enforce the balance between demanded power by all loads and the overall power delivered by all synchronous generators. In our subsequent discussion, we shall ignore this technical detail and for the sake of conceptual simplicity deal with generator and load buses characterized by formulas (5.2) and (5.3), respectively.

Now, the problem of power flow analysis can be stated as follows: given the quantities specified in formulas (5.2) and (5.3), find

$$Q_k \text{ and } \varphi_k, \quad (k = 1, 2, ..., G), \tag{5.4}$$

for generator buses and

$$V_{mk} \text{ and } \varphi_k, \quad (k = G+1, G+2, ..., N), \tag{5.5}$$

for load buses.

To solve the stated problem, power flow equations will be derived. The derivation is based on the nodal analysis. The first step of the derivation is to write KCL for each bus:

$$\hat{I}_k = \sum_{n=1}^{N} \hat{I}_{kn}, \quad (k = 1, 2, ..., N), \tag{5.6}$$

where $\hat{I}_k$ is the phasor of the current from the bus number k to the neutral, while $\hat{I}_{kn}$ is the phasor of the current between buses number k and n.

According to the nodal analysis, currents $\hat{I}_{kn}$ can be expressed in terms of branch admittances and bus (nodal) voltages $\hat{V}_n$. This leads to the following formulas:

$$\hat{I}_k = \sum_{n=1}^{N} Y_{kn}\hat{V}_n, \quad (k = 1, 2, ..., N). \tag{5.7}$$

It is convenient to represent these formulas in the matrix form

$$\begin{pmatrix} \hat{I}_1 \\ \hat{I}_2 \\ \vdots \\ \hat{I}_N \end{pmatrix} = \begin{pmatrix} Y_{11} & Y_{12} & \cdots & Y_{1N} \\ Y_{21} & Y_{22} & \cdots & Y_{2N} \\ \vdots & \vdots & \ddots & \vdots \\ Y_{N1} & Y_{N2} & \cdots & Y_{NN} \end{pmatrix} \begin{pmatrix} \hat{V}_1 \\ \hat{V}_2 \\ \vdots \\ \hat{V}_N \end{pmatrix}. \tag{5.8}$$

The following expressions are valid for the elements of the admittance matrix in (5.8):

$$Y_{kk} = \sum_i Y_i, \tag{5.9}$$

where the sum is taken over all admittances connected to the bus number k, while

$$Y_{kn} = -\sum_i Y_i, \tag{5.10}$$

where the sum is taken over all admittances connected between buses number k and n. These expressions are usually derived in the discussion of nodal analysis (see [35]).

Since not all buses are interconnected with one another, this implies in accordance with formula (5.10) that many off-diagonal elements of the admittance matrix in equation (5.8) are equal to zero. In other words, this admittance matrix is usually quite sparse.

The further derivation of power flow equations proceeds as follows. The complex power $\hat{S}_k$ at each bus can be written in the form

$$\hat{S}_k = \frac{1}{2}\hat{V}_k\hat{I}_k^*, \quad (k = 1, 2, ..., N). \tag{5.11}$$

According to formulas (5.7), we find

$$\hat{I}_k^* = \sum_{n=1}^{N} Y_{kn}^* \hat{V}_n^*, \quad (k = 1, 2, ..., N). \tag{5.12}$$

By substituting the last formulas into equations (5.11), we get

$$\hat{S}_k = \frac{1}{2}\hat{V}_k \sum_{n=1}^{N} Y_{kn}^* \hat{V}_n^*, \quad (k = 1, 2, ..., N), \tag{5.13}$$

which is equivalent to

$$\hat{S}_k = \frac{1}{2} \sum_{n=1}^{N} \hat{V}_k \hat{V}_n^* Y_{kn}^*, \quad (k = 1, 2, ..., N). \tag{5.14}$$

Equations (5.14) are written for complex quantities. We shall next transform them into equations for real quantities P_k, Q_k, V_{mk} and φ_k. To this end, we shall first recall that

$$\hat{V}_k = V_{mk} e^{j\varphi_k}, \tag{5.15}$$

$$\hat{V}_n^* = V_{mn} e^{-j\varphi_n}. \tag{5.16}$$

Consequently,

$$\hat{V}_k \hat{V}_n^* = V_{mk} V_{mn} e^{j(\varphi_k - \varphi_n)}, \tag{5.17}$$

which is tantamount to

$$\hat{V}_k \hat{V}_n^* = V_{mk} V_{mn} \left[\cos(\varphi_k - \varphi_n) + j \sin(\varphi_k - \varphi_n)\right]. \tag{5.18}$$

Furthermore,

$$Y_{kn} = G_{kn} + jB_{kn}, \tag{5.19}$$

where G_{kn} are conductances, while B_{kn} are susceptances.

Formula (5.19) implies that

$$Y_{kn}^* = G_{kn} - jB_{kn}. \tag{5.20}$$

Finally,

$$\hat{S}_k = P_k + jQ_k, \quad (k = 1, 2, ..., N). \tag{5.21}$$

Now, by substituting formulas (5.18), (5.20) and (5.21) into equations (5.14), we obtain

$$P_k + jQ_k = \frac{1}{2} \sum_{n=1}^{N} V_{mk} V_{mn} \left[\cos(\varphi_k - \varphi_n) + j \sin(\varphi_k - \varphi_n)\right] (G_{kn} - jB_{kn}),$$

$$(k = 1, 2, ..., N). \tag{5.22}$$

After simple algebraic transformation, we find

$$P_k + jQ_k = \frac{1}{2}\sum_{n=1}^{N} V_{mk}V_{mn}\left[G_{kn}\cos(\varphi_k - \varphi_n) + B_{kn}\sin(\varphi_k - \varphi_n)\right]$$
$$+ \frac{j}{2}\sum_{n=1}^{N} V_{mk}V_{mn}\left[G_{kn}\sin(\varphi_k - \varphi_n) - B_{kn}\cos(\varphi_k - \varphi_n)\right],$$
$$(k = 1, 2, ..., N). \tag{5.23}$$

Now, by separating real and imaginary parts in the last equations, we derive

$$P_k = \frac{1}{2}\sum_{n=1}^{N} V_{mk}V_{mn}\left[G_{kn}\cos(\varphi_k - \varphi_n) + B_{kn}\sin(\varphi_k - \varphi_n)\right], \tag{5.24}$$

$$Q_k = \frac{1}{2}\sum_{n=1}^{N} V_{mk}V_{mn}\left[G_{kn}\sin(\varphi_k - \varphi_n) - B_{kn}\cos(\varphi_k - \varphi_n)\right], \tag{5.25}$$
$$(k = 1, 2, ..., N).$$

Finally, by introducing rms values V_k and V_n instead of peak values V_{mk} and V_{mn}, equations (5.24) and (5.25) can be written as follows:

$$\boxed{P_k = \sum_{n=1}^{N} V_kV_n\left[G_{kn}\cos(\varphi_k - \varphi_n) + B_{kn}\sin(\varphi_k - \varphi_n)\right],} \tag{5.26}$$

$$\boxed{Q_k = \sum_{n=1}^{N} V_kV_n\left[G_{kn}\sin(\varphi_k - \varphi_n) - B_{kn}\cos(\varphi_k - \varphi_n)\right],} \tag{5.27}$$
$$(k = 1, 2, ..., N).$$

This is the final set of power flow equations. The power flow equations can also be written in another (more compact) form. To arrive at this form, we shall use the following expression

$$Y_{kn}^* = |Y_{kn}|e^{-j\gamma_{kn}} \tag{5.28}$$

instead of (5.20).

Then, by substituting formulas (5.17), (5.21) and (5.28) into equations (5.14), we find

$$P_k + jQ_k = \frac{1}{2}\sum_{n=1}^{N} V_{mk}V_{mn}|Y_{kn}|e^{j(\varphi_k - \varphi_n - \gamma_{kn})},$$
$$(k = 1, 2, ..., N). \tag{5.29}$$

By separating real and imaginary parts in the last equations and replacing peak values of voltages by their rms values, we end up with power flow equations in the form

$$P_k = \sum_{n=1}^{N} V_k V_n |Y_{kn}| \cos(\varphi_k - \varphi_n - \gamma_{kn}), \tag{5.30}$$

$$Q_k = \sum_{n=1}^{N} V_k V_n |Y_{kn}| \sin(\varphi_k - \varphi_n - \gamma_{kn}), \tag{5.31}$$

$$(k = 1, 2, ..., N).$$

It is apparent that this is a set of $2N$ nonlinear simultaneous equations. These equations are nonlinear with respect to voltages because of product terms $V_k V_n$ and with respect to initial phases φ_k because of cosine and sine terms. In these equations, $|Y_{kn}|$ and γ_{kn} are regarded as known and they are determined from the connectivity of buses in the power transmission network. In these equations, P_k and V_k for $k = 1, 2, ..., G$ as well as P_k and Q_k for $k = G + 1, G + 2, ..., N$ are also known because the corresponding buses are generator and load buses, respectively. Thus, it can be concluded that the $2N$ nonlinear equations (5.26) and (5.27) have $2N$ unknowns, which are Q_k and φ_k for $k = 1, 2, ..., G$ as well as V_{mk} and φ_k for $k = G+1, G+2, ..., N$. By solving these nonlinear equations numerically, these unknowns can be found. It must be remarked that not all these equations are fully coupled. Namely, the so-called P-equations for real powers P_k at all buses (equations (5.30) for $k = 1, 2, ..., N$) and the so-called Q-equations for reactive powers Q_k at all *load* buses (equations (5.31) for $k = G+1, G+2, ..., G + L = N$) are coupled and constitute $N + L$ nonlinear equations with respect to φ_k ($k = 1, 2, ..., N$) and V_k ($k = G+1, G+2, ..., G+L = N$). As soon as these $N + L$ coupled nonlinear equations are solved, reactive power Q_k at generator buses can be computed by using formulas (5.31) for $k = 1, 2, ..., G$.

One of the most powerful techniques for the solution of these $N + L$ nonlinear algebraic equations is the Newton-Raphson method which is discussed in the next section. Here, however, it is worthwhile to mention some intrinsic difficulties associated with power flow equations (5.30) and (5.31) which stem from their nonlinear nature. First, these real nonlinear equations may not always have solutions. Physically, it means that not all regimes of power systems specified by the conditions at generator and load buses are realizable. A simple example of this situation is when the total power demand exceeds the total power generating capacity. Second, nonlinear equations usually have not a single but multiple solutions. This raises an immediate question which of these solutions are relevant to the operation of power systems and under what circumstances one or another of these solutions is physically realizable. Third, numerical solution of nonlinear equations is carried out by using iterative techniques. These techniques usually have local convergence. The latter means that iterations converge to one of the possible solutions if an initial guess (i.e., a starting point of iterations) is sufficiently close to this solution. How to choose the initial guess in order to converge to a desired (relevant) solution is still by and large an open question. Finally, if a specific solution of the power flow equations is found, it is not always clear that this solution may correspond to *stable* operation of the power system. Interesting research on the aspects mentioned above has been carried out and published in such papers as [3], [28], [56].

5.2 Newton-Raphson and Continuation Methods

The Newton-Raphson method is a very powerful iterative technique for the solution of nonlinear equations. It has a quadratic rate of convergence. Conceptually, it is based on local linearization of nonlinear equations on each iteration step. For pedagogical reasons, we shall first discuss this technique for the one-dimensional case, i.e., for a single nonlinear equation with one unknown and shall prove its quadratic rate of convergence. Subsequently, we shall discuss the m-dimensional case, i.e., the case of m simultaneous nonlinear equations with m unknowns.

Consider the nonlinear equation

$$f(x) = 0, \tag{5.32}$$

where f is a nonlinear function of a single variable x. We are looking for a solution x_* of this equation. This means that we want to find such a number x_* that

$$f(x_*) = 0. \tag{5.33}$$

The essence of the Newton-Raphson method is that we start from some initial guess x_0 and linearize function $f(x)$ around x_0 by using the first two (linear) terms of the Taylor expansion:

$$f(x) \approx f(x_0) + f'(x_0)(x - x_0). \tag{5.34}$$

Since we want to solve equation (5.32), we consider the solution x_1 of linear equation

$$f(x_0) + f'(x_0)(x_1 - x_0) = 0 \tag{5.35}$$

as the next (hopefully better than x_0) approximation to the actual solution x_*. From the last formula we find

$$x_1 = x_0 - \frac{f(x_0)}{f'(x_0)}. \tag{5.36}$$

Having found x_1, we shall linearize function $f(x)$ around x_1 by using the first two terms of the Taylor expansion,

$$f(x) \approx f(x_1) + f'(x_1)(x - x_1), \tag{5.37}$$

and consider the solution x_2 of linear equation

$$f(x_1) + f'(x_1)(x_2 - x_1) = 0 \tag{5.38}$$

as the next approximation to the actual solution x_*.

From the last formula we find

$$x_2 = x_1 - \frac{f(x_1)}{f'(x_1)}. \tag{5.39}$$

Now, we shall linearize function $f(x)$ around x_2 by using the first two terms of the Taylor expansion and solve the corresponding linear equation. We shall continue local linearizations for $f(x)$ around each iteration and find the solution of the corresponding linear equations until the convergence of such iterations is achieved with desired accuracy. It is apparent that on iteration number k we have to solve linear equation

$$f(x_k) + f'(x_k)(x_{k+1} - x_k) = 0, \tag{5.40}$$

which leads to

$$\boxed{x_{k+1} = x_k - \frac{f(x_k)}{f'(x_k)}.} \tag{5.41}$$

Thus, the same formula (5.41) has to be repeatedly used to find new approximations.

The geometric interpretation of the Newton-Raphson method is illustrated by Figure 5.1. It is clear that linear equation (5.34) is the equation of the tangent line to the graph of function $f(x)$ at the point $(x_0, f(x_0))$ and that x_1 is the point of intersection of this tangent line with the x-axis. The same geometric interpretation is valid for any Newton-Raphson iteration. This geometric interpretation is the reason why the Newton-Raphson method is also called the method of tangents.

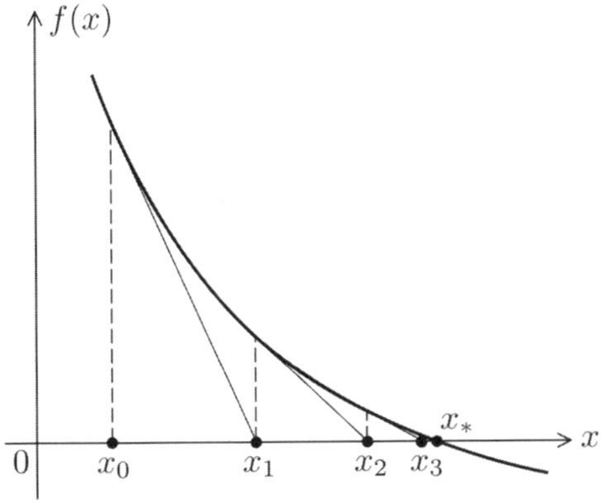

Fig. 5.1

We shall next prove the convergence of the Newton-Raphson method and establish its quadratic rate of convergence. To this end, by using formula (5.33), we shall make the following equivalent transformation of formula (5.41):

$$x_{k+1} - x_* = x_k - x_* - \frac{f(x_k) - f(x_*)}{f'(x_k)}. \tag{5.42}$$

The next step is to rewrite the last formula as

$$x_{k+1} - x_* = \frac{f'(x_k)(x_k - x_*)}{f'(x_k)} - \frac{f(x_k) - f(x_*)}{f'(x_k)}. \tag{5.43}$$

Now, we shall use the *mean value theorem* to represent the difference $f(x_k) - f(x_*)$ in the form

$$f(x_k) - f(x_*) = f'(\tilde{x}_k)(x_k - x_*), \tag{5.44}$$

where $\tilde{x}_k$ is between x_k and x_*. The latter implies the inequality

$$|x_k - \tilde{x}_k| < |x_k - x_*|. \tag{5.45}$$

By substituting formula (5.44) into equation (5.43), we obtain

$$x_{k+1} - x_* = \frac{f'(x_k) - f'(\tilde{x}_k)}{f'(x_k)}(x_k - x_*), \tag{5.46}$$

which leads to

$$|x_{k+1} - x_*| = \frac{|f'(x_k) - f'(\tilde{x}_k)|}{|f'(x_k)|}|x_k - x_*|. \tag{5.47}$$

Next, we have

$$|f'(x_k) - f'(\tilde{x}_k)| \leq \max|f''(x)||x_k - \tilde{x}_k|. \tag{5.48}$$

From formula (5.45) and the last inequality follows that

$$|f'(x_k) - f'(\tilde{x}_k)| \leq \max |f''(x)||x_k - x_*|. \tag{5.49}$$

By taking into account the last inequality in formula (5.47), we derive

$$\boxed{|x_{k+1} - x_*| \leq C|x_k - x_*|^2,} \tag{5.50}$$

where

$$C = \frac{\max |f''(x)|}{\min |f'(x)|}. \tag{5.51}$$

By using inequality (5.50) the *local* convergence of Newton-Raphson iterations can be established. Indeed, let us suppose that we can find such an initial guess x_0 that

$$C|x_0 - x_*| = q < 1. \tag{5.52}$$

Then, from inequality (5.50) written for $k = 0$, we find

$$|x_1 - x_*| \leq \frac{q^2}{C}. \tag{5.53}$$

Next, by using inequality (5.50) for $k = 1$, we obtain

$$|x_2 - x_*| \leq C|x_1 - x_*|^2. \tag{5.54}$$

The last two inequalities imply that

$$|x_2 - x_*| \leq \frac{q^4}{C}. \tag{5.55}$$

Now, by using the induction argument, it is easy to establish that

$$\boxed{|x_k - x_*| \leq \frac{q^{2k}}{C}.} \tag{5.56}$$

Since $q < 1$ (see formula (5.52)), the last inequality implies that

$$\lim_{k \to \infty} |x_k - x_*| = 0. \tag{5.57}$$

Thus, the convergence of the Newton-Raphson iterations is established. This is very fast convergence because according to (5.50) the deviation of new iteration x_{k+1} from the solution x_* is of the second order of smallness in comparison with the deviation of the previous iteration x_k from the same solution x_*. This is the reason why it is said that the Newton-Raphson method has the quadratic rate of convergence. However, it must be stressed that the convergence of the Newton-Raphson iterations has been established under the condition that the initial guess x_0 is chosen in such a way that the inequality (5.52) is satisfied. In other words, the convergence

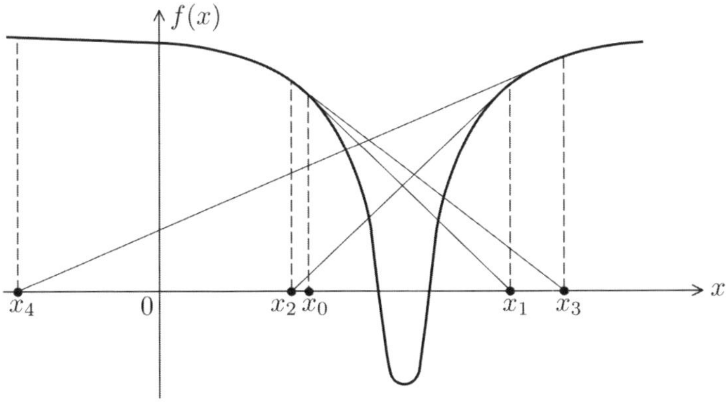

Fig. 5.2

is established when the initial guess x_0 is sufficiently close to an actual solution x_*. It is quite possible that the convergence may occur even when the inequality (5.52) is not satisfied and that this possibility is not captured by the presented proof of convergence. Nevertheless, it can be graphically illustrated that in general Newton-Raphson iterations do not converge for any choice of initial guess x_0. This illustration is presented in Figure 5.2. Thus, it can be concluded that in general the Newton-Raphson technique has *local* convergence. This is especially true when equation (5.32) has many solutions.

Now, we shall proceed to the discussion of the multi-dimensional case, that is, the case when we deal with many simultaneous equations with respect to many unknowns. Conceptually, our discussion will replicate the presented discussion of the one-dimensional case.

Consider m simultaneous nonlinear equations with m unknowns:

$$\begin{cases} F_1(x_1, x_2, ..., x_m) = 0, \\ F_2(x_1, x_2, ..., x_m) = 0, \\ \qquad \vdots \\ F_m(x_1, x_2, ..., x_m) = 0. \end{cases} \tag{5.58}$$

These simultaneous equations can be written in concise vector form as

$$\mathbf{F}(\mathbf{x}) = 0, \tag{5.59}$$

where $\mathbf{F}$ and $\mathbf{x}$ are m-dimensional vectors

$$\mathbf{F}(\mathbf{x}) = \begin{pmatrix} F_1(\mathbf{x}) \\ F_2(\mathbf{x}) \\ \vdots \\ F_m(\mathbf{x}) \end{pmatrix}, \qquad \mathbf{x} = \begin{pmatrix} x_1 \\ x_2 \\ \vdots \\ x_m \end{pmatrix}. \tag{5.60}$$

We are interested in numerical solution of simultaneous equations (5.59) by using the Newton-Raphson iterations. To start these iterations, we choose some initial guess $\mathbf{x}_0$ and we linearize the nonlinear vector function $\mathbf{F}(\mathbf{x})$ around this guess. This is done by using the first two (linear) terms of the Taylor expansion of $\mathbf{F}(\mathbf{x})$ at $\mathbf{x}_0$. In vector form these two terms of the Taylor expansion can be written as

$$\mathbf{F}(\mathbf{x}) \approx \mathbf{F}(\mathbf{x}_0) + \hat{J}(\mathbf{x}_0)(\mathbf{x} - \mathbf{x}_0), \tag{5.61}$$

where $\hat{J}(\mathbf{x}_0)$ is the Jacobian matrix of $\mathbf{F}(\mathbf{x})$ at $\mathbf{x}_0$. The elements of this matrix are the first derivatives of functions $F_i(\mathbf{x})$ evaluated at $\mathbf{x}_0$. Namely,

$$\hat{J}(\mathbf{x}_0) = \begin{pmatrix} \frac{\partial F_1}{\partial x_1}(\mathbf{x}_0) & \frac{\partial F_1}{\partial x_2}(\mathbf{x}_0) & \cdots & \frac{\partial F_1}{\partial x_m}(\mathbf{x}_0) \\ \frac{\partial F_2}{\partial x_1}(\mathbf{x}_0) & \frac{\partial F_2}{\partial x_2}(\mathbf{x}_0) & \cdots & \frac{\partial F_2}{\partial x_m}(\mathbf{x}_0) \\ \vdots & \vdots & \ddots & \vdots \\ \frac{\partial F_m}{\partial x_1}(\mathbf{x}_0) & \frac{\partial F_m}{\partial x_2}(\mathbf{x}_0) & \cdots & \frac{\partial F_m}{\partial x_m}(\mathbf{x}_0) \end{pmatrix}. \tag{5.62}$$

The first approximation (the first iteration) $\mathbf{x}_1$ to the solution of simultaneous equations (5.59) can be found by equating the right-hand side of formula (5.61) to zero. This leads to

$$\hat{J}(\mathbf{x}_0)(\mathbf{x}_1 - \mathbf{x}_0) = -\mathbf{F}(\mathbf{x}_0). \tag{5.63}$$

By introducing vector

$$\mathbf{X}_1 = \mathbf{x}_1 - \mathbf{x}_0, \tag{5.64}$$

the last formula can be written as

$$\hat{J}(\mathbf{x}_0)\mathbf{X}_1 = -\mathbf{F}(\mathbf{x}_0). \tag{5.65}$$

These are linear simultaneous equations for the components of $\mathbf{X}_1$. By solving these linear equations, we can find $\mathbf{X}_1$ and then according to formula (5.64) we can compute the new iteration vector $\mathbf{x}_1$:

$$\mathbf{x}_1 = \mathbf{x}_0 + \mathbf{X}_1. \tag{5.66}$$

Having found $\mathbf{x}_1$, we shall linearize nonlinear vector function $\mathbf{F}(\mathbf{x})$ around $\mathbf{x}_1$ by using the first two terms of the Taylor expansion. Then, we equate these two linear terms to zero to find the new iteration $\mathbf{x}_2$. Afterwards, the process is repeated to find subsequent iterations. These Newton-Raphson iterations can be summarized by the following recurrent relations:

$$\boxed{\begin{aligned} \hat{J}(\mathbf{x}_k)\mathbf{X}_{k+1} &= -\mathbf{F}(\mathbf{x}_k), \\ \mathbf{x}_{k+1} &= \mathbf{x}_k + \mathbf{X}_{k+1}. \end{aligned}} \qquad \begin{aligned} (5.67) \\ (5.68) \end{aligned}$$

Thus, numerical realization of Newton-Raphson iterations requires the solution of simultaneous linear equations (5.67) with the Jacobian matrix $\hat{J}(\mathbf{x}_k)$ and the right-hand side $-\mathbf{F}(\mathbf{x}_k)$ being computed by using the previous iteration $\mathbf{x}_k$. After simultaneous linear equations (5.67) are solved, the new iteration is computed by using formula (5.68).

Formulas (5.67) and (5.68) are convenient for performing computations. However, for the convergence study of the Newton-Raphson technique it is convenient to represent these formulas in another equivalent form,

$$\hat{J}(\mathbf{x}_k)(\mathbf{x}_{k+1} - \mathbf{x}_k) = -\mathbf{F}(\mathbf{x}_k), \tag{5.69}$$

which leads to

$$\mathbf{x}_{k+1} = \mathbf{x}_k - \hat{J}^{-1}(\mathbf{x}_k)\mathbf{F}(\mathbf{x}_k), \tag{5.70}$$

where $\hat{J}^{-1}(\mathbf{x}_k)$ is the inverse of the Jacobian $\hat{J}(\mathbf{x}_k)$. Let $\mathbf{x}_*$ be a solution of equations (5.59), i.e.,

$$\mathbf{F}(\mathbf{x}_*) = 0. \tag{5.71}$$

By using the last formula, the recurrent relations (5.70) can be represented in the following equivalent form:

$$\mathbf{x}_{k+1} - \mathbf{x}_* = \mathbf{x}_k - \mathbf{x}_* - \hat{J}^{-1}(\mathbf{x}_k)[\mathbf{F}(\mathbf{x}_k) - \mathbf{F}(\mathbf{x}_*)]. \tag{5.72}$$

Next, we shall derive the integral formula for the difference $\mathbf{F}(\mathbf{x}_k) - \mathbf{F}(\mathbf{x}_*)$. To this end, we introduce the vector

$$\mathbf{x}(\lambda) = \mathbf{x}_* + \lambda(\mathbf{x}_k - \mathbf{x}_*), \quad 0 \le \lambda \le 1. \tag{5.73}$$

It is clear that

$$\mathbf{x}(0) = \mathbf{x}_* \quad \text{and} \quad \mathbf{x}(1) = \mathbf{x}_k. \tag{5.74}$$

Now, it is clear that

$$\mathbf{F}(\mathbf{x}_k) - \mathbf{F}(\mathbf{x}_*) = \int_0^1 \frac{d\mathbf{F}(\mathbf{x}(\lambda))}{d\lambda} d\lambda. \tag{5.75}$$

By performing differentiation, we find

$$\frac{d\mathbf{F}(\mathbf{x}(\lambda))}{d\lambda} = \hat{J}(\mathbf{x}(\lambda))(\mathbf{x}_k - \mathbf{x}_*). \tag{5.76}$$

By substituting the last formula into equation (5.75), we obtain

$$\mathbf{F}(\mathbf{x}_k) - \mathbf{F}(\mathbf{x}_*) = \left[\int_0^1 \hat{J}(\mathbf{x}(\lambda))d\lambda\right](\mathbf{x}_k - \mathbf{x}_*). \tag{5.77}$$

The last formula is the substitute for the mean value relation (5.44) in the multi-dimensional vectorial case.

Next, we insert the last formula into equation (5.72),

$$\mathbf{x}_{k+1} - \mathbf{x}_* = \left[\hat{I} + \hat{J}^{-1}(\mathbf{x}_k)\int_0^1 \hat{J}(\mathbf{x}(\lambda))d\lambda\right](\mathbf{x}_k - \mathbf{x}_*), \tag{5.78}$$

where $\hat{I}$ is the identity matrix, which can be represented as

$$\hat{I} = \hat{J}^{-1}(\mathbf{x}_k)\hat{J}(\mathbf{x}_k). \tag{5.79}$$

By inserting the last formula into equation (5.78), we derive

$$\mathbf{x}_{k+1} - \mathbf{x}_* = \hat{J}^{-1}(\mathbf{x}_k)\left[\hat{J}(\mathbf{x}_k) - \int_0^1 \hat{J}(\mathbf{x}(\lambda))d\lambda\right](\mathbf{x}_k - \mathbf{x}_*). \tag{5.80}$$

From the last formula follows the inequality

$$\|\mathbf{x}_{k+1} - \mathbf{x}_*\| \leq \left\|\hat{J}^{-1}(\mathbf{x}_k)\right\| \left\|\hat{J}(\mathbf{x}_k) - \int_0^1 \hat{J}(\mathbf{x}(\lambda))d\lambda\right\| \|\mathbf{x}_k - \mathbf{x}_*\|, \tag{5.81}$$

where $\|\cdot\|$ is a norm (for instance, ℓ_2-norm) in m-dimensional space. Next, we find that

$$\left\|\hat{J}(\mathbf{x}_k) - \int_0^1 \hat{J}(\mathbf{x}(\lambda))d\lambda\right\| = \left\|\int_0^1 \left[\hat{J}(\mathbf{x}_k) - \hat{J}(\mathbf{x}(\lambda))\right]d\lambda\right\|. \tag{5.82}$$

The elements of matrix $\hat{J}(\mathbf{x}_k) - \hat{J}(\mathbf{x}(\lambda))$ are

$$\frac{\partial F_i}{\partial x_j}(\mathbf{x}_k) - \frac{\partial F_i}{\partial x_j}(\mathbf{x}(\lambda)). \tag{5.83}$$

By assuming that the second-order derivatives of functions $F_i(\mathbf{x})$ are bounded, it can be concluded that

$$\left|\frac{\partial F_i}{\partial x_j}(\mathbf{x}_k) - \frac{\partial F_i}{\partial x_j}(\mathbf{x}(\lambda))\right| < L\|\mathbf{x}_k - \mathbf{x}(\lambda)\|, \tag{5.84}$$

where L is some constant. This, according to (5.73), implies that

$$\left|\frac{\partial F_i}{\partial x_j}(\mathbf{x}_k) - \frac{\partial F_i}{\partial x_j}(\mathbf{x}(\lambda))\right| \leq (1-\lambda)L\|\mathbf{x}_k - \mathbf{x}_*\|. \tag{5.85}$$

From formula (5.83) and the last inequality follows that

$$\left\|\hat{J}(\mathbf{x}_k) - \int_0^1 \hat{J}(\mathbf{x}(\lambda))d\lambda\right\| \leq D\|\mathbf{x}_k - \mathbf{x}_*\|, \tag{5.86}$$

where D is some constant.

Now, by taking into account the last inequality in formula (5.81) and assuming that norms $\left\| \hat{J}^{-1}(\mathbf{x}_k) \right\|$ are bounded, we conclude that for some constant C the following inequality is valid:

$$\boxed{\|\mathbf{x}_{k+1} - \mathbf{x}_*\| \le C\|\mathbf{x}_k - \mathbf{x}_*\|^2.} \tag{5.87}$$

This inequality is similar to inequality (5.50) derived for the one-dimensional case. Consequently, by using the same line of reasoning as before, it can be established that if the initial guess $\mathbf{x}_0$ is chosen in such a way that

$$C\|\mathbf{x}_0 - \mathbf{x}_*\| = q < 1, \tag{5.88}$$

then

$$\|\mathbf{x}_k - \mathbf{x}_*\| < \frac{q^{2k}}{C}, \tag{5.89}$$

which implies the convergence of the Newton-Raphson iterations. This establishes *local quadratic* convergence of iterations (5.67)-(5.68).

Now, we shall apply the Newton-Raphson technique (5.67) and (5.68) to the solution of the power flow equations (5.30)-(5.31). As was pointed out in the previous section, not all these equations are fully coupled. Namely, all P-type equations for active power at all buses and Q-type equations for reactive power at all load buses constitute $N + L$ coupled nonlinear equations for $N + L$ unknowns which are the initial phases of voltages at all (N) buses and rms values of voltages at all (L) load buses. We shall write these equations again in the form (5.58) by using different notations for indices and numerating first all load buses instead of generator buses. Then, these nonlinear equations are

$$F_i(\mathbf{x}) = P_i - \sum_{j=1}^{N} V_i V_j |Y_{ij}| \cos(\varphi_i - \varphi_j - \gamma_{ij}) = 0, \tag{5.90}$$

$$(i = 1, 2, ..., N),$$

$$F_{N+i}(\mathbf{x}) = Q_i - \sum_{j=1}^{N} V_i V_j |Y_{ij}| \sin(\varphi_i - \varphi_j - \gamma_{ij}) = 0, \tag{5.91}$$

$$(i = 1, 2, ..., L),$$

where the unknown vector $\mathbf{x}$ has the components

$$x_j = \varphi_j, \qquad (j = 1, 2, ..., N), \tag{5.92}$$

$$x_{N+j} = V_j, \qquad (j = 1, 2, ..., L). \tag{5.93}$$

It is clear now that the Jacobian $\hat{J}(\mathbf{x})$ has four major blocks:

$$\hat{J} = \left(\begin{array}{cc|c} \dfrac{\partial F_i}{\partial \varphi_j} & N & \dfrac{\partial F_i}{\partial V_j} \\ N & & L \\ \hline \dfrac{\partial F_{N+i}}{\partial \varphi_j} & L & \dfrac{\partial F_{N+i}}{\partial V_j} \end{array}\right). \tag{5.94}$$

From formulas (5.90) and (5.91) we easily derive the elements of the Jacobian matrix:

$$\frac{\partial F_i}{\partial \varphi_j} = -V_i V_j |Y_{ij}| \sin(\varphi_i - \varphi_j - \gamma_{ij}) \quad \text{if } i \neq j, \tag{5.95}$$

$$\frac{\partial F_i}{\partial \varphi_i} = \sum_{j=1}^{N} V_i V_j |Y_{ij}| \sin(\varphi_i - \varphi_j - \gamma_{ij}), \tag{5.96}$$

$$\frac{\partial F_{N+i}}{\partial \varphi_j} = V_i V_j |Y_{ij}| \cos(\varphi_i - \varphi_j - \gamma_{ij}) \quad \text{if } i \neq j, \tag{5.97}$$

$$\frac{\partial F_{N+i}}{\partial \varphi_i} = -\sum_{j=1}^{N} V_i V_j |Y_{ij}| \cos(\varphi_i - \varphi_j - \gamma_{ij}), \tag{5.98}$$

$$\frac{\partial F_i}{\partial V_j} = -V_i |Y_{ij}| \cos(\varphi_i - \varphi_j - \gamma_{ij}) \quad \text{if } i \neq j, \tag{5.99}$$

$$\frac{\partial F_i}{\partial V_i} = -\sum_{j \neq i}^{N} V_j |Y_{ij}| \cos(\varphi_i - \varphi_j - \gamma_{ij}) - 2V_i |Y_{ii}| \cos \gamma_{ii}, \tag{5.100}$$

$$\frac{\partial F_{N+i}}{\partial V_j} = -V_i |Y_{ij}| \sin(\varphi_i - \varphi_j - \gamma_{ij}) \quad \text{if } i \neq j, \tag{5.101}$$

$$\frac{\partial F_{N+i}}{\partial V_i} = -\sum_{j \neq i}^{N} V_j |Y_{ij}| \sin(\varphi_i - \varphi_j - \gamma_{ij}) + 2V_i |Y_{ii}| \sin \gamma_{ii}. \tag{5.102}$$

Having specified the above formulas for the computation of the matrix elements of the Jacobian, the numerical realization of the Newton-Raphson iterations (5.67)-(5.68) becomes transparent.

Next, we shall discuss another efficient method for numerical solution of nonlinear algebraic equations which is often used in combination with the Newton-Raphson method. This is the continuation method.* The central idea of this method can be briefly described as follows. Consider two separate sets of simultaneous nonlinear equations

$$\mathbf{F}_0(\mathbf{x}) = 0 \tag{5.103}$$

and

$$\mathbf{F}_1(\mathbf{x}) = 0. \tag{5.104}$$

Suppose that a solution $\mathbf{x}_0$ of nonlinear equations (5.103) is known (or found), i.e., $\mathbf{F}(\mathbf{x}_0) = 0$. We want to find a solution of equation (5.104). For this purpose, we introduce a new vector-function $\mathbf{F}(\mathbf{x}, \lambda)$, where λ is a parameter (i.e., a real number) which varies (for instance) between 0 and 1,

$$0 \le \lambda \le 1. \tag{5.105}$$

This parameter is introduced in such a way that

$$\mathbf{F}(\mathbf{x}, 0) = \mathbf{F}_0(\mathbf{x}) \tag{5.106}$$

and

$$\mathbf{F}(\mathbf{x}, 1) = \mathbf{F}_1(\mathbf{x}). \tag{5.107}$$

In other words, by continuously varying the parameter λ, the vector-function $\mathbf{F}(\mathbf{x}, \lambda)$ is continuously deformed from $\mathbf{F}_0(\mathbf{x})$ into $\mathbf{F}_1(\mathbf{x})$.

One example (just an example) of such introduction of parameter λ is illustrated by the formula

$$\mathbf{F}(\mathbf{x}, \lambda) = (1 - \lambda)\mathbf{F}_0(\mathbf{x}) + \lambda \mathbf{F}_1(\mathbf{x}). \tag{5.108}$$

Next, for each value of λ we consider simultaneous nonlinear equations

$$\mathbf{F}(\mathbf{x}(\lambda), \lambda) = 0. \tag{5.109}$$

Here, we use the notation $\mathbf{x}(\lambda)$ for a solution of nonlinear equations (5.109) because this solution varies with variations of λ and, consequently, this solution is a function of λ.

Now, by differentiating both sides of formula (5.109) with respect to λ, we obtain

$$\hat{J}_{\mathbf{x}}(\mathbf{x}(\lambda), \lambda)\frac{d\mathbf{x}(\lambda)}{d\lambda} + \frac{\partial \mathbf{F}(\mathbf{x}(\lambda), \lambda)}{\partial \lambda} = 0, \tag{5.110}$$

*This method was proposed by Russian mathematician D. F. Davidenko for numerical solution of nonlinear algebraic equations, although for analysis of PDEs it was much earlier developed by S. N. Bernstein.

where $\hat{J}_{\mathbf{x}}$ is, as before, the Jacobian matrix of $\mathbf{F}(\mathbf{x}(\lambda), \lambda)$ whose elements are the first-order partial derivatives of $\mathbf{F}$ with respect to the Cartesian components of vector $\mathbf{x}(\lambda)$ (see formula (5.62)).

The last formula can be written as

$$\hat{J}_{\mathbf{x}}(\mathbf{x}(\lambda), \lambda)\frac{d\mathbf{x}(\lambda)}{d\lambda} = -\frac{\partial \mathbf{F}(\mathbf{x}(\lambda), \lambda)}{\partial \lambda} \tag{5.111}$$

and treated as the differential equation for $\mathbf{x}(\lambda)$ with the initial condition

$$\mathbf{x}(0) = \mathbf{x}_0, \tag{5.112}$$

where, as mentioned above, $\mathbf{x}_0$ is a solution of equation (5.103).

By numerically integrating the initial value problem (5.111)-(5.112) in the closed interval (5.105), vector $\mathbf{x}(1)$ may be found. According to formulas (5.107) and (5.109), this vector is a solution of nonlinear equations (5.104). This is, in a nutshell, the essence of the continuation method, which reduces the numerical solution of algebraic equations to numerical integration of specific differential equations.

If the Jacobian in equation (5.111) is invertible for all λ (which is not always the case), then this equation can be written in the form

$$\frac{d\mathbf{x}(\lambda)}{d\lambda} = -\hat{J}_{\mathbf{x}}^{-1}(\mathbf{x}(\lambda), \lambda)\frac{\partial \mathbf{F}(\mathbf{x}(\lambda), \lambda)}{\partial \lambda} \tag{5.113}$$

and numerically integrated with the initial condition (5.112). This integration can be accomplished by introducing some mesh

$$0 < \lambda_1 < \lambda_2 < \cdots < \lambda_n = 1 \tag{5.114}$$

and by replacing the differential equation (5.113) by the finite-difference equation

$$\frac{\mathbf{x}(\lambda_{k+1}) - \mathbf{x}(\lambda_k)}{\lambda_{k+1} - \lambda_k} \simeq -\hat{J}_{\mathbf{x}}^{-1}(\mathbf{x}(\lambda_k), \lambda_k)\frac{\partial \mathbf{F}}{\partial \lambda}(\mathbf{x}(\lambda_k), \lambda_k). \tag{5.115}$$

The last equation can also be written in the form

$$\mathbf{x}(\lambda_{k+1}) \simeq \mathbf{x}(\lambda_k) - (\lambda_{k+1} - \lambda_k)\hat{J}_{\mathbf{x}}^{-1}(\mathbf{x}(\lambda_k), \lambda_k)\frac{\partial \mathbf{F}}{\partial \lambda}(\mathbf{x}(\lambda_k), \lambda_k). \tag{5.116}$$

By using the last formula, $\mathbf{x}(\lambda_k)$ can be consecutively computed until $\mathbf{x}(\lambda_n) = \mathbf{x}(1)$ is found. However, more accurate results can be obtained if the accuracy of $\mathbf{x}(\lambda_k)$ is improved (for each λ_k) by using the Newton-Raphson iterations. In this case, $\mathbf{x}(\lambda_k)$ computed by means of formula (5.116) is used as an initial guess (i.e., as a starting point) for the Newton-Raphson iterations.

It is clear from the above description of the general scheme of the continuation method that there is freedom in choices of function $\mathbf{F}_0(\mathbf{x})$ and the continuation parameter λ. Function $\mathbf{F}_0(\mathbf{x})$ is usually chosen to simplify numerical solution of equations (5.103), i.e., to simplify the computing of initial condition $\mathbf{x}_0$ in (5.112). $\mathbf{F}_0(\mathbf{x})$ can even be, for instance, a linear vector-function. As far as λ is concerned, it is beneficial to introduce parameter λ in the load terms of the power flow equations. Such a parameter can naturally be called a "load parameter." Then, by using the continuation method (i.e., by integrating the appropriate differential equations), the nonlinear power flow equations can be solved for different loading conditions. In other words, in this case, solutions of equations (5.109) for various values of λ (not only for $\lambda = 1$) become physically and practically meaningful. The detailed discussion of this and other issues is beyond the scope of this book and can be found in [2], where the continuation technique is extensively used for voltage stability assessment and control in power systems. It is worthwhile to mention that the continuation method has been widely used for solution of various nonlinear problems. For instance, it has been used in [31] for solution of nonlinear magnetostatic problems when saturation of ferromagnetic cores must be accounted for.

5.3 Stability of Power Systems

In the previous chapter (see section 4.4), we have discussed static stability of the synchronous generator under the condition of complete balance between loads and generation, that is, when synchronism is maintained. In this section, our discussion will be concerned with the transient stability of synchronous generators when this balance is perturbed. Transient stability implies the ability of the synchronous generator to attain synchronism (or close to synchronous performance) after sudden disturbances. These disturbances create imbalances between generation and loads which cause rotor speed deviations from synchronous speed. Thus, it is clear that the transient stability study should be based on the analysis of dynamics of mechanical motion of rotors of synchronous generators. For this reason, we shall start with the derivation of the so-called "swing" equation which describes nonsynchronous dynamics of the rotor. We shall limit our discussion of transient stability to the case of a single generator. The case of multiple generators is very difficult from the mathematical point of view and will not be addressed in this text. Even in the case of a single generator, our

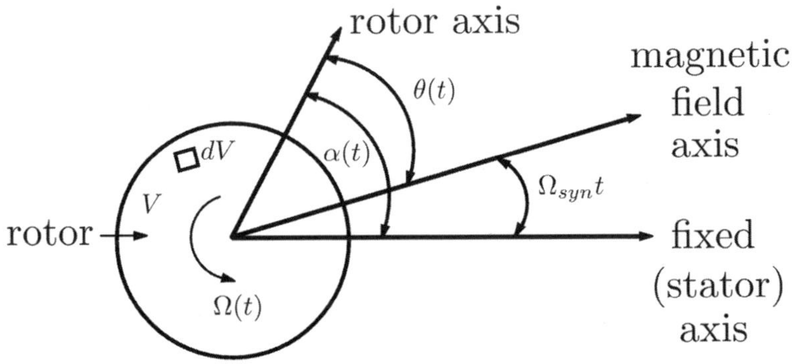

Fig. 5.3

analysis will be limited to the discussion of the very basic facts and many important details will not be covered.

First, we shall demonstrate that the dynamics of the rotor is described by the following equation:

$$\boxed{I\frac{d\Omega(t)}{dt} = T_{ap},} \tag{5.117}$$

where I is the moment of inertia of the rotor, T_{ap} is the torque applied to the rotor, while $\Omega(t)$ is the instantaneous angular speed of the rotor. This speed is given by the formula

$$\Omega(t) = \frac{d\alpha(t)}{dt} \tag{5.118}$$

with $\alpha(t)$ being an angle between some chosen axis of the rotating rotor and the fixed (stator) axis (see Figure 5.3). To derive equation (5.117), we shall start with the second Newton's law for infinitesimally small volume dV of the rotor,

$$adm = r\beta dm = dF, \tag{5.119}$$

where dF is a force applied to this volume, dm is its mass, a is its acceleration, while β is its angular acceleration.

It is apparent that

$$\beta = \frac{d\Omega(t)}{dt}. \tag{5.120}$$

Next, we shall multiply all sides of equation (5.119) by r and integrate over the volume V of the rotating rotor. This yields

$$\beta \int_V r^2 dm = \int_V r dF. \tag{5.121}$$

It is apparent that

$$\int_V r dF = T_{ap}, \tag{5.122}$$

while

$$\int_V r^2 dm = I. \tag{5.123}$$

Now, by substituting formulas (5.120), (5.122) and (5.123) into equation (5.121), we shall arrive at equation (5.117).

By using formula (5.118), the dynamics equation (5.117) can be written as the following second-order differential equation:

$$I\frac{d^2\alpha(t)}{dt^2} = T_{ap}. \tag{5.124}$$

It is very convenient to replace the angle $\alpha(t)$ between the rotor and stator axes by the angle $\theta(t)$ between the rotor axis and the axis of the in-gap magnetic field rotating with constant synchronous speed Ω_{syn} with respect to the stator. This can be done by using the formula (see Figure 5.3)

$$\alpha(t) = \theta(t) + \Omega_{syn}t. \tag{5.125}$$

By twice differentiating the last equation with respect to time, we find

$$\frac{d^2\alpha(t)}{dt^2} = \frac{d^2\theta(t)}{dt^2}, \tag{5.126}$$

and, consequently, equation (5.124) can be written as

$$I\frac{d^2\theta(t)}{dt^2} = T_{ap}. \tag{5.127}$$

Next, we shall multiply both sides of equation (5.127) by $\Omega(t)$:

$$I\Omega(t)\frac{d^2\theta(t)}{dt^2} = T_{ap}\Omega(t). \tag{5.128}$$

The right-hand side of the last equation has the meaning of total power applied to the rotor. This power is the difference between the mechanical shaft power P_{mech} supplied by the prime mover, which drives the rotor, and the generated electric power P supplied to the power network. P_{mech} corresponds to the mechanical shaft torque applied by the prime mover, while P corresponds to electromagnetic torque caused by interaction of armature reaction magnetic field and rotor currents. Thus,

$$T_{ap}\Omega(t) = P_{ap} = P_{mech} - P. \tag{5.129}$$

The product $I\Omega(t)$ in the left-hand side of equation (5.128) is called angular momentum M,

$$M = I\Omega(t). \tag{5.130}$$

This angular momentum varies with time. However, it is usually assumed in transient stability studies that deviations of rotor angular speed are small in comparison with synchronous speed, i.e.,

$$\Omega(t) \approx \Omega_{syn}. \tag{5.131}$$

For this reason, it can be assumed that

$$M \approx const. \tag{5.132}$$

By substituting formulas (5.129) and (5.130) into equation (5.128), we end up with

$$M\frac{d^2\theta(t)}{dt^2} = P_{mech} - P. \tag{5.133}$$

As discussed in the last section of the previous chapter, geometric angle $\theta(t)$ can be construed as a load (or power) angle which is the phase shift in time between the induced internal voltage $\hat{\mathcal{E}}_s$ and terminal voltage $\hat{V}_s$. This fact is used as a justification for using the following expression for P in the case of cylindrical rotor turbine generators:

$$P = P_m \sin\theta, \tag{5.134}$$

where as before P_m stands for the maximum of P. By substituting the last relation into formula (5.133), we arrive at

$$\boxed{M\frac{d^2\theta(t)}{dt^2} = P_{mech} - P_m \sin\theta.} \tag{5.135}$$

This is the so-called "swing" equation which has been traditionally used in the analysis of transient stability. It is worthwhile to stress here that the derivation of this equation has been based on some approximations. First, it has been assumed that the total in-gap magnetic field rotates with constant angular synchronous speed Ω_{syn} despite the fact that the rotor and its magnetic field rotate with variable angular speed $\Omega(t)$ different from Ω_{syn}. This assumption can be justified by accepting the approximation (5.131) which limits the analysis of transient stabilities to relatively small deviations of rotor angular speed from synchronous angular speed. Second, it has been assumed in the derivation of the "swing" equation (5.135) that formula (5.134) is valid for electric power generated by the synchronous machine. However, this formula has been derived in the last section of the previous chapter for steady-state operation of the synchronous generator, that is, when the rotor and the armature reaction magnetic field rotate in synchronism. When this synchronism is maintained, no eddy currents are induced in the solid rotors of turbine generators. This is not the case for transient performance of synchronous machines when appreciable eddy currents may be induced in solid rotors. Furthermore, these eddy currents may result (among other things) in additional mechanical torque which is not accounted for in the "swing" equation (5.135). This torque usually has a damping effect on the rotor dynamics. Phenomenologically, the stabilizing effect of such damping torques is discussed at the end of this section. The approximate nature of the "swing" equation (5.135) implies that the results of transient stability study obtained by using this equation are suggestive in nature rather than being unquestionably valid.

The swing equation (5.135) is written for cylindrical rotor machines and the following discussion deals exclusively with these machines. However, a similar equation can be written for salient pole rotor machines, and many results of our subsequent discussions can be extended to those machines.

It is convenient to carry out the study of transient stability by writing the "swing" equation (5.135) in the "state space" form, that is, as coupled first-order differential equations. This can be done by introducing a new variable

$$\gamma(t) = \frac{d\theta(t)}{dt}, \tag{5.136}$$

which according to formulas (5.118) and (5.125) is the measure of deviation from synchronism

$$\gamma(t) = \Omega(t) - \Omega_{syn}. \tag{5.137}$$

Now the "swing" equation can be written as the following two coupled equations:

$$\boxed{\begin{aligned} \frac{d\theta(t)}{dt} &= \gamma(t), && (5.138)\\ \frac{d\gamma(t)}{dt} &= \frac{1}{M}\left[P_{mech} - P_m \sin\theta\right]. && (5.139)\end{aligned}}$$

Next, we introduce the function

$$\boxed{H(\gamma,\theta) = \frac{\gamma^2}{2} - \frac{1}{M}\left[\int_0^{\theta} (P_{mech} - P_m \sin u)du\right],} \tag{5.140}$$

or, after the integration,

$$\boxed{H(\gamma,\theta) = \frac{\gamma^2}{2} + \frac{1}{M}[P_m(1-\cos\theta) - P_{mech}\theta].} \tag{5.141}$$

It is clear that

$$\frac{\partial H}{\partial \gamma} = \gamma, \tag{5.142}$$

and

$$-\frac{\partial H}{\partial \theta} = \frac{1}{M}[P_{mech} - P_m \sin\theta]. \tag{5.143}$$

The last two formulas imply that the dynamics equations (5.138) and (5.139) can be written as

$$\frac{d\theta(t)}{dt} = \frac{\partial H}{\partial \gamma}, \tag{5.144}$$

$$\frac{d\gamma(t)}{dt} = -\frac{\partial H}{\partial \theta}. \tag{5.145}$$

The last two relations are Hamiltonian equations and the function $H(\gamma,\theta)$ is the Hamiltonian, which is why the letter H is used for its notation. Equations of this type are encountered in many different areas of physics. Usually, function H in Hamiltonian equations has the physical meaning of energy. Hamiltonian equations play a central role in modern physics because they reveal that the underlying dynamics is controlled by energy. This fundamental physical feature is replicated in the structure of quantum mechanics where the time evolution of a wave function is controlled by the Hamiltonian (energy) operator. The Hamiltonian equations (5.144)-(5.145) are highly symmetric and this aspect is extensively utilized in the mathematical theory of Hamiltonian equations. We shall use this symmetry to demonstrate that the Hamiltonian function is an integral (i.e., conserved quantity) of rotor dynamics. Indeed, suppose that $\gamma(t)$ and $\theta(t)$ are a solution of equations (5.144)-(5.145), and consider the following function of time:

$$H(t) = H[\gamma(t), \theta(t)]. \tag{5.146}$$

Then,

$$\frac{dH(t)}{dt} = \frac{\partial H}{\partial \gamma}\frac{d\gamma(t)}{dt} + \frac{\partial H}{\partial \theta}\frac{d\theta(t)}{dt}. \tag{5.147}$$

Now, by using equations (5.144) and (5.145) in the last formula, we find

$$\frac{dH(t)}{dt} = -\frac{\partial H}{\partial \gamma}\frac{\partial H}{\partial \theta} + \frac{\partial H}{\partial \theta}\frac{\partial H}{\partial \gamma} = 0. \tag{5.148}$$

The latter implies that

$$\boxed{H[\gamma(t), \theta(t)] = const.} \tag{5.149}$$

This means that the quantity $H[\gamma(t), \theta(t)]$ is conserved along any solution of equations (5.144)-(5.145) or equations (5.138)-(5.139). In other words, $H[\gamma(t), \theta(t)]$ is an integral of rotor dynamics. The practical significance of this fact is that any solution trajectory of equations (5.138)-(5.139) on the (γ, θ)-plane coincides with one constant level line of function $H(\gamma, \theta)$ given by formula (5.141); and the other way around, each constant level line of $H(\gamma, \theta)$ coincides with a specific solution trajectory of equations (5.138)-(5.139). This fact allows one to compute the solution trajectories for differential equations (5.138)-(5.139). Indeed, we can choose any value of H. Then, we can choose θ and compute the last term in formula (5.141). If the value of this term is smaller than the chosen value of H, then by using formula (5.141) we can compute positive and negative values of γ corresponding to the chosen values of H and θ. Doing this for different values of θ but the same value of H, we construct point-by-point a specific solution trajectory corresponding to the chosen value of H. It is clear from the above discussion that if for a chosen value of θ the value of the second term in formula (5.141) is larger than the chosen value of H, then this value of θ is not reachable by this specific trajectory corresponding to the chosen value of H. Performing these computations for different values of H, we can construct a set of solution trajectories which represent a phase portrait of the dynamical system described by equations (5.138) and (5.139). It is also clear from the presented discussion that solution trajectories have even symmetry with respect to the line $\gamma = 0$ (which is the θ-axis on the (γ, θ)-plane). This is so because for any value of θ reachable by a specific trajectory there are positive and negative values of γ of the same magnitude (with only one exception when $\gamma = 0$).

It is also very helpful for the construction of phase portraits to consider the critical points of rotor dynamics described by equations (5.138) and (5.139). The critical points are defined by equations

$$\frac{d\theta(t)}{dt} = 0, \qquad \frac{d\gamma(t)}{dt} = 0. \tag{5.150}$$

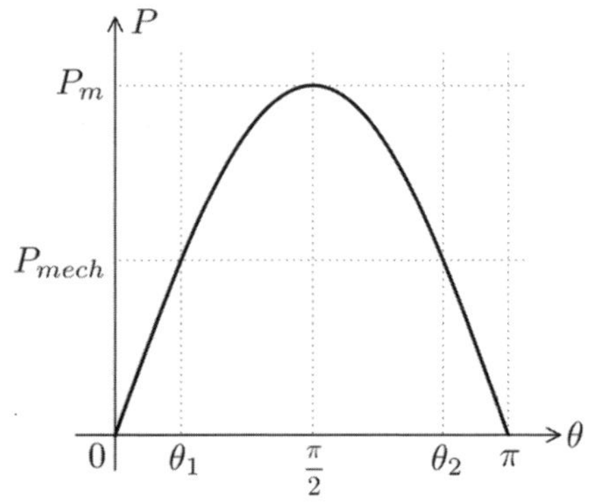

Fig. 5.4

From the first equation in (5.150) and formulas (5.136) and (5.137) we find that at the critical points

$$\Omega(t) = \Omega_{syn} \tag{5.151}$$

and

$$\gamma = 0. \tag{5.152}$$

From the second equation in (5.150) and equation (5.139) we conclude that at critical points there exists balance of the total power:

$$P_{mech} - P_m \sin\theta = 0. \tag{5.153}$$

This balance occurs for the following values of θ, $(0 < \theta < \pi)$:

$$\theta_1 = \sin^{-1}\left(\frac{P_{mech}}{P_m}\right) < \frac{\pi}{2} \tag{5.154}$$

and

$$\theta_2 = \pi - \theta_1 > \frac{\pi}{2}. \tag{5.155}$$

This is illustrated by Figure 5.4. Thus, it can be concluded that the points $(\gamma = 0,\ \theta = \theta_1)$ and $(\gamma = 0,\ \theta = \theta_2)$ on the $(\gamma,\ \theta)$-plane are the critical points of rotor dynamics and they correspond to the physical conditions when synchronism (see (5.151)) and balance of the total power (see (5.153)) occur.

Next, we shall demonstrate that the critical point $(0,\ \theta_1)$ corresponds to the minimum of Hamiltonian H, while the critical point $(0,\ \theta_2)$ is the saddle point of H. It is clear from formula (5.141) that for any fixed θ, H is a quadratic function of γ which achieves its minimum at $\gamma = 0$. In other words, function $H(\gamma, \theta)$ describes a parabolic well with its bottom (its floor) given by the equation

$$H(0,\theta) = \frac{1}{M}[P_m(1-\cos\theta) - P_{mech}\theta]. \tag{5.156}$$

We shall now analyze the extremum points of this floor. First, we find that

$$\frac{dH(0,\theta)}{d\theta} = \frac{1}{M}[P_m \sin\theta - P_{mech}] = 0, \tag{5.157}$$

which is the same as equation (5.153). Thus, the extremum points of the floor coincide with the critical points of rotor dynamics. Next,

$$\frac{d^2H(0,\theta)}{d\theta^2} = \frac{P_m}{M}\cos\theta, \tag{5.158}$$

and according to (5.154)

$$\left.\frac{d^2H(0,\theta)}{d\theta^2}\right|_{\theta=\theta_1} = \frac{P_m}{M}\cos\theta_1 > 0, \tag{5.159}$$

while according to (5.155)

$$\left.\frac{d^2H(0,\theta)}{d\theta^2}\right|_{\theta=\theta_2} = \frac{P_m}{M}\cos\theta_2 < 0. \tag{5.160}$$

It is clear from formulas (5.159) and (5.160) that the critical points $(0,\ \theta_1)$ and $(0,\ \theta_2)$ are the points at which the floor function $H(0,\theta)$ achieves its minimum and maximum values, respectively. A sketch of function $H(0,\theta)$ is shown in Figure 5.5. Since the second-order mixed derivative $\frac{\partial^2 H(\gamma,\theta)}{\partial\gamma\partial\theta}$ is (identically) equal to zero, it is easy to prove that the critical point $(0,\ \theta_1)$ is the point at which $H(\gamma,\theta)$ achieves its minimum value, while the critical point $(0,\ \theta_2)$ is the saddle point. The latter is transparent from the geometric point of view because $H(\gamma,\theta_2)$ achieves its minimum at $\gamma = 0$ and $H(0,\theta)$ achieves its maximum at $\theta = \theta_2$. The fact that point $(0,\ \theta_1)$ is the point of minimum of H and the point $(0,\ \theta_2)$ is the saddle point of H simplifies the construction of the phase portrait of the rotor dynamics described by the equations (5.138) and (5.139). Indeed, this fact implies that there are level lines of $H(\gamma,\theta)$ that enclose the critical point $(0,\ \theta_1)$ at which the minimum of H is achieved. These level lines correspond to the values of H which are between $H_{min} = H(0,\theta_1)$ and $H_{sad} = H(0,\theta_2)$. The structure of these level lines is changed for level values above H_{sad} as illustrated by Figure 5.6, where the level lines (and consequently, the phase portrait of rotor dynamics) are presented for some particular values of P_{mech} and P_m. In this figure, θ_1 is labeled as θ_{eq}, while θ_2 is labeled as θ_{sad}.

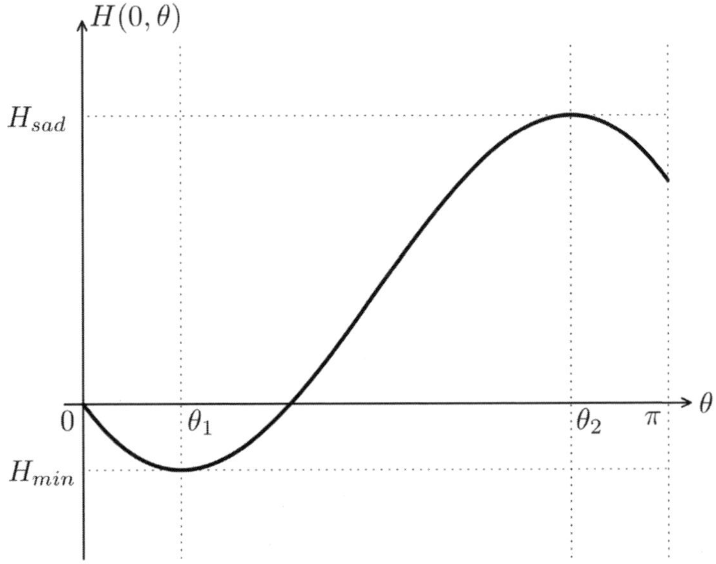

Fig. 5.5

The first labeling is justified because it is clear from the phase portrait that the critical point $(0,\ \theta_1 = \theta_{eq})$ corresponds to stable (equilibrium) synchronous performance of the generator. Indeed, small perturbations can only result in rotor dynamics (rotor swings) around this critical point along closed trajectories (trajectory 1, for instance) and no "run away" (unstable) rotor dynamics can be observed. The second labeling reflects that $(0,\ \theta_2 = \theta_{sad})$ is the saddle point of the dynamics, which is unstable equilibrium. Indeed, small perturbations may result in values of H above H_{sad} and this will cause the "run away" rotor dynamics along trajectory 2, for instance. The solution trajectory (level line of H) that goes through the saddle point separates stable and unstable regions. For this reason, it is called the "separatrix." Thus, it can be concluded that the region on the $(\gamma,\ \theta)$-plane specified by inequality

$$\boxed{H(\gamma,\theta) < H_{sad}} \tag{5.161}$$

is the region of stability, while the region specified by inequality

$$\boxed{H(\gamma,\theta) > H_{sad}} \tag{5.162}$$

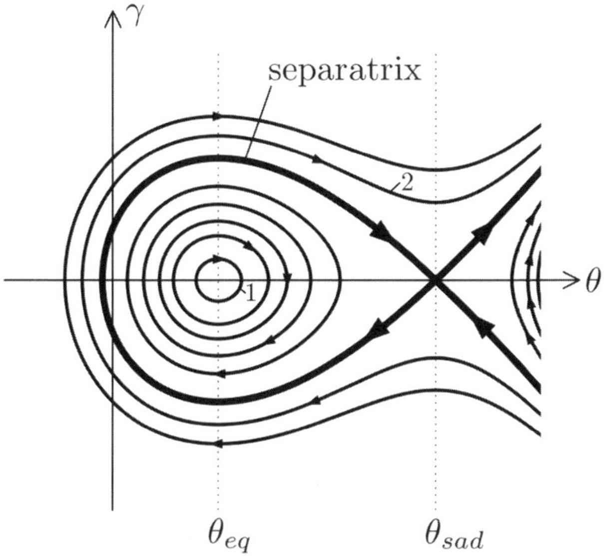

Fig. 5.6

is the region of instability.

The "size" of the stability region (or the depth of the "energy" well) which is measured by the difference $H_{sad} - H_{min}$ depends on the values of P_{mech} and P_m. Indeed, by using formulas (5.154), (5.155) and (5.156), it can be shown that

$$H_{sad} - H_{min} = \frac{1}{M}[2P_m \cos\theta_1 + P_{mech}(2\theta_1 - \pi)]. \tag{5.163}$$

It is clear from the last equation and formula (5.154) that the difference $H_{sad} - H_{min}$ goes to zero as P_{mech} approaches P_m. This fact has an important practical implication: since the stability is compromised when P_{mech} is close to P_m and since at synchronism $P_{mech} = P$, generators are usually operated at powers P substantially below P_m.

In power systems, the transient stability analysis is usually performed for sudden changes of P_{mech} and/or P_m. Consider one example of such an analysis. Assume that before time instant $t = 0$ a generator was delivering power under the conditions that mechanical shaft power was equal to $P_{mech}^{(0)}$ and P_m was equal to $P_m^{(0)}$. The synchronous performance of the generator under these conditions is characterized by a load (power) angle $\theta_{eq}^{(0)}$ at which $P_m^{(0)} \sin\theta$ is equal to $P_{mech}^{(0)}$ (see Figure 5.7). Then, at time $t = 0$ the mechanical shaft power is abruptly (very quickly) changed to a new

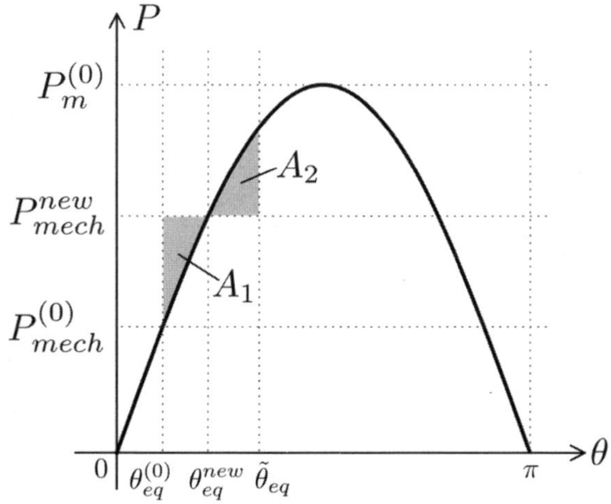

Fig. 5.7

value P_{mech}^{new}. It is required to find out if the performance of the generator will remain stable after this change. To solve this problem, we shall first find the initial conditions for the rotor dynamics caused by the change in P_{mech}. It is apparent that these initial conditions are the values of γ and θ immediately before the change in P_{mech}, namely,

$$\gamma(0) = 0, \qquad \theta(0) = \theta_{eq}^{(0)} = \sin^{-1}\left(\frac{P_{mech}^{(0)}}{P_m^{(0)}}\right). \tag{5.164}$$

Then, we shall use the Hamiltonian H^{new} for new value P_{mech}^{new},

$$H^{new}(\gamma, \theta) = \frac{\gamma^2}{2} + \frac{1}{M}\left[P_m^{(0)}(1 - \cos\theta) - P_{mech}^{new}\theta\right], \tag{5.165}$$

and evaluate this Hamiltonian at initial conditions (5.164) and at the saddle point $(0, \theta_{sad}^{new})$, respectively:

$$H^{new}\left(0, \theta_{eq}^{(0)}\right) = \frac{1}{M}\left[P_m^{(0)}(1 - \cos\theta_{eq}^{(0)}) - P_{mech}^{new}\theta_{eq}^{(0)}\right], \tag{5.166}$$

$$H_{sad}^{new}\left(0, \theta_{sad}^{new}\right) = \frac{1}{M}\left[P_m^{(0)}(1 - \cos\theta_{sad}^{new}) - P_{mech}^{new}\theta_{sad}^{new}\right], \tag{5.167}$$

where (see formulas (5.154) and (5.155))

$$\theta_{sad}^{new} = \pi - \sin^{-1}\left(\frac{P_{mech}^{new}}{P_m^{(0)}}\right). \tag{5.168}$$

Next, we check the stability condition (5.161). Namely, we check the validity of inequality

$$H^{new}\left(0, \theta_{eq}^{(0)}\right) < H_{sad}^{new} \tag{5.169}$$

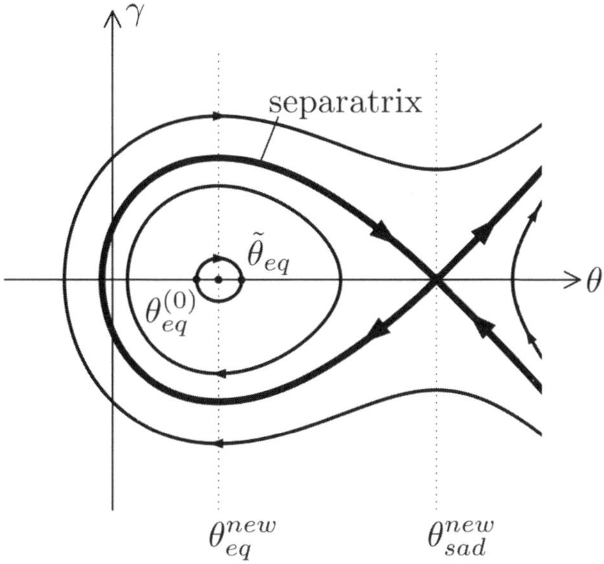

Fig. 5.8

for $H^{new}(0, \theta_{eq}^{(0)})$ and H_{sad}^{new} computed by using formulas (5.166), (5.167) and (5.168). If this inequality is satisfied, then the rotor dynamics after the sudden change in P_{mech} will be stable. Otherwise, it is unstable. It is apparent that the verification of inequality (5.169) is tantamount to the verification that the initial condition (5.164) belongs to the new region of stability formed after the change in P_{mech}.

Next, we shall demonstrate that the stability condition (5.169) is equivalent to the celebrated equal area criterion for stability of rotor dynamics. To this end, we consider the phase portrait of the rotor dynamics after the change in P_{mech} (see Figure 5.8). On this phase portrait, θ_{eq}^{new} corresponds to synchronous operation of the generator for P_{mech}^{new} (see Figure 5.7). It is clear that for rotor dynamics to be stable, initial condition $(0,\ \theta_{eq}^{(0)})$ must be in the region of stability, namely, it should be on some closed trajectory in the region inside the separatrix. Consider another point $(0,\ \tilde{\theta}_{eq})$ on the same solution trajectory (see Figure 5.8). Since any solution trajectory is a level line for the Hamiltonian, we find

$$H^{new}\left(0, \tilde{\theta}_{eq}\right) = H^{new}\left(0, \theta_{eq}^{(0)}\right). \tag{5.170}$$

Now, by using formula (5.140) for the Hamiltonian, we find from the last equation that

$$\int_{\theta_{eq}^{(0)}}^{\tilde{\theta}_{eq}} \left[P_{mech}^{new} - P_m^{(0)} \sin u\right] du = 0. \tag{5.171}$$

Since $\theta_{eq}^{(0)} < \theta_{eq}^{new} < \tilde{\theta}_{eq}$, the last equality can be transformed as follows:

$$\int_{\theta_{eq}^{(0)}}^{\theta_{eq}^{new}} \left[P_{mech}^{new} - P_m^{(0)} \sin u\right] du = \int_{\theta_{eq}^{new}}^{\tilde{\theta}_{eq}} \left[P_m^{(0)} \sin u - P_{mech}^{new}\right] du. \tag{5.172}$$

It is apparent that the integral in the left-hand side of equality (5.172) is equal to the shaded area A_1 on Figure 5.7, while the integral in the right-hand side of equality (5.172) is equal to the shaded area A_2 on the same figure. Consequently,

$$\boxed{A_1 = A_2.} \tag{5.173}$$

This is the equal area stability criterion.

It is worthwhile to stress that the reasoning used in the discussed example can be easily extended to a general case of changes in P_{mech} and P_m which may be required in the case of fault situations or other emergencies. *Namely, the first step is always to find the initial condition for γ and θ from the operational conditions of the synchronous generator before the sudden changes in P_{mech} and P_m occurred. The next step is to form the new Hamiltonian by using new (changed) values of P_{mech} and P_m in formula (5.141) and to find θ_{sad}^{new} by using formulas (5.154) and (5.155) (please remember that θ_2 in these formulas is θ_{sad}). Finally, evaluate the new Hamiltonian at the initial conditions found on the first step and at the saddle point ($\gamma = 0$, $\theta = \theta_{sad}^{new}$) and verify the validity of inequality (5.161).* The outlined procedure of stability verification is quite simple and purely algebraic in nature. It can always be recast in terms of the equal area criterion as has been demonstrated above.

In our previous discussion, stable rotor dynamics has consisted of periodic swings around an equilibrium corresponding to synchronous operation of the generator. These undamped swings occur because the structure of the "swing" equation (5.135) does not account for any damping which may occur due to the induced eddy currents in solid conducting rotors or mechanical (frictional) losses. One possible way to account for such damping is by modifying equation (5.135) through the introduction of a damping term:

$$M\frac{d^2\theta(t)}{dt^2} + D\frac{d\theta(t)}{dt} = P_{mech} - P_m \sin\theta, \tag{5.174}$$

where it is assumed that $D > 0$. As before, the last equation can be written in the following state space form:

$$\frac{d\theta(t)}{dt} = \gamma(t), \tag{5.175}$$

$$\frac{d\gamma(t)}{dt} = -\frac{D}{M}\gamma + \frac{1}{M}[P_{mech} - P_m \sin\theta]. \tag{5.176}$$

It can be easily shown that the introduction of damping does not change the critical points of the dynamics. Namely, the critical points of the dynamics described by equations (5.175)-(5.176) are the same as the critical points of the dynamics described by equations (5.138)-(5.139). Let $\gamma(t)$ and $\theta(t)$ be a solution of equations (5.175)-(5.176) starting from some initial condition in the region inside the separatrix. Consider $H[\gamma(t), \theta(t)])$, where H is the Hamiltonian given by the formula (5.141). Then, we find

$$\frac{dH[\gamma(t), \theta(t)]}{dt} = \frac{\partial H}{\partial \gamma}\frac{d\gamma(t)}{dt} + \frac{\partial H}{\partial \theta}\frac{d\theta(t)}{dt}. \tag{5.177}$$

By using equations (5.175) and (5.176) as well as formula (5.141), we obtain

$$\frac{dH[\gamma(t), \theta(t)]}{dt} = -\frac{D}{M}\gamma^2 + \frac{\gamma}{M}[P_{mech} - P_m \sin\theta] - \frac{\gamma}{M}[P_{mech} - P_m \sin\theta], \tag{5.178}$$

which is reduced to

$$\frac{dH[\gamma(t), \theta(t)]}{dt} = -\frac{D}{M}\gamma^2 \leq 0, \tag{5.179}$$

where the equality is reached only at synchronism when $\gamma = 0$. Thus, the damped rotor dynamics is such that it leads to the continuous decay of the Hamiltonian function. This continuous decay may eventually bring the damped dynamics to the minimum of H which is the equilibrium corresponding to synchronous operation of the generator.

Chapter 6

Induction Machines

6.1 Design and Principle of Operation of Induction Machines

In this chapter, induction machines are discussed. Induction machines can be operated as motors as well as generators. Induction motors have been the workhorse of industry since the very inception of ac power. One of the reasons is that the induction motors have been extensively used in various power tools, which have dramatically increased the productivity of labor. Induction motors achieved and maintained this unique position due to the simplicity of their design and relatively low cost. During the past thirty years, this position has been somewhat challenged by the advent of permanent magnet synchronous motors. However, during the same time span, induction machines have found new applications as generators in wind energy systems.

Now, we shall discuss the basic design of induction machines. Induction machines have two major parts (see Figure 6.1): *stator* and *rotor* separated by a *very small air gap*. The design of the stator is conceptually identical to the design of the stator of synchronous machines. Namely, it has a laminated structure. This means that it is assembled of a very large number of very thin varnished (or oxidized) silicon steel laminations. The latter is done to substantially reduce eddy current power losses. The stator has slots uniformly distributed over its interior (gap-facing) surface. The distributed three-phase windings are embedded in these slots. Furthermore, these phase windings are shifted with respect to one another along the interior circumference of the stator by $120°$ in the case of two-pole machines or by $240°/p$ in the case of p-pole machines. These three stationary phase windings are usually connected to a three-phase power network. As a result,

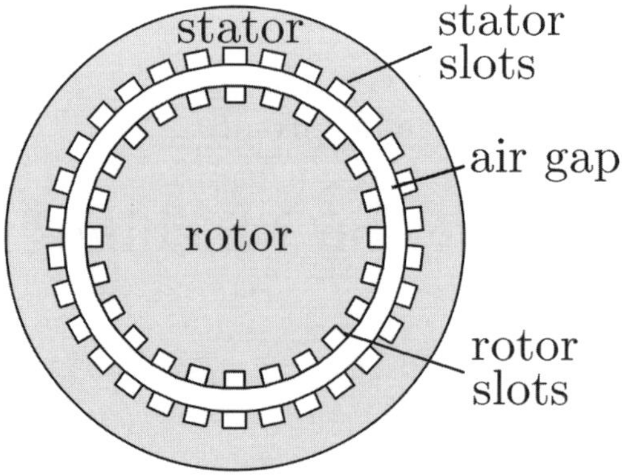

Fig. 6.1

they are excited by currents of the same peak value and the same frequency but phase-shifted in time with respect to one another by $2\pi/3$ (or $120°$). It has been demonstrated in Chapter 4 that the stationary distributed three-phase windings energized in this way create uniformly rotating magnetic field (when higher-order spatial magnetic field harmonics are neglected). It has also been shown in Chapter 4 that the speed of rotation of this magnetic field measured in terms of revolutions per minute is given by the following formula:

$$n_{syn} = \frac{120f}{p}. \tag{6.1}$$

This rotating magnetic field created by the stator winding is at the very foundation of the principle of operation of the induction machine.

The rotor is the rotating part of an induction machine. It also has laminated structure to reduce eddy current power losses as well as many slots uniformly distributed over the exterior (gap-facing) surface of the rotor (see Figure 6.1). The number of slots in the rotor (N_r) and stator (N_s) should be carefully chosen to avoid the appearance of very strong parasitic torques. Especially strong parasitic torques occur when

$$N_s = N_r \quad \text{or} \quad N_s - N_r = 2p, \tag{6.2}$$

and these torques are detrimental to the operation of induction machines. The detailed discussion of the physical origin of these parasitic torques is beyond the scope of this text. We shall only mention that slots are sometimes skewed slightly along the length of the rotor to suppress parasitic torques caused by the slotted structure of rotors and stators.

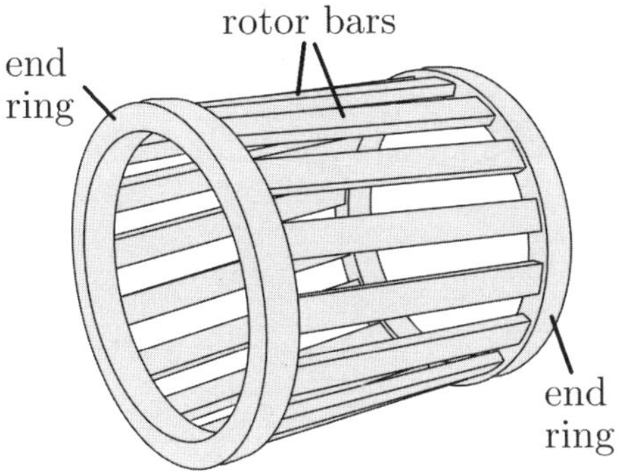

Fig. 6.2

Rotor windings are embedded in slots. As far as the design of these windings is concerned, there are two distinct types of rotors of induction machines: 1) *wound* rotors and 2) *squirrel cage* rotors. In the case of wound rotors, distributed three-phase windings are embedded in rotor slots. These windings are similar in design to those in stators and they are usually wound for the same number of poles as the stator windings. These rotor windings are connected to three slip rings mounted on the rotor shaft. Carbon brushes ride over these slip rings, and they electrically connect the rotor windings to three equal external resistances. These external resistances are usually needed during the start of induction motors in order to increase overall rotor winding resistances which result in the increase of the starting torque. After the starting of an induction motor is accomplished, the carbon brushes are externally short-circuited.

In the case of *squirrel cage* rotors, the rotor slots are occupied by copper or aluminum bars (known as rotor bars) which are short-circuited by two conducting end rings of the same material as the rotor bars. As a result, the conducting rotor windings have the structure of a squirrel cage shown in Figure 6.2. In the case of squirrel cage rotors, the high resistance during the start and, consequently, high starting torques can be achieved by using double-cage or deep-bar rotor designs illustrated in Figures 6.3a and 6.3b, respectively. These designs exploit the *skin effect* phenomenon to increase the rotor winding resistance during the start. Indeed, as is intuitively clear and will be discussed in detail later, the rotor frequency f_r is substantially higher during the start, i.e., when the rotor is originally at standstill, than during the normal operation when rotor speed n is close to the synchronous

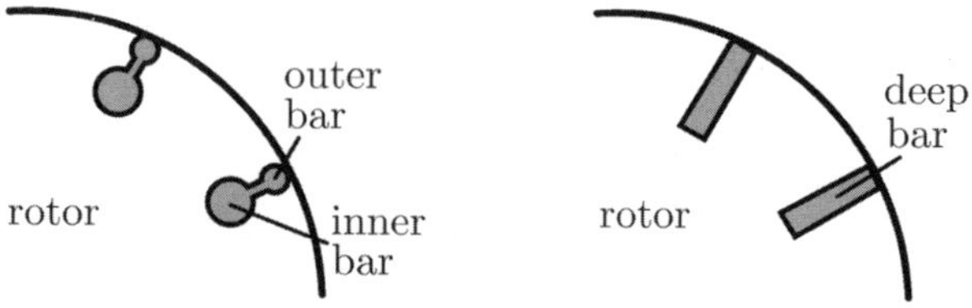

(a) (b)

Fig. 6.3

speed of the rotating magnetic field n_{syn}. For this reason, the skin effect phenomenon occurs during the start and only outer bars or outer sections of deep bars are utilized for current conduction, which results in large resistances of rotor windings. During the normal operation with rotor speed close to synchronous speed and f_r close to zero, both outer and inner bars or entire deep bars are utilized for current conduction and this leads to small resistances of rotor windings. It must be noted that the manifestation of the skin effect phenomenon in deep bars (or double-cage bars) is strongly enhanced by the embedding of these bars in slots surrounded by high magnetic permeability ferromagnetic material.

Having described the basic design of induction machines, we shall turn our attention to the discussion of the principle of operation of induction motors. As mentioned before, the stator winding creates a uniformly rotating magnetic field as soon as this winding is electrically connected to a conventional three-phase power network. This rotating magnetic field induces currents in the rotor winding. These induced currents interact with the rotating magnetic field of the stator, and as a result of this interaction, electromagnetic forces and torque appear that cause the rotor to rotate in the direction of the stator magnetic field. The latter is the case because according to Lenz's law currents are always induced in such a way as to counteract (to reduce) the cause of induction. The latter is achieved when the rotor rotates in the same direction as the stator magnetic field because this reduces the relative speed of the stator magnetic field with respect to the rotor. It is apparent that the speed n of rotation of the rotor of the induction motor is always below the speed of rotation of the stator magnetic field n_{syn}:

$$\boxed{n < n_{syn}.} \tag{6.3}$$

Indeed, if hypothetically mechanical speed n of the rotor of the induction motor reaches the speed n_{syn} of the stator magnetic field, then this magnetic field is static (motionless) with respect to the rotor and, consequently, no currents are induced in the rotor winding and no torque is created. Thus, this hypothetical situation is not possible. However, as will be discussed in the last section of this chapter, the mechanical speed of the rotor is actually very close to the speed of the stator magnetic field n_{syn} defined by formula (6.1). Consequently,

$$\boxed{n \approx n_{syn} = \frac{120f}{p}.} \tag{6.4}$$

The last formula clearly suggests that the mechanical speed of the induction motor can be controlled by controlling the frequency of stator ac currents. The latter is the foundation for the frequency control of speed of induction motors that can be achieved by using ac-to-ac power electronic converters. These converters are discussed in the last part of this book. The integration of these power converters with induction machines results in so-called "power semiconductor drives" which have wide-ranging applications in industry.

It is clear from the presented discussion that the principle of operation of induction motors is based on the induction of currents in rotor windings of these motors by stator rotating magnetic fields. This explains why the word "induction" is used in the name of these motors. It is also clear that the rotor of induction motors and the stator magnetic field do not rotate in synchronism (see formula (6.3)). For this reason, induction motors are often called "asynchronous" motors. There is always some finite "slip" between the rotating rotor and the rotating stator magnetic field. This slip is denoted s and mathematically defined by the formula

$$\boxed{s = \frac{n_{syn} - n}{n_{syn}}.} \tag{6.5}$$

The last formula can also be written in terms of angular speeds of the rotor (Ω) and stator magnetic field (Ω_{syn}) measured in terms of radians per second instead of rpm. Namely,

$$s = \frac{\Omega_{syn} - \Omega}{\Omega_{syn}}. \tag{6.6}$$

Since the rotor moves in the same direction as the stator magnetic field ($n > 0$) and since inequality (6.3) is valid, we conclude that for induction motors

$$\boxed{0 < s < 1.} \tag{6.7}$$

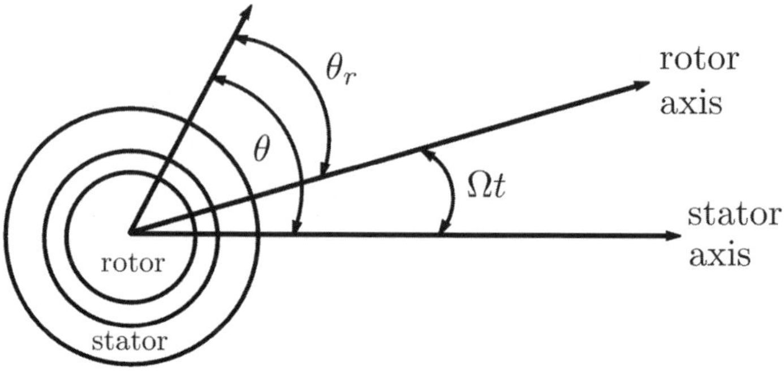

Fig. 6.4

Now, we shall discuss the important relation between the frequency f_r of currents induced in the rotor windings and the frequency f of the stator currents. To this end, we shall recall that the vector magnetic potential of the rotating stator magnetic field is given by the formula (see (4.69))

$$A(r, \theta, t) = \frac{3\mu_0 b}{2\nu\delta} F_m \sin(\omega t - \nu\theta), \tag{6.8}$$

where θ is a polar angle measured in the stator reference frame (see Figure 6.4), ν is (as before) the number of pole pairs ($\nu = p/2$) of the stator winding, while F_m can be construed as the fundamental spatial harmonic of mmf created by the entire three-phase stator winding. It is clear from Figure 6.4 that the angle θ is related to the polar angle θ_r measured in the rotor reference frame by the formula

$$\theta = \theta_r + \Omega t. \tag{6.9}$$

By substituting the last relation into equation (6.8), we find the following expression for the vector magnetic potential in the rotor reference frame:

$$A(r, \theta_r, t) = \frac{3\mu_0 b}{2\nu\delta} F_m \sin[(\omega - \nu\Omega)t - \nu\theta_r]. \tag{6.10}$$

Next, we shall derive the following formula:

$$\omega = \nu\Omega_{syn}. \tag{6.11}$$

Indeed, from equation (6.1) we have

$$\omega = 2\pi f = 2\pi \frac{n_{syn} p}{120} = \frac{\pi n_{syn}}{30} \nu. \tag{6.12}$$

On the other hand,

$$\Omega_{syn} = \frac{n_{syn} 2\pi}{60} = \frac{\pi n_{syn}}{30}. \tag{6.13}$$

From the last two formulas, we easily obtain relation (6.11). By using this relation, we derive

$$\omega - \nu\Omega = \nu(\Omega_{syn} - \Omega) = \nu\Omega_{syn}\frac{\Omega_{syn} - \Omega}{\Omega_{syn}}. \tag{6.14}$$

Now, by recalling formulas (6.6) and (6.11), we conclude that

$$\omega - \nu\Omega = s\omega. \tag{6.15}$$

By using the last equation in formula (6.10), we find

$$A(r, \theta_r, t) = \frac{3\mu_0 b}{2\nu\delta}F_m \sin[s\omega t - \nu\theta_r], \tag{6.16}$$

which means that in the rotor reference frame the vector magnetic potential of the stator magnetic field and, consequently, the magnetic field itself, vary with the angular frequency

$$\omega_r = s\omega. \tag{6.17}$$

This implies that the rotating stator magnetic field will induce electric currents in the rotor winding whose frequency f_r is given by the formula

$$\boxed{f_r = sf.} \tag{6.18}$$

It is apparent from the last equation that the rotor frequency is the largest ($f_r = f$) when the rotor is at standstill and that $f_r \approx 0$ during the normal operation when $n \approx n_{syn}$. This fact has already been used in our discussion of operation of induction motors with double-cage and deep-bar rotors.

It is mentioned in the beginning of this section that the air gap lengths δ in induction motors are quite small. One of the important reasons is that the smaller the air gap length, the less reactive power needed for the performance of an induction motor and the larger its power factor. To illustrate this, the following expression for the power factor can be invoked:

$$\cos\varphi = \frac{P}{\left|\hat{S}\right|} = \frac{P}{\sqrt{P^2 + Q^2}}. \tag{6.19}$$

As was demonstrated in Chapter 3 of Part I, in magnetic systems with air gaps most of the magnetic field energy is localized in the air gaps. This means that when the same peak value of magnetic flux density is maintained in the air gaps, the smaller the air gap, the smaller the reactive power that is needed to be supplied and, according to formula (6.19), the larger the power factor. This explains why the air gap length in the induction motor is very small and it is usually between 0.5 and 3 millimeters. One may even say that "an air gap in an induction motor is so small that snakes can hardly crawl through it." This is clearly in contrast with the large air gaps in synchronous generators, which for two-pole machines may approach 15 cm. Large air gaps in synchronous generators result in reduction of their synchronous impedances, which is beneficial (among other things) for static and transient stability of these generators.

It is clear from the presented discussion of the design and principle of operation of induction motors that in these motors there exists a very strong electromagnetic coupling between stator and rotor windings. In this sense, induction motors are quite similar to transformers. This very strong electromagnetic coupling makes many elements of the theory of induction motors (in particular, the structure and the derivation of their equivalent circuits) resemble those of transformers.

Up to this point, the operation of induction machines as motors has been discussed. However, induction machines can be operated as generators as well. This can be achieved by connecting a rotor of an induction machine to a prime mover (turbine, for instance) and by driving the rotor above the synchronous speed:

$$\boxed{n > n_{syn}.} \tag{6.20}$$

This, according to formula (6.5), results in negative slip:

$$\boxed{s < 0.} \tag{6.21}$$

It can be shown (see the last section of this chapter) that the change in the sign of slip results in the change in the sign of active electric power at the induction machine terminals. This change in sign implies that the induction machine instead of being operated as a motor for $s > 0$ is operated as a generator for $s < 0$. It is remarkable that it generates active power of *the frequency of the power network* to which it is connected despite the non-synchronous (and possibly variable) speed of its rotor. This makes this regime of induction generator attractive for utilization in wind energy systems. Indeed, the speed of wind turbines is not constant in time. However, by connecting wind turbines to rotor shafts of induction machines through proper gearboxes, it is possible to maintain the rotor speed above synchronous and to operate induction machines as wind turbine-driven generators.

There exists another way to operate wound rotor induction machines as variable-speed generators of ac electric power of constant frequency. In this case, the induction machines are operated as *doubly-fed* machines. The latter implies that both stator and rotor windings of such machines are connected to power networks. For stator windings, these are *direct* connections, while rotor windings are connected to the same power networks *through slip rings, brushes and ac-to-ac power electronic converters.* These converters supply ac currents to the distributed three-phase rotor windings of such controllable frequency that these excited rotor windings create rotor magnetic field rotating with respect to the rotor at the speed n' equal to $n_{syn} - n$. This implies that these rotor magnetic fields rotate with respect to stators with synchronous speed n_{syn}. In this sense, the rotors of such doubly-fed induction machines create the same magnetic fields as rotors of synchronous generators excited by dc currents and driven with constant-in-time synchronous speed. Such doubly-fed induction generators are now extensively used in wind power generation. It is worthwhile to note that induction generators, in contrast with synchronous generators, are not stand-alone generators; they require the energized power network for their operation. This may eventually limit the penetration of induction machine-based wind power generation in existing utility power systems.

6.2 Coupled Circuit Equations and Equivalent Circuits for Induction Machines

In induction machines, there are stator and rotor windings which are strongly electromagnetically coupled. For instance, in the case of wound rotor machines, there are three-phase windings on the stator and three-phase windings on the rotor which are all coupled to one another. This, in general, results in six coupled circuit equations. However, these coupled equations can be appreciably simplified and reduced to two coupled equations by taking into account that *the electromagnetic coupling is mostly actualized through rotating magnetic fields.* Indeed, the stator currents have the same peak values, the same frequency and they are phase-shifted in time by 120°. These currents create the rotating magnetic field of the stator, which induces three voltages in the three-phase distributed windings of the rotor which are of the same magnitude and phase-shifted in time by 120°. These voltages result in three-phase rotor currents of the same magnitude and phase-shifted in time by 120°. These currents in the rotor windings also create rotating magnetic fields in the air gaps of induction machines, and *the coupling between various windings is realized through these rotating magnetic fields of the stator and rotor.* Magnetic fields created by stator windings rotate with speed n_{syn} given by formula (6.1). Similarly, magnetic fields created by the rotor windings rotate with respect to the rotor with speed n'_{syn} given by the formula

$$n'_{syn} = \frac{120f_r}{p}. \tag{6.22}$$

By taking into account that rotor frequency f_r is related to stator frequency f by expression (6.18), from the last equation we find

$$n'_{syn} = \frac{120f}{p}s = sn_{syn}, \tag{6.23}$$

which, according to the definition of slip s (see (6.5)), means that

$$n'_{syn} = n_{syn} - n, \tag{6.24}$$

where as before n is the rotational speed of the rotor. From the last equation immediately follows that the magnetic fields created by the rotor windings rotate with respect to the stator with speed n_{syn}. In other words, magnetic fields of the stator and rotor windings rotate in synchronism.

In section 4.3, we derived the following expression for the reactance of the stator winding (see formula (4.109)):

$$X_{11}^{(m)} = \frac{12\mu_0 b\ell}{\delta} f N_1^2 k_{w1}^2, \tag{6.25}$$

where all notations have the same meaning as before, and subscript "1" indicates that the quantities are related to the stator windings, which are regarded as the primary windings. It is discussed in section 4.3 that the reactance $X_{11}^{(m)}$ accounts for voltage induced in one phase of the stator winding by the rotating magnetic field created by three-phase currents in the stator winding. In this sense, this reactance accounts for self and mutual inductances of all three stator phase windings. It is also discussed in section 4.3 that this is the main reactance (as indicated by superscript "(m)") because it is derived for the ideal machine where end parts, slots and higher-order spatial harmonics of stator magnetic field are neglected. These neglected (small) contributions to the overall reactance are accounted for by the leakage reactance X_1^ℓ, which is added to the main reactance $X_{11}^{(m)}$ to get the total reactance of the stator (primary) winding:

$$X_{11} = X_{11}^{(m)} + X_1^\ell. \tag{6.26}$$

By using the same line of reasoning that has been used for the derivation of "rotating field" reactance $X_{11}^{(m)}$, the following formula can be derived for the main reactance of the rotor (secondary) winding:

$$X_{22}^{(m)} = \frac{12\mu_0 b\ell}{\delta} f N_2^2 k_{w2}^2, \tag{6.27}$$

where it is tacitly assumed that the rotor is at standstill and $f_r = f$. For the rotating rotor, f should be replaced by $f_r = sf$.

As before, formula (6.27) is valid for the ideal machine. The total reactance of the rotor winding which accounts for contributions coming from end parts, slots and higher-order spatial harmonics of rotor magnetic field can be written as

$$X_{22} = X_{22}^{(m)} + X_2^{\ell}, \tag{6.28}$$

where X_2^{ℓ} is the secondary leakage reactance.

Next, we shall discuss the coupling reactance between the stator and rotor windings. In doing so, we neglect coupling due to leakage magnetic fields as a small effect and consider only the coupling due to rotating magnetic fields created by the stator and rotor windings. By almost literally repeating the line of reasoning used in section 4.3 for the derivation of formula (6.25) for reactance $X_{11}^{(m)}$, the following formula can be obtained for the "rotating magnetic field" coupling reactance:

$$X_{12} = \frac{12\mu_0 b\ell}{\delta} f N_1 N_2 k_{w1} k_{w2}. \tag{6.29}$$

This reactance accounts for coupling between one phase of the stator winding and three phases of the rotor winding. And the other way around, reactance X_{12} given by formula (6.29) accounts for coupling of one phase of the rotor winding with three phases of the stator winding. This aggregate nature of coupling represented by reactance X_{12} is the consequence of it being computed by using rotating magnetic fields, i.e., fields created by all three-phase stator (or rotor) currents. It is clear as before that, in the case of the rotating rotor, frequency f in formula (6.29) should be replaced by $f_r = sf$. Finally, it must be remarked that formulas (6.25), (6.27) and (6.29) are valid for two-pole machines. In the case when $\nu = \frac{p}{2} > 1$, δ must be replaced by $\nu\delta$ in the mentioned formulas.

From formulas (6.25), (6.27) and (6.29), we find

$$\frac{X_{11}^{(m)}}{X_{22}^{(m)}} = a^2, \tag{6.30}$$

$$\frac{X_{11}^{(m)}}{X_{12}} = a, \tag{6.31}$$

$$\frac{X_{22}^{(m)}}{X_{12}} = \frac{1}{a}, \tag{6.32}$$

where

$$\boxed{a = \frac{N_1 k_{w1}}{N_2 k_{w2}}.} \tag{6.33}$$

Now, we shall use per-phase analysis and write the coupled circuit equations for stator and rotor windings in the form

$$\begin{cases} \hat{V}_1 = R_1\hat{I}_1 + jX_{11}\hat{I}_1 - jX_{12}\hat{I}_2, & (6.34) \\ 0 = R_2\hat{I}_2 + jsX_{22}\hat{I}_2 - jsX_{12}\hat{I}_1. & (6.35) \end{cases}$$

In the written equation (6.34), the term $R_1\hat{I}_1$ accounts for voltage drop across the resistance R_1 of one phase (phase a, for instance) of the stator winding; the term $jX_{11}\hat{I}_1$ accounts for the voltage induced in the same phase by the rotating magnetic field of the stator; the term $-jX_{12}\hat{I}_2$ accounts for the voltage induced in the same phase by the rotating magnetic field of the rotor; finally, $\hat{V}_1$ is the phasor of the applied terminal voltage for the same phase. The three terms in equation (6.35), written for one phase (phase a, for instance) of the rotor windings have similar meaning. Factor s in the last two terms of equation (6.35) indicates that this equation is written for the rotating rotor when the rotor currents have frequency $f_r = sf$, while reactances X_{22} and X_{12} are given by formulas (6.27) and (6.29) for frequency f. Thus, by using the fact that the coupling between six stator and rotor phase windings is *caused by rotating magnetic fields* of the rotor and stator, this coupling can be fully described by two coupled circuit equations (6.34) and (6.35). All subsequent discussion deals with mathematical transformations of these equations. The first step in this direction is to divide the equation (6.35) by s and to rewrite these equations as follows:

$$\begin{cases} \hat{V}_1 = R_1\hat{I}_1 + jX_{11}\hat{I}_1 - jX_{12}\hat{I}_2, & (6.36) \\ 0 = \dfrac{R_2}{s}\hat{I}_2 + jX_{22}\hat{I}_2 - jX_{12}\hat{I}_1. & (6.37) \end{cases}$$

The last two equations can be physically interpreted as coupled circuit equations for a standstill induction machine whose secondary (rotor) resistance is equal to R_2/s. Thus, any rotating induction machine can be reduced to an equivalent standstill induction machine by the proper adjustment of its secondary resistance.

The next step in the mathematical transformations of the coupled circuit equations is to introduce the scaled rotor (secondary) current

$$\hat{I}_2' = \frac{1}{a}\hat{I}_2, \qquad \hat{I}_2 = a\hat{I}_2'. \tag{6.38}$$

By replacing $\hat{I}_2$ by $\hat{I}_2'$ in equations (6.36) and (6.37) and by multiplying the equation (6.37) by a, we find

$$\begin{cases} \hat{V}_1 = R_1\hat{I}_1 + jX_{11}\hat{I}_1 - jaX_{12}\hat{I}_2', & (6.39) \\ 0 = \dfrac{a^2R_2}{s}\hat{I}_2' + ja^2X_{22}\hat{I}_2' - jaX_{12}\hat{I}_1. & (6.40) \end{cases}$$

The next step of the transformations is to subtract and add the term $jaX_{12}\hat{I}_1$ in equation (6.39) as well as to subtract and add the term $jaX_{12}\hat{I}_2'$ in equation (6.40). This leads to

$$\begin{cases} \hat{V}_1 = R_1\hat{I}_1 + j(X_{11} - aX_{12})\hat{I}_1 + jaX_{12}\left(\hat{I}_1 - \hat{I}_2'\right), & (6.41) \\ 0 = \dfrac{a^2R_2}{s}\hat{I}_2' + ja^2\left(X_{22} - \dfrac{1}{a}X_{12}\right)\hat{I}_2' + jaX_{12}\left(\hat{I}_2' - \hat{I}_1\right). & (6.42) \end{cases}$$

From formulas (6.26) and (6.31) we obtain

$$X_{11} - aX_{12} = X_{11} - X_{11}^{(m)} = X_1^{\ell}. \tag{6.43}$$

Furthermore, from equation (6.31) follows

$$aX_{12} = X_{11}^{(m)}. \tag{6.44}$$

From relations (6.28) and (6.32) we derive

$$X_{22} - \frac{1}{a}X_{12} = X_{22} - X_{22}^{(m)} = X_2^{\ell}. \tag{6.45}$$

Now, we introduce the scaled secondary resistance and scaled secondary leakage reactance by using the following formulas, respectively:

$$R_2' = a^2R_2, \tag{6.46}$$

$$a^2\left(X_{22} - \frac{1}{a}X_{12}\right) = a^2X_2^{\ell} = \left(X_2^{\ell}\right)'. \tag{6.47}$$

By substituting formulas (6.43) and (6.44) into equation (6.41) and at the same time substituting formulas (6.44), (6.46) and (6.47) into equation (6.42), we arrive at the following form of the coupled circuit equations:

$$\begin{cases} \hat{V}_1 = R_1\hat{I}_1 + jX_1^{\ell}\hat{I}_1 + jX_{11}^{(m)}\left(\hat{I}_1 - \hat{I}_2'\right), & (6.48) \\ 0 = \dfrac{R_2'}{s}\hat{I}_2' + j\left(X_2^{\ell}\right)'\hat{I}_2' + jX_{11}^{(m)}\left(\hat{I}_2' - \hat{I}_1\right). & (6.49) \end{cases}$$

Thus, by using *equivalent mathematical transformations*, we have reduced the original coupled circuit equations (6.34)-(6.35) to the coupled equations (6.48)-(6.49). The important outcome of these transformations is the fact that equations (6.48)-(6.49) coincide with KVL equations for the electric

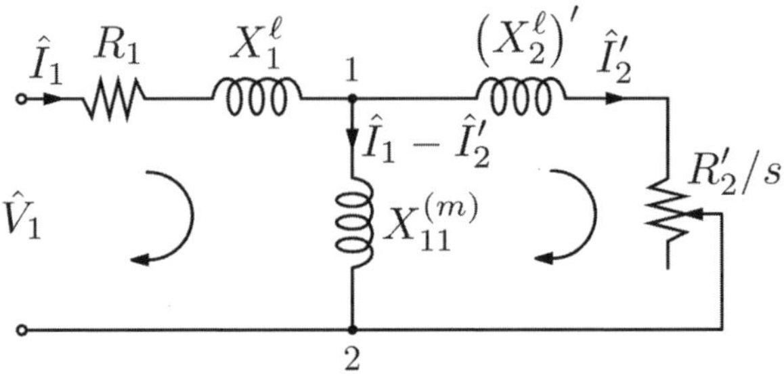

Fig. 6.5

circuit shown in Figure 6.5. This circuit can be considered as an equivalent circuit for an induction machine. This is so because this circuit and the induction machine are described by mathematically equivalent sets of equations. For this reason, the induction machine and the electric circuit shown in Figure 6.5 are indistinguishable as far as the relations between terminal voltages and terminal currents are concerned. In other words, we can replace in a power network an induction machine by its equivalent circuit without affecting currents and voltages in the network because in both cases the network is described by mathematically equivalent sets of equations.

It is interesting to point out that the equivalent circuit for an induction machine is not unique. Indeed, coupled equations (6.41) and (6.42) have been derived from the original coupled circuit equations (6.34) and (6.35) by using equivalent mathematical transformations and the scaling (6.38). Furthermore, by using formula (6.46) and notation

$$X_{22}' = a^2 X_{22}, \tag{6.50}$$

equations (6.41) and (6.42) can be rewritten as follows:

$$\begin{cases} \hat{V}_1 = R_1\hat{I}_1 + j(X_{11} - aX_{12})\hat{I}_1 + jaX_{12}\left(\hat{I}_1 - \hat{I}_2'\right), & (6.51) \\ 0 = \dfrac{R_2'}{s}\hat{I}_2' + j\left(X_{22}' - aX_{12}\right)\hat{I}_2' + jaX_{12}\left(\hat{I}_2' - \hat{I}_1\right). & (6.52) \end{cases}$$

The last two equations are valid for any value of scaling factor a, not only when a is given by formula (6.33). The coupled equations (6.51) and (6.52) coincide with the KVL equations for the electric circuit shown in Figure 6.6. Consequently, this electric circuit can be considered as an equivalent circuit for the induction machine as well, and this is true for any choice of scaling parameter a. The choice of a defined by formula (6.33) leads

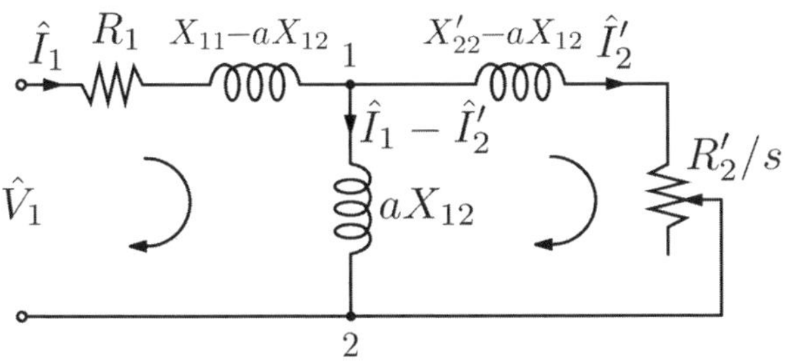

Fig. 6.6

to the exposure of the leakage reactances and this facilitates the further development of the equivalent circuit as discussed below.

Now, we shall return to the discussion of the equivalent circuit shown in Figure 6.5. In deriving this equivalent circuit, eddy current losses in the stator and rotor cores were neglected. Usually, eddy current losses in the rotor are quite small in comparison with eddy current losses in the stator. This is so because classical eddy current losses are proportional to the square of the frequency of the time-varying magnetic field. This frequency in the rotor reference frame is $f_r = sf$, and it is quite small compared to the stator frequency f for normal operation of the induction machine when slip s is quite small. Thus, we shall discuss only how eddy current losses in the stator core can be accounted for in the structure of the equivalent circuit shown in Figure 6.5. It has been shown in section 3.5 of Part I that the eddy current losses are proportional to the square of peak value of magnetic flux density B_m^2 in the ferromagnetic core:

$$P_e \sim B_m^2. \tag{6.53}$$

On the other hand, B_m^2 is proportional to the square of magnetic flux peak value which, in turn, is proportional to the square of peak value of the voltage induced by this flux. In the equivalent circuit shown in Figure 6.5, this voltage can be identified with the voltage V_{m12} across terminals 1 and 2. Thus, it can be concluded that

$$P_e \sim V_{m12}^2. \tag{6.54}$$

The last formula suggests the idea of modeling eddy current losses in the stator core of the induction machine as ohmic losses P_{R_e} in some equivalent resistor R_e connected across the terminals 1 and 2, that is, in parallel with $X_{11}^{(m)}$ (see Figure 6.7). The rationale behind this idea is the fact that the

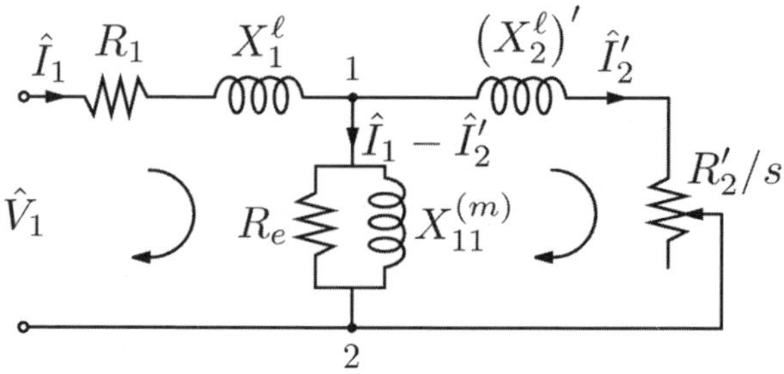

Fig. 6.7

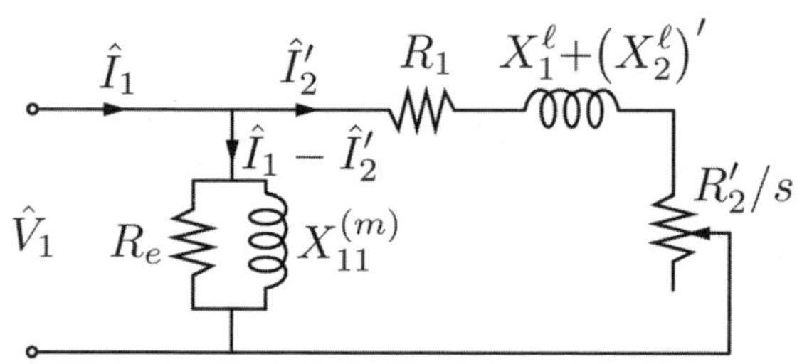

Fig. 6.8

ohmic losses in R_e are also proportional to V_{m12}^2:

$$P_{R_e} = \frac{V_{m12}^2}{2R_e}. \tag{6.55}$$

The resistor R_e can be chosen from the condition

$$P_e = P_{R_e}, \tag{6.56}$$

which leads to

$$R_e = \frac{V_{m12}^2}{2P_e}. \tag{6.57}$$

The electric circuit shown in Figure 6.7 is the complete equivalent circuit for an induction machine.

It is often convenient to deal with the approximate equivalent circuit obtained from the circuit shown in Figure 6.7 by moving the parallel-connected R_e and $X_{11}^{(m)}$ directly across the primary terminals. Such an equivalent circuit is shown in Figure 6.8. Such a transformation of the equivalent circuit is usually justified on the grounds that R_e and $X_{11}^{(m)}$ are quite large. Consequently, current $\hat{I}_1 - \hat{I}_2'$ (see Figure 6.7) is small in magnitude. For this reason, current $\hat{I}_1$ through R_1 and X_1^ℓ is almost equal to current $\hat{I}_2'$. Moreover, since R_1 and X_1^ℓ are small, the voltage across terminals 1 and 2 is almost equal to $\hat{V}_1$. The mentioned facts are consistent with the structure of the equivalent electric circuit shown in Figure 6.8. The parameters of the approximate equivalent circuit can be experimentally determined by using the no-load test ($s \approx 0$) and locked-rotor test ($s = 1$). These tests are similar to the open- and short-circuit tests for the transformer, respectively, and the details are left as a useful problem for the reader.

In our discussion, we dealt with an induction machine with a wound rotor. The case of a squirrel cage rotor is theoretically more complicated. The reason is that the squirrel cage cannot be treated as a three-phase winding but rather must be treated as a polyphase winding. The latter treatment is based on the fact that the stator rotating magnetic field induces the currents in slot bars of the squirrel cage which all have the same magnitude but are incrementally shifted in time with respect to one another by the same angle dependent on the number of rotor slots. It can be shown that these induced bar currents create uniformly rotating magnetic fields (as far as their fundamental spatial harmonic is concerned), and the speed of this field is given by the formula (6.22). This eventually leads to coupled circuit equations similar to equations (6.34)-(6.35) and to an equivalent electric circuit of the type shown in Figure 6.7. The detailed discussion of all these issues is beyond the scope of this text.

6.3 Torque-Speed Characteristics of the Induction Motor

An induction motor is an electromechanical device which is used to convert electrical energy supplied by a power network into mechanical energy of the rotating rotor. In this section, we shall discuss mechanical characteristics of the induction motor such as the mechanical torque on its rotor shaft and the dependence of this torque on the rotational speed of the rotor. The derivation of the mathematical expression for the mechanical torque will be based on the equivalent circuit for the induction machine shown in Figure 6.8. This equivalent circuit suggests that the total active power P_2 transferred across the air gap to the rotor can be written as

$$P_2 = 3\left(I_2'\right)^2 \frac{R_2'}{s}, \tag{6.58}$$

where the factor 3 is used to account for the three phases of the stator winding and I_2' is the rms value of $\hat{I}_2'$.

By using formulas (6.38) and (6.46) in the last equation, we obtain

$$P_2 = 3I_2^2 \frac{R_2}{s}. \tag{6.59}$$

The active power transferred to the rotor has two distinct components: 1) an active power P_2^{heat} which covers ohmic losses (heat dissipation) in the rotor winding and 2) an active power $\tilde{P}_2$ which is converted into the mechanical power of the rotating rotor. Thus,

$$P_2 = P_2^{heat} + \tilde{P}_2. \tag{6.60}$$

It is apparent that

$$P_2^{heat} = 3I_2^2 R_2, \tag{6.61}$$

and, consequently,

$$\tilde{P}_2 = P_2 - P_2^{heat} = 3I_2^2 \frac{R_2}{s} - 3I_2^2 R_2, \tag{6.62}$$

which leads to

$$\tilde{P}_2 = 3I_2^2 R_2 \frac{1-s}{s}. \tag{6.63}$$

It is clear from the last formula that $\tilde{P}_2 > 0$ in the motor regime when $0 < s < 1$. It is also clear from the last formula that $\tilde{P}_2 < 0$ if $s < 0$, that is, when the rotor of the induction machine is driven by a prime mover above synchronous speed n_{syn}. This change in the sign of $\tilde{P}_2$ suggests that the power is transferred from the rotor to the stator and that the induction machine operates as a generator.

By using again formulas (6.38) and (6.46), equation (6.63) can be written as

$$\tilde{P}_2 = 3\left(I_2'\right)^2 R_2' \frac{1-s}{s}. \tag{6.64}$$

Next, by using the equivalent electric circuit shown in Figure 6.8, we find

$$\left(I_2'\right)^2 = \frac{V_1^2}{\left[R_1 + \frac{R_2'}{s}\right]^2 + \left[X_1^\ell + \left(X_2^\ell\right)'\right]^2}, \tag{6.65}$$

where V_1 stands for the rms value of the stator voltage.

By substituting the last formula into equation (6.64), we obtain

$$\tilde{P}_2 = 3V_1^2 \frac{R_2' \frac{1-s}{s}}{\left[R_1 + \frac{R_2'}{s}\right]^2 + \left[X_1^\ell + \left(X_2^\ell\right)'\right]^2}. \tag{6.66}$$

It is known from classical mechanics that the mechanical power $\tilde{P}_2$ of the rotating rotor is related to the torque T applied to the rotor by the formula

$$\tilde{P}_2 = T\Omega, \tag{6.67}$$

where, as before, Ω stands for the angular speed of the rotor.

Invoking formula (6.6), we find

$$\Omega = (1 - s)\Omega_{syn}. \tag{6.68}$$

Furthermore, from formulas (6.1) and (6.13) follows that

$$\Omega_{syn} = \frac{4\pi f}{p}, \tag{6.69}$$

which leads to

$$\Omega = (1 - s)\frac{4\pi f}{p}. \tag{6.70}$$

By substituting the last formula into equation (6.67), we obtain

$$\tilde{P}_2 = T(1 - s)\frac{4\pi f}{p}. \tag{6.71}$$

By using relation (6.66) in the last equation and solving for T, we derive

$$\boxed{T(s) = \frac{3p}{4\pi f}V_1^2 \frac{R_2'/s}{[R_1 + R_2'/s]^2 + \left[X_1^\ell + \left(X_2^\ell\right)'\right]^2}.} \tag{6.72}$$

This is the sought expression for the mechanical torque on the rotor shaft of the induction machine. What follows next is the analytical study of this expression and its physical interpretation.

It is apparent from the last formula that for very small s we have

$$T(s) \approx \frac{3p}{4\pi f}V_1^2 \frac{s}{R_2'}, \tag{6.73}$$

and in the limit of $s \to 0$ we find

$$T(0) = 0. \tag{6.74}$$

The last equality is transparent from the physical point of view and mathematically confirms that there is no rotor torque at synchronous speed n_{syn}.

From formula (6.72) we also find

$$T_{start} = T(1) \tag{6.75}$$

and

$$T_{start} = \frac{3p}{4\pi f} V_1^2 \frac{R_2'}{[R_1 + R_2']^2 + \left[X_1^\ell + \left(X_2^\ell\right)'\right]^2}, \tag{6.76}$$

where T_{start} stands for the starting torque when $n = 0$ and, consequently, $s = 1$.

Usually, resistances of the rotor and stator windings are quite small. For this reason,

$$R_1 + R_2' \ll X_1^\ell + \left(X_2^\ell\right)', \tag{6.77}$$

and formula (6.76) can be simplified as follows:

$$\boxed{T_{start} \approx \frac{3p}{4\pi f} V_1^2 \frac{R_2'}{\left[X_1^\ell + \left(X_2^\ell\right)'\right]^2}.} \tag{6.78}$$

We shall next examine the torque $T(s)$ as a function of s. To this end, we introduce a new variable

$$y = \frac{1}{s} \tag{6.79}$$

and represent the equation (6.72) in the form

$$T(y) = \frac{Ay}{By^2 + Cy + D}, \tag{6.80}$$

where the following notations are introduced:

$$A = \frac{3p}{4\pi f} V_1^2 R_2', \tag{6.81}$$

$$B = \left(R_2'\right)^2, \tag{6.82}$$

$$C = 2R_1 R_2', \tag{6.83}$$

and

$$D = \left[X_1^\ell + \left(X_2^\ell\right)'\right]^2 + R_1^2. \tag{6.84}$$

From formula (6.80) we find

$$\frac{dT(y)}{dy} = \frac{A(By^2 + Cy + D) - Ay(2By + C)}{(By^2 + Cy + D)^2}, \tag{6.85}$$

which is reduced to

$$\frac{dT(y)}{dy} = \frac{A(D - By^2)}{(By^2 + Cy + D)^2}. \tag{6.86}$$

To find the extremum values of the torque, we consider the equation

$$\frac{dT(y)}{dy} = 0. \tag{6.87}$$

It is apparent from formula (6.86) that the last equality is satisfied for such values of y_m that

$$D - By_m^2 = 0. \tag{6.88}$$

This leads to

$$y_m = \pm\sqrt{\frac{D}{B}}. \tag{6.89}$$

Taking into account formula (6.79), we conclude that the torque $T(s)$ achieves its extrema at the following values of s:

$$s_m = \pm\sqrt{\frac{B}{D}}. \tag{6.90}$$

By recalling formulas (6.82) and (6.84), we find

$$s_m = \pm\frac{R_2'}{\sqrt{R_1^2 + \left[X_1^\ell + (X_2^\ell)'\right]^2}}. \tag{6.91}$$

By taking into account the inequality (6.77), from the last formula we conclude

$$\boxed{s_m \approx \pm\frac{R_2'}{X_1^\ell + (X_2^\ell)'}.} \tag{6.92}$$

It can be easily seen from formula (6.72) that the torque $T(s)$ assumes its maximum value at $s = s_m$ and its minimum value at $s = -s_m$. Indeed, $T(s) > 0$ for $s > 0$ and $T(0) = T(\infty) = 0$. Since $T(s)$ assumes only one extremum value for $s > 0$, this extremum is the maximum of $T(s)$. Since $T(s) < 0$ for $s < 0$, similar reasoning leads to the conclusion that $T(s)$ assumes its minimum value at $s = -s_m$. By using the last formula in equation (6.72) and by taking into account that R_1 is quite small in comparison with the leakage reactances, we easily derive the following expression for the maximum T_m of T:

$$\boxed{T_m = T(|s_m|) \approx \frac{3p}{8\pi f}\frac{V_1^2}{X_1^\ell + (X_2^\ell)'}.} \tag{6.93}$$

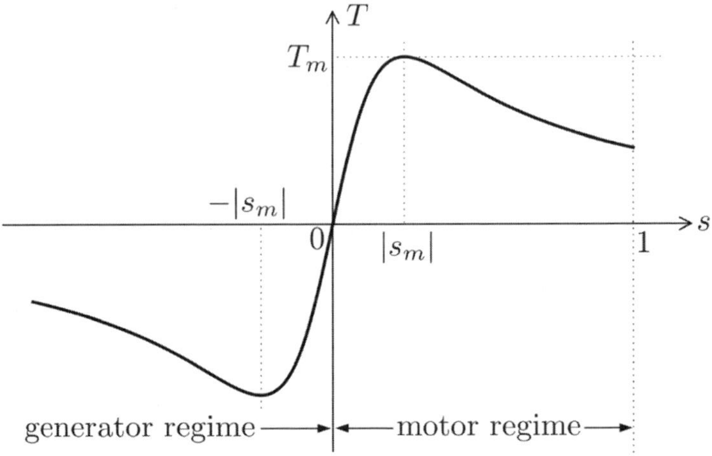

Fig. 6.9

After the presented discussion, it is easy to plot the torque $T(s)$ as a function of s. This plot is presented in a qualitative way in Figure 6.9. In reality, $|s_m|$ is quite small because R_2' is much smaller than $X_1^\ell + \left(X_2^\ell\right)'$. As is clear from this plot and formula (6.72), $T(s)$ changes its sign with the change in sign of s. This is consistent with the fact that for $s < 0$, an induction machine operates as a generator and the torque T changes its nature from being a driving torque in the motor regime ($s > 0$) to being an impeding torque in the generator regime ($s < 0$).

It is customary to plot the torque T as a function of rotor speed n. To do this, we shall use the relation

$$n = (1 - s)n_{syn} \tag{6.94}$$

which follows from formula (6.5). By using this relation and Figure 6.9, we can easily replot T as a function of n for the motor regime as shown in Figure 6.10. In this figure, n_m is the rotor speed at which the torque $T(n)$ achieves its maximum value. According to (6.94),

$$n_m = (1 - |s_m|)n_{syn}. \tag{6.95}$$

Since, as discussed before, $|s_m|$ is quite small, we find that

$$n_m \approx n_{syn}. \tag{6.96}$$

Now, suppose that the starting torque T_{start} of the induction motor is larger than the load torque T_{load}: $T_{start} > T_{load}$. Then, the induction motor will

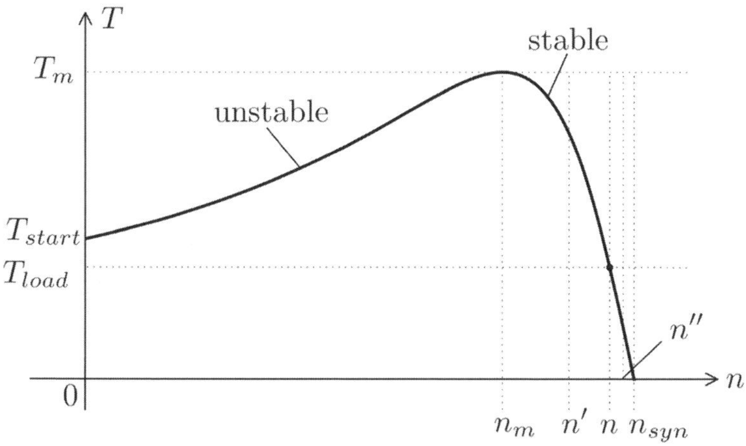

Fig. 6.10

get started and its speed will be increased until the equality between the motor torque and the load torque is achieved at some speed n. It is clear from Figure 6.10 that

$$n_m < n < n_{syn} \tag{6.97}$$

and this implies according to (6.96) that

$$n \approx n_{syn} = \frac{120f}{p}. \tag{6.98}$$

The last formula is the foundation of frequency control of speed of the induction motor. By varying f, we shall vary n_{syn} and, consequently, according to formula (6.98) we shall vary n.

Next, we shall point out that the operation point, where the equality between $T(n)$ and T_{load} is achieved, is *stable*. Indeed, suppose that as a result of some temporary disturbance the speed of the induction motor is reduced to n'. Then, since the motor torque $T(n')$ is larger than the load torque, the rotor will be accelerated until speed n is achieved where the motor torque and the load torque are the same. Similarly, suppose that as a result of some temporary perturbations the speed of the induction motor is increased to n''. Then, since the motor torque $T(n'')$ is smaller than the load torque, the rotor will be slowed down and the speed n is achieved where the motor torque and load torque are the same. It is clear that this reasoning is valid for any operational speed between n_m and n_{syn} and, consequently, the part of the mechanical characteristics $T(n)$ between n_m

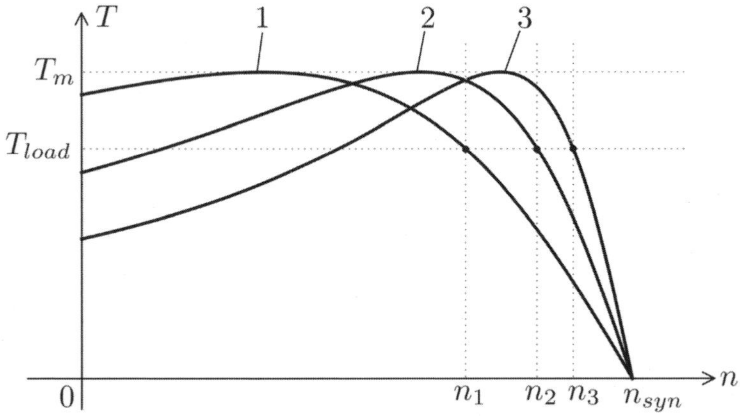

Fig. 6.11

and n_{syn} is stable. By using similar reasoning, it can be shown that the part of the mechanical characteristics $T(n)$ between $n = 0$ and $n = n_m$ is *unstable*.

Up to this point, we have discussed the operation of the induction motor when the starting torque is above the load torque. In the case when $T_{start} < T_{load}$, the induction motor cannot start. To remedy this situation, the secondary resistance R_2 (and consequently, R_2') is increased. For wound rotor machines, this is done by using external resistance which can be connected to the rotor winding through brushes and slip rings. It is clear from formula (6.78) that T_{start} is increased with the increase in R_2'. It is also clear from formula (6.93) that the maximum torque T_m does not change with the increase of R_2', however the speed n_m at which this maximum is achieved is reduced as is evident from formulas (6.92) and (6.95). This is illustrated by Figure 6.11 where the mechanical characteristics $T(n)$ are plotted for different values of the secondary resistance R_2'. It is clear from this figure that for sufficiently large R_2' (curve 1) starting torque can be achieved which is larger than the load torque. As a result, the induction motor can be started and achieve the speed n_1. As the external resistance (and, consequently, R_2') is reduced, the torque-speed characteristics are changed (see curves 2 and 3), and the induction motor speed will be increased to n_2 and, finally, to n_3. For squirrel cage rotor machines with double-cage or deep-bar designs the same result is achieved due to the manifestation of the skin effect (as discussed in the first section of this chapter). Indeed, for such machines, the mechanical characteristics of torque versus speed deviate from what is shown in Figure 6.10 where the case of constant secondary resistance R_2' is illustrated. The above deviations are usually pronounced at low rotor speeds when the skin effect in rotor conductors is strongly manifested and results in the increase of the starting torque. For sufficiently high rotor speeds, the skin effect is not exhibited and the mechanical characteristics practically coincide with the curve in Figure 6.10.

In conclusion, the main features of the torque of induction machines can be summarized as follows:

- All torques of induction machines are determined by leakage reactances. Thus, as in the case of transformers, leakage reactances determine the quality of induction machines.
- All torques are proportional to the square of rms (or peak) value of the primary voltage.
- The larger R_2', the larger the starting torque.
- The maximum torque does not depend on R_2'.
- The operational speed of induction motors is very close to the synchronous speed and this fact can be utilized for frequency control of speed.
- The operational point always belongs to the stable part of the torque-speed characteristics located between the synchronous speed n_{syn} and the speed n_m at which the maximum torque is achieved.

Problems

(1) What are the three main components of conventional utility power systems?
(2) What are the main types of power plants and what are the energy conversion processes performed at these power plants?
(3) Explain why high voltages are used for power transmission.
(4) Explain what the essence of utility industry deregulation is.
(5) Give a brief description of the structure of three-phase circuits and define phase and line voltages. Explain what a neutral wire is useful for.
(6) (a) Suppose that the lines and neutral of a three-phase circuit are not marked but accessible for measurements. Explain how by knowing the value of line voltage and by using only two voltmeter measurements the neutral can be identified. (b) Consider the split-phase method currently often used in the US for final distribution of ac power to residential loads. In this method, the primary winding of a load single-phase transformer is fed by an ac voltage from the utility distribution system. The transformer's secondary winding is tapped in the center, which makes available a center-tap connection in addition to the two terminals of the secondary winding. Typically, the voltage measured between either of the two secondary terminals to the center tap is about 120 V rms and in such case the voltage between the two secondary terminals is 240 V rms. In the case when the three wires are not marked, explain how the center-tap wire can be identified by using two voltage measurements. What is the phase shift between the voltages measured between each secondary winding terminal and the center tap?
(7) Describe the essence of per-phase analysis of three-phase circuits with balanced loads.

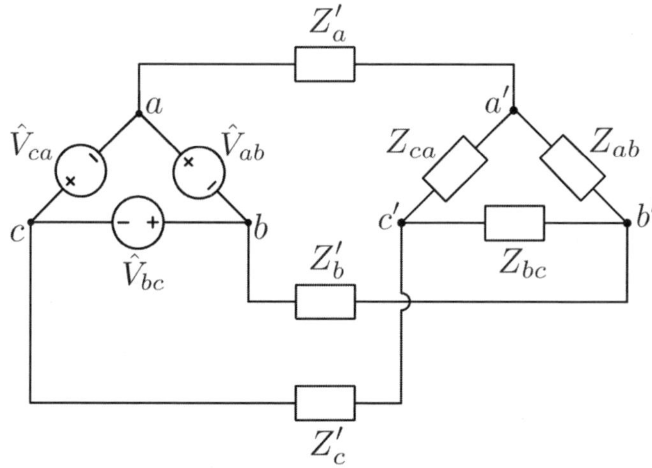

Fig. P.1

(8) Describe how the delta connection of three-phase voltages can be equivalently transformed into star connection of three-phase voltages.

(9) Perform the analysis (i.e., find all branch currents) of a three-phase circuit with delta connections of sources and loads (see Figure P.1).

(10) Draw the plots of $p(t)$ (see formula (1.64)) for two cases: 1) when the power factor ($\cos\varphi$) is adjusted to one and 2) when the power factor is not adjusted to one. By using these two plots explain the difference in energy consumption for the above two cases.

(11) Draw the phasor diagrams for the electric circuit in Figure 1.13b in the cases when the power factor is adjusted and not adjusted to one.

(12) By using the phasor diagram from the previous example, derive formula (1.83) for the capacitance that results in adjustment of power factor to one.

(13) Explain two beneficial effects of leading power factor ($\varphi < 0$).

(14) Prove that the sum of two wattmeter measurements (see Figure P.2) gives the total active power supplied to the load.

(15) What are the most typical power line faults and why is their analysis important?

(16) Describe the algorithm of using the Thevenin theorem for fault analysis.

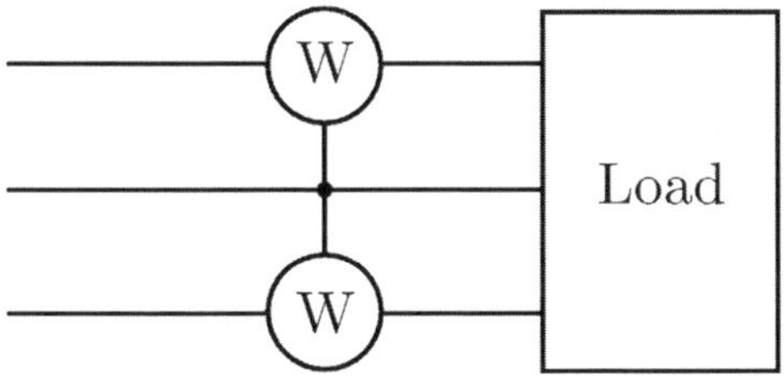

Fig. P.2

(17) By using the Thevenin theorem, analyze the SLG fault in a three-phase circuit (see Figure 2.4) when the node O' is grounded and the grounding impedance is neglected (assumed to be equal to zero).
(18) Solve problem 17 when the grounding impedance Z'_n of node O' is not equal to zero. (Hint: use delta-to-star transformation on the second step.)
(19) By using the Thevenin theorem, carry out the analysis of LL fault (see Figure 2.7a) when nodes O' and O are grounded through impedances Z'_n and Z_n. (Hint: use delta-to-star transformation on the second step.)
(20) By using the Thevenin theorem, carry out the analysis of DLG fault (see Figure 2.10a) by using the steps described in the text.
(21) Summarize concisely the definition of symmetrical components and their mathematical relations with the three-phase quantities they represent.
(22) Prove formula (2.83).
(23) Prove formula (2.86).
(24) Draw sequence networks for two distinct cases, when the center O' of star connection of three-phase voltage sources is grounded, and when it is not grounded.
(25) State the general algorithm of using sequence networks in fault analysis.
(26) By using the sequence networks, carry out the analysis of SLG fault in the case when node O' is grounded through zero impedance.
(27) By using the sequence networks, carry out the analysis of DLG fault (see Figure 2.24) in the case when node O' is grounded through zero impedance.
(28) Complete the analysis of LL fault (see Figure 2.26) described in the text.

(29) Describe the basic design and the principle of operation of single-phase transformers. Explain for what transformers are used.
(30) Define an ideal transformer and describe its terminal relations and equivalent circuit.
(31) Consider an ac transmission line with a characteristic impedance Z_0 of 100 Ω to which you want to connect a resistive load with $R = 25$ Ω. How could you use a transformer to maximize the power transfer to the load?
(32) Explain what leakage inductances are and why they are important as far as the operation of transformers is concerned.
(33) Summarize the main steps used in the mathematical transformation of the coupled circuit equations in order to arrive at the equivalent circuit of the transformer.
(34) Draw the equivalent circuit of the nonideal transformer and describe the meaning of each parameter in the equivalent circuit.
(35) Is the equivalent circuit unique?
(36) Draw the simplified equivalent circuit of the transformer that is used in the power system analysis.
(37) Describe the open- and short-circuit tests used for the identification of parameters of the (approximate) equivalent circuit of the transformer.
(38) Suppose the open- and short-circuit tests were performed for a transformer with a rated rms voltage of 120 V and rated rms current of 10 A and the following values were obtained, respectively: $P_{oc} = 10$ W, $I_1 = 1$ A, $V_2 = 240$ V; $P_{sc} = 60$ W, $V_1 = 7$ V. Find the approximate equivalent circuit model parameters. Explain if the assumptions made in obtaining the approximate equivalent circuit model are reflected in the values you obtain.
(39) Suppose you have characterized the approximate equivalent circuit corresponding to an actual transformer. What power losses are described by this circuit and how are they modeled? How would you compute these losses using the circuit model if the transformer drives some arbitrary load within its rating?
(40) How could you compute a transformer's efficiency (the ratio of secondary to primary power) for a given load from the knowledge of the transformer's equivalent circuit and the losses it models?
(41) Describe various designs (i.e., various core configurations and various connectivities of primary and secondary windings) of three-phase transformers.

(42) Derive the expressions for the ratio of primary and secondary line voltage phasors for each of the following three-phase transformer connectivities: a) star-star, b) star-delta, c) delta-star and d) delta-delta.
(43) Which of the connectivities in problem 42 gives the highest secondary line voltage for a given primary line voltage and turns ratio? Explain.
(44) Describe the design and principle of operation of synchronous (cylindrical rotor and salient pole) generators.
(45) What is synchronism? Define and explain the significance of synchronous speed for the performance of synchronous generators. Why is a synchronous generator a (P, V)-source?
(46) Explain how the rotating magnetic field can be created by the stationary three-phase stator winding.
(47) Demonstrate that negative-sequence three-phase stator currents create mmf and magnetic fields rotating with speed n_{syn} in the direction opposite to the rotor rotation.
(48) By using the language of symmetrical components, explain what happens to synchronous generators in the case of unbalanced loads. Why are such loads undesirable?
(49) Explain how in the case of unbalanced loads the negative effect of the zero sequence of stator currents can be eliminated.
(50) Describe the main principles of stator winding design. Explain how it is achieved that the stator winding serves as a filter of higher-order spatial and temporal (time) harmonics.
(51) Construct an occupation diagram for a two-layer full-pitch stator winding with 18 slots on the stator.
(52) How is the occupation diagram in the previous example modified in the case of fractional-pitch winding?
(53) Explain what the (main) synchronous reactance of the stator winding is and how it is related to self and mutual reactances of the phase windings. How does this reactance depend on the air gap length?
(54) Describe and explain the open- and short-circuit tests for experimental identification of synchronous reactance.
(55) Describe the essence of the two-reactance theory for salient pole machines.
(56) What is the load angle? Explain its significance.

(57) What is the limit of static stability of a synchronous generator? Explain why the air-gap length of synchronous generators is usually large.
(58) Consider a cylindrical rotor synchronous generator connected to the three-phase power grid, which maintains the constant terminal voltage (infinite bus). Describe what happens to the magnitude of the internal voltage (emf) as the rotor dc winding current is increased from a small value up to some maximum allowed value.
(59) For the situation described in problem 58, describe what happens to the stator current and the power factor as the rotor dc winding current is changed and the generated active power is maintained constant. Draw and explain the corresponding phasor diagrams.
(60) Define generator buses and load buses and state the essence of power flow analysis.
(61) State two equivalent forms of the final power flow equations. Explain the cause of nonlinearity of these equations. Explain if all of these equations are coupled or not.
(62) State the algorithm of Newton-Raphson iterations for the solution of general nonlinear equations. What is the main mathematical idea of these iterations? Is the convergence of these iterations local or global? What is the rate of convergence?
(63) Construct a graphical example (different than in the text) when Newton-Raphson iterations do not converge.
(64) Describe the Newton-Raphson iterations in the case of power flow equations.
(65) Describe the central idea and main formulas of the continuation technique.
(66) Describe the continuation technique for power flow equations when the parameter is introduced in active and reactive powers of load buses to solve the problems for various load configurations.
(67) State and explain the "swing" equation for rotor dynamics of a synchronous generator.
(68) Explain how the "swing" equation can be written in the Hamiltonian form and how this form can be used for the construction of the phase portrait.
(69) Explain what the "saddle" point and the separatrix of rotor dynamics are and what their significance is.
(70) State the algebraic criterion of stability of rotor dynamics in terms of the values of the Hamiltonian at initial conditions and at the saddle point.

(71) Demonstrate the equivalence of the algebraic stability criterion of problem 70 and the classical "equal area" stability condition.

(72) What is the physical effect of eddy currents induced in generator solid rotors on the stability of rotor dynamics? How can these currents be accounted for and what is their effect on time variations of the Hamiltonian?

(73) Describe the basic design and principle of operation of induction machines as motors and generators.

(74) What is "slip" and what is the relation between the frequencies of rotor and stator currents? Give the range of slip variations in the motor and generator regimes.

(75) Explain why the air gap length in induction machines is usually very small.

(76) Explain how a doubly-fed induction machine can be operated as a generator. In what type of power plants are induction generators utilized?

(77) Explain how the electromagnetic coupling between stator and rotor windings realized through rotating magnetic fields leads to the simplification of the coupled circuit equations. Explain the physical meaning of the reactances in the coupled circuit equations.

(78) Describe the main steps of mathematical transformation of the coupled circuit equations that are used in the derivation of the equivalent circuit for the induction machine. Is the equivalent circuit unique?

(79) Describe how the no-load test and locked-rotor test can be used for the identification of parameters of the approximate equivalent circuit.

(80) Give the formula and draw the graph for the torque-speed characteristics of induction machines for unlimited variation of speed in both directions. What parts of these characteristics are stable and unstable?

(81) Explain what the relation between the mechanical speed of the induction motor and the speed of the rotating magnetic field of the stator winding is. What does this relationship imply for the control of induction motor speed?

(82) What is the starting torque and how can it be controlled for wound rotor induction motors?

(83) Explain what physical phenomenon is utilized for the increase in the starting torque in the case of the squirrel cage rotor induction

motor. Explain what kinds of designs of squirrel cage rotors are used for starting torque increase.

(84) Suppose you are told that a particular induction machine has a speed of 1100 RPM at its rated load torque. Assuming the line frequency is 60 Hz, how many poles does this machine have?

(85) Suppose you have a three-phase induction machine at rest ($n = 0$) with rated line-to-line voltage of 208 V and you have connected a variable magnitude three-phase ac supply to its stator terminals. As you steadily increase the stator voltage from 0 V, you notice that the rotor starts to spin when you reach about 50 V. Why might this occur? What does the stator voltage magnitude control in the induction machine?

(86) Suppose that two distributed stator windings are angularly shifted with respect to one another along the stator circumference by 90° and driven by ac currents of the same peak value and frequency but shifted in time by 90°. Show that this two-phase winding produces a uniformly rotating magnetic field.

(87) A single-phase winding when driven by ac current produces a pulsing magnetic field. Prove that this pulsing magnetic field can be equivalently described as the superposition of two uniformly rotating magnetic fields with the same speed but opposite rotation directions. What is the speed of rotation of these fields?

(88) Induction motors can also be constructed to be driven by single-phase ac power and these machines are called single-phase induction motors. Using the fact described in the previous problem, graphically derive a generic plot of the single-phase induction motor torque-speed characteristics from those of the three-phase induction machine. How do the single-phase induction motor mechanical characteristics compare with the three-phase induction motor characteristics as far as the starting torque is concerned?

(89) One common design of a single-phase induction machine incorporates another (auxiliary) stator winding which is electrically excited in parallel with the main single-phase winding. This auxiliary winding is designed in such a way as to give the single-phase induction machine a nonzero starting torque. Describe the geometrical and electrical characteristics of the auxiliary starting winding necessary for the single-phase induction machine to start itself when the main and auxiliary windings are excited in parallel by a single-phase ac voltage source. Explain the usefulness of constructing the auxiliary winding with a series capacitor.

PART III

Power Electronics

Chapter 1

Power Semiconductor Devices

1.1 Introduction; Basic Facts Related to Semiconductor Physics

This part of the book is concerned with power electronics. Power electronics can be defined as the area of electrical engineering which deals with the design of electric circuits that use semiconductor devices as switches to convert electric power from one (*available*) form into another (*desired*) form. Such circuits are called power converters. It is clear that these converters are switching-mode devices. In this sense, power electronics is similar to digital electronics where semiconductor devices are also used as switches but for different purposes such as storage, processing and transmission of digital information. Furthermore, another important difference is that in power electronics semiconductor devices are used as switches with high current and high voltage-handling capabilities. Switching of semiconductor devices in power converters inevitably results in ripples in voltages and currents, and these ripples are suppressed by using energy storage elements such as inductors and capacitors. It turns out that there exists a trade-off between the switching speed of semiconductor devices and the size of energy storage elements. Namely, the faster the switching of semiconductor devices, the smaller the energy storage elements needed for suppression of ripples. This, in turn, results in more compact, lighter and cheaper power electronics converters. Thus, the progress in semiconductor device technology leading to faster semiconductor switches of high currents and high voltages is very beneficial to the progress in power electronics.

It is well known that ac and dc are the two most prevalent forms of electric power. For this reason,the following four types of power converters are often encountered in power electronics applications:

- ac-to-dc converters (called rectifiers),
- dc-to-ac converters (called inverters),
- dc-to-dc converters (called choppers),
- ac-to-ac converters.

In this book we shall study these converters and we will be mostly concerned with their steady-state performance, that is, the performance for which the power converters were designed. Due to the switching-mode nature of power converters, the analysis of their steady-state performance is reduced to the analysis of steady states in electric circuits driven (excited) by periodic non-sinusoidal voltage sources. The methods of analysis of such electric circuits are discussed in detail in Chapter 2 of Part I of this book, and these methods will be extensively used in our forthcoming study of power converters. In this study, we shall also use some facts from magnetics that are presented in Chapter 3 of Part I as well as some background material related to three-phase circuits and transformers discussed in Part II.

Power electronics is an enabling technology which has already found numerous applications in very diverse areas of engineering. Just to name but a few, power electronics is indispensable in integrating such renewable energy sources as wind and solar into existing power systems; it plays the central role in the construction of HVDC (high voltage dc) transmission lines; it is at the very foundation of the development of semiconductor drives; it is the critical technology in the design of new and efficient hybrid and electric cars; it enables the development of uninterruptible power supplies (UPS) to provide backup electric power to various loads in the case of emergency; power electronics converters can be found in many consumer devices such as televisions, personal computers, battery chargers, etc.

As mentioned above, the use of semiconductor devices as switches is at the very foundation of power electronics. For this reason, this chapter is concerned with the discussion of the designs and principles of operation of basic semiconductor devices and their utilization as switches. To proceed with this discussion, we shall first review in this section the basic facts related to semiconductor physics.

We shall begin with the discussion of intrinsic (pure) semiconductors. They are made up of a huge number of atoms which consist of heavy positive nuclei surrounded by light negative electrons. These nuclei form a rigid periodic lattice which is essentially "frozen" at very low temperatures. As temperature is raised, the atoms exhibit thermal vibrations about their equilibrium (mean) positions. All electrons can be subdivided into two

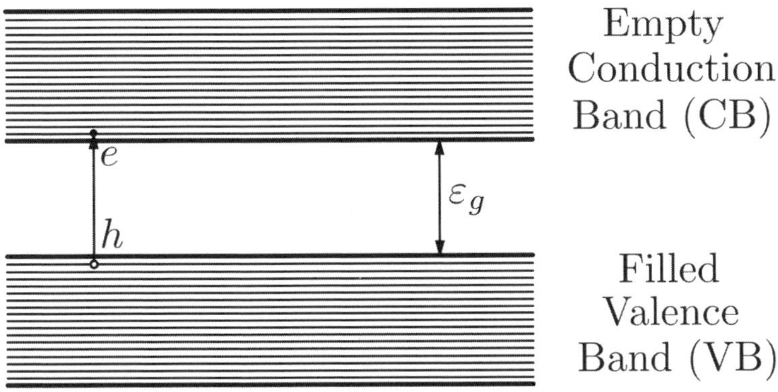

Fig. 1.1

distinct groups. There are electrons which are tightly bound to nuclei and can be naturally regarded as part of the lattice. They are referred to as "core" electrons. There are, however, electrons that are spread out over the entire semiconductor. They are referred to as "valence" or mobile electrons. The transport of these mobile electrons results in electric current conduction.

According to quantum mechanics, the energy spectrum of valence electrons in perfectly periodic (frozen) crystals can be described in terms of so-called Bloch states, and this spectrum exhibits band structure. This means that not all energy values (levels) are permissible and those which are permissible are grouped into *bands*. In each band there are very many permissible energy levels (states) which are separated from one another by very small energy increments, while the bands may be separated by appreciable *energy gaps*. As far as the transport of electrons in semiconductors is concerned, two bands of high energy are most important. They are the *valence band* (VB) all of whose energy levels are occupied by electrons at zero temperature, and the next in increasing energy and completely empty of electrons is the *conduction band* (CB). These two bands are separated by energy gap ε_g. This is schematically represented by Figure 1.1.

For silicon (Si), which is still the main material for semiconductor devices,

$$\varepsilon_g = 1.1 \text{ eV}. \tag{1.1}$$

For applications in power electronics, wide bandgap semiconductors are very attractive and promising. Examples of wide bandgap semiconductors include silicon carbide (SiC) with

$$\varepsilon_g = 3.2 \text{ eV} \tag{1.2}$$

and gallium nitride (GaN) with

$$\varepsilon_g = 3.4 \text{ eV}. \tag{1.3}$$

Such high energy gaps lead to appreciably higher breakdown electric fields, which is beneficial for the operation of semiconductor devices at high voltages. Furthermore, large energy bandgaps also result in much higher operating temperatures and higher radiation hardness. The former is important for the operation of semiconductor devices at high currents and voltages.

When the temperature of semiconductors is gradually increased above zero, then some electrons in the valence band may acquire sufficient energy from the thermally vibrating lattice to make transitions across the energy gap into the conduction band. As a result, some "vacancies" are formed in the valence band. These vacancies are usually referred to as *holes.* Such electron-hole pair production in a pure semiconductor is called intrinsic electron-hole pair generation and is schematically shown in Figure 1.1. It is apparent that the simultaneous production of electrons and holes results in equal density of electrons and holes,

$$n = p = n_i, \tag{1.4}$$

where n stands for electron density, p stands for hole density, while n_i is called intrinsic density (concentration). This density depends on temperature and for Si at room temperature

$$n_i = 1.45 \cdot 10^{10} \text{ cm}^{-3}. \tag{1.5}$$

It must be remarked that the electric current conduction in semiconductors is due to the transport of conduction band and valence band electrons. However, the transport properties of conduction band electrons are quite different from transport properties of valence band electrons. To distinguish between these two transports, the notion of an imaginary and positively charged particle, a hole, is introduced in semiconductor physics, and the actual transport of valence band electrons is described (is modeled) as transport of holes.

In the design of semiconductor devices, extrinsic semiconductors are used. These semiconductors are *doped.* The latter means that specific impurities are intentionally introduced in semiconductors by means of ion implantation (or other fabrication techniques). Introduced impurities result in localized energy levels within the energy gap. Usually, "*shallow*"

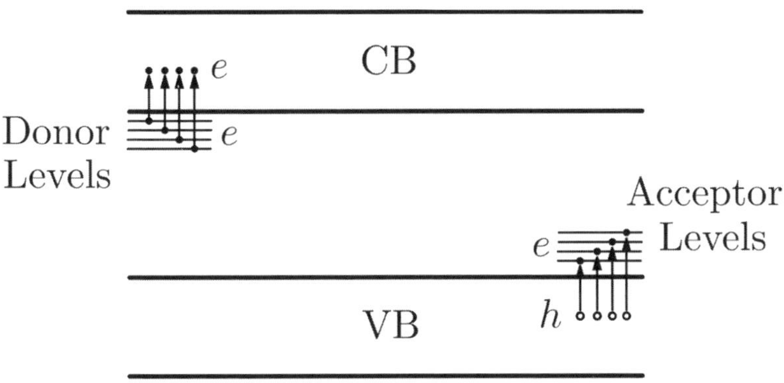

Fig. 1.2

impurities are used for doping which result in energy levels close to the boundaries of the energy bandgap. When the impurity energy levels are close to the lower edge of the conduction band (see Figure 1.2), they are called "*donor*" levels and at very low temperatures these energy levels are completely occupied by electrons. Impurities whose implantation results in such energy levels are called donors, and for Si such impurities are phosphorus (P) and arsenic (As). As temperature is slightly increased, electrons occupying donor energy levels acquire enough energy from thermal lattice vibrations to make transitions to the conduction band. This results in the appearance of mobile conduction band electrons and positively charged (ionized) immobile impurities. On the other hand, when impurity energy levels are close to the upper edge of the valence band (see again Figure 1.2), they are called *acceptor* levels and at very low temperatures these energy levels are not occupied by electrons. Impurities whose implantation results in such energy levels are called acceptors, and for Si such impurities are boron (B) and aluminum (Al). As temperature is slightly increased, electrons from the valence band acquire enough energy from thermal lattice vibrations to make transitions to acceptor energy levels. This results in appearance of mobile holes in the valence band and negatively charged (ionized) immobile impurities. It is clear from the presented discussion that doping and thermal ionization of impurities may result in production of a specific type of carriers in chosen regions. Indeed, if some regions of semiconductors are doped by donor impurities, then in these regions as a result of thermal ionization we have inequality

$$n > p, \tag{1.6}$$

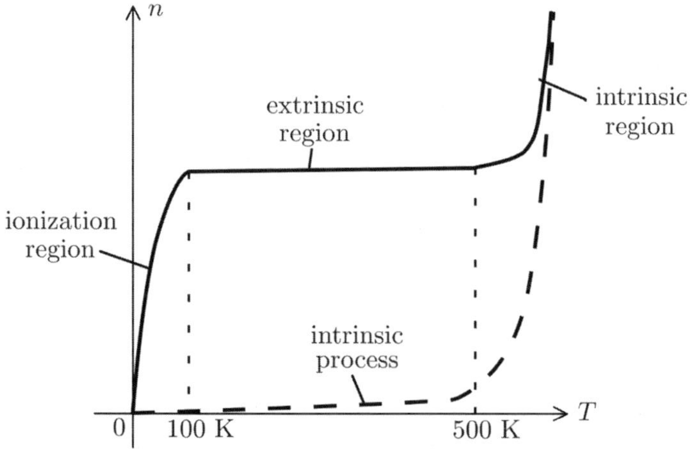

Fig. 1.3

and such regions are called n-type (or n^+-type if $n \gg p$). On the other hand, if some regions of semiconductors are doped by acceptors, then in these regions

$$p > n \tag{1.7}$$

and such regions are called p-type (or p^+-type if $p \gg n$).

It can be shown that, regardless of the nature of doping, at equilibrium

$$np = n_i^2. \tag{1.8}$$

It is clear from the previous discussion that in doped semiconductors there are two distinct mechanisms of mobile carrier (electron or hole) production: an *intrinsic* mechanism associated with transitions of electrons across the entire energy bandgap and resulting in simultaneous and equal production of electrons and holes; and an *extrinsic* mechanism associated with thermal ionization of shallow impurities resulting in production of electrons *or* holes. The interplay of these two mechanisms is illustrated by a plot of electron density versus temperature for n-type Si shown in Figure 1.3. A similar plot can be drawn for hole density in the case of p-type Si. It is apparent from this plot that there are three distinct regions: the ionization region (0 K $< T <$ 100 K) with rapid growth of electron density in the conduction band due to thermal ionization of donor impurities, the extrinsic region (100 K $< T <$ 500 K) with more or less constant electron density and the intrinsic region ($T >$ 500 K) with rapid growth in electron density due to the intensified intrinsic process for sufficiently large temperatures. The extrinsic region where electron density is practically constant and does not depend on temperature is usually a desired region for operation of semiconductor devices. It is apparent on physical grounds that this region will be extended to appreciably higher temperatures for wide bandgap semiconductors, which is one of their attractive features.

It is clear from the presented discussion that in semiconductors there are two types of mobile carriers and two types of immobile ionized impurities. Consequently, we can talk about volume charge density ρ inside semiconductors,

$$\rho = q(p - n + N), \tag{1.9}$$

where q is the absolute value of electron charge, while

$$N = N_n - N_p, \tag{1.10}$$

with N_n being the density of ionized donors and N_p being the density of ionized acceptors. By neglecting magnetic field effects associated with the transport of mobile carriers, we shall characterize the effect of charges only by electric field $\mathbf{E}$. This field can be represented as the gradient of scalar potential φ,

$$\mathbf{E} = -\nabla\varphi, \tag{1.11}$$

and this leads in the usual way to the Poisson equation

$$\nabla^2\varphi = -\frac{\rho}{\varepsilon}, \tag{1.12}$$

or, taking into account formula (1.9),

$$\nabla^2\varphi = \frac{q}{\varepsilon}(n - p - N), \tag{1.13}$$

where $\varepsilon = 11.7\varepsilon_0$ for Si.

It must be noted that electron (n) and hole (p) densities in the Poisson equation (1.13) are not known beforehand and these densities are dependent on transport of mobile carriers in semiconductor devices. As far as this transport is concerned, electron and hole currents in semiconductors are due to two distinct physical mechanisms: *drift* and *diffusion*. Accordingly, the electron and hole current densities can be written as follows:

$$\mathbf{J}_n = \mathbf{J}_n^{drift} + \mathbf{J}_n^{dif}, \tag{1.14}$$

$$\mathbf{J}_p = \mathbf{J}_p^{drift} + \mathbf{J}_p^{dif}, \tag{1.15}$$

where the meanings of subscripts and superscripts are self-explanatory.

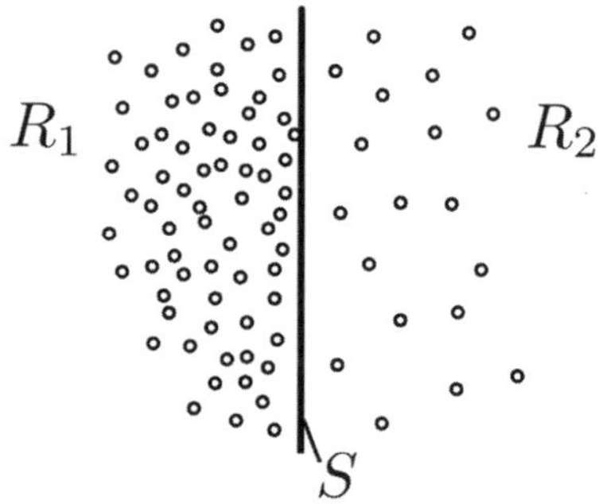

Fig. 1.4

The drift mechanism of current conduction in semiconductors is quite similar to the mechanism of current conduction in conductors. Namely, an existing electric field accelerates holes and electrons along the field direction and opposite to it, respectively. This accelerated motion is impeded by scattering from random lattice vibrations and impurities. The average effect of this scattering results in electrons and holes attaining some average velocity proportional to the local electric field $\mathbf{E}$. This leads to the following expressions for drift currents:

$$\mathbf{J}_n^{drift} = q\mu_n n\mathbf{E}, \tag{1.16}$$

$$\mathbf{J}_p^{drift} = q\mu_p p\mathbf{E}, \tag{1.17}$$

where μ_n and μ_p are mobilities of electrons and holes, respectively. The positive sign in equation (1.16) can be explained as follows. Since electrons are negatively charged, they move in the direction opposite to electric field $\mathbf{E}$. However, the motion of negative charges in the direction opposite to $\mathbf{E}$ is equivalent to positive electric current in the direction of $\mathbf{E}$.

The diffusion mechanism of current conduction is of stochastic origin and it is caused by the random nature of scattering. Macroscopically, it manifests itself in the motion of carriers from high density regions to low density regions. This mechanism can be illustrated as follows. Consider two adjacent regions R_1 and R_2 of semiconductor (see Figure 1.4) with different hole densities p_1 and p_2, assuming for certainty that $p_1 > p_2$. Consider also that there exists electric field $\mathbf{E}$ parallel to the interface S between R_1 and R_2 that causes drift transport of holes parallel to S. Then, due to the random component of hole motion in the direction perpendicular to $\mathbf{E}$, some holes from region R_1 cross S into region R_2 and some holes from R_2 cross S into R_1. However, since $p_1 > p_2$, a larger number of holes (on average) cross S from R_1 to R_2 than the other way around. This results in a net influx of holes into R_2 and produces the component of hole current through S in the direction from high hole density region to low hole density region. This is the diffusion current, and it is clear that it is controlled by the gradient in hole density. Mathematically, it is expressed as

$$\mathbf{J}_p^{dif} = -qD_p\nabla p, \tag{1.18}$$

where D_p is the hole diffusion coefficient (diffusivity), while the minus sign indicates that the hole diffusion current is directed opposite to the gradient direction, that is, from high density to low density regions.

Similarly, electron diffusion current density can be written as

$$\mathbf{J}_n^{dif} = qD_n\nabla n, \tag{1.19}$$

where D_n is the electron diffusion coefficient and positive sign indicates that the electron diffusion *current* is in the direction of the gradient because *negatively* charged electrons diffuse in the opposite direction to ∇n. Diffusion currents play a crucial role in the performance of many semiconductor devices. These currents are engineered by doping differently adjacent parts of semiconductor devices.

Now, by combining formulas (1.14)-(1.19), we arrive at the following expressions for total electron and hole current densities:

$$\mathbf{J}_n = q\mu_n n\mathbf{E} + qD_n\nabla n, \tag{1.20}$$

$$\mathbf{J}_p = q\mu_p p\mathbf{E} - qD_p\nabla p. \tag{1.21}$$

It turns out that there is the following remarkable relation between mobilities and diffusivities called the Einstein relation:

$$\frac{D_n}{\mu_n} = \frac{D_p}{\mu_p} = \frac{k_B T}{q}, \tag{1.22}$$

where $k_B = 1.38 \cdot 10^{-23}$ joule/K is the Boltzmann constant and, as before, $q = 1.6 \cdot 10^{-19}$ coulomb is the absolute value of electron charge.

The quantity

$$V_T = \frac{k_B T}{q} \tag{1.23}$$

has the dimension of voltage, and it is called thermal voltage. At room temperatures,

$$V_T = 0.026 \text{ V}. \tag{1.24}$$

Thermal voltage plays an important role in the theory of semiconductor devices.

The Einstein relation can be derived as follows. Consider equilibrium conditions in semiconductors when net electron and hole current densities are equal to zero:

$$\mathbf{J}_n = 0, \tag{1.25}$$

$$\mathbf{J}_p = 0. \tag{1.26}$$

From equations (1.11), (1.21) and (1.26) we derive

$$\mu_p p \nabla\varphi + D_p \nabla p = 0, \tag{1.27}$$

which leads to

$$\frac{\mu_p}{D_p}\nabla\varphi + \frac{\nabla p}{p} = 0 \tag{1.28}$$

or

$$\nabla\left(\frac{\mu_p}{D_p}\varphi + \ln p\right) = 0. \tag{1.29}$$

By integrating the last equation, we find

$$p = Ce^{-\frac{\mu_p}{D_p}\varphi}. \tag{1.30}$$

On the other hand, the Boltzmann distribution is valid for p at equilibrium conditions:

$$p = Ce^{-\frac{\varepsilon_p}{k_B T}} = Ce^{-\frac{q\varphi}{k_B T}}. \tag{1.31}$$

The last two equations will be identical if

$$\frac{D_p}{\mu_p} = \frac{k_B T}{q}. \tag{1.32}$$

By using the same line of reasoning, the Einstein relation can be established for electrons. Finally, it is worthwhile to note that the Einstein relations are often called in literature fluctuation-dissipation relations. The reason is that these relations establish connection between fluctuations in the motion of mobile carriers described by diffusivity (D) and dissipation described by mobility (μ).

Other important physical phenomena which occur in semiconductors are *generation* and *recombination* of mobile carriers, i.e., electrons and holes. There are several mechanisms of recombination and generation. Below, we briefly consider only two of them: indirect or Shockley-Read-Hall (SRH) recombination-generation and Auger (or three-particle) recombination-generation.

SRH recombination-generation is dominant for indirect bandgap semiconductors such as Si and SiC. The term "indirect bandgap" means that

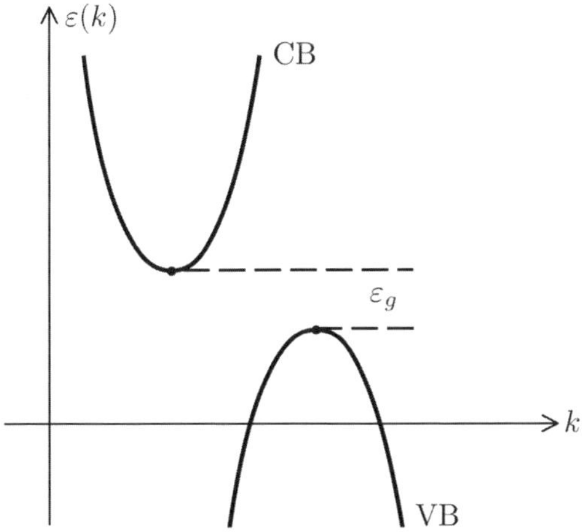

Fig. 1.5

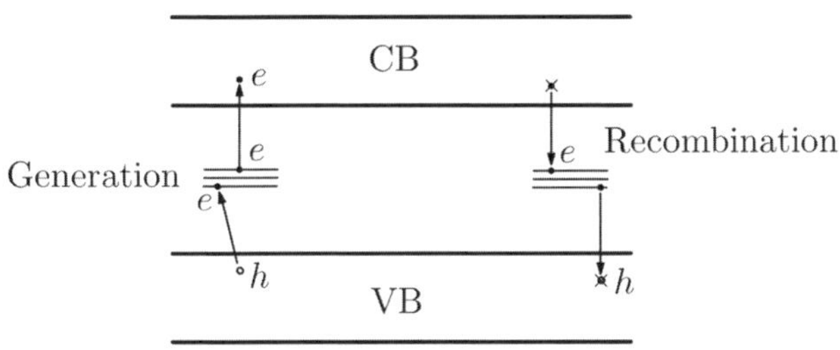

Fig. 1.6

the electron energy $\varepsilon(k)$ as a function of its momentum (crystal momentum, to be precise) k achieves its minimum for the conduction band (CB) and its maximum for the valence band (VB) for different values of k. This is illustrated by Figure 1.5. The latter means that the most probable direct transitions (i.e., transitions with small changes in energy) from the states near the lower energy edge of the conduction band to the states near the upper energy edge of the valence band are prohibited because they cannot be realized with the conservation of momentum k. For this reason, the most probable transitions between the valence and conduction bands in indirect semiconductors occur through localized energy states (energy levels) created in the middle of the bandgap by implantation of so-called "deep" impurities (gold, for instance). The physical mechanism of such transitions is illustrated by Figure 1.6. During the recombination process, an electron

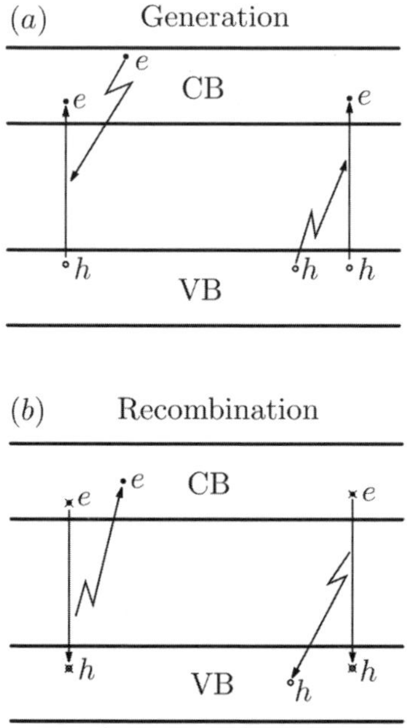

Fig. 1.7

is captured by an empty trap and almost simultaneously a hole is captured by an electron from a filled trap. The inverse process, generation, consists of almost simultaneous transitions of an electron from the valence band to an empty trap and an electron from an occupied trap to the conduction band. The following formula can be derived for the rate $R^{(SRH)}(n,p)$ of this recombination-generation process:

$$R^{(SRH)}(n,p) = \frac{np - n_i^2}{t_p(n+n_1) + t_n(p+p_1)}, \tag{1.33}$$

where t_p and t_n are hole and electron lifetimes, while n_1 and p_1 are some constants.

Now, we consider Auger recombination-generation, which is quite active for high doping levels. The physical mechanism of Auger recombination-generation is illustrated by Figures 1.7a and 1.7b. It is clear from these figures that three particles participate in each elementary process. In the case of generation, an electron from the valence band absorbs energy emitted by an energetic electron (or hole) and makes a transition to the conduction band. This generation process can also be viewed as an impact ionization process where energetic mobile carriers cause generation of electron-hole pairs. In the case of recombination, an electron in the conduction band makes a transition to the valence band and the released energy is transferred to an electron (in n-type material) or to a hole (in p-type material). The following formula can be derived for the rate $R^{(Au)}(n,p)$ of the Auger recombination-generation process:

$$R^{(Au)}(n,p) = (np - n_i^2)(\alpha_n n + \alpha_p p), \tag{1.34}$$

where α_n and α_p are Auger constants.

The transport of electrons and holes in semiconductors is governed by the following continuity equations which express the balance of electrons and holes:

$$\frac{\partial n}{\partial t} = \frac{1}{q}\text{div } \mathbf{J}_n - R(n,p), \tag{1.35}$$

$$\frac{\partial p}{\partial t} = -\frac{1}{q}\text{div } \mathbf{J}_p - R(n,p). \tag{1.36}$$

These equations state that in any infinitesimally small volume the time variation of mobile carrier density is due to local inflow (or outflow) of carriers due to their drift and diffusion currents as well as due to local recombination-generation of carriers.

These current continuity equations together with the Poisson equation (1.13) and formulas (1.20) and (1.21) for electron and hole currents constitute three coupled partial differential equations for n, p and φ. These coupled equations are the essence of the *drift-diffusion* model for electron and hole transport in semiconductors. This drift-diffusion model is widely used for analytical and numerical analysis of semiconductor devices. For very small (nanoscale) devices this drift-diffusion model is replaced by more accurate and relevant models such as semiclassical transport or quantum transport models. Discussion of these models is beyond the scope of this text.

In summary, there are two types of mobile carriers in semiconductors, electrons and holes, whose transport occurs within conduction and valence bands, respectively. There are two distinct mechanisms of current conduction, drift and diffusion. There are phenomena of recombination-generation caused by transition of mobile carriers across energy bandgaps or some parts of them. The transport of carriers is described by coupled continuity and Poisson equations which constitute the drift-diffusion model.

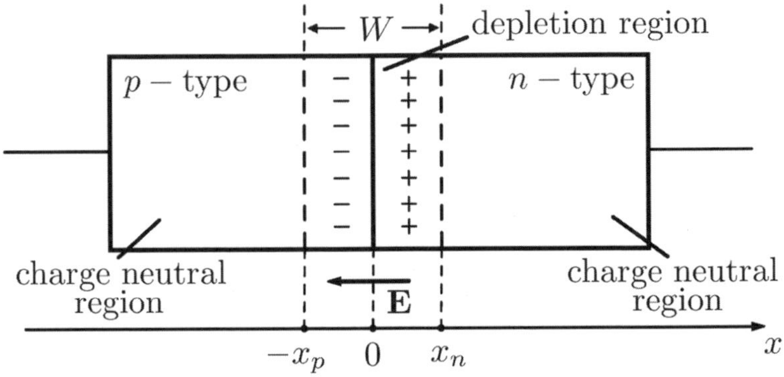

Fig. 1.8

1.2 P-N Junctions and Diodes

In this section, we shall discuss *p*-*n* junctions and their use as diodes. These junctions are ubiquitous in semiconductor electronics because most semiconductor devices utilize at least one junction between *n*-type and *p*-type materials. For this reason, one may say that *p*-*n* junctions are among the main building blocks of semiconductor devices. These *p*-*n* junctions are fundamental in carrying out such functions as rectification, switching, amplification, etc.

A simple *p*-*n* junction can be viewed as a piece of semiconductor with two adjacent *p*-type and *n*-type regions (see Figure 1.8). It will be assumed in our discussion that these *p*-type and *n*-type regions are uniformly doped (abrupt junction). Namely, the densities of ionized impurities can be plotted as shown in Figure 1.9. Since these two differently doped regions are adjacent to one another, electrons tend to diffuse from the *n*-type region to the *p*-type region, while holes tend to diffuse from the *p*-type region into the *n*-type region resulting in nonzero diffusion currents. Immediately the question can be asked how equilibrium conditions (i.e., when electron and hole current densities are equal to zero) can be realized in such junctions. It is clear that such equilibrium conditions can be achieved only if these diffusion currents are counterbalanced by drift currents. These drift currents can be created when the depletion region is formed. The latter means that a narrow region around the interface between *n*-type and *p*-type materials is depleted of mobile carriers. As a result, the positive and negative charges of ionized impurities are exposed and these charges create nonzero electric

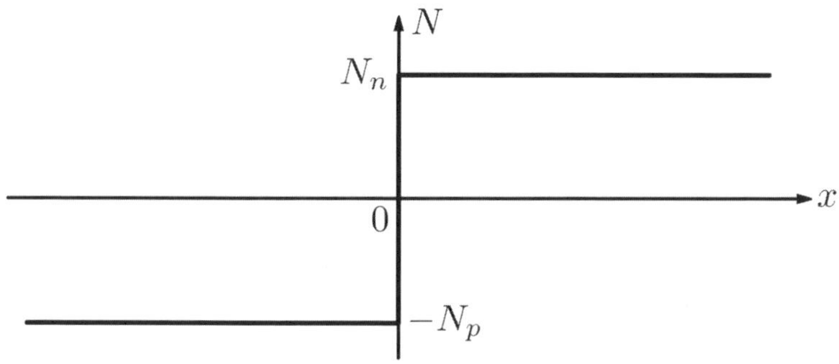

Fig. 1.9

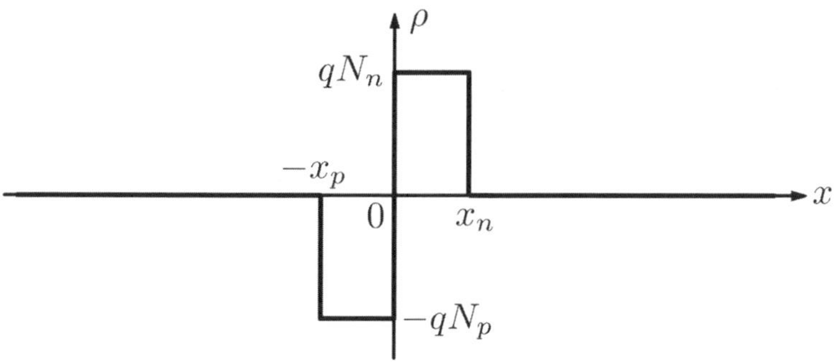

Fig. 1.10

field **E** that drives electrons and holes in directions opposite to the directions of their diffusion. Indeed, as seen from Figure 1.8, holes are driven by electric field **E** back to the p-type region, while electrons are driven by the same field back to the n-type region.

It is important to stress that the physical mechanism of depletion region formation is diffusion of mobile carriers. Indeed, holes and electrons will continue to diffuse into n-type and p-type regions, respectively, and recombine there until sufficiently strong electric field **E** is established as a result of mobile carrier depletion and this field counteracts the above diffusion.

Outside the depletion region, there are two charge neutral regions (see Figure 1.8) where

$$\rho = q(p - n + N) = 0. \tag{1.37}$$

Thus, the plot of volume charge density can be drawn as shown in Figure 1.10. This plot corresponds to the *depletion approximation* when it is

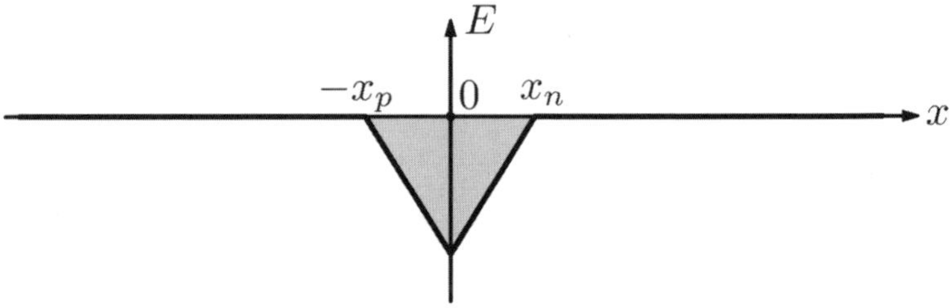

Fig. 1.11

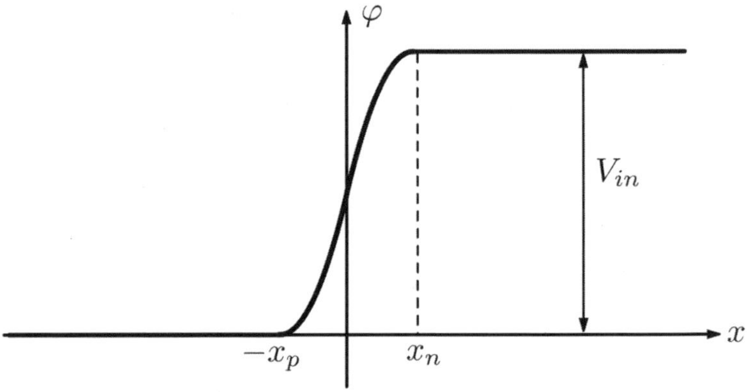

Fig. 1.12

assumed that the region $-x_p < x < x_n$ is completely depleted of mobile carriers. Next, we shall plot the graphs for electric field E and electric potential φ. In doing so, we shall use the relations

$$\frac{dE}{dx} = \frac{\rho}{\varepsilon}, \tag{1.38}$$

$$E = -\frac{d\varphi}{dx}. \tag{1.39}$$

From formula (1.38) and the plot shown in Figure 1.10 follows that E is piecewise linear in the depletion region and it is equal to zero outside this region (see Figure 1.11). From formula (1.39) and the plot shown in the last figure follows that electric potential φ is piecewise quadratic within the depletion region, equal to zero for $x < -x_p$ and assumes constant value V_{in} for $x > x_n$ (see Figure 1.12). Thus, it can be concluded that a specific potential difference V_{in}, called built-in potential, appears across the depletion region. This potential is central to the performance of p-n junctions. For this reason, we shall next derive the expression for V_{in}.

At equilibrium, we have

$$\mathbf{J}_p = 0, \qquad \mathbf{J}_n = 0. \tag{1.40}$$

The first equation in (1.40) can be written as follows:

$$D_p \nabla p + \mu_p p \nabla \varphi = 0. \tag{1.41}$$

In the one-dimensional case when all physical quantities vary only with respect to one variable x, the last equation leads to

$$D_p \frac{dp}{dx} = -\mu_p p \frac{d\varphi}{dx}. \tag{1.42}$$

By using the Einstein relation (see formulas (1.22) and (1.23)), we derive from (1.42) that

$$\frac{d\varphi}{dx} = -V_T \frac{d}{dx}(\ln p). \tag{1.43}$$

By integrating from $-x_p$ to x_n both sides of the last equality, we find

$$\varphi(x_n) - \varphi(-x_p) = -V_T[\ln p(x_n) - \ln p(-x_p)]. \tag{1.44}$$

The built-in potential V_{in} is defined as

$$\varphi(x_n) - \varphi(-x_p) = V_{in}. \tag{1.45}$$

The latter means that formula (1.44) can be written as

$$V_{in} = V_T \ln \frac{p(-x_p)}{p(x_n)}. \tag{1.46}$$

Now, by using the charge neutrality condition (1.37) at $x = -x_p$ and taking into account that in the p region the density of electrons n is very small in comparison with the density of holes p and density of ionized donors N_p, we find

$$p(-x_p) \approx N_p. \tag{1.47}$$

Then, according to formula (1.8), we have

$$p(x_n)n(x_n) = n_i^2, \tag{1.48}$$

and

$$p(x_n) = \frac{n_i^2}{n(x_n)}. \tag{1.49}$$

Now, by using the charge neutrality condition (1.37) at $x = x_n$ and the same reasoning as before, we conclude

$$n(x_n) \approx N_n. \tag{1.50}$$

By using the last relation in formula (1.49), we find that

$$p(x_n) = \frac{n_i^2}{N_n}. \tag{1.51}$$

By substituting formulas (1.47) and (1.51) into equation (1.46), we derive the following important expression:

$$\boxed{V_{in} = V_T \ln \frac{N_n N_p}{n_i^2}.} \tag{1.52}$$

The starting point in our derivation of formula (1.52) was the first equation in (1.40). It can be shown that the same formula (1.52) can be obtained by using the second equation in (1.40) as the starting point of derivation.

Typically, the range of built-in potential variations is

$$0.5\,\text{V} \leq V_{in} \leq 0.85\,\text{V}. \tag{1.53}$$

For instance, if

$$N_n = N_p = 10^{16}\ \text{cm}^{-3}, \tag{1.54}$$

then

$$V_{in} = 0.699\ \text{V}. \tag{1.55}$$

Such a relatively small range of variation of V_{in} is due to the smallness of V_T (see formula (1.24)) and the presence of "ln" in equation (1.52).

It is also quite interesting to find the expression for the overall width W of the depletion region as well as for its one-sided width x_n (or x_p). This can be done by using the following reasoning. By using formula (1.38) in the depletion region, we find

$$\frac{dE(x)}{dx} = -\frac{qN_p}{\varepsilon} \quad \text{for} \ -x_p < x < 0, \tag{1.56}$$

$$\frac{dE(x)}{dx} = \frac{qN_n}{\varepsilon} \quad \text{for} \quad 0 < x < x_n. \tag{1.57}$$

Furthermore, at the boundary of the depletion region, we have

$$E(-x_p) = E(x_n) = 0. \tag{1.58}$$

From the last three formulas, we obtain

$$\max_x E(x) = E(0) = -\frac{qN_n}{\varepsilon} x_n = -\frac{qN_p}{\varepsilon} x_p. \tag{1.59}$$

The last equation yields the following relation between one-sided widths of the depletion region:

$$N_n x_n = N_p x_p. \tag{1.60}$$

On the other hand, we have

$$V_{in} = \varphi(x_n) - \varphi(-x_p) = -\int_{-x_p}^{x_n} E(x)dx. \tag{1.61}$$

It is clear that the absolute value of the integral in the last formula is equal to the shaded area in Figure 1.11. Consequently,

$$\int_{-x_p}^{x_n} E(x)dx = \frac{1}{2}E(0)W, \tag{1.62}$$

where

$$W = x_n + x_p. \tag{1.63}$$

Now, by combining formulas (1.59), (1.61) and (1.62), we derive

$$V_{in} = \frac{qN_n}{2\varepsilon} x_n W. \tag{1.64}$$

Next, by solving equations (1.60) and (1.63) we find the following expression for x_n:

$$x_n = \frac{N_p}{N_n + N_p} W. \tag{1.65}$$

By substituting the last formula into equation (1.64), we obtain

$$V_{in} = \frac{qN_n N_p}{2\varepsilon(N_n + N_p)} W^2, \tag{1.66}$$

which leads to

$$W = \left[\frac{2\varepsilon(N_n + N_p)}{qN_n N_p} V_{in}\right]^{\frac{1}{2}}. \tag{1.67}$$

Finally, by using formula (1.52) for V_{in} in the last equation, we arrive at

$$\boxed{W = \left[\frac{2\varepsilon V_T(N_n + N_p)}{qN_n N_p} \ln \frac{N_n N_p}{n_i^2}\right]^{\frac{1}{2}}.} \tag{1.68}$$

The last formula gives the expression for the depletion width in terms of doping densities. It is apparent that the higher the doping densities, the narrower the depletion width. Indeed, in the particular case when $N_p = N_n = N$, from the last formula we find

$$W = 2\left[\frac{2\varepsilon V_T}{qN} \ln \frac{N}{n_i}\right]^{\frac{1}{2}} \tag{1.69}$$

and it is clear that $W \to 0$ as $N \to \infty$.

From the equations (1.65) and (1.68), we derive the following expression for one-sided width of the depletion region in terms of doping densities:

$$\boxed{x_n = \left[\frac{2\varepsilon V_T N_p}{q N_n (N_n + N_p)} \ln \frac{N_n N_p}{n_i^2} \right]^{\frac{1}{2}}.} \tag{1.70}$$

It is clear from the last formula that for very large N_n we have the asymptotic relation

$$x_n \sim \frac{(\ln N_n)^{\frac{1}{2}}}{N_n}, \tag{1.71}$$

which suggests that the substantial one-sided narrowing of the depletion region occurs with increase in N_n.

Overall narrowing of the depletion region may present some problems related to avalanche breakdown of p-n junctions as discussed later in this section. On the other hand, the one-sided narrowing of the depletion region is practically utilized in many semiconductor devices for constructing ohmic contacts. Indeed, typically, ohmic contacts are metal-semiconductor contacts with heavily doped semiconductors in the contact regions. This makes depletion widths very narrow to allow carriers to tunnel through.

The built-in potential can be viewed as a potential (or energy) barrier which prevents mobile carrier transport through the depletion region at equilibrium. This property of the built-in potential is the key to understanding the rectifying functions of the p-n junction. When an external voltage is applied to the p-n junction, then depending on its polarity this voltage may decrease or increase the potential (energy) barrier already existing due to the built-in potential. If the applied voltage causes a decrease in the potential barrier, this will result in current conduction. It is said in this case that the p-n junction is forward biased and the polarity of the applied voltage is treated as positive. If, on the other hand, the applied voltage causes an increase in the potential barrier, this will further impede the current conduction. It is said in this case that the p-n junction is reverse biased and the polarity of the applied voltage is treated as negative.

The dependence of the p-n junction current on polarity and magnitude of applied voltage is given by the following equation derived by W. Shockley:

$$I = I_s \left(e^{\frac{qV}{k_B T}} - 1 \right), \tag{1.72}$$

where I_s is the so-called saturation current. By recalling the definition of thermal voltage V_T (see formula (1.23)), the last equation can be written

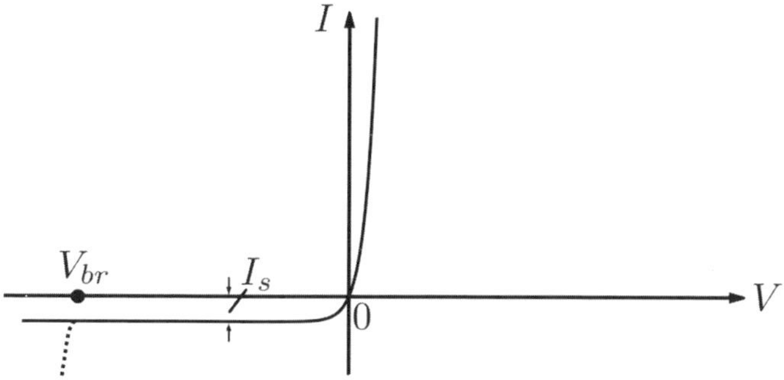

Fig. 1.13

as follows:

$$I = I_s \left(e^{\frac{V}{V_T}} - 1 \right). \tag{1.73}$$

A typical plot of this equation is shown in Figure 1.13. It is apparent from this figure that the current is very rapidly and exponentially increased for positive (forward) bias of the junction. This rapid increase in junction current for small forward bias voltages is due to the smallness of V_T, which is equal to about 0.026 V at room temperatures (see (1.24)). For negative bias voltages, the junction current rapidly reaches the value $-I_s$. The question can be immediately asked why there exists nonzero current for negative voltages when the junction is reverse biased, and what the physical origin of this current is. The answer is that the saturation current I_s is due to electron-hole pair (EHP) generation within the depletion region. Indeed, such generation (however small) always exists (due, for instance, to an intrinsic process). Electrons and holes generated in the depletion region are swept by an electric field in this region in opposite directions (i.e., toward n-type and p-type regions, respectively). This results in negative current I_s. At equilibrium (zero bias voltage) this current I_s is fully compensated by the positive diffusion current which is due to a small number of high energy electrons and holes that are able to surmount the energy barrier created by the built-in potential. As negative bias is increased, this results in an increase in the energy barrier and in an appreciable decrease in the number of high energy mobile carriers able to surmount this increased energy barrier. Consequently, the positive diffusion current due to high energy carriers is practically reduced to zero and the negative current I_s due to EHP generation is exposed, as can be seen from Figure 1.13.

Mobile carriers have exponential (or close to exponential) distribution in energy. When the forward (positive) bias voltage is applied to the junction,

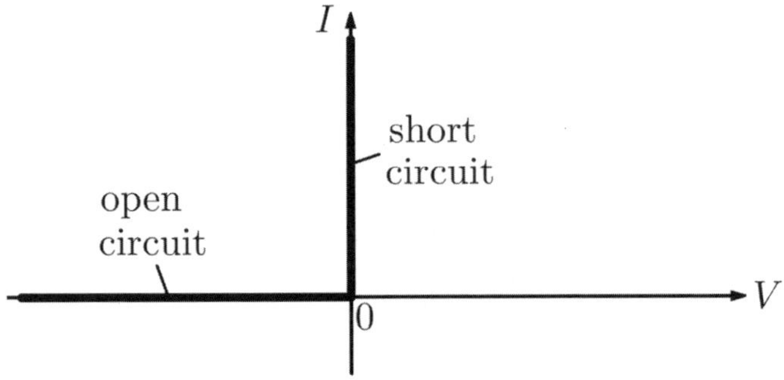

Fig. 1.14

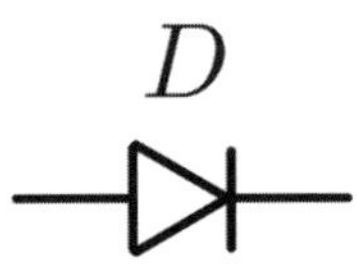

Fig. 1.15

the potential (energy) barrier is decreased. This results in exponential increase in the number of mobile carriers able to surmount the reduced energy barrier, and this leads to the exponential growth in junction diffusion current as consistent with formulas (1.72) and (1.73).

The I-V curve presented in Figure 1.13 suggests that the resistance of the p-n junction for positive bias voltages is very small, while for negative bias voltages this resistance is very large. Using this fact, the actual I-V curves can be idealized and represented by the plot shown in Figure 1.14. It is apparent from the previous discussion and Figure 1.14 that a p-n junction can be used as a diode, i.e., a circuit element that can be switched from open-circuit state to short-circuit state and vice versa by the change in polarity of applied voltage. The circuit notation for the diode element is presented in Figure 1.15.

Power diodes used in power electronics are required to sustain large negative (reverse bias) voltages. Such voltages may result in large electric fields across depletion regions and may cause the breakdown of p-n junctions at some voltages V_{br}. One of the physical mechanisms of such breakdown is impact ionization when EHPs generated in depletion regions are appreciably accelerated by strong depletion region electric fields due to large reverse

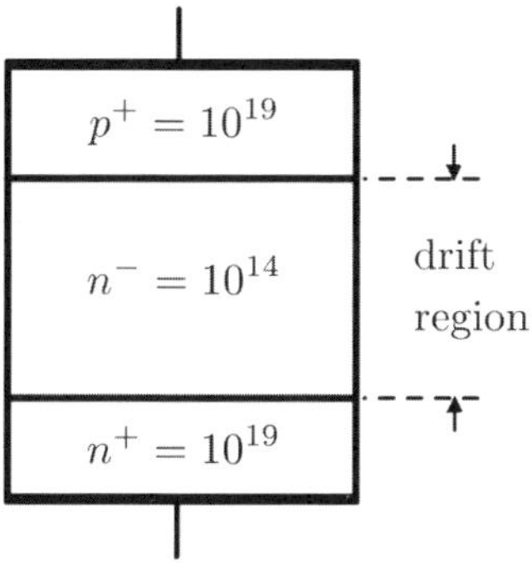

Fig. 1.16

biases. These high energy mobile carriers cause new generation of EHPs through the Auger generation process. These newly generated EHPs are also accelerated by the electric field and also generate new EHPs through the Auger process. This cascading generation leads to an avalanche of EHPs and large breakdown current.

To increase the breakdown voltage in power diodes, special designs of such diodes are used. In these designs, a lightly doped (almost intrinsic in p-i-n diodes) semiconductor region is placed between heavily doped p^+-type and n^+-type regions used for ohmic contacts. An example of such design is shown in Figure 1.16. It must be stressed that the presence of lightly doped and sufficiently thick n^- layer is a typical structural feature of power diodes. To demonstrate how lightly doped n^- regions lead to the increase in breakdown voltage, we shall start with the formula similar to (1.61):

$$V_{br} = \varphi(x_n) - \varphi(-x_p) = -\int_{-x_p}^{x_n} E(x)dx, \tag{1.74}$$

where x_n and $-x_p$ are the boundary coordinates of the depletion region when the reverse bias voltage is equal to V_{br}. In the case of depletion approximation, $E(x)$ is a piecewise linear function (see Figure 1.11). Consequently,

$$-\int_{-x_p}^{x_n} E(x)dx = \frac{1}{2}E_{br}W, \tag{1.75}$$

where E_{br} is the breakdown value of electric field whose magnitude is equal to $E(0)$, while $W = x_n + x_p$. From the last two formulas, we derive

$$V_{br}^2 = \frac{1}{4}E_{br}^2 W^2. \tag{1.76}$$

By using the same line of reasoning which led to the derivation of formula (1.66), it can be shown that

$$V_{br} = \frac{qN_nN_p}{2\varepsilon(N_n + N_p)} W^2. \tag{1.77}$$

By dividing formula (1.76) by formula (1.77), we find

$$V_{br} = \frac{\varepsilon(N_n + N_p)}{2qN_nN_p} E_{br}^2. \tag{1.78}$$

Since

$$N_p \gg N_n, \tag{1.79}$$

from equation (1.78) we finally obtain

$$V_{br} \approx \frac{\varepsilon E_{br}^2}{2qN_n}. \tag{1.80}$$

The last formula clearly reveals that for the same breakdown electric field E_{br} the breakdown voltage can be appreciably increased by reducing density N_n. The latter justifies using lightly doped n^- regions in power diodes.

In the conclusion of this section, it is worthwhile to mention that *p-n* junctions are used in solar cells for conversion of solar energy into electric energy. The principle of operation of these cells is based on optical generation of EHPs in the depletion region of the *p-n* junction. This generation occurs because in the depletion region exposed to optical radiation valence electrons may absorb optical energy from incident light sufficient for their transition to the conduction band. These optically generated electrons and holes are then swept by the electric field in the depletion region in opposite directions resulting in junction electric current. This conversion of energy of optical radiation into the energy of electric currents in junctions is the physical foundation of solar cell operation.

1.3 BJT and Thyristor

In this section, we shall first discuss the basic design and the principle of operation of the bipolar junction transistor (BJT) as well as how this transistor can be utilized as a switch. Then, we shall consider the basic structure of the thyristor, which is often called a semiconductor-controlled rectifier (SCR), and discuss its principle of operation by using a two-BJT model of the thyristor.

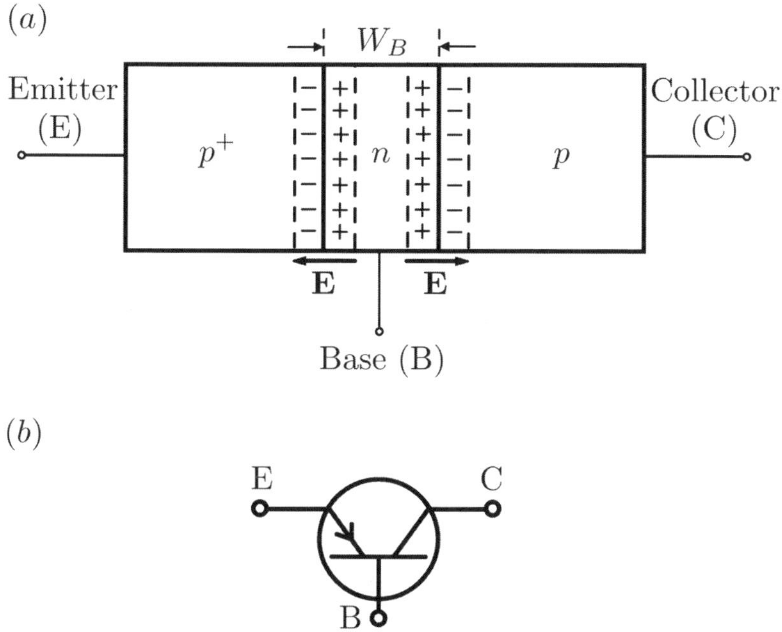

Fig. 1.17

The BJT is a three-terminal device and its three contacts are connected to three differently doped regions which are called *emitter*, *base* and *collector*. Schematics of the BJT are shown in Figure 1.17a for a p^+np transistor, while its circuit symbol is presented in Figure 1.17b. The BJT can also be designed as a n^+pn transistor as shown in Figure 1.18a, and its circuit symbol is depicted in Figure 1.18b. Below, we shall discuss only the p^+np transistor; the treatment of the n^+pn device is very similar and will be left as an exercise.

There are two very important features of BJT design. First, the emitter is *heavily doped*, which implies that

$$p^+ \gg n. \tag{1.81}$$

Second, the base region of the BJT is *very narrow*. The latter statement is usually characterized by the inequality

$$W_B \ll L_p, \tag{1.82}$$

where W_B stands for the width of the base, while L_p is the diffusion length of holes in the n-type region, i.e., the average distance through which the holes can diffuse without recombination.

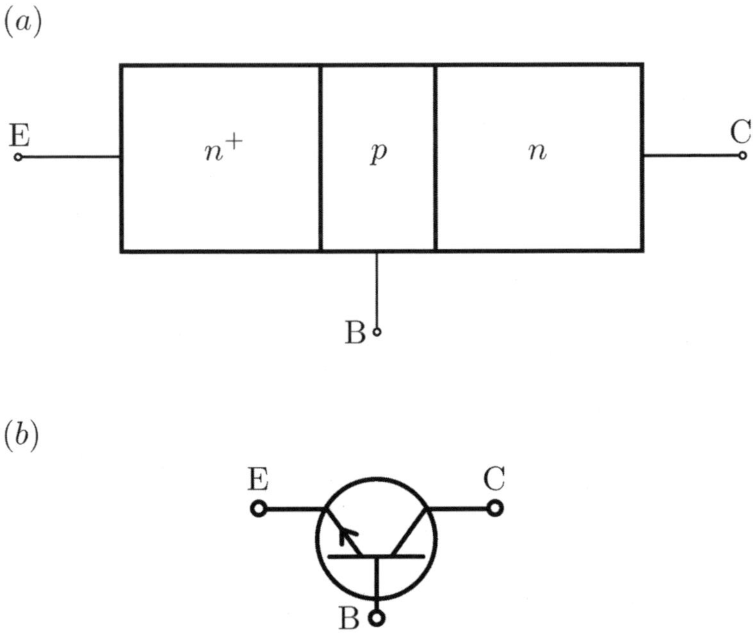

Fig. 1.18

The above two features define the quality of the BJT.

Next, we shall discuss the operation of the BJT as a current-controlled device. In doing so, we shall emphasize only the main features of this operation without covering the complete theory of this device.

A BJT has two junctions, the emitter junction between the emitter and the base regions, and the collector junction between the base and the collector regions. At equilibrium, that is, when the current through the base terminal is zero, there are two depletion regions corresponding to the emitter and collector junctions. The electric fields in these depletion regions are shown in Figure 1.17a, and these electric fields prevent the transport of mobile carriers through the emitter and collector junctions. Hence, there is no net current flow from emitter to collector. Also, at equilibrium the part of the base between these two depletion regions is *charge neutral.*

Now, consider what happens when a small electron current I_B through the base is introduced. This current results in the small excess of negatively charged electrons in the previously charge neutral region of the base. The electric field of this small net negative charge in the base is directed opposite to the electric field in the depletion region of the emitter junction. This

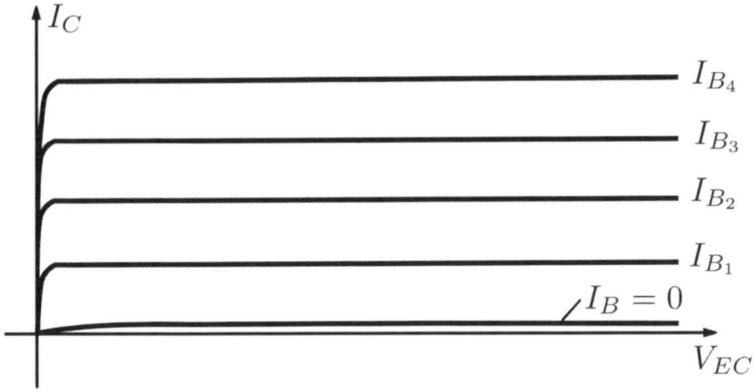

Fig. 1.19

leads to the small reduction in the electric field in the emitter junction, which is equivalent to *small forward bias* of this junction. Since the current through a *p-n* junction is *exponentially* dependent on bias voltages (see equation (1.73)), this small forward bias results in a large current through the emitter junction called the emitter current. Since the emitter is heavily doped (see inequality (1.81)), the emitter current consists *predominantly of holes* injected into the base. These holes diffuse through the base with very *little recombination* because the base is narrow (see formula (1.82)). When the injected holes reach as a result of diffusion the depletion region of the collector junction, they are swept by the electric field in this depletion region into the collector region, resulting in large collector current I_C. Thus, a small base current results in large collector current. The latter can be mathematically written as

$$I_C = \beta I_B, \tag{1.83}$$

where β is the amplification factor (or current gain). For properly designed BJTs, β is quite large and typically

$$\beta > 100. \tag{1.84}$$

It is clear from (1.83) that the collector current is mostly controlled by the base current and does not depend much on the voltage V_{EC} across the transistor, i.e., the voltage between the emitter and collector terminals. This is reflected in the family of curves $I_C(V_{EC})$ shown in Figure 1.19 for different values of the base current I_B, namely $I_{B_1} < I_{B_2} < I_{B_3} < I_{B_4}$. Now, we shall discuss how the BJT can be used as a *current-controlled* switch. This is done by using the common emitter configuration where

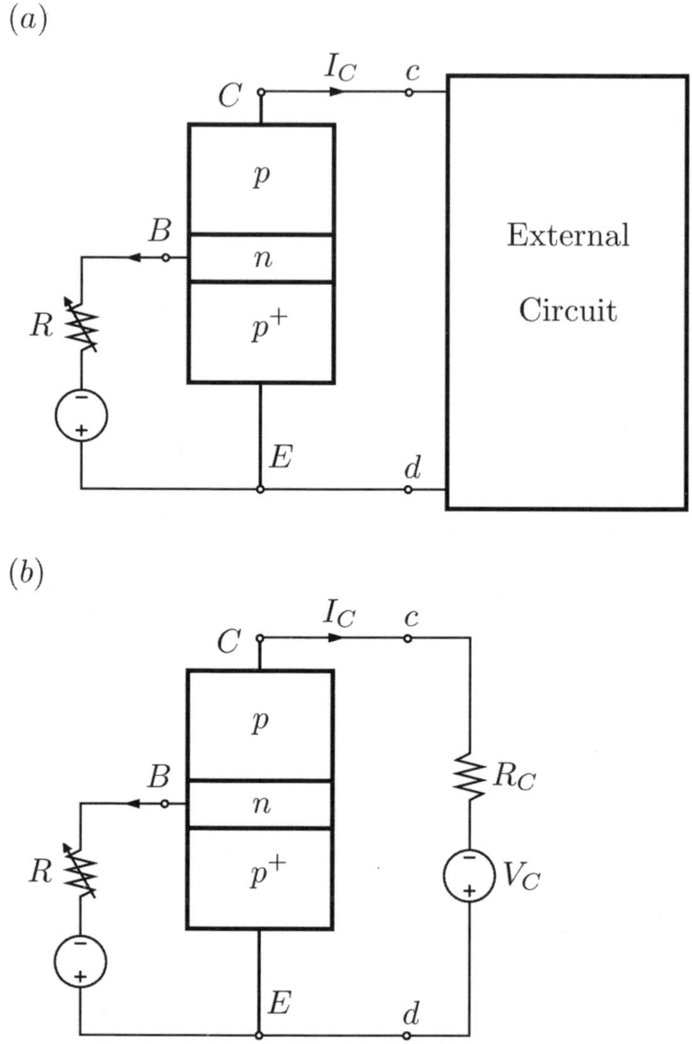

Fig. 1.20

the emitter terminal is common to the circuit controlling the base current and to an external circuit for which the BJT serves as a switch. This configuration is shown in Figure 1.20a. By assuming that the switching of the BJT occurs sufficiently fast, the external circuit can be represented by the Thevenin-equivalent resistance R_C and voltage source V_C (see Figure 1.20b). Indeed, during the fast switching, the voltages across capacitors

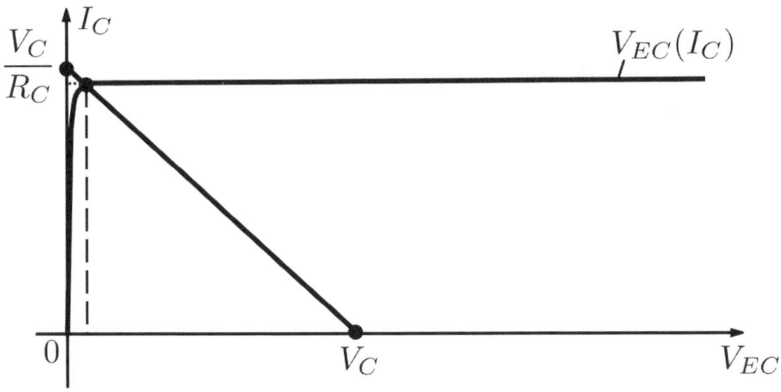

Fig. 1.21

and currents through inductors in the external circuit do not change much. This means that, during the switching, capacitances and inductances can be replaced by voltage and current sources, respectively. For this reason, the external circuit can be treated as a resistive circuit with sources and can be replaced by its Thevenin equivalent.

By using KVL for the loop consisting of the BJT, R_C and voltage source V_C, we find

$$V_{EC} + I_C R_C = V_C, \tag{1.85}$$

or

$$V_{EC} = V_C - I_C R_C. \tag{1.86}$$

It is apparent that the voltage V_{EC} is a function of I_C (see the curves in Figure 1.19). Consequently, the last equation can be written as follows:

$$V_{EC}(I_C) = V_C - I_C R_C. \tag{1.87}$$

This nonlinear equation can be solved graphically by plotting the nonlinear function $V_{EC}(I_C)$ and the straight line representing the right-hand side of the last formula, which is usually called a "load line." This is done in Figure 1.21, where V_C and V_C/R_C are the intercepts of the load line with the V_{EC}-axis and the I_C-axis, respectively. It is apparent that the solution of equation (1.87) corresponds to the intersection point of the above two graphs, because for the value of I_C corresponding to this intersection point the equality of both sides of equation (1.87) is achieved. This graphical representation of the operational condition of the BJT is very useful for the description of BJT performance as a switch. Indeed, consider a set of

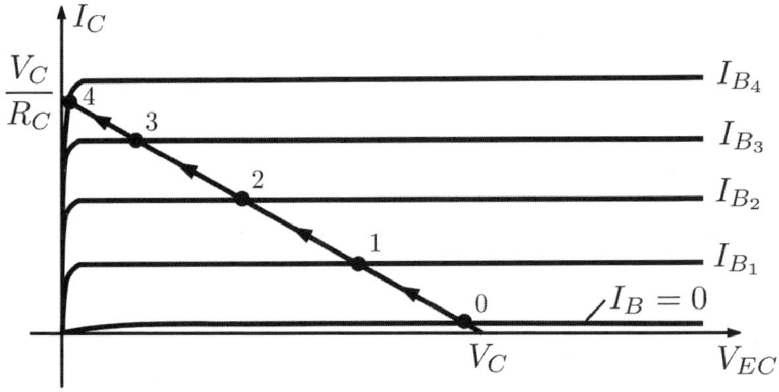

Fig. 1.22

curves representing $V_{EC}(I_C)$ for different values of the base current I_B and the load line representing the external circuit during the switching process (see Figure 1.22). The switching of the BJT by means of the appropriate increase in the base current can be explained as follows. For zero base current, there is practically no current through the transistor and the operating point of the BJT is marked as point "0" on Figure 1.22. As the base current I_B is increased, the operating points will be, in succession, points "1," "2," "3" and "4" on the same figure, and they represent intersection points between the load line and the set of $V_{EC}(I_C)$ curves corresponding to monotonically increased values of I_B. Thus, for zero (or negative) value of I_B, there is very small (or no) current through the BJT for large values of V_{EC}. This clearly corresponds to the high resistive (so-called "off") state of the transistor. On the other hand, for sufficiently large base current I_{B_4} (at operating point "4"), there exists sufficiently large current through the BJT for negligibly small value of V_{EC}. This clearly corresponds to the low resistive (or so-called "on") state of the transistor. By neglecting small currents in the "off" states and small voltages in "on" states, the BJT can be viewed as an ideal current-controlled switch characterized by the plot shown in Figure 1.23. It is clear from the presented discussion that the BJT has high-quality "on" and "off" states. The latter means that in these states electric power losses are very small because voltages are very small in the "on" state and currents are very small in the "off" state. That is not true for the intermediate states that the BJT goes through during the switching process. In these intermediate states, $V_{EC}(t)$ and $I_C(t)$ are appreciable and the total energy loss during the switching $\mathcal{E}_{sw}$ is given by the

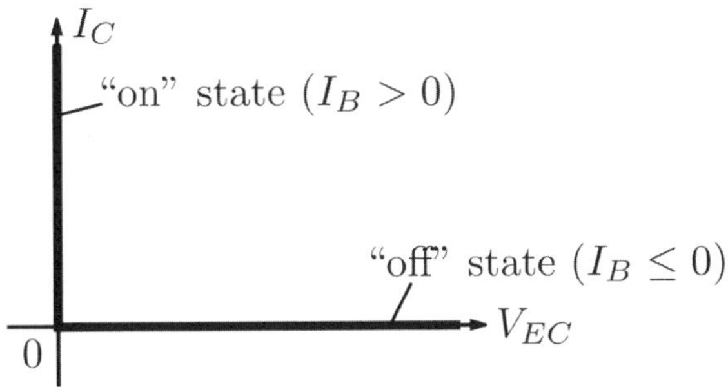

Fig. 1.23

formula

$$\mathcal{E}_{sw} = \int_0^{\tau_{sw}} V_{EC}(t) I_C(t) dt, \tag{1.88}$$

where τ_{sw} is the switching time. It is clear from the last formula that the switching losses can be made small if the switching is very fast. Unfortunately, the BJT is intrinsically a relatively slow switching device. This is because the "turn-on" and "turn-off" switching of the BJT is controlled by how fast charges in the base region (so-called stored charges) can be injected or removed. Charge removing may be especially slow because it is naturally accomplished through the recombination process. To expedite the charge removal at turn-off, the base current is reversed, i.e., it is driven in the direction opposite to that during the turn-on process.

Power BJTs usually have vertical four-layer structures with a collector lightly doped (drift) region to increase breakdown voltages. Vertical structures are also desirable because they lead to large cross-sectional areas for transistor currents. This is beneficial for reduction of on-state resistance and on-state power dissipation.

Now, we shall proceed to the discussion of the thyristor (SCR). This is a three-terminal device and its terminals are marked as anode (A), cathode (K) and gate (G). The circuit symbol of this device is presented in Figure 1.24. The switching of this device is controlled by the polarity of voltage V_{AK} applied between the anode and cathode as well as by a short current pulse through the gate. This device is designed to achieve the switching performance which is (in an idealized manner) illustrated by Figure 1.25. This switching performance can be described in words as follows. When a negative voltage is applied between the anode and cathode ($V_{AK} < 0$), no (appreciable) current conduction is possible; the SCR is in the "off" state. When the polarity of the applied voltage is reversed ($V_{AK} > 0$), the SCR

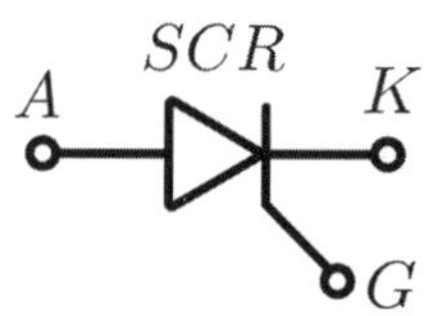

Fig. 1.24

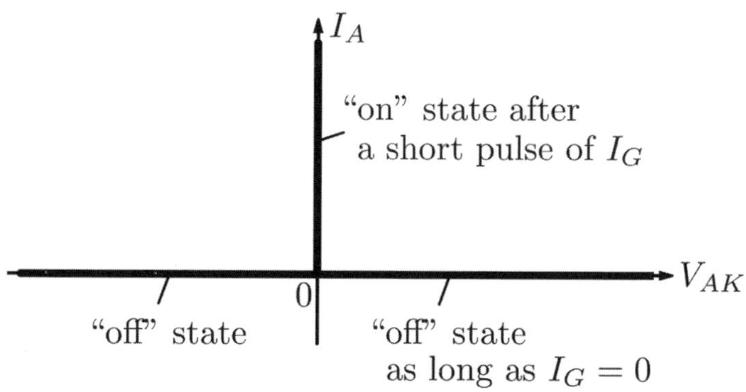

Fig. 1.25

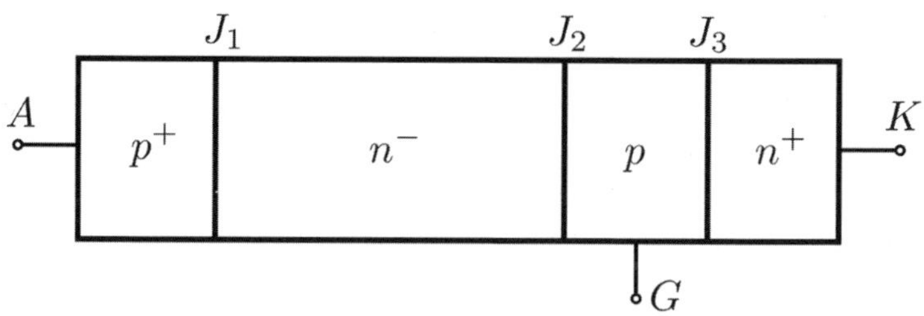

Fig. 1.26

still remains in the "off" state as long as no current I_G is pulsed through the gate terminal. At the timing of proper choice, a short pulse of current I_G turns the SCR into the "on" state, and the device remains in this self-sustaining state with low on-state voltage and high on-state current after the end of the gate current pulse. To turn off the SCR, the polarity of the voltage V_{AK} must be reversed.

The described switching performance can be achieved by using the design shown in Figure 1.26. It is clear from this figure that the structure of the thyristor consists of four semiconductor regions doped in alternating

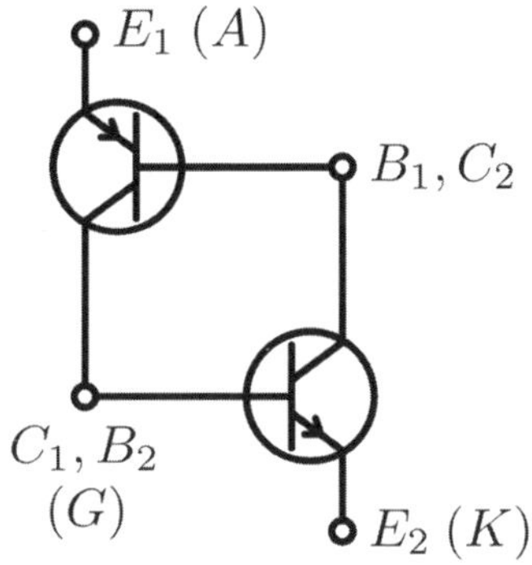

Fig. 1.27

manner. One of these regions is lightly doped (n^- region) and serves as the drift region that supports (without breakdown) high voltages when the device is in the nonconducting (blocking) state. It is clear that the device has three junctions J_1, J_2 and J_3 marked in Figure 1.26. It is also clear that the device design is such as if it contains two BJTs: a p^+n^-p BJT and a n^+pn^- BJT. It can be easily observed that the collector (p region) of the p^+n^-p BJT serves as the base for the n^+pn^- BJT; and the other way around, the collector (n^- region) of the n^+pn^- BJT serves as the base of the p^+n^-p BJT. This leads to the two-transistor model of the thyristor shown in Figure 1.27, which can be used to explain the operation of the thyristor.

First, assume that the polarity of the applied voltage V_{AK} is such that junctions J_1 and J_3 are reverse biased. This polarity is defined as negative ($V_{AK} < 0$). It is clear that there is no current through the thyristor at these biasing conditions. This is the reverse blocking mode (state).

Next, assume that the polarity of V_{AK} is reversed ($V_{AK} > 0$). Under this biasing condition, junctions J_1 and J_3 are forward biased, while junction J_2 is reverse biased. Again, there is no current through the device; this is the forward blocking mode (state).

Now, assume that as the device is in the forward blocking mode a short pulse of current I_G is sent through the gate terminal. This pulse triggers the n^+pn^- BJT and sets into motion electrons from the n^+ region through the p region into the n^- region. This flow of electrons into the n^- region triggers the p^+n^-p BJT, which results in the motion of holes from the p^+ region through the n^- region into the p region. This keeps the n^+pn^- BJT triggered even if the gate current is completely diminished. This mutual triggering mechanism maintains the "on" (current conducting) state of the thyristor that can only be switched off by changing the polarity of the applied voltage V_{AK}.

1.4 MOSFET, Power MOSFET, IGBT

In this section, we shall discuss MOSFET-type devices whose principle of operation is based on electric field-induced inversion phenomena rather than on proper biasing of junctions. These devices offer some clear advantages over BJT devices. Indeed, bipolar transistors are operated as current-controlled switches. As a result, appreciable base currents are required to maintain them in on-states, and even larger reverse currents are needed to speed up their turn-off. This makes the base drive circuits quite complicated and expensive. Furthermore, BJT switches are intrinsically slow, which results in large switching losses. In contrast, MOSFET devices are operated as voltage-controlled switches and no delays occur due to storage or recombination of mobile carriers during the turn-off process. As a result, the switching speed of MOSFET devices is orders of magnitude faster than for bipolar transistors. This fast switching of MOSFET devices may lead to lower overall (total) losses in comparison with bipolar transistors despite the fact that on-state losses for MOSFET devices are larger. The fast intrinsic switching of MOSFETs is also beneficial for ripple suppression that can be accomplished by using smaller energy storage elements (inductors and capacitors).

We shall start with the discussion of the design and the principle of operation of the *lateral* MOSFET, which is the workhorse of digital electronics. A schematic depiction of the structure of this MOSFET is shown in Figure 1.28a. There are many circuit symbols for MOSFETs which are currently in use and which reflect different specific features of their designs and/or operation. In this text, we shall use the circuit symbol shown in Figure 1.28b. The term "MOSFET" is an abbreviation that reflects the main features of the design and the principle of operation of such transistors. The first three letters "MOS" stand for the metal-oxide-semiconductor structure which is evident from Figure 1.28a. The last three letters "FET" stand for "field-effect transistor," which captures the main feature of the principle of operation of the MOSFET. In the MOSFET shown in Figure 1.28a there are four terminals: gate (G), source (S), drain (D) and substrate (Sub), and usually the "S" and "Sub" terminals are connected together. When a

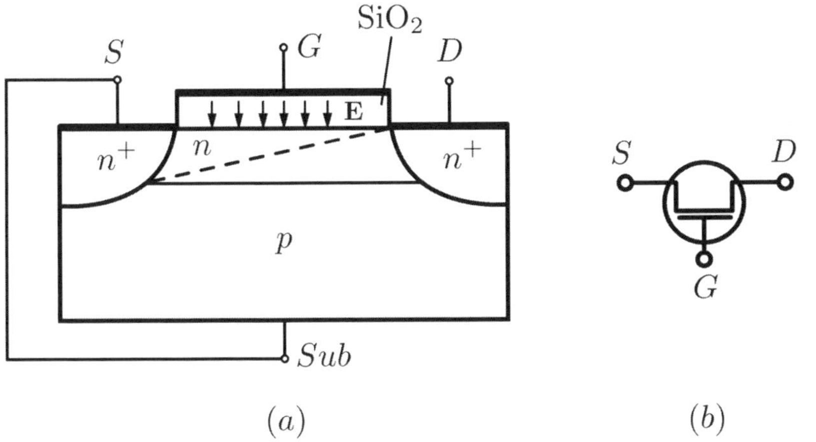

Fig. 1.28

positive voltage V_G is applied to the gate terminal, it creates a vertical electric field **E** directed down from the gate terminal. This electric field forces holes in the p-doped silicon substrate to move in the direction of the field and away from the silicon dioxide (SiO_2) and silicon interface. This results in a thin layer under the dioxide interface depleted of holes. When the gate voltage V_G and electric field **E** are further increased, electrons moving in the direction opposite to **E** are brought close to the dioxide interface. As a result, the thin layer under the dioxide interface is changed from p-type to n-type. This process is called *inversion*, and it leads to the formation of an *n-channel* that connects the two n^+ regions of the source and drain contacts (see Figure 1.28a). Now, if a small voltage V_{DS} between the drain (D) and source (S) is applied, it results in current I_D from drain to source. This current is linearly increased with small increases in V_{DS} because the n-channel serves as a resistor (see Figure 1.29a). As V_{DS} and I_D are further increased, this results in appreciable gradual increase in potential along the dioxide interface in the direction from source to drain contacts. This gradual increase in potential results in gradual decrease in vertical electric field **E** which, in turn, leads to gradual n-channel narrowing as one moves from source to drain. This narrowing causes an increase in the n-channel resistance and manifests itself in gradual decrease in the local slope of the curve $I_D(V_{DS})$ (see Figure 1.29a). As voltage V_{DS} is further increased, the point is reached when the thickness of the n-channel near the drain contact is reduced to zero (see dashed line in Figure 1.28a). In other words, the

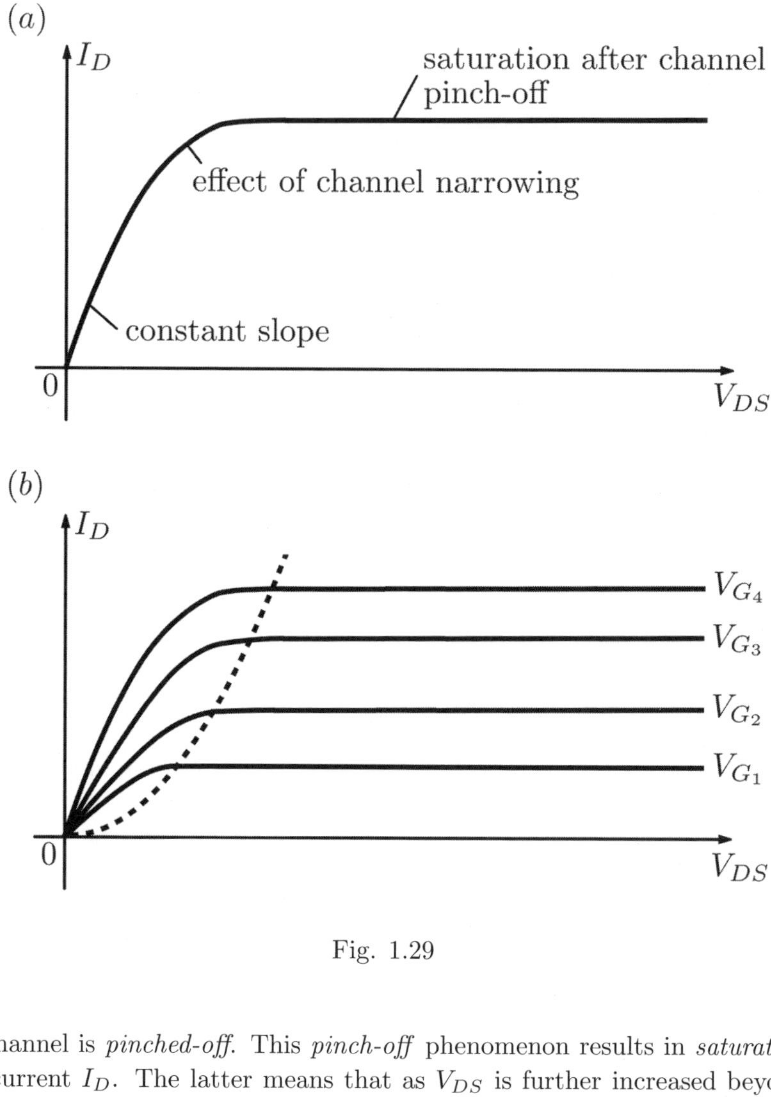

Fig. 1.29

n-channel is *pinched-off*. This *pinch-off* phenomenon results in *saturation* of current I_D. The latter means that as V_{DS} is further increased beyond its pinch-off value practically no significant increase in I_D is observed (see Figure 1.29a). It is clear that the shape of the curve $I_D(V_{DS})$ depends on the value of voltage V_G. Indeed, the larger V_G, the larger is the vertical electric field emanating from the gate and the larger is the thickness of the inversion layer (i.e., n-channel). The latter results in smaller resistance of this n-channel and the larger current I_D for the same value of V_{DS}. This also results in the increase of pinch-off value of V_{DS}. A set of I_D versus

V_{DS} curves for different values of V_G is presented in Figure 1.29b. It can be shown theoretically that in the saturation state the drain current I_D has a "square-law" dependence on the gate voltage V_G. This is indicated by the dashed line in Figure 1.29b.

Next, we shall discuss how a MOSFET device can be utilized as a *voltage-controlled* switch. This is done by connecting source and drain terminals to an external circuit as shown in Figure 1.30a. Since the switching of the MOSFET is quite fast, capacitances and inductances can be replaced during the switching by voltage and current sources, respectively. This means that the external circuit can be represented with respect to source and drain terminals by the Thevenin-equivalent resistance R_D and voltage source V_D (see Figure 1.30b). By using KVL for the loop consisting of the MOSFET, R_D and voltage source V_D, we find

$$V_{DS}(I_D) + I_D R_D = V_D, \tag{1.89}$$

or

$$V_{DS}(I_D) = V_D - I_D R_D. \tag{1.90}$$

The last nonlinear equation can be solved graphically by using the concept of the "load line" (see the previous section). This graphical solution is illustrated in Figure 1.31, which also clarifies the operation of the MOSFET as a switch. Indeed, for zero (or slightly negative) voltage V_G, there is practically no current through the MOSFET. This operating point is marked as point "0" on Figure 1.31, and it corresponds to the "off" state of the transistor. As the gate voltage V_G is monotonically increased, achieving successively the values V_{G_1}, V_{G_2}, V_{G_3} and V_{G_4}, the operating points achieved consecutively are the points "1," "2," "3" and "4" (see again Figure 1.31). Thus, for sufficiently large gate voltage V_{G_4}, the corresponding operating point is point "4" where there exists a large current through the MOSFET for a relatively small voltage V_{DS}. This operating point can be treated as the "on" state of the transistor. It is clear from the above figure that in the on-state of the MOSFET there are non-negligible losses. This is the main shortcoming of using the MOSFET as a switch.

Now, we proceed to the discussion of the power MOSFET. As with most power semiconductor devices, the structure of this device is vertical. A schematic depiction of this structure is presented in Figure 1.32. By using a vertical structure and lightly doped (drift) n^- region, it is possible for power MOSFETs to sustain high blocking voltages and high currents. The principle of operation of the power MOSFET is essentially the same as

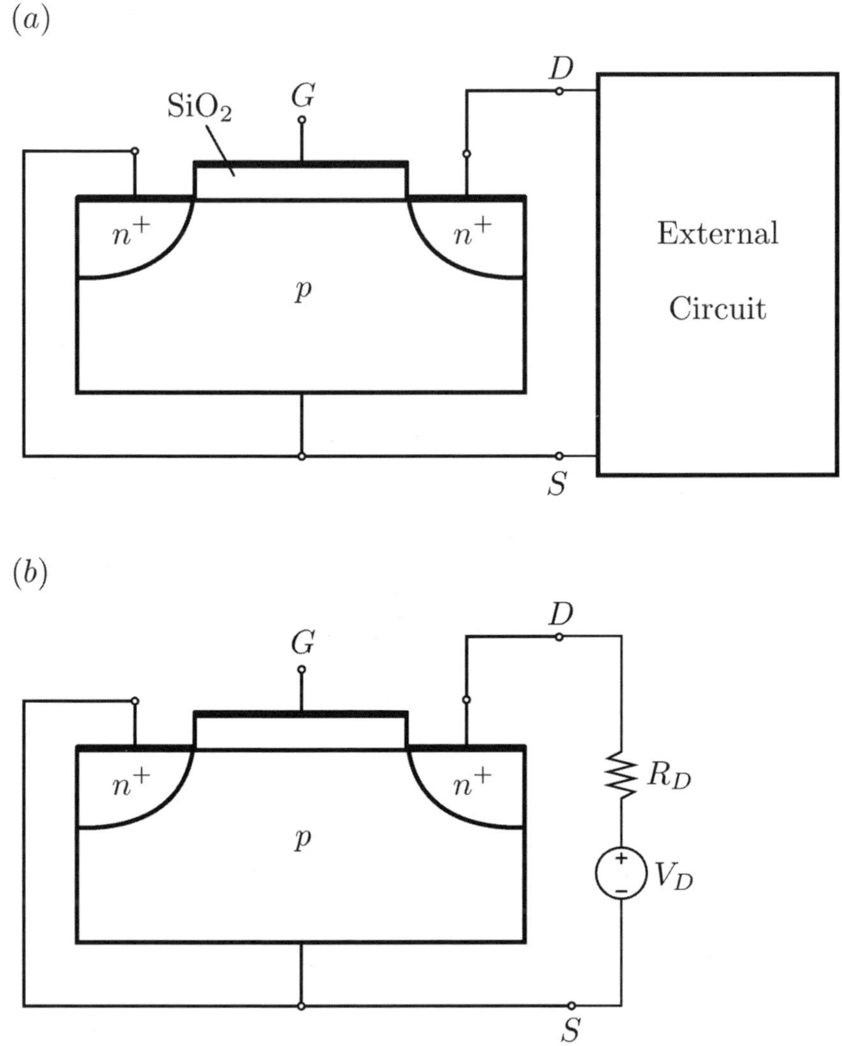

Fig. 1.30

the principle of operation of the lateral MOSFET. Namely, when a positive gate voltage is applied, an n-channel is formed in the p regions through the inversion process. This n-channel connects the n^+ regions of the source contacts with the n^- region. As the voltage between the drain and source is applied, the electron flow from source to drain is established resulting in drain current I_D. The curves I_D versus V_{DS} are basically the same as

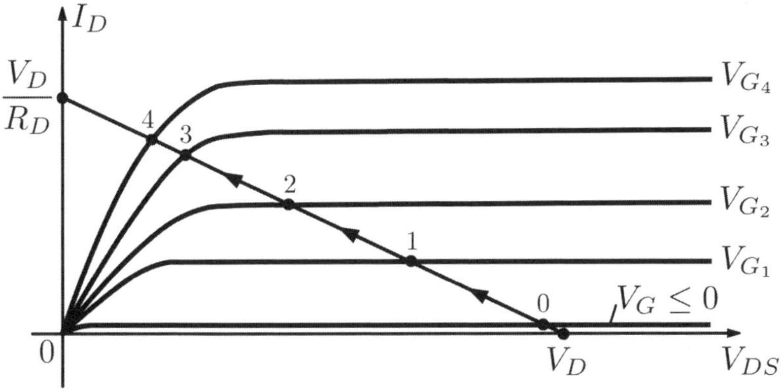

Fig. 1.31

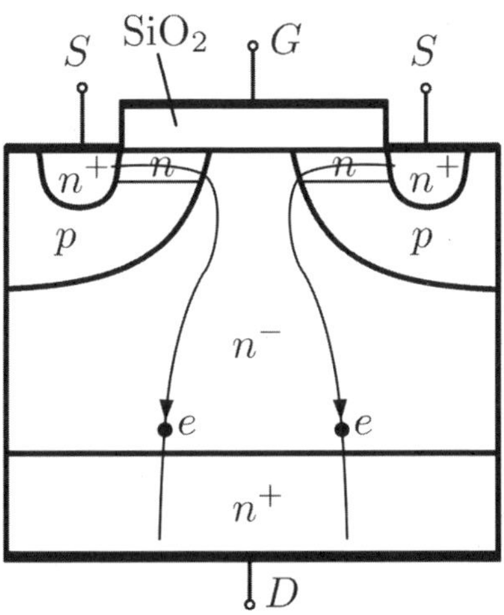

Fig. 1.32

those shown in Figure 1.29b for the lateral MOSFET. Power MOSFETs are fabricated by using a vertical double diffusion process to create n^+ and p regions. For this reason, these power MOSFETs are sometimes called VDMOSFETs or DMOSFETs. Power MOSFETs are fabricated as multi-cell devices and Figure 1.32 represents the schematics of one cell. A large number of such cells are closely packed in a single silicon chip and all these cells are connected in parallel. The number of parallel-connected cells varies (depending on the geometric dimensions of the chip) from several

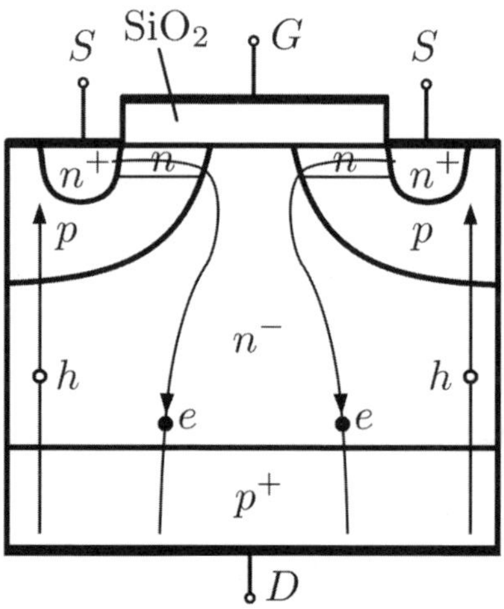

Fig. 1.33

thousand to more than twenty thousand. As a result of parallel connectivity of the cells, the overall on-state resistance (i.e., resistance between drain and source terminals in the conducting state) is substantially reduced in comparison with the on-state resistance of an individual cell.

It turns out that the structure of the power MOSFET can be modified to create another power semiconductor device called the IGBT or COMFET. The first abbreviation stands for "insulated-gate bipolar transistor," while the second abbreviation stands for "conductivity modulated field-effect transistor." A schematic depiction of one cell of the IGBT is shown in Figure 1.33. In actual IGBT devices a very large number of such cells are connected in parallel.

It is evident from this figure that the main structural difference between the power MOSFET and IGBT is the replacement of the n^+ region of the power MOSFET by a p^+ region in the IGBT. This replacement has important consequences. Indeed, on two sides of each cell of the IGBT two p^+n^-p bipolar transistors are formed as a result of the above replacement. When a positive voltage is applied to the gate resulting in the formation of two n channels in the p regions, the flow of electrons will be caused by the application of drain-to-source voltage. This electron current entering the n^- region (which is the base region for the bipolar transistors on the sides) will trigger these bipolar transistors, resulting in the side flow of holes from drain to source. Thus, the current in the IGBT has two distinct components, the electron current due to the MOSFET action and the hole current due to the BJT action:

$$I_{\text{IGBT}} = I_{\text{MOSFET}} + I_{\text{BJT}}. \tag{1.91}$$

This leads to the increase in IGBT current for the same value of V_{DS} in comparison with the power MOSFET, where only the electron component of the drain current is present. This increase in the drain current results in the reduction of on-state resistance and on-state losses.

It must be remarked that the introduction of the p^+ region in the IGBT creates a four-layer vertical structure $p^+n^-pn^+$ similar to the one used in the design of thyristors (see the previous section). This may lead to parasitic thyristor action in the IGBT which is usually called thyristor latch-up. This parasitic thyristor action may compromise the gate control over the drain current. Special techniques have been developed to achieve non-latch-up operation of the IGBT. The discussion of these techniques is outside the scope of this text.

1.5 Snubbers and Resonant Switches

It has been repeatedly emphasized in our discussion that in power electronics semiconductor devices are used as switches and that it is desirable to use fast switching devices in order to reduce ripples in power converters as well as their overall size, weight and cost. Fast switchings may result in fast time variations of voltages (i.e., large $\frac{dv}{dt}$) and currents (i.e., large $\frac{di}{dt}$), which is a cause of electromagnetic interference (EMI). The fast switchings may also result in large voltages across semiconductor devices during turn-off transients and large currents through devices during turn-on transients. To ameliorate these adverse effects of fast switchings, special snubber circuits are used in combination with semiconductor devices. These snubber circuits are not fundamental to the understanding of the principle of operation and the main properties of power converters. For this reason, in subsequent chapters the snubber circuits are neglected and switches are assumed to be ideal.

To illustrate the central idea of snubber circuits, consider the effects of a parallel capacitor (Figure 1.34a) and a series inductor (Figure 1.34b) during turn-off and turn-on of semiconductor switches, respectively. It is apparent that during the turn-off of the switch in Figure 1.34a the capacitor tends to maintain the voltage across the switch close to zero, i.e., close to its value immediately before switching. Similarly, it is apparent that during the turn-on of the switch in Figure 1.34b the inductor tends to

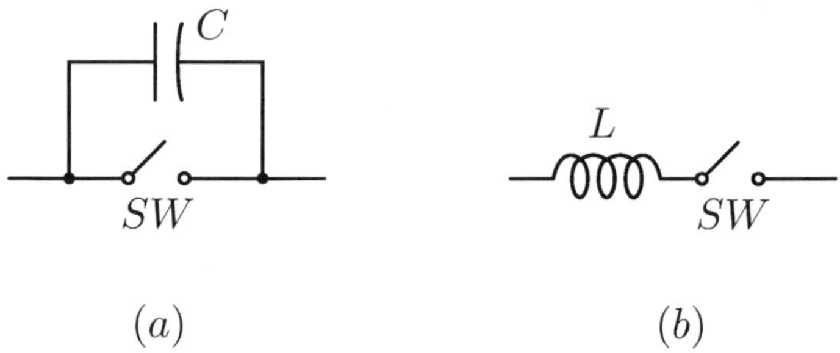

Fig. 1.34

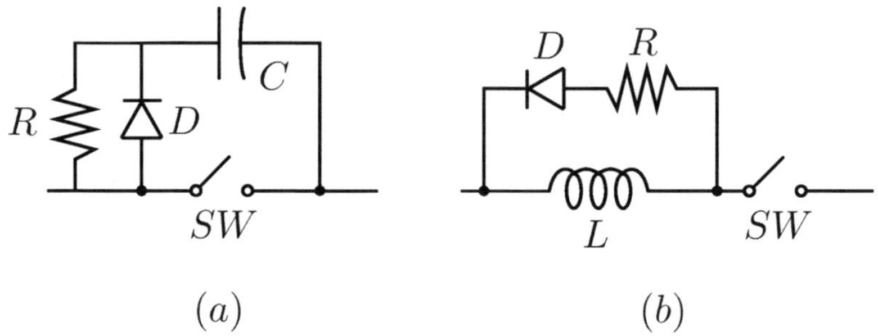

Fig. 1.35

maintain the current through the switch close to zero, i.e., close to its value immediately before switching. In this way, parallel capacitors and series inductors tend to reduce the rates of time variations of voltages across the switches and currents through the switches immediately after their turn-off or turn-on, respectively. However, the circuits shown in Figure 1.34 are not fully satisfactory. Indeed, at turn-on of the switch shown in Figure 1.34a, the charge accumulated in the capacitor is dissipated through the switch and may result in a high current overshoot and large transistor losses. Similarly, at turn-off of the switch shown in Figure 1.34b, high voltage overshoot may occur across the switch as well as large switching losses. To avoid these shortcomings, the above circuits are modified as shown in Figure 1.35 and they represent so-called "turn-on" and "turn-off" snubber circuits, respectively. The operation of the circuit shown in Figure 1.35a can be briefly described as follows. During turn-off, the diode turns on and the capacitor is being gradually charged. This slows the rate of voltage increase across the semiconductor switch. The next time the transistor

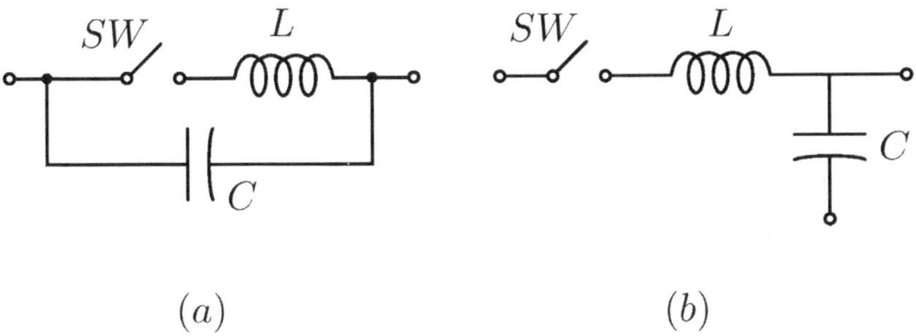

Fig. 1.36

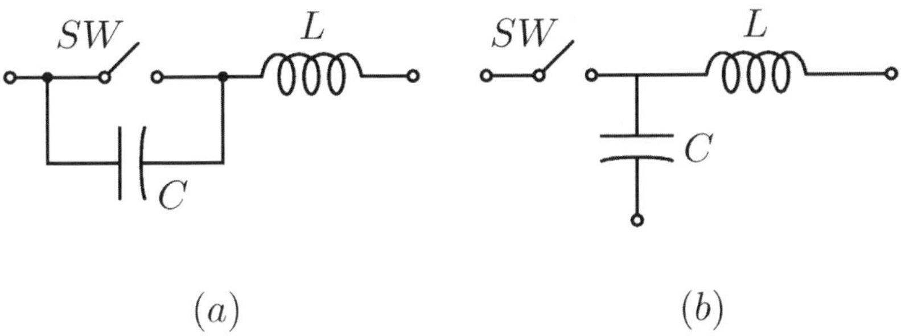

Fig. 1.37

is turned on, the capacitor discharges through the resistor and this limits the discharge current through the switch. This also suggests that in the circuits shown in Figure 1.35 the energy losses in the resistors may result in reduction of energy losses in the semiconductor switches if proper selections of capacitors, inductors and resistors are made. This and other issues such as proper combining of turn-on and turn-off snubbers to form "unified" snubbers as well as the design of "energy recovery" snubbers are outside the scope of this book. The discussion of such issues can be found in more comprehensive books on power electronics (see, for instance, [29] and [43]).

A very viable and promising alternative to snubber circuits has been developed and it is based on the concept of a resonant switch. Two techniques of resonant switching have been advanced. The first one is the *zero-current-switching* (ZCS) technique that can be accomplished by using, for instance, the resonant switches shown in Figure 1.36. The other one is the *zero-voltage-switching* (ZVS) technique that can be accomplished by using, for instance, the resonant switches shown in Figure 1.37. It is clear from the above figures that in resonant switches semiconductor devices are used in combination with LC resonant circuits (LC tanks). These LC circuits are needed to shape the current or voltage waveforms of semiconductor switches. Namely, the addition of these resonance circuits results in waveforms with zero crossings of transistor currents or transistor voltages. These zero crossings are used for on-to-off (or off-to-on) switch transitions. We shall briefly illustrate this by considering the operation of the ZCS resonant switch shown in Figure 1.36a. This operation can be roughly described as follows. Since the switch is connected in series with an inductor, during the turning-on of this switch its current remains close to zero. As soon as the switch is turned on, the parallel resonant LC circuit is formed. This usually results in the switch current waveform with zero crossing. This current zero crossing is used for turning the switch off. It is clear from the above description that the switch turns on and off at close to zero current. This substantially reduces the switching losses in comparison with conventional transistor switching discussed in the previous sections of this chapter. Such switching with almost zero switch current (or almost zero switch voltage) is called *soft switching*, while the previously discussed non-resonant switching of transistors (accompanied by appreciable power losses) is called *hard switching*. It is also clear from the schematic description of the operation of resonant switching that the realization of this switching requires the proper control of transistors to guarantee that their switching is accomplished at zero current (or voltage) crossings. This has important implications concerning the principle of operation and the structure of power converters. For instance, in the case of using resonant switches in dc-to-dc converters (see the last chapter of the book), the duty factor can no longer serve as the primary control parameter to regulate the converter output dc voltage. In resonant (or quasi-resonant) dc-to-dc converters, the output dc voltage is often controlled by the switching frequency (or by the ratio of switching and resonance frequencies) rather than by the duty factor. High frequency resonance switching and resonant converters are currently a very active area of research in power electronics, and the detailed discussion of the many issues related to these converters is beyond the scope of this book. This discussion can be found in [30].

Chapter 2

Rectifiers

2.1 Single-Phase Rectifiers with RL Loads

In this chapter, we shall discuss rectifiers, which are ac-to-dc converters. The latter means that the input of these converters is an ac voltage, while the output is a dc voltage. These converters use diodes and thyristors for rectification. We shall first consider diode rectifiers and then conclude the chapter with the detailed discussion of controlled rectifiers, which are thyristor (SCR)-based rectifiers.

We start with the discussion of the single-phase full-wave diode bridge rectifier shown in Figure 2.1a. Another (equivalent) drawing of this rectifier circuit is shown in Figure 2.1b. Both of these equivalent drawings will be used in our subsequent discussions in this chapter. In the figure, there is a four-shoulder bridge with diodes D_1, D_2, D_3 and D_4, one respectively in each of its shoulders. Across one diagonal of the bridge, a known ac voltage source

$$v_s(t) = V_{ms} \sin \omega t \tag{2.1}$$

is connected, while across the other bridge diagonal an RL branch is connected. The purpose of the design of this circuit is to achieve voltage $v_{out}(t)$ across the resistor terminals that is constant in its polarity and almost constant in its magnitude. Such a voltage can then be regarded as a dc voltage source for an external circuit connected across the resistor terminals. It is shown below that the constant polarity of $v_{out}(t)$ is achieved due to the diode bridge, while an almost constant magnitude of $v_{out}(t)$ is obtained by using an inductor with sufficiently large inductance L.

We shall now proceed to the detailed analysis of the rectifier shown in Figure 2.1 and present this analysis as the sequence of the following specific steps.

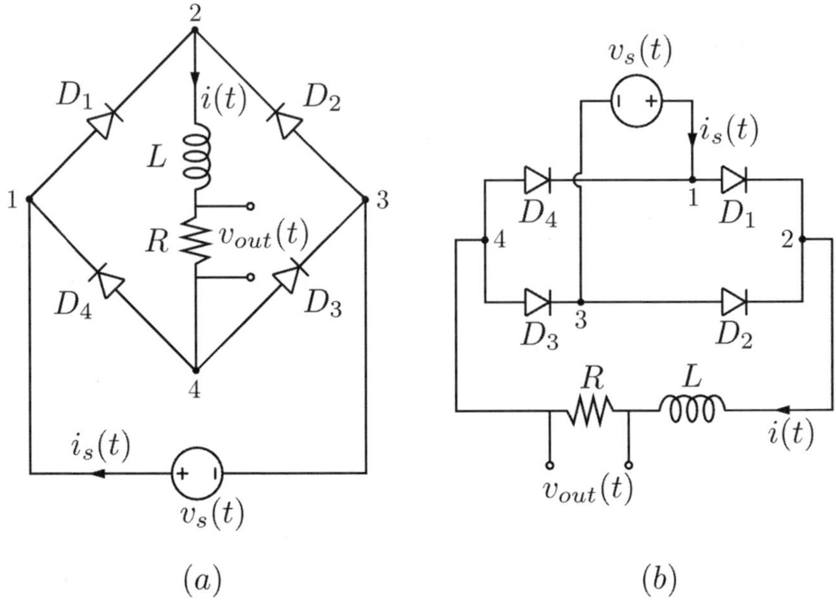

Fig. 2.1

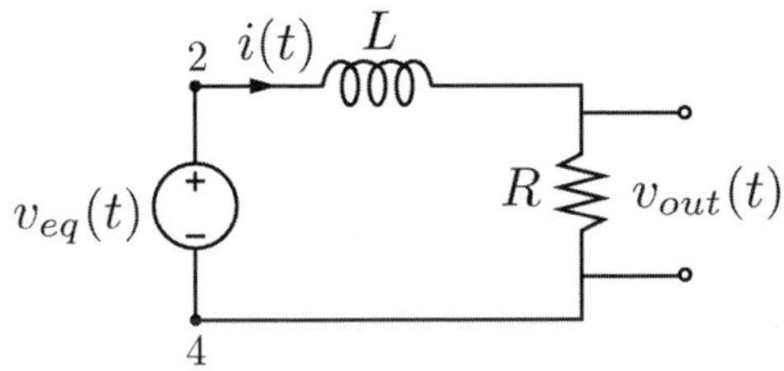

Fig. 2.2

Step 1. The purpose of this step is to replace the diode bridge and the ac voltage source $v_s(t)$ by the equivalent voltage source $v_{eq}(t)$. This replacement leads to the reduction of the electric circuits shown in Figure 2.1 to the equivalent electric circuit shown in Figure 2.2. It must be remarked that this type of first step where given sources and semiconductor switches are replaced by equivalent voltage sources is generic in the analysis of all power electronics converters. A step of this kind will be used time and again in our subsequent discussions. This step requires "proper reading"

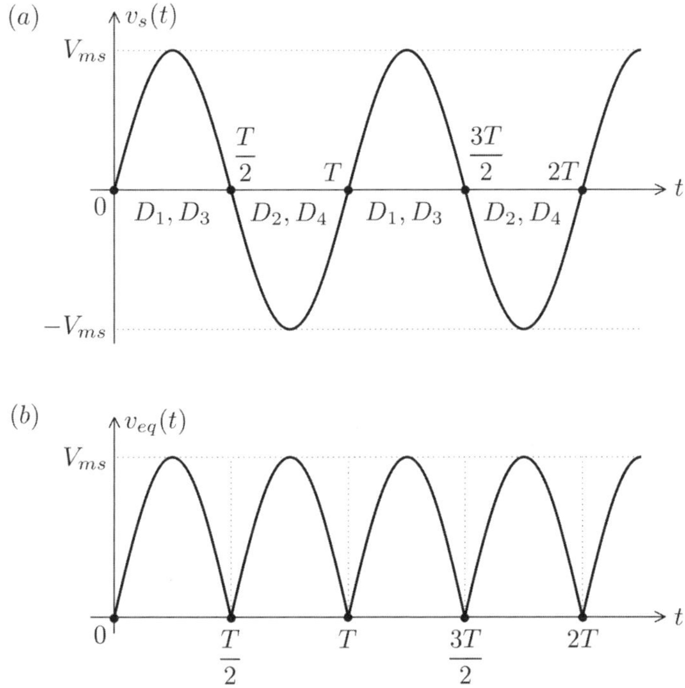

Fig. 2.3

of electric circuits of power converters. This "reading" is exactly what we should do next.

It is clear from the comparison of the electric circuits shown in Figures 2.1a and 2.2 that $v_{eq}(t)$ is the equivalent voltage source if and only if at any time t this voltage source is equal to the voltage between nodes 2 and 4 in the original circuit shown in Figure 2.1a. Thus, in order to find $v_{eq}(t)$ we have to find the voltage between the nodes 2 and 4. To this end, consider the positive half-cycle of $v_s(t)$ (see Figure 2.3a). It is apparent that during this half-cycle the anode of diode D_1 is at the highest positive potential in the circuit, while the cathode of diode D_3 is at the most negative potential in the circuit. This implies that during the positive half-cycle of $v_s(t)$ diodes D_1 and D_3 are in "on" (i.e., conducting) states. It is also clear that diodes D_2 and D_4 are in "off" states because the highest positive potential is at the cathode of diode D_4 and the most negative potential is at the anode of diode D_2. Thus, it can be concluded that during the positive half-cycle the voltage between nodes 2 and 4 (and, consequently, $v_{eq}(t)$) is equal to

$v_s(t)$ as shown in Figure 2.3b. Next, consider the negative half-cycle of $v_s(t)$ when node 3 is at the highest positive potential while node 1 is at the most negative potential. By using the same line of reasoning as before, it is easy to conclude that during this half-cycle diodes D_2 and D_4 are in "on" states while diodes D_1 and D_3 are in "off" states. This means that during the negative half-cycle of $v_s(t)$ the voltage between the nodes 2 and 4 (and, consequently, voltage $v_{eq}(t)$) is equal to $-v_s(t)$ as illustrated in Figure 2.3b. Thus, the diode bridge enforces the constant polarity of voltage between nodes 2 and 4. It is also clear from Figure 2.3 that

$$v_{eq}(t) = V_{ms}|\sin \omega t|. \tag{2.2}$$

It is apparent that $v_{eq}(t)$ is a periodic function with period

$$T' = \frac{T}{2} = \frac{\pi}{\omega}. \tag{2.3}$$

This concludes the first step.

Step 2. Having found $v_{eq}(t)$, we can proceed to the steady-state analysis of the equivalent circuit shown in Figure 2.2. We shall carry out this analysis by using the time-domain technique developed in Chapter 2 of Part I for the analysis of an electric circuit excited by a periodic non-sinusoidal voltage source. In accordance with this technique, we shall consider one period, namely,

$$0 < t < \frac{\pi}{\omega} \tag{2.4}$$

of $v_{eq}(t)$ and the steady-state response of the electric circuit in Figure 2.2 during this time period. This requires finding the solution of the following differential equation with periodic boundary conditions:

$$L\frac{di(t)}{dt} + Ri(t) = V_{ms} \sin \omega t, \tag{2.5}$$

$$i(0) = i\left(\frac{\pi}{\omega}\right). \tag{2.6}$$

Indeed, equation (2.5) is the KVL equation for the circuit shown in Figure 2.2, and the form of the right-hand side of this equation accounts for the fact that $v_{eq}(t) = v_s(t)$ in the time interval specified by formula (2.4). The boundary condition (2.6) is used because we are interested in the steady-state (i.e., periodic) solution of equation (2.5). As soon as the solution of the boundary value problem (2.5)-(2.6) is found for the time interval (2.4), this solution can be periodically extended for any time interval by using the formula

$$i(t) = i\left(t + k\frac{T}{2}\right), \tag{2.7}$$

where k is any integer.

Step 3. Next, we shall find the mathematical form of the general solution of equation (2.5). This solution has two distinct components: a particular solution $i_p(t)$ of the inhomogeneous equation (2.5) and a general solution $i_h(t)$ of the corresponding homogeneous equation. Namely,

$$i(t) = i_p(t) + i_h(t). \tag{2.8}$$

We shall treat $i_p(t)$ as the ac steady state of the electric circuit shown in Figure 2.2 excited by the sinusoidal voltage source $v_s(t)$ rather than $v_{eq}(t)$. It is clear that this ac steady state is a solution of equation (2.5). This ac steady state is given by the formula

$$i_p(t) = I_m \sin(\omega t - \varphi), \tag{2.9}$$

where

$$I_m = \frac{V_{ms}}{\sqrt{R^2 + \omega^2 L^2}}, \tag{2.10}$$

$$\boxed{\tan\varphi = \frac{\omega L}{R}.} \tag{2.11}$$

Thus,

$$i_p(t) = \frac{V_{ms}}{\sqrt{R^2 + \omega^2 L^2}} \sin(\omega t - \varphi). \tag{2.12}$$

The component $i_h(t)$ of $i(t)$ is a general solution of the homogeneous equation

$$L\frac{di_h(t)}{dt} + Ri_h(t) = 0. \tag{2.13}$$

It is apparent that this solution is given by the formula

$$i_h(t) = Ae^{-\frac{R}{L}t} \tag{2.14}$$

where A is an arbitrary constant. By combining formulas (2.8), (2.12) and (2.14), we find that the mathematical form of the general solution of equation (2.5) is given by the expression

$$i(t) = \frac{V_{ms}}{\sqrt{R^2 + \omega^2 L^2}} \sin(\omega t - \varphi) + Ae^{-\frac{R}{L}t}. \tag{2.15}$$

Step 4. Now, we shall find the constant A by using the periodic boundary condition (2.6). Indeed, from formula (2.15) we find

$$i(0) = -\frac{V_{ms}\sin\varphi}{\sqrt{R^2+\omega^2L^2}} + A, \tag{2.16}$$

$$i\left(\frac{\pi}{\omega}\right) = \frac{V_{ms}\sin(\pi-\varphi)}{\sqrt{R^2+\omega^2L^2}} + Ae^{-\frac{\pi R}{\omega L}}. \tag{2.17}$$

By using the last two formulas in the boundary condition (2.6), we obtain

$$-\frac{V_{ms}\sin\varphi}{\sqrt{R^2+\omega^2L^2}} + A = \frac{V_{ms}\sin\varphi}{\sqrt{R^2+\omega^2L^2}} + Ae^{-\frac{\pi R}{\omega L}}, \tag{2.18}$$

which leads to

$$A\left(1-e^{-\frac{\pi R}{\omega L}}\right) = \frac{2V_{ms}\sin\varphi}{\sqrt{R^2+\omega^2L^2}}, \tag{2.19}$$

and

$$A = \frac{2V_{ms}\sin\varphi}{\left(1-e^{-\frac{\pi R}{\omega L}}\right)\sqrt{R^2+\omega^2L^2}}. \tag{2.20}$$

By substituting this expression for A into formula (2.15), we derive

$$i(t) = \frac{V_{ms}}{\sqrt{R^2+\omega^2L^2}}\sin(\omega t-\varphi) + \frac{2V_{ms}\sin\varphi}{\left(1-e^{-\frac{\pi R}{\omega L}}\right)\sqrt{R^2+\omega^2L^2}}e^{-\frac{R}{L}t}. \tag{2.21}$$

Finally, since

$$v_{out}(t) = Ri(t), \tag{2.22}$$

we conclude that

$$\boxed{v_{out}(t) = \frac{V_{ms}R}{\sqrt{R^2+\omega^2L^2}}\sin(\omega t-\varphi) + \frac{2V_{ms}R\sin\varphi}{\left(1-e^{-\frac{\pi R}{\omega L}}\right)\sqrt{R^2+\omega^2L^2}}e^{-\frac{R}{L}t}.} \tag{2.23}$$

The last formula together with equation (2.11) for φ gives the explicit analytical expression for the output voltage of the power converter shown in Figure 2.1.

Step 5. Next, we shall demonstrate that for sufficiently large L, that is, when

$$\omega L \gg R, \tag{2.24}$$

the voltage $v_{out}(t)$ has almost constant value and we shall find this value. From the last inequality we have

$$\sqrt{R^2 + \omega^2 L^2} \approx \omega L \tag{2.25}$$

and

$$\frac{R}{\sqrt{R^2 + \omega^2 L^2}} \approx 0. \tag{2.26}$$

This means that the first term in the right-hand side of formula (2.23) is quite small and can be neglected. We turn now to the analysis of the second term of the right-hand side of the same formula. It is clear from inequality (2.24) and formula (2.11) that

$$\tan\varphi = \frac{\omega L}{R} \gg 1. \tag{2.27}$$

This implies that

$$\varphi \approx \frac{\pi}{2} \quad \text{and} \quad \sin\varphi \approx 1. \tag{2.28}$$

Furthermore, by retaining only the first two terms in the power series expansion of the exponential, we find

$$1 - e^{-\frac{\pi R}{\omega L}} \approx 1 - \left(1 - \frac{\pi R}{\omega L}\right) = \frac{\pi R}{\omega L}. \tag{2.29}$$

Next, we shall evaluate $e^{-\frac{R}{L}t}$ in the time interval (2.4):

$$1 > e^{-\frac{R}{L}t} > e^{-\frac{\pi R}{\omega L}} \approx 1 - \frac{\pi R}{\omega L} \approx 1. \tag{2.30}$$

The last formula implies that

$$e^{-\frac{R}{L}t} \approx 1. \tag{2.31}$$

By using formulas (2.25), (2.28), (2.29) and (2.31) in equation (2.23), we find

$$\boxed{v_{out}(t) \approx \frac{2}{\pi} V_{ms} \approx 0.637 V_{ms}.} \tag{2.32}$$

The plot of $v_{out}(t)$ is illustrated by Figure 2.4, that is, the output voltage $v_{out}(t)$ is indeed almost constant.

Step 6. Now, we shall demonstrate that this (almost) constant value of $v_{out}(t)$ can be found directly by using the averaging technique. To this end, by using formula (2.22), we shall rewrite the equation (2.5) as follows:

$$L\frac{di(t)}{dt} + v_{out}(t) = V_{ms}\sin\omega t. \tag{2.33}$$

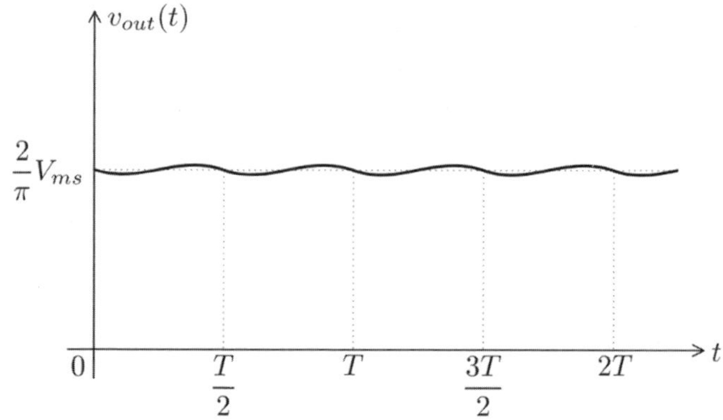

Fig. 2.4

We shall average both sides of the last equation over one period $\frac{\pi}{\omega}$:

$$L\frac{\omega}{\pi}\int_0^{\frac{\pi}{\omega}}\frac{di(t)}{dt}dt+\frac{\omega}{\pi}\int_0^{\frac{\pi}{\omega}}v_{out}(t)dt=\frac{\omega V_{ms}}{\pi}\int_0^{\frac{\pi}{\omega}}\sin\omega t dt. \tag{2.34}$$

By using the boundary condition (2.6), we find

$$\int_0^{\frac{\pi}{\omega}}\frac{di(t)}{dt}dt=i\left(\frac{\pi}{\omega}\right)-i(0)=0. \tag{2.35}$$

By definition,

$$\overline{v_{out}(t)}=\frac{\omega}{\pi}\int_0^{\frac{\pi}{\omega}}v_{out}(t)dt, \tag{2.36}$$

where $\overline{v_{out}(t)}$ stands for the average value of $v_{out}(t)$.

Finally,

$$\int_0^{\frac{\pi}{\omega}}\sin\omega t dt=-\frac{1}{\omega}\cos\omega t\Big|_0^{\frac{\pi}{\omega}}=\frac{2}{\omega}. \tag{2.37}$$

By combining formulas (2.34), (2.35), (2.36) and (2.37), we find

$$\boxed{\overline{v_{out}(t)}=\frac{2}{\pi}V_{ms}.} \tag{2.38}$$

The following four remarks are in order. First, the derivation of formula (2.38) for the average value of the output voltage was done without using inequality (2.24). This implies that formula (2.38) is valid for any value of inductance. Second, the output voltage will almost coincide with its average value if the inductance is sufficiently large. Third, by subtracting formulas (2.38) and (2.23), the analytical expression for the ripple $v_{out}^{rip}(t)$ in the output voltage is obtained:

$$v_{out}^{rip}(t) = \frac{V_{ms}R}{\sqrt{R^2+\omega^2L^2}}\sin(\omega t-\varphi) + \frac{2V_{ms}R\sin\varphi}{\left(1-e^{-\frac{\pi R}{\omega L}}\right)\sqrt{R^2+\omega^2L^2}}e^{-\frac{R}{L}t} - \frac{2}{\pi}V_{ms}. \tag{2.39}$$

The last formula can be used for the evaluation of ripple for any values of rectifier parameters. Fourth, for sufficiently large inductance L, that is, when inequality (2.24) is satisfied, the dc output voltage is given by formula (2.32) and it does not depend on the value of resistance R. This implies that a resistive load across the terminals of the output voltage will change the overall resistance but will not affect the value of the output voltage. In this sense, this dc output voltage can be treated as an ideal dc voltage source with respect to the external load. In fact, the external resistive load will reduce the overall resistance and, in this way, it will strengthen the inequality (2.24) and, consequently, reduce the ripple in the dc output voltage.

The current $i(t)$ is an almost dc current and is equal to

$$i(t) \approx \frac{2V_{ms}}{\pi R}, \tag{2.40}$$

when the inductance is sufficiently large. It is clear that this current coincides with the current $i_s(t)$ through the sinusoidal voltage source during its positive half-cycle (that is, when diodes D_1 and D_3 conduct). During the negative half-cycle of $v_s(t)$, that is, when diodes D_2 and D_4 conduct, these two currents have opposite directions. This implies that the current $i_s(t)$ has the time variations shown in Figure 2.5. This clearly results in higher-order harmonics contaminating the ac network supplying power to the rectifier. This type of harmonics contamination is typical for power electronics converters.

The dc voltage output of the rectifier shown in Figure 2.1 is fixed and not controllable. This output voltage may be above or may be below the desired dc voltage. A certain level of adjustment of dc output voltage can be achieved by using a center-tapped transformer rectifier shown in Figure 2.6, where the primary voltage of the transformer is the input ac voltage $v_s(t)$ of the rectifier:

$$v_1(t) = v_s(t) = V_{ms}\sin\omega t. \tag{2.41}$$

As in the previous analysis of the bridge rectifier, the first step in the analysis of the center-tapped transformer rectifier is to replace the ac voltage

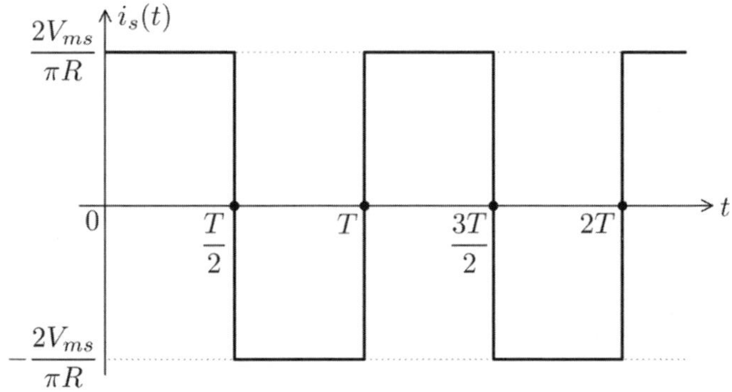

Fig. 2.5

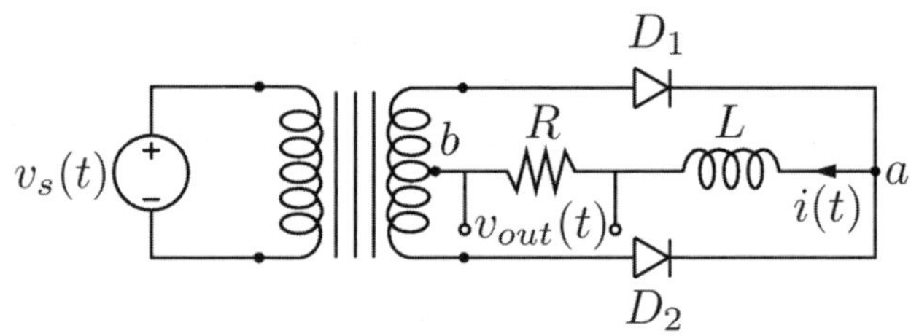

Fig. 2.6

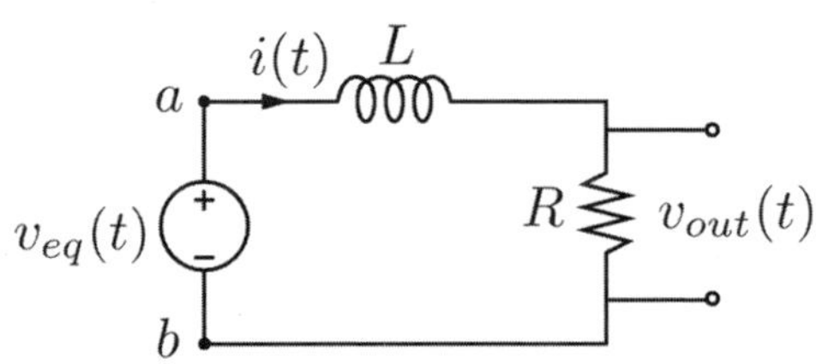

Fig. 2.7

source, transformer and two diodes by the equivalent voltage source $v_{eq}(t)$. This leads to the reduction of the circuit shown in Figure 2.6 to the equivalent circuit shown in Figure 2.7. The equivalent voltage source $v_{eq}(t)$ is equal at any instant of time to the actual voltage between nodes "a" and "b" in the actual rectifier circuit. To find this voltage between nodes "a" and "b," we shall use the model of the ideal transformer (see Chapter 3 of

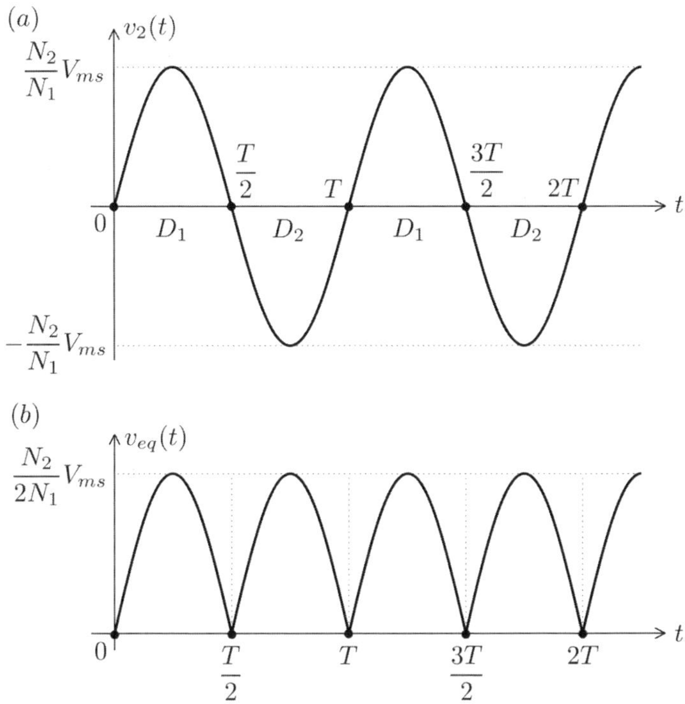

Fig. 2.8

Part II). According to this model, we find

$$\frac{v_1(t)}{v_2(t)} = a = \frac{N_1}{N_2}. \tag{2.42}$$

By using formulas (2.41) and (2.42), we conclude that the voltage across the terminals of the secondary winding of the transformer is equal to

$$v_2(t) = \frac{N_2}{N_1} V_{ms} \sin \omega t. \tag{2.43}$$

This voltage is plotted in Figure 2.8a. It is clear from Figure 2.6 that during the positive half-cycle of $v_2(t)$ the diode D_1 is in the "on" state, while the diode D_2 is in the "off" state. This implies that the voltage between nodes "a" and "b" has positive polarity and it is equal to one half of the secondary voltage $v_2(t)$. During the negative half-cycle of $v_2(t)$, the diode D_2 is in the "on" state, while the diode D_1 is in the "off" state. This implies that the voltage between nodes "a" and "b" retains the same positive polarity (i.e., the polarity opposite to $v_2(t)$) and it is equal to one half of $-v_2(t)$. Thus, we can conclude that

$$v_{eq}(t) = v_{ab}(t) = \frac{N_2}{2N_1} V_{ms} |\sin \omega t|, \tag{2.44}$$

as shown in Figure 2.8b. Having found $v_{eq}(t)$, we can proceed to the analysis of the equivalent circuit shown in Figure 2.7. It is clear that this analysis is mostly identical to the analysis of the electric circuit shown in Figure 2.2 where $v_{eq}(t)$ is defined by formula (2.2). The only difference is in the peak value of $v_{eq}(t)$. This means that the expression for $v_{out}(t)$ for the center-tapped transformer rectifier can be found by using formula (2.23) and by replacing V_{ms} in this formula by $\frac{N_2}{2N_1} V_{ms}$. This also means that for sufficiently large inductance L in the circuit shown in Figure 2.6, the output voltage $v_{out}(t)$ is practically constant and it is given by the formula

$$\boxed{v_{out}(t) \approx \frac{N_2}{\pi N_1} V_{ms}.} \tag{2.45}$$

It is apparent from the last formula that by properly choosing the ratio N_2/N_1 the desired level of dc output voltage $v_{out}(t)$ can be achieved.

2.2 Single-Phase Rectifiers with *RC* and *RLC* Loads

In the previous section, we have discussed single-phase full-wave rectifiers with RL loads and have stressed that sufficiently large values of inductances are usually needed to effectively suppress ripples in the output voltage and to make this voltage practically constant in time. Hardware realizations of large inductances may lead to expensive and heavy rectifier designs. For this reason, it may be of interest to use capacitors for ripple suppression in the design of rectifiers. Such designs are discussed in this section.

We shall begin with the discussion of full-wave diode bridge rectifiers with RC loads shown in Figure 2.9. The only design difference in comparison with rectifiers shown in Figure 2.1 is the replacement of the RL branch by an RC branch with parallel connection of the capacitor and resistor. Here, as before,

$$v_s(t) = V_{ms} \sin \omega t \tag{2.46}$$

and the intent of this design is to achieve practically constant output voltage $v_{out}(t)$ across the terminals of the resistor. The first step in the analysis of this rectifier is to replace the given ac voltage source and four diodes by the equivalent voltage source $v_{eq}(t)$. This is done by using the same reasoning

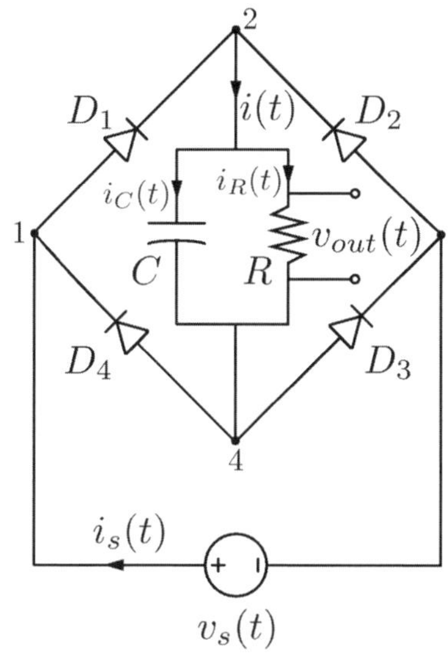

Fig. 2.9

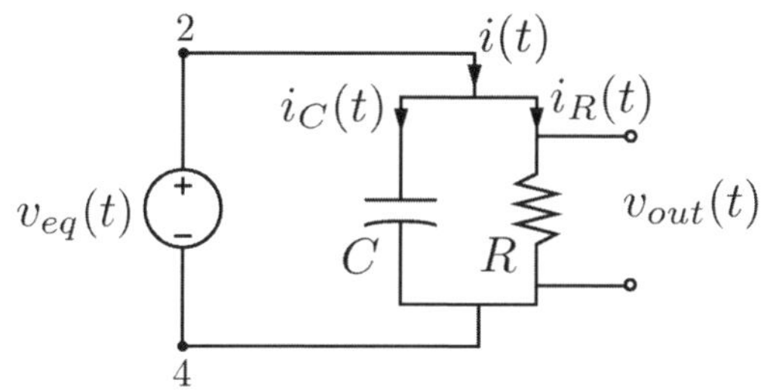

Fig. 2.10

as in the previous section and the resulting equivalent electric circuit is shown in Figure 2.10 with

$$v_{eq}(t) = V_{ms}|\sin \omega t|. \tag{2.47}$$

At first glance it may seem from this circuit that the capacitor has no effect on ripple suppression because the output voltage $v_{out}(t)$ is equal to $v_{eq}(t)$ and, consequently, the output voltage has the same (100%) level of ripple as $v_{eq}(t)$. However, this is not the case. The reason is that the circuit shown in Figure 2.10 is not valid for all times. It is only valid for time intervals when

$$i(t) > 0. \tag{2.48}$$

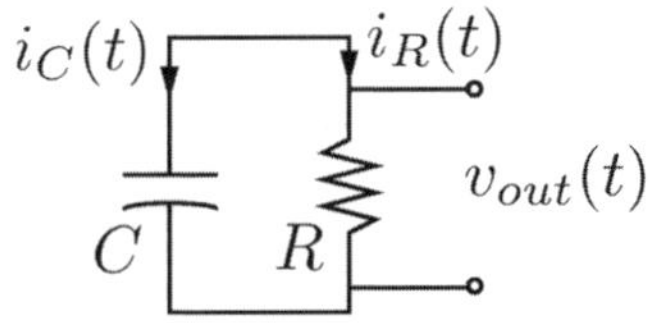

Fig. 2.11

Indeed, the negative current flow (i.e., the current $i(t)$ flow in the direction opposite to the one marked on Figures 2.9 and 2.10) is prohibited by the diodes D_1 and D_2 which may conduct only in the directions from nodes 1 and 3, respectively, towards the node 2. This implies that for time intervals when

$$i(t) = 0 \tag{2.49}$$

the performance of the rectifier is described by the circuit shown in Figure 2.11. Indeed, when the current $i(t)$ in the circuit shown in Figure 2.10 reaches zero, the current $i_C(t)$ is negative, while the current $i_R(t) = v_{eq}(t)/R$ is always positive. The negative current through the capacitor implies that for $i(t) = 0$ the capacitor is discharged through the resistor as represented by the circuit shown in Figure 2.11.

The presented discussion can be summarized as follows. The operation of the rectifier shown in Figure 2.9 consists of two distinct and periodically repeated regimes: regime 1, when the current $i(t)$ is positive and the capacitor is charged according to the circuit shown in Figure 2.10; and regime 2, when the current $i(t)$ is equal to zero and the capacitor is discharged through the resistor as shown in Figure 2.11. These two regimes are periodically repeated as graphically illustrated by Figure 2.12. In this figure, the first regime occurs during the time interval

$$t_0 < t < t_1 < \frac{\pi}{\omega} \tag{2.50}$$

when (see Figure 2.10)

$$v_{out}(t) = v_C(t) = v_{eq}(t), \tag{2.51}$$

while the second regime occurs during the time interval

$$t_1 < t < t_2 = t_0 + \frac{\pi}{\omega} \tag{2.52}$$

and (see Figure 2.11)

$$v_{out}(t) = v_C(t). \tag{2.53}$$

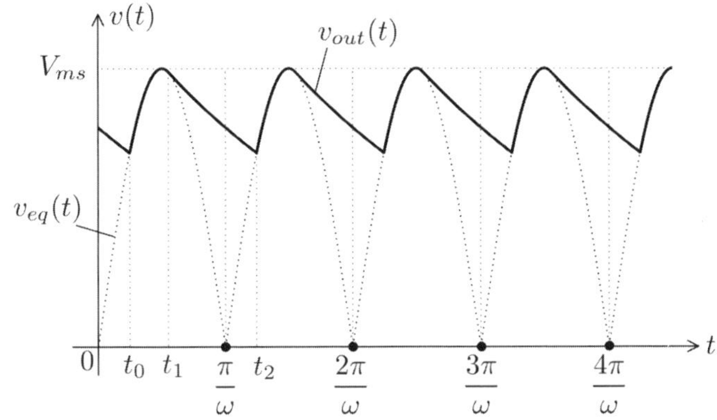

Fig. 2.12

Thus, in order to find $v_{out}(t)$ we need to find expressions for t_0, t_1 and for $v_C(t)$ during the time interval specified by formula (2.52). To do this, we shall initially consider the first regime. According to Figure 2.10, we find

$$i(t) = i_C(t) + i_R(t). \tag{2.54}$$

It is clear that

$$i_C(t) = C\frac{dv_C(t)}{dt}, \tag{2.55}$$

$$i_R(t) = \frac{v_C(t)}{R}. \tag{2.56}$$

For the time interval defined in (2.50), we have

$$v_C(t) = v_{eq}(t) = V_{ms}\sin\omega t. \tag{2.57}$$

By combining formulas (2.54), (2.55), (2.56) and (2.57) we derive

$$i(t) = \omega C V_{ms}\cos\omega t + \frac{V_{ms}}{R}\sin\omega t. \tag{2.58}$$

At time t_1 the current $i(t)$ reaches zero,

$$i(t_1) = 0. \tag{2.59}$$

Consequently,

$$\omega C V_{ms}\cos\omega t_1 + \frac{V_{ms}}{R}\sin\omega t_1 = 0, \tag{2.60}$$

which yields

$$\tan\omega t_1 = -\omega CR. \tag{2.61}$$

The last equation can be solved for t_1:

$$\boxed{t_1 = \frac{\pi}{\omega} - \frac{1}{\omega}\arctan \omega CR.} \tag{2.62}$$

Now, we consider the second regime. According to Figure 2.11, we find

$$v_C(t) - v_R(t) = 0, \tag{2.63}$$

$$i_C(t) = -i_R(t), \tag{2.64}$$

which leads to

$$\frac{dv_C(t)}{dt} + \frac{1}{RC}v_C(t) = 0. \tag{2.65}$$

A general solution of the last equation is given by the formula

$$v_C(t) = Ae^{-\frac{t}{RC}}, \tag{2.66}$$

where the constant A can be found from the initial condition

$$v_C(t_1) = V_{ms} \sin \omega t_1 = Ae^{-\frac{t_1}{RC}}. \tag{2.67}$$

This leads to

$$A = V_{ms}e^{\frac{t_1}{RC}} \sin \omega t_1. \tag{2.68}$$

By substituting the last expression for A into formula (2.66), we find

$$v_C(t) = V_{ms}e^{\frac{t_1-t}{RC}} \sin \omega t_1. \tag{2.69}$$

Next, we shall discuss how to find t_0. Since $v_C(t)$ is periodic with period $\frac{T}{2} = \frac{\pi}{\omega}$, we find

$$v_C(t_0) = v_C(t_2) = v_C\left(t_0 + \frac{\pi}{\omega}\right). \tag{2.70}$$

It is clear that

$$v_C(t_0) = V_{ms} \sin \omega t_0, \tag{2.71}$$

while

$$v_C\left(t_0 + \frac{\pi}{\omega}\right) = V_{ms}e^{\frac{t_1-t_0-\frac{\pi}{\omega}}{RC}} \sin \omega t_1. \tag{2.72}$$

By substituting the last two formulas into equation (2.70), we obtain

$$V_{ms} \sin \omega t_0 = V_{ms}e^{\frac{t_1-t_0-\frac{\pi}{\omega}}{RC}} \sin \omega t_1. \tag{2.73}$$

This leads to

$$\boxed{e^{\frac{t_0}{RC}} \sin \omega t_0 = e^{\frac{t_1-\frac{\pi}{\omega}}{RC}} \sin \omega t_1.} \tag{2.74}$$

Since t_1 can be determined by using the relation (2.62), the last formula can be regarded as an equation for t_0. As soon as t_0 and t_1 are found, the following expression can be used for the computation of $v_{out}(t)$:

$$\boxed{v_{out}(t) = \begin{cases} V_{ms} \sin \omega t, & \text{if } t_0 < t < t_1, \\ V_{ms} e^{\frac{t_1 - t}{RC}} \sin \omega t_1, & \text{if } t_1 < t < t_0 + \frac{\pi}{\omega}. \end{cases}} \tag{2.75}$$

Finally, we shall demonstrate that under the condition

$$\boxed{RC \gg \frac{\pi}{\omega},} \tag{2.76}$$

the output voltage v_{out} is almost constant and

$$\boxed{v_{out}(t) \approx V_{ms}.} \tag{2.77}$$

Indeed, since t_0 and t_1 belong to the time interval $(0, \pi/\omega)$, from inequality (2.76) we find

$$e^{\frac{t_0}{RC}} \approx 1, \tag{2.78}$$

$$e^{\frac{t_1 - \frac{\pi}{\omega}}{RC}} \approx 1. \tag{2.79}$$

This implies that equation (2.74) can be reduced to

$$\sin \omega t_0 \approx \sin \omega t_1, \tag{2.80}$$

which leads to the conclusion that

$$\omega t_0 \approx \pi - \omega t_1. \tag{2.81}$$

On the other hand, from inequality (2.76) follows that

$$\arctan \omega RC \approx \frac{\pi}{2}, \tag{2.82}$$

which, according to equation (2.62), yields

$$t_1 \approx \frac{\pi}{2\omega}. \tag{2.83}$$

From formulas (2.81) and (2.83) we find

$$t_0 \approx \frac{\pi}{2\omega}. \tag{2.84}$$

Thus, it follows from formulas (2.75), (2.83) and (2.84) that under the condition (2.76) the first regime of capacitor charging is quite short, while during the second regime the capacitor discharges very slowly. This is illustrated by Figure 2.13, which implies that under the condition (2.76) the output voltage $v_{out}(t)$ is given by formula (2.77). A shortcoming of the

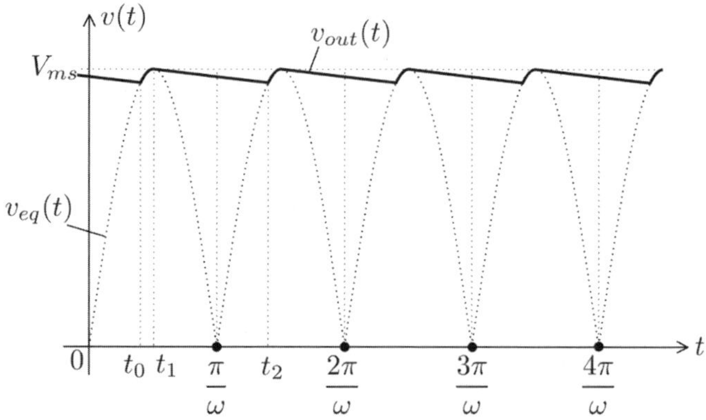

Fig. 2.13

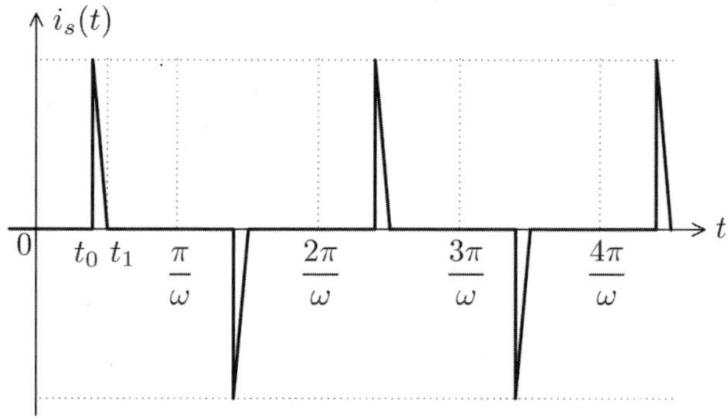

Fig. 2.14

rectifier shown in Figure 2.9 is that the current $i_s(t)$ through the voltage source $v_s(t)$ occurs during short time intervals as illustrated by Figure 2.14. This clearly results in strong higher-order harmonics contamination of the ac network supplying power to the rectifier. Another shortcoming of this rectifier is that its output dc voltage is fixed and it is equal to the peak value of the ac voltage source, and this level of dc voltage may not be desired. It turns out that a desired value of dc output voltage can be obtained by using the center-tapped transformer rectifier shown in Figure 2.15. By replacing the ac voltage source, transformer and two diodes by the equivalent voltage source $v_{eq}(t)$, the analysis of this rectifier is reduced to the analysis of the equivalent circuits for "charging" and "discharging" regimes shown in Figures 2.16a and 2.16b, respectively. By using the same

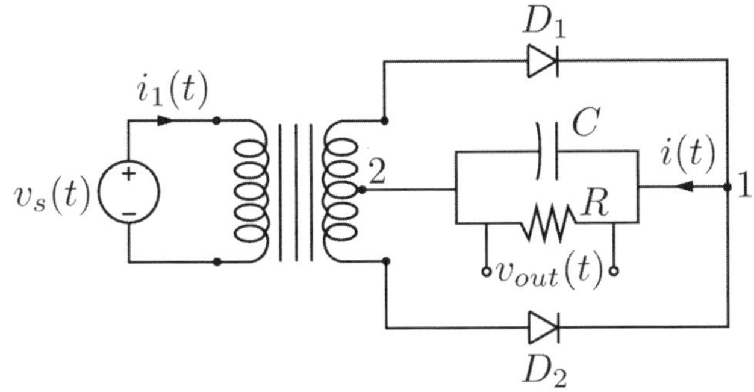

Fig. 2.15

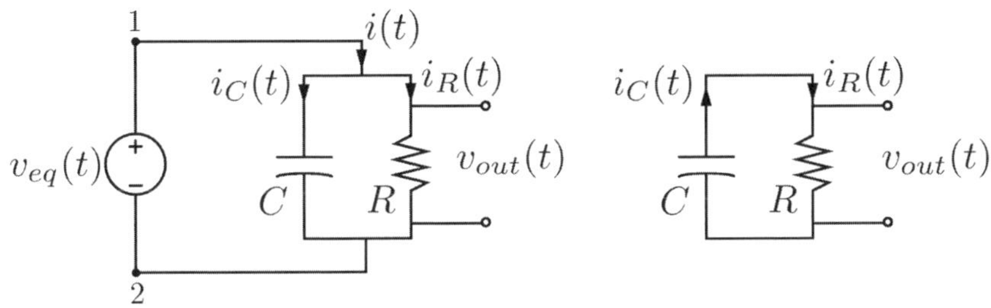

(a) (b)

Fig. 2.16

line of reasoning as in the discussion of the rectifier presented in Figure 2.6 (see the previous section), we conclude that

$$v_{eq}(t) = \frac{N_2}{2N_1} V_{ms} |\sin \omega t|, \tag{2.85}$$

where N_1 and N_2 are the numbers of turns of the primary and secondary windings of the transformer, respectively. Now, it is apparent that the analysis of the rectifier shown in Figure 2.15 is mostly identical to the analysis of the bridge rectifier presented in Figure 2.9; the only difference is that the peak value V_{ms} in the previous relevant formulas must be replaced by the value $\frac{N_2}{2N_1} V_{ms}$. This leads to the conclusion that under the condition (2.76) the output dc voltage of the rectifier shown in Figure 2.15 is given by the formula

$$v_{out}(t) \approx \frac{N_2}{2N_1} V_{ms}. \tag{2.86}$$

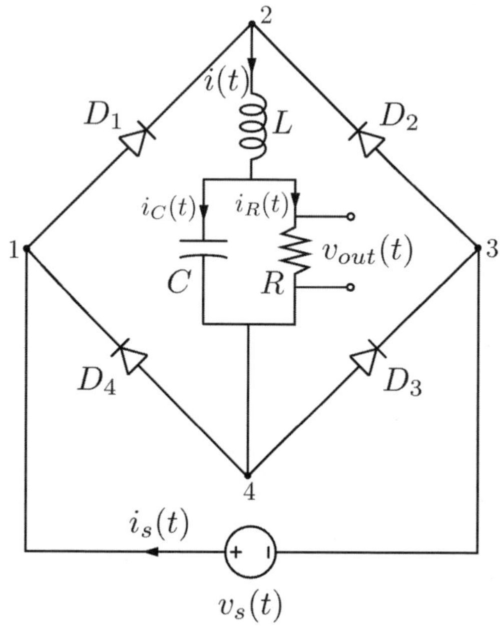

Fig. 2.17

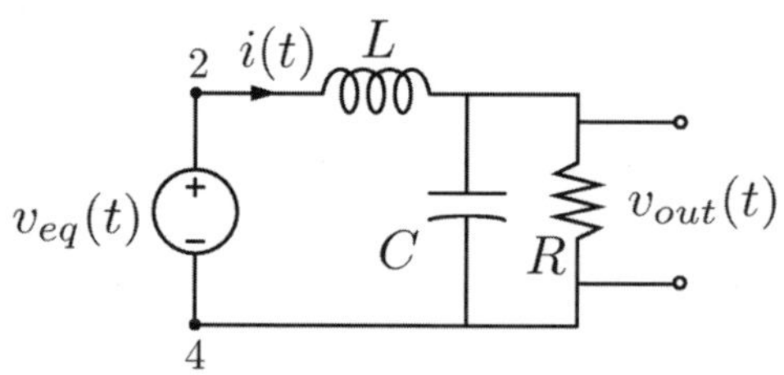

Fig. 2.18

It is apparent from the last formula that the desired value of dc output voltage can be achieved by the proper choice of turns ratio N_2/N_1.

In the conclusion of this section, we consider the bridge rectifier with RLC load shown in Figure 2.17. By using the same line of reasoning as before, the four diodes and ac voltage source can be replaced by the equivalent voltage source $v_{eq}(t)$. This leads to the equivalent circuit shown in Figure 2.18 with

$$v_{eq}(t) = V_{ms}|\sin \omega t|, \tag{2.87}$$

which is graphically illustrated in Figure 2.3b. The steady-state analysis of this circuit is carried out in section 2.3 of Part I (see Example 2), where it is demonstrated that this analysis can be reduced to the solution of the differential equation

$$LC\frac{d^2 v_C(t)}{dt^2} + \frac{L}{R}\frac{dv_C(t)}{dt} + v_C(t) = V_{ms}\sin\omega t \tag{2.88}$$

subject to periodic boundary conditions

$$v_C(0) = v_C\left(\frac{\pi}{\omega}\right), \tag{2.89}$$

$$\frac{dv_C}{dt}(0) = \frac{dv_C}{dt}\left(\frac{\pi}{\omega}\right). \tag{2.90}$$

It is demonstrated in section 2.3 of Part I that the solution of boundary value problem (2.88), (2.89) and (2.90) is given by the following formulas:

$$v_{out}(t) = v_C(t) = A_1 e^{s_1 t} + A_2 e^{s_2 t} + \frac{V_{ms}R}{\sqrt{(\omega^2 LC - 1)^2 R^2 + \omega^2 L^2}}\sin(\omega t - \varphi), \tag{2.91}$$

where

$$\tan\varphi = \frac{\omega L}{R - \omega^2 LCR}, \tag{2.92}$$

$$s_1 = -\frac{1}{2RC} + \sqrt{\frac{1}{4R^2C^2} - \frac{1}{LC}}, \tag{2.93}$$

$$s_2 = -\frac{1}{2RC} - \sqrt{\frac{1}{4R^2C^2} - \frac{1}{LC}}, \tag{2.94}$$

$$A_1 = \frac{2RV_{ms}(s_2\sin\varphi + \omega\cos\varphi)}{(s_2 - s_1)\left(1 - e^{\frac{s_1\pi}{\omega}}\right)\sqrt{(\omega^2 LC - 1)^2 R^2 + \omega^2 L^2}}, \tag{2.95}$$

$$A_2 = \frac{2RV_{ms}(s_1\sin\varphi + \omega\cos\varphi)}{(s_1 - s_2)\left(1 - e^{\frac{s_2\pi}{\omega}}\right)\sqrt{(\omega^2 LC - 1)^2 R^2 + \omega^2 L^2}}. \tag{2.96}$$

By averaging all terms of equation (2.88) over one period $\frac{\pi}{\omega}$ and by using the periodic boundary conditions (2.89) and (2.90), it can be shown (in the same way as in the derivation of formula (2.38)) that

$$\boxed{\overline{v_{out}(t)} = \overline{v_C(t)} = \frac{2}{\pi}V_{ms}.} \tag{2.97}$$

This formula for the average value of output voltage is valid for any parameters R, L and C. Subtracting this formula from equation (2.91), the explicit analytical expression for the ripple in the output voltage is obtained. It can

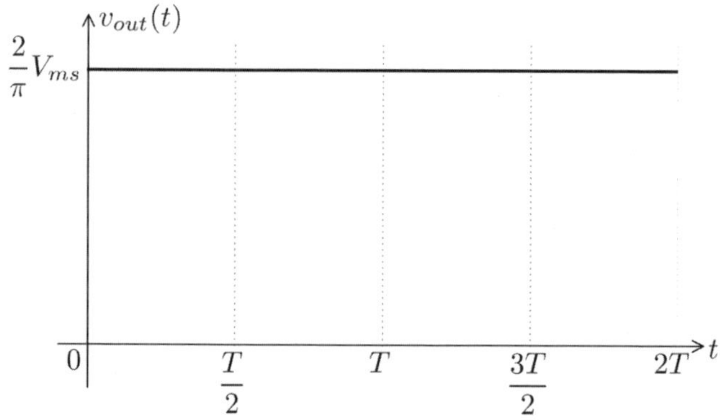

Fig. 2.19

be shown that this ripple will be quite small if the parameters R, L and C are chosen in such a way that

$$\omega^2 LC \gg 1, \tag{2.98}$$

$$\omega RC \gg 1. \tag{2.99}$$

Indeed, under these conditions

$$\sqrt{(\omega^2 LC - 1)^2 R^2 + \omega^2 L^2} \approx \omega^2 LCR, \tag{2.100}$$

$$\tan\varphi \approx 0 \tag{2.101}$$

and

$$\left|e^{s_1 t}\right| \approx \left|e^{s_2 t}\right| \approx 1 \tag{2.102}$$

for $0 < t < \frac{\pi}{\omega}$. By using formulas (2.98)-(2.102) and the same line of reasoning as in the derivation of relation (2.32), it can be shown that from equations (2.91)-(2.96) follows that

$$v_{out}(t) \approx \frac{2}{\pi} V_{ms}. \tag{2.103}$$

An example of calculation of $v_{out}(t)$ under conditions (2.98) and (2.99) is shown in Figure 2.19.

It can also be remarked that, in the case of small ripple, the current $i_s(t)$ through the voltage source $v_s(t)$ is the same as shown in Figure 2.5. This implies that higher-order harmonics contamination of the ac power network is not as severe as in the case of the rectifier shown in Figure 2.9 for which $i_s(t)$ is represented by Figure 2.14.

2.3 Three-Phase Diode Rectifiers

In the previous two sections, single-phase diode rectifiers have been discussed. In these rectifiers, the equivalent voltage sources $v_{eq}(t)$ which replace diodes and single-phase ac sources have constant polarity but large (100%) ripple. The latter means that the range of variation of $v_{eq}(t)$ is from zero to some peak value. For this reason, sufficiently large energy storage elements are needed to effectively suppress this ripple and to achieve more or less constant in time output voltage. It turns out that equivalent voltage sources $v_{eq}(t)$ with appreciably smaller ripples can be achieved by using three-phase rectifiers.

We shall start with the discussion of the three-phase rectifiers with three diodes shown in Figure 2.20. Such rectifiers are also called half-wave or three-pulse rectifiers. In this figure, $v_a(t)$, $v_b(t)$ and $v_c(t)$ are three-phase star-connected voltage sources that have the same peak values V_{ms} and the same frequency ω, but are phase-shifted in time with respect to one another by $\frac{2\pi}{3}$:

$$v_a(t) = V_{ms} \cos \omega t, \tag{2.104}$$

$$v_b(t) = V_{ms} \cos \left(\omega t - \frac{2\pi}{3} \right), \tag{2.105}$$

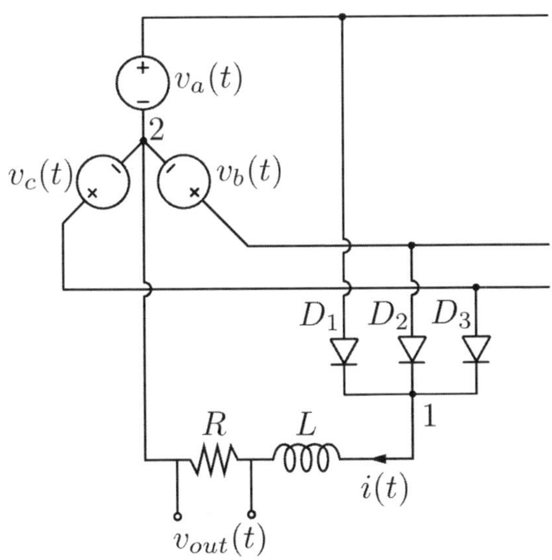

Fig. 2.20

$$v_c(t) = V_{ms} \cos\left(\omega t - \frac{4\pi}{3}\right). \qquad (2.106)$$

The first step in the analysis of this rectifier is to make equivalent replacement of the three ac voltage sources and three diodes by one equivalent voltage source $v_{eq}(t)$ and to reduce the circuit shown in Figure 2.20 to the equivalent circuit shown in Figure 2.21. It is clear that the circuit shown in Figure 2.21 is equivalent to the circuit shown in Figure 2.20 if and only if at any instant of time t the equivalent voltage $v_{eq}(t)$ is equal to the actual voltage between nodes 1 and 2 in the original circuit shown in Figure 2.20. To find this actual voltage, it is necessary to figure out the time intervals when these three diodes are in "on" or "off" states. To this end, the plots of the three-phase voltages $v_a(t)$, $v_b(t)$ and $v_c(t)$ are presented in Figure 2.22a.

It is apparent from this figure that for the time interval $-\frac{\pi}{3\omega} < t < \frac{\pi}{3\omega}$ the anode of diode D_1 is at the highest positive potential in the circuit. For this reason, the diode D_1 is in the "on" state in the above time interval, while the diodes D_2 and D_3 are in the "off" state. Any other option results in contradiction. Indeed, assuming, for instance, that during the above time interval diode D_2 is in the "on" state while diode D_1 is in the "off" state implies that the potential of node 1 and, consequently, the potential of the cathode of diode D_1, is equal to $v_b(t)$, which is less than $v_a(t)$. The latter means that the diode D_1 must be in the "on" state, and this precludes the diode D_2 being in the "on" state. By using the same line of reasoning, it can be concluded that during the time interval $\frac{\pi}{3\omega} < t < \frac{\pi}{\omega}$ the diode D_2 is in the "on" state, while the diodes D_1 and D_3 are in the "off" state. Similarly, during the time interval $\frac{\pi}{\omega} < t < \frac{5\pi}{3\omega}$ the diode D_3 is in the "on" state, while the diodes D_1 and D_2 are in the "off" state. The above discussion implies that the voltage between nodes 1 and 2 in the circuit shown in Figure 2.20 is equal sequentially to the phase voltages $v_a(t)$, $v_b(t)$ and $v_c(t)$ for the time intervals $-\frac{\pi}{3\omega} < t < \frac{\pi}{3\omega}$, $\frac{\pi}{3\omega} < t < \frac{\pi}{\omega}$ and $\frac{\pi}{\omega} < t < \frac{5\pi}{3\omega}$,

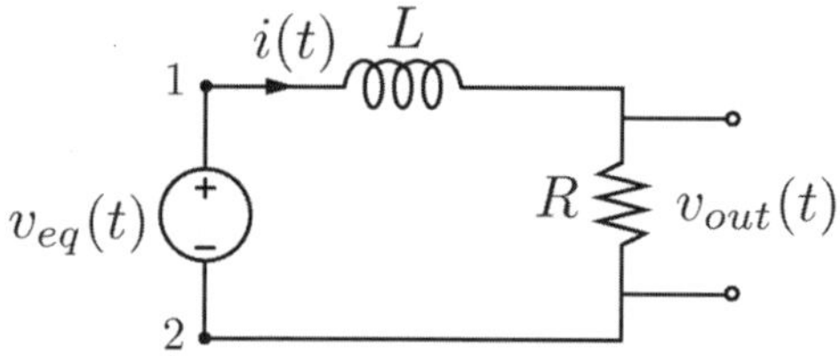

Fig. 2.21

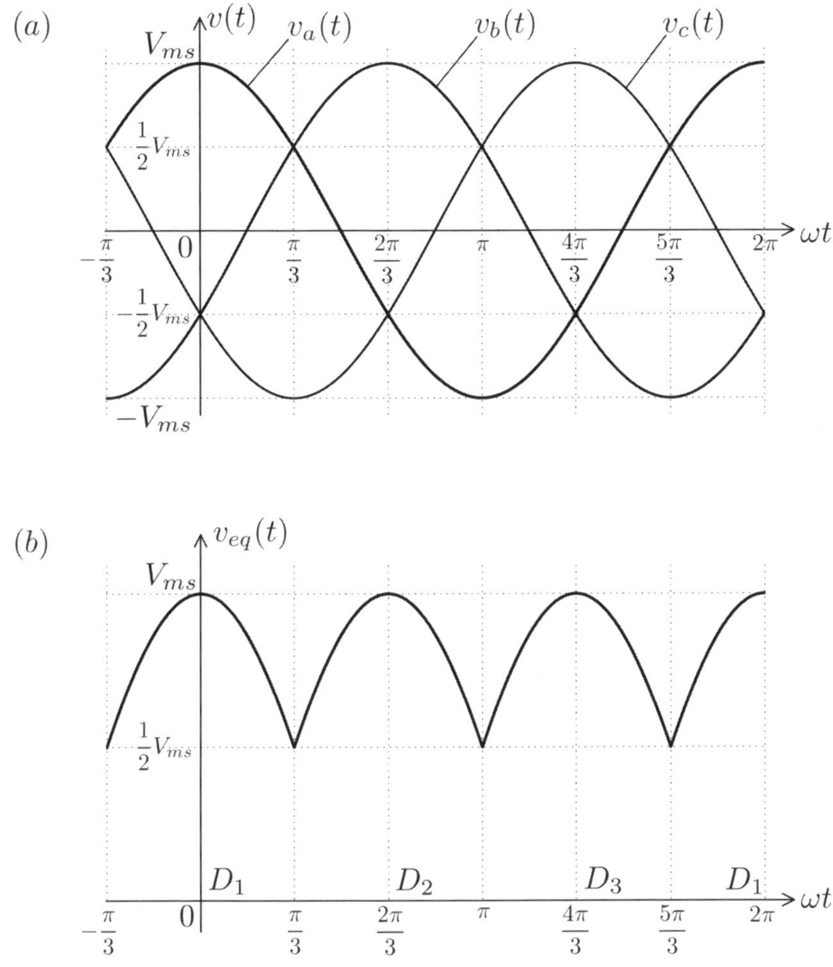

Fig. 2.22

respectively. This means that the equivalent voltage source $v_{eq}(t)$ can be represented by the plot shown in Figure 2.22b.

It is clear that the ripple of $v_{eq}(t)$ shown in Figure 2.22b is 50% if this ripple is measured as

$$\text{ripple } v_{eq}(t) = \frac{\max v_{eq}(t) - \min v_{eq}(t)}{\max v_{eq}(t)} \cdot 100\%. \tag{2.107}$$

To conclude the analysis of the three-phase rectifier shown in Figure 2.20, it is necessary to find the steady state in the equivalent circuit shown in Figure 2.21 with $v_{eq}(t)$ given by the plot in Figure 2.22b. Since $v_{eq}(t)$ is periodic with period $\frac{2\pi}{3\omega}$, we consider only one period

$$-\frac{\pi}{3\omega} < t < \frac{\pi}{3\omega} \tag{2.108}$$

and write KVL equation

$$L\frac{di(t)}{dt} + Ri(t) = v_{eq}(t). \tag{2.109}$$

For the period defined by formula (2.108),

$$v_{eq}(t) = v_a(t) = V_{ms}\cos\omega t \tag{2.110}$$

and equation (2.109) can be represented in the form

$$L\frac{di(t)}{dt} + Ri(t) = V_{ms}\cos\omega t. \tag{2.111}$$

The steady state satisfies the periodic boundary conditions

$$i\left(-\frac{\pi}{3\omega}\right) = i\left(\frac{\pi}{3\omega}\right). \tag{2.112}$$

As before, a general solution of equation (2.111) has two distinct components,

$$i(t) = i_p(t) + i_h(t), \tag{2.113}$$

where $i_p(t)$ is a particular solution of the inhomogeneous equation (2.111) which can be identified with the ac steady state in the circuit shown in Figure 2.21 excited by the voltage $V_{ms}\cos\omega t$. This steady state is given by the formulas

$$i_p(t) = \frac{V_{ms}}{\sqrt{R^2 + \omega^2 L^2}}\cos(\omega t - \varphi), \tag{2.114}$$

$$\boxed{\tan\varphi = \frac{\omega L}{R}.} \tag{2.115}$$

The second component $i_h(t)$ in formula (2.113) is a general solution of the corresponding homogeneous equation. It is easy to see that

$$i_h(t) = Ae^{-\frac{R}{L}t}, \tag{2.116}$$

where A is some constant.

By substituting formulas (2.114) and (2.116) into equation (2.113), we find

$$i(t) = \frac{V_{ms}}{\sqrt{R^2 + \omega^2 L^2}} \cos(\omega t - \varphi) + Ae^{-\frac{R}{L}t}. \tag{2.117}$$

The constant A can be found from the periodic boundary condition (2.112), which leads to the following equation for A:

$$\frac{V_{ms}\cos\left(\frac{\pi}{3} + \varphi\right)}{\sqrt{R^2 + \omega^2 L^2}} + Ae^{\frac{\pi R}{3\omega L}} = \frac{V_{ms}\cos\left(\frac{\pi}{3} - \varphi\right)}{\sqrt{R^2 + \omega^2 L^2}} + Ae^{-\frac{\pi R}{3\omega L}}. \tag{2.118}$$

By taking into account that

$$\cos\left(\frac{\pi}{3} - \varphi\right) - \cos\left(\frac{\pi}{3} + \varphi\right) = \sqrt{3}\sin\varphi, \tag{2.119}$$

from equation (2.118) we derive

$$A = \frac{\sqrt{3}V_{ms}\sin\varphi}{2\sinh\frac{\pi R}{3\omega L}\sqrt{R^2 + \omega^2 L^2}}. \tag{2.120}$$

By substituting the last relation into formula (2.117), we find

$$i(t) = \frac{V_{ms}}{\sqrt{R^2 + \omega^2 L^2}} \cos(\omega t - \varphi) + \frac{\sqrt{3}V_{ms}\sin\varphi}{2\sinh\frac{\pi R}{3\omega L}\sqrt{R^2 + \omega^2 L^2}} e^{-\frac{R}{L}t}. \tag{2.121}$$

It is clear that

$$v_{out}(t) = i(t)R, \tag{2.122}$$

which leads to

$$\boxed{v_{out}(t) = \frac{V_{ms}R}{\sqrt{R^2 + \omega^2 L^2}} \cos(\omega t - \varphi) + \frac{\sqrt{3}V_{ms}R\sin\varphi}{2\sinh\frac{\pi R}{3\omega L}\sqrt{R^2 + \omega^2 L^2}} e^{-\frac{R}{L}t}.} \tag{2.123}$$

Next, we shall demonstrate that under the condition

$$\omega L \gg R \tag{2.124}$$

the output voltage $v_{out}(t)$ is practically constant and equal to

$$\boxed{v_{out}(t) \approx \frac{3\sqrt{3}}{2\pi} V_{ms}.} \tag{2.125}$$

Indeed, from (2.124) follows that

$$\sqrt{R^2 + \omega^2 L^2} \approx \omega L, \tag{2.126}$$

$$\frac{R}{\sqrt{R^2+\omega^2L^2}} \approx \frac{R}{\omega L} \approx 0. \tag{2.127}$$

This means that the first term in the right-hand side of formula (2.123) is negligible. Now, we shall evaluate the second term of this right-hand side. From formulas (2.115) and (2.124), we conclude

$$\varphi \approx \frac{\pi}{2} \quad \text{and} \quad \sin\varphi \approx 1. \tag{2.128}$$

Furthermore, the inequality (2.124) implies that

$$\sinh\frac{\pi R}{3\omega L} \approx \frac{\pi R}{3\omega L}. \tag{2.129}$$

It is clear that

$$e^{\frac{\pi R}{3\omega L}} > e^{-\frac{R}{L}t} > e^{-\frac{\pi R}{3\omega L}}, \tag{2.130}$$

which, according to the inequality (2.124), means that in the time interval (2.108) we have

$$e^{-\frac{R}{L}t} \approx 1. \tag{2.131}$$

By using formulas (2.126), (2.128), (2.129) and (2.131), it is easy to conclude that the second term in the right-hand side of equation (2.123) is practically constant and equal to $\frac{3\sqrt{3}}{2\pi}V_{ms}$. Thus, formula (2.125) is established.

It is easy to demonstrate that the average value $\overline{v_{out}(t)}$ is given by the formula

$$\boxed{\overline{v_{out}(t)} = \frac{3\sqrt{3}}{2\pi}V_{ms} \approx 0.827V_{ms}.} \tag{2.132}$$

Indeed, by writing equation (2.111) in the form

$$L\frac{di(t)}{dt} + v_{out}(t) = V_{ms}\cos\omega t, \tag{2.133}$$

by averaging all terms of the last equation over one period $\left(-\frac{\pi}{3\omega}, \frac{\pi}{3\omega}\right)$ and taking into account the periodic boundary conditions (2.112), we derive

$$\overline{v_{out}(t)} = V_{ms}\frac{3\omega}{2\pi}\int_{-\frac{\pi}{3\omega}}^{\frac{\pi}{3\omega}} \cos\omega t dt, \tag{2.134}$$

which leads to formula (2.132).

It is worthwhile to point out that formula (2.132) is valid for any values of parameters in the rectifier circuit in Figure 2.20. By subtracting formula (2.132) from formula (2.123), we obtain the analytical expression for the

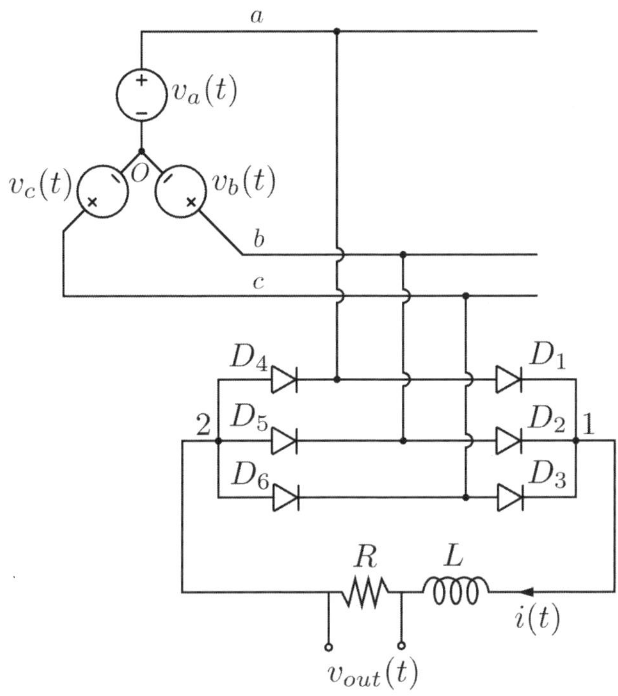

Fig. 2.23

output voltage ripple, which can be used for ripple evaluation for any values of L and R.

As discussed above, the ripple of $v_{eq}(t)$ for the three-phase rectifier shown in Figure 2.20 is 50%. It turns out that this ripple can be substantially reduced by using the three-phase bridge rectifier shown in Figure 2.23. Such rectifiers are also called six-pulse rectifiers. As before, the first step in the analysis of this rectifier is to make the equivalent replacement of the three-phase voltage sources and six diodes by the equivalent voltage source $v_{eq}(t)$. This replacement leads to the equivalent electric circuit shown in Figure 2.24. The equivalent voltage source $v_{eq}(t)$ must be equal at any instant of time t to the actual voltage between nodes 1 and 2 in the original circuit shown in Figure 2.23. To find this actual voltage, the conduction pattern of the six diodes must first be understood. By using the same line of reasoning as in the discussion of the three-diode rectifier shown in Figure 2.20, it can be shown that at any instant of time t only one of the three diodes D_1, D_2 and D_3 is in the "on" state, while the other

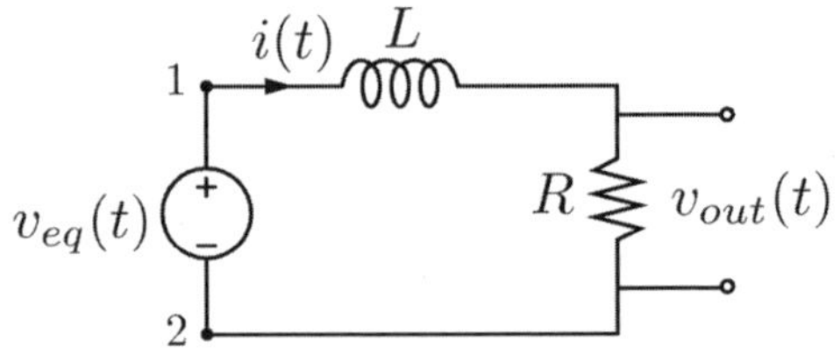

Fig. 2.24

two diodes are in the "off" state. The diode which is in the "on" state at time t is connected to the line (i.e., voltage source) with the highest positive potential at time t. Similarly, at any instant of time t only one of the three diodes D_4, D_5 and D_6 is in the "on" state, while the other two diodes are in the "off" state. The diode which is in the "on" state at time t is connected to the line (i.e., voltage source) with the most negative potential at time t. Thus, it is clear that the voltage between nodes 1 and 2 in the circuit in Figure 2.23 is always equal to one of the line voltages, namely, to the line voltage with largest positive value at time t. The specific line voltage appearing across nodes 1 and 2 at time t is determined by the conduction pattern of the six diodes at this instant of time t. This diode conduction pattern can now be discerned by using the plots of voltages $v_a(t)$, $v_b(t)$ and $v_c(t)$ shown in Figure 2.22a. It is clear according to this figure that during the time interval $0 < t < \frac{\pi}{3\omega}$ diodes D_1 and D_6 are connected to the lines with the most positive (line a) and most negative (line c) potentials, respectively. For this reason, these diodes are in the "on" state and the line voltage $v_{ac}(t)$ appears between nodes 1 and 2. During the time interval $\frac{\pi}{3\omega} < t < \frac{2\pi}{3\omega}$, diodes D_2 and D_6 are in the "on" state and the line voltage $v_{bc}(t)$ appears between nodes 1 and 2. The diode conduction patterns at other time intervals are similarly determined and they are illustrated in Figure 2.25 along with the plot of $v_{eq}(t)$. By using the phasor diagram for phase and line voltages, it is easy to find that

$$v_{ac}(t) = -v_{ca}(t) = \sqrt{3}V_{ms} \cos\left(\omega t - \frac{\pi}{6}\right), \tag{2.135}$$

$$v_{bc}(t) = -v_{cb}(t) = \sqrt{3}V_{ms} \cos\left(\omega t - \frac{\pi}{2}\right), \tag{2.136}$$

$$v_{ba}(t) = -v_{ab}(t) = \sqrt{3}V_{ms} \cos\left(\omega t - \frac{5\pi}{6}\right). \tag{2.137}$$

It is clear that

$$v_{ac}(0) = v_{ac}\left(\frac{\pi}{3\omega}\right) = \sqrt{3}V_{ms} \cos\frac{\pi}{6} = \frac{3}{2}V_{ms}, \tag{2.138}$$

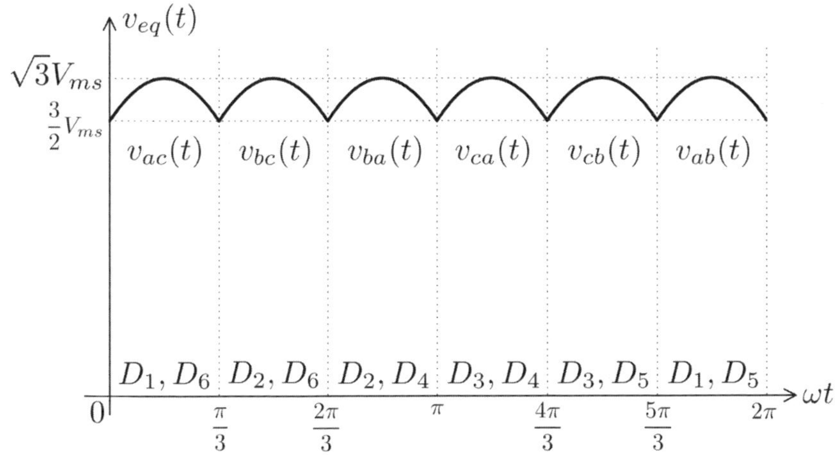

Fig. 2.25

which implies that

$$\text{ripple } v_{eq}(t) = \frac{\sqrt{3}V_{ms} - \frac{3}{2}V_{ms}}{\sqrt{3}V_{ms}} \cdot 100\% \approx 13.4\%. \tag{2.139}$$

Thus, in the case of the three-phase bridge rectifier the ripple of $v_{eq}(t)$ is appreciably reduced.

Having found $v_{eq}(t)$, we can now proceed to the analysis of the steady state in the electric circuit shown in Figure 2.24. We consider one period of $v_{eq}(t)$,

$$0 < t < \frac{\pi}{3\omega}. \tag{2.140}$$

For this time interval, the analysis of the steady state can be reduced to the following boundary value problem:

$$L\frac{di(t)}{dt} + Ri(t) = \sqrt{3}V_{ms}\cos\left(\omega t - \frac{\pi}{6}\right), \tag{2.141}$$

$$i(0) = i\left(\frac{\pi}{3\omega}\right). \tag{2.142}$$

A general solution of equation (2.141) can be written in the form

$$i(t) = \frac{\sqrt{3}V_{ms}}{\sqrt{R^2 + \omega^2 L^2}}\cos\left(\omega t - \frac{\pi}{6} - \varphi\right) + Ae^{-\frac{R}{L}t}, \tag{2.143}$$

$$\boxed{\tan\varphi = \frac{\omega L}{R}.} \tag{2.144}$$

By using the periodic boundary conditions (2.142), we end up with the following equation for A:

$$\frac{\sqrt{3}V_{ms}\cos\left(\frac{\pi}{6}+\varphi\right)}{\sqrt{R^2+\omega^2L^2}}+A=\frac{\sqrt{3}V_{ms}\cos\left(\frac{\pi}{6}-\varphi\right)}{\sqrt{R^2+\omega^2L^2}}+Ae^{-\frac{\pi R}{3\omega L}}. \tag{2.145}$$

By solving this equation for A and taking into account that

$$\cos\left(\frac{\pi}{6}-\varphi\right)-\cos\left(\frac{\pi}{6}+\varphi\right)=\sin\varphi, \tag{2.146}$$

we find

$$A=\frac{\sqrt{3}V_{ms}\sin\varphi}{\left(1-e^{-\frac{\pi R}{3\omega L}}\right)\sqrt{R^2+\omega^2L^2}}. \tag{2.147}$$

By substituting the last formula into equation (2.143), we end up with

$$i(t)=\frac{\sqrt{3}V_{ms}}{\sqrt{R^2+\omega^2L^2}}\cos\left(\omega t-\frac{\pi}{6}-\varphi\right)+\frac{\sqrt{3}V_{ms}\sin\varphi}{\left(1-e^{-\frac{\pi R}{3\omega L}}\right)\sqrt{R^2+\omega^2L^2}}e^{-\frac{R}{L}t}. \tag{2.148}$$

This leads to the following final expression for the output voltage:

$$\boxed{\begin{aligned}v_{out}(t)=&\frac{\sqrt{3}V_{ms}R}{\sqrt{R^2+\omega^2L^2}}\cos\left(\omega t-\frac{\pi}{6}-\varphi\right)\\&+\frac{\sqrt{3}V_{ms}R\sin\varphi}{\left(1-e^{-\frac{\pi R}{3\omega L}}\right)\sqrt{R^2+\omega^2L^2}}e^{-\frac{R}{L}t}.\end{aligned}} \tag{2.149}$$

It is remarkable that the complicated circuit shown in Figure 2.23 admits such a simple analytical solution. By using this solution, it can be shown that under the condition

$$\omega L\gg R \tag{2.150}$$

the output voltage is almost constant in time and it is given by the formula

$$\boxed{v_{out}(t)\approx\frac{3\sqrt{3}}{\pi}V_{ms}.} \tag{2.151}$$

The derivation is performed in the same way as in the case of the rectifier shown in Figure 2.20 and is left as a useful exercise for the reader.

By averaging both sides of equation (2.141) over one period $\left(0,\frac{\pi}{3\omega}\right)$, it can be found that, as expected,

$$\boxed{\overline{v_{out}(t)}=\frac{3\sqrt{3}}{\pi}V_{ms}\approx1.65V_{ms}.} \tag{2.152}$$

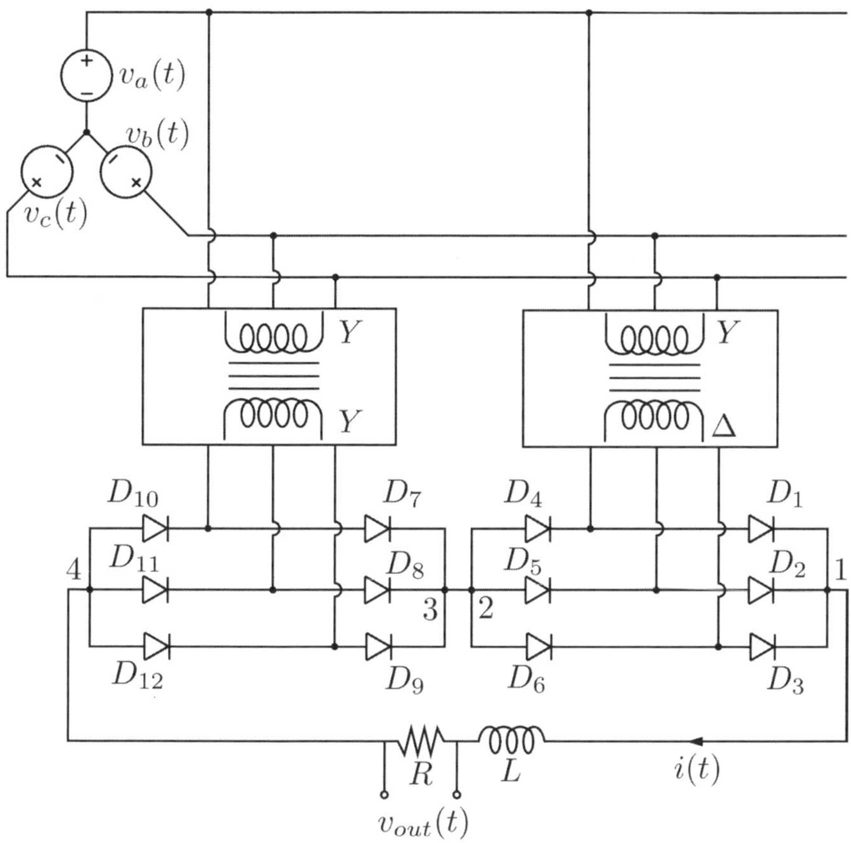

Fig. 2.26

By subtracting formula (2.152) from formula (2.149), the analytical expression for the ripple of the output voltage is found and can be used for evaluation of this ripple for any values of L and R. The clear advantage of the three-phase rectifiers shown in Figures 2.20 and 2.23 is that they lead to appreciable reduction in the ripple of $v_{eq}(t)$. A shortcoming of these rectifiers is that the levels of the output dc voltages are fixed and given by formulas (2.132) and (2.152), respectively. These levels can be adjusted by using three-phase transformers in the rectifier designs. It turns out that the use of three-phase transformers opens the opportunities for further reduction of ripple of $v_{eq}(t)$. We shall present below one such rectifier design shown in Figure 2.26. This rectifier is also called a twelve-pulse rectifier. In this rectifier, there are two three-phase transformers with Y-Y and Y-Δ connectivity of primary and secondary windings as well as two three-phase diode bridges connected in series. Across nodes 1 and 2 as well as nodes 3 and 4 the line voltages from the secondary windings of the Y-Δ and Y-Y transformers appear, respectively. By using the proper choice of turns ratios of these transformers, the desired and equal peak values $V_{ms}^{(2)}$ of their secondary line voltages can be achieved. However, these line voltages will be relatively phase-shifted in time by $\frac{\pi}{6}$ (see Chapter 3 of Part II). Thus, at any time instant t the voltage between nodes 1 and 4 (i.e., $v_{eq}(t)$) is equal to the sum of the line voltages of equal peak values but phase-shifted in time by $\frac{\pi}{6}$. This implies that $v_{eq}(t)$ is periodic with period $\frac{\pi}{6\omega}$ and for the time interval

$$0 < t < \frac{\pi}{6\omega} \tag{2.153}$$

the equivalent voltage $v_{eq}(t)$ is equal to

$$v_{eq}(t) = V_{ms}^{(2)} \cos \omega t + V_{ms}^{(2)} \cos \left(\omega t - \frac{\pi}{6} \right). \tag{2.154}$$

By using simple trigonometry, the last formula can be transformed as follows:

$$v_{eq}(t) = 2V_{ms}^{(2)} \cos \frac{\pi}{12} \cos \left(\omega t - \frac{\pi}{12} \right). \tag{2.155}$$

It can be found from the last equation that

$$\text{ripple } v_{eq}(t) = \frac{v_{eq}\left(\frac{\pi}{12\omega}\right) - v_{eq}(0)}{v_{eq}\left(\frac{\pi}{12\omega}\right)} \cdot 100\% = \left(1 - \cos \frac{\pi}{12} \right) \cdot 100\% < 4\%. \tag{2.156}$$

Thus, the ripple of $v_{eq}(t)$ is quite small. By using the same technique as frequently employed in this chapter, a simple analytical expression for $v_{out}(t)$ can be derived. This derivation is left as a useful exercise for the reader.

The rectifier presented in Figure 2.26 has two three-phase transformers. In practice, a single three-phase transformer with two sets of secondary windings connected in Y and Δ can be used. These two sets of secondary windings are connected to two bridge rectifiers and the basic operation of such rectifiers is identical to the operation of the rectifier shown in Figure 2.26.

2.4 Phase-Controlled Rectifiers

In the previous sections of this chapter, diode rectifiers have been discussed. In these rectifiers, the output dc voltages are fixed and non-controllable.

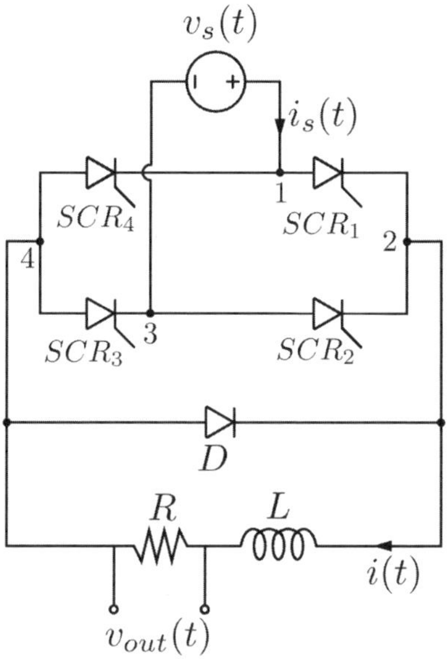

Fig. 2.27

The controllability of output dc voltages can be achieved by replacing diodes by thyristors (SCRs). This results in so-called phase-controlled rectifiers, which are discussed in this section.

We start with the discussion of the full-wave phase-controlled rectifier whose electric circuit is presented in Figure 2.27. This circuit is obtained from the electric circuit of the full-wave diode bridge rectifier (see Figure 2.1b) by replacing the four diodes by thyristors (SCRs) and by introducing a diode across the nodes 2 and 4. This diode is called a freewheeling diode (FWD). This diode prevents the appearance of negative voltages across nodes 2 and 4 which otherwise may occur in this circuit. Indeed, this negative voltage would correspond to forward bias of this diode and result in its switching into the "on" (conduction) state with zero voltage across it. This freewheeling diode also provides an alternative path for the load current $i(t)$ and, in this way, it is instrumental in maintaining the continuity in time of this current. It is furthermore clear that in the case of positive polarity of the voltage across nodes 2 and 4 the freewheeling diode is reverse biased, and it is in the "off" state. The current conduction path in this case goes through SCR_1 and SCR_3 or SCR_2 and SCR_4.

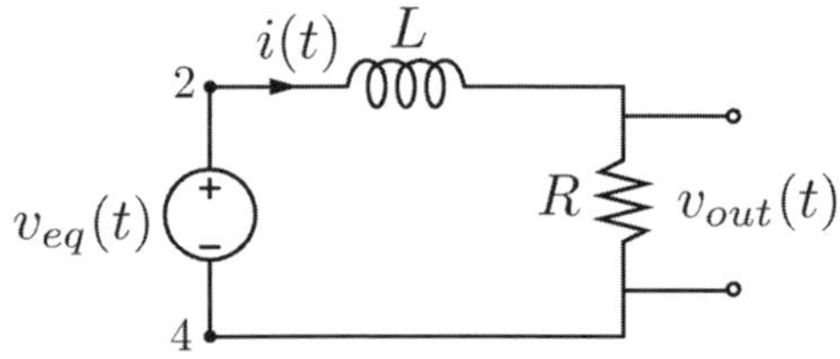

Fig. 2.28

As before, the first step in the analysis of the rectifier shown in Figure 2.27 is to make the equivalent replacement of the ac voltage $v_s(t)$, the four SCRs and the freewheeling diode by the voltage source $v_{eq}(t)$. This replacement results in the equivalent electric circuit shown in Figure 2.28. It is clear that the equivalent voltage source $v_{eq}(t)$ must be equal at any instant of time t to the actual voltage between nodes 2 and 4 in the original circuit shown in Figure 2.27. To find this voltage, the conduction patterns of the SCRs and the freewheeling diode must be first determined. To this end, we turn to the plot of sinusoidal voltage of the source $v_s(t)$ shown in Figure 2.29a and consider the positive half-cycle of this voltage in the time interval $0 < t < \frac{\pi}{\omega}$. In the case of the diode bridge rectifier (see Figure 2.1b), diodes D_1 and D_3 become immediately forward biased with the advent of this half-cycle and are immediately turned on. This is not the case for SCR_1 and SCR_3 in the electric circuit shown in Figure 2.27. These SCRs will remain in the "off" (forward-blocking) state until triggering currents are pulsed through their gates at some time $t_0 = \frac{\alpha}{\omega}$. This implies that SCR_2 and SCR_4 triggered during the preceding negative half-cycle of $v_s(t)$ tend to remain in the "on" state with the advent of the positive half-cycle and this leads to the appearance of negative polarity voltage across the nodes 2 and 4. As a result, the freewheeling diode becomes forward biased and starts conducting, resulting in zero voltage across nodes 2 and 4 as shown in Figure 2.29b. As soon as SCR_1 and SCR_3 are turned on by current pulses through their gates at time $t_0 = \frac{\alpha}{\omega}$, the voltage $v_s(t)$ of positive polarity appears across the nodes 2 and 4, and the freewheeling diode is turned off. This means that $v_{eq}(t)$ replicates $v_s(t)$ during the time interval $t_0 < t < \frac{\pi}{\omega}$ as shown in Figure 2.29b. With the advent of the next negative half-cycle, the freewheeling diode is again turned on to prevent the appearance of negative polarity voltage across the nodes 2 and 4 and it remains in the "on" state until SCR_2 and SCR_4 are triggered by current pulses through their gates at the time $t_0 + \frac{\pi}{\omega}$. The discussed conduction pattern of SCRs

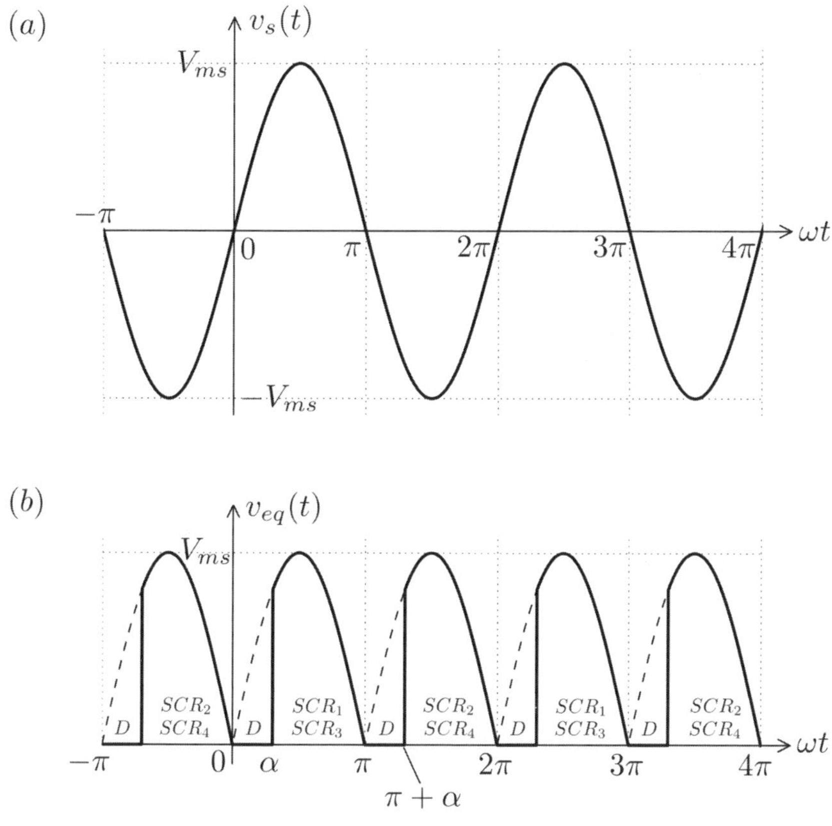

Fig. 2.29

and the freewheeling diode is periodically repeated resulting in $v_{eq}(t)$ shown in Figure 2.29b.

Now, having determined $v_{eq}(t)$, we shall proceed to the analysis of the steady state in the equivalent circuit shown in Figure 2.28. As discussed before in this chapter, it is sufficient to carry out this analysis for one period

$$0 < t < \frac{\pi}{\omega} \tag{2.157}$$

and then periodically extend the found expression for $v_{out}(t)$ to other time intervals.

The KVL equation for the circuit in Figure 2.28 can be written as

$$L\frac{di(t)}{dt} + Ri(t) = v_{eq}(t). \tag{2.158}$$

By using the plot of $v_{eq}(t)$ shown in Figure 2.29b, we conclude that the last equation can be written as the following two equations:

$$L\frac{di(t)}{dt} + Ri(t) = 0, \qquad \text{if } 0 < t < t_0, \tag{2.159}$$

$$L\frac{di(t)}{dt} + Ri(t) = V_{ms}\sin\omega t, \qquad \text{if } t_0 < t < \frac{\pi}{\omega}. \tag{2.160}$$

The steady-state solution of the above equations must satisfy the periodic boundary condition

$$i(0) = i\left(\frac{\pi}{\omega}\right) \tag{2.161}$$

as well as the continuity condition

$$i(t_{0-}) = i(t_{0+}). \tag{2.162}$$

A general solution of equation (2.159) can be written as

$$i(t) = A_1 e^{-\frac{R}{L}t}, \qquad (0 < t < t_0), \tag{2.163}$$

where A_1 is some constant. A general solution of equation (2.160) has the form

$$i(t) = \frac{V_{ms}}{\sqrt{R^2+\omega^2L^2}}\sin(\omega t - \varphi) + A_2 e^{-\frac{R}{L}t}, \qquad \left(t_0 < t < \frac{\pi}{\omega}\right), \tag{2.164}$$

where

$$\tan\varphi = \frac{\omega L}{R} \tag{2.165}$$

and A_2 is some constant.

To find constants A_1 and A_2, the periodicity condition (2.161) and continuity condition (2.162) can be used. This leads to the following simultaneous equations for A_1 and A_2:

$$A_1 = A_2 e^{-\frac{\pi R}{\omega L}} + \frac{V_{ms}\sin\varphi}{\sqrt{R^2+\omega^2L^2}}, \tag{2.166}$$

$$A_1 e^{-\frac{Rt_0}{L}} = A_2 e^{-\frac{Rt_0}{L}} + \frac{V_{ms}\sin(\omega t_0 - \varphi)}{\sqrt{R^2+\omega^2L^2}}. \tag{2.167}$$

By solving the above equations, we find

$$A_1 = \frac{V_{ms}\left[\sin\varphi - e^{\frac{Rt_0}{L} - \frac{\pi R}{\omega L}}\sin(\omega t_0 - \varphi)\right]}{\left(1 - e^{-\frac{\pi R}{\omega L}}\right)\sqrt{R^2+\omega^2L^2}}, \tag{2.168}$$

$$A_2 = \frac{V_{ms}\left[\sin\varphi - e^{\frac{Rt_0}{L}}\sin(\omega t_0 - \varphi)\right]}{\left(1 - e^{-\frac{\pi R}{\omega L}}\right)\sqrt{R^2+\omega^2L^2}}. \tag{2.169}$$

By substituting expressions (2.168) and (2.169) into formulas (2.163) and (2.164), respectively, and taking into account that

$$v_{out}(t) = Ri(t), \tag{2.170}$$

we derive that for $0 < t < t_0$,

$$\boxed{v_{out}(t) = \frac{V_{ms}R\left[\sin\varphi - e^{\frac{Rt_0}{L}-\frac{\pi R}{\omega L}}\sin(\omega t_0 - \varphi)\right]}{\left(1 - e^{-\frac{\pi R}{\omega L}}\right)\sqrt{R^2+\omega^2 L^2}}e^{-\frac{R}{L}t},} \tag{2.171}$$

while for $t_0 < t < \frac{\pi}{\omega}$,

$$\boxed{\begin{aligned} v_{out}(t) = {} & \frac{V_{ms}R}{\sqrt{R^2+\omega^2L^2}}\sin(\omega t - \varphi) \\ & + \frac{V_{ms}R\left[\sin\varphi - e^{\frac{Rt_0}{L}}\sin(\omega t_0 - \varphi)\right]}{\left(1 - e^{-\frac{\pi R}{\omega L}}\right)\sqrt{R^2+\omega^2 L^2}}e^{-\frac{R}{L}t} \end{aligned}} \tag{2.172}$$

and φ is defined by formula (2.165).

The last two formulas provide explicit analytical expressions for the output voltage of the phase-controlled rectifier shown in Figure 2.27. Next, we demonstrate that under the condition

$$\omega L \gg R, \tag{2.173}$$

this output voltage is practically constant in time. The demonstration is in the same way as before.

From the inequality (2.173), we find

$$\sqrt{R^2+\omega^2L^2} \approx \omega L, \tag{2.174}$$

$$\frac{R}{\sqrt{R^2+\omega^2L^2}} \approx 0, \tag{2.175}$$

$$\varphi \approx \frac{\pi}{2} \quad \text{and} \quad \sin\varphi \approx 1, \tag{2.176}$$

$$\sin(\omega t_0 - \varphi) \approx -\cos\omega t_0 = -\cos\alpha, \tag{2.177}$$

$$1 - e^{-\frac{\pi R}{\omega L}} \approx \frac{\pi R}{\omega L}, \tag{2.178}$$

$$e^{\frac{Rt_0}{L}-\frac{\pi R}{\omega L}} \approx 1, \tag{2.179}$$

$$e^{-\frac{R}{L}t} \approx 1 \quad \text{for} \quad 0 < t < \frac{\pi}{\omega}. \tag{2.180}$$

By using the last seven formulas to simplify the expressions (2.171) and (2.172), it can be easily demonstrated that under the condition (2.173) we have

$$\boxed{v_{out}(t) \approx \frac{V_{ms}}{\pi}(1 + \cos\alpha).} \tag{2.181}$$

As expected, this approximate value of $v_{out}(t)$ coincides with its average value. Indeed, by writing equations (2.159) and (2.160) in the form

$$L\frac{di(t)}{dt} + v_{out}(t) = \begin{cases} 0, & \text{if } 0 < t < t_0, \\ V_{ms}\sin\omega t, & \text{if } t_0 < t < \frac{\pi}{\omega}, \end{cases} \tag{2.182}$$

then by averaging all terms of the last equation over the period $\left(0, \frac{\pi}{\omega}\right)$ and taking into account the periodicity condition (2.161), we derive

$$\overline{v_{out}(t)} = V_{ms}\frac{\omega}{\pi}\int_{t_0}^{\frac{\pi}{\omega}} \sin\omega t dt, \tag{2.183}$$

which leads to

$$\boxed{\overline{v_{out}(t)} = \frac{V_{ms}}{\pi}(1 + \cos\alpha),} \tag{2.184}$$

where as before $\alpha = \omega t_0$. It is clear from formulas (2.181) and (2.184) that by varying the timing t_0 of triggering the SCRs the value of the output dc voltage can be continuously controlled.

It can also be remarked that by using formulas (2.171), (2.172) and (2.184) the ripple level in $v_{out}(t)$ can be computationally evaluated for any values of R and L.

It is evident from formulas (2.181) and (2.184) that voltage $v_{out}(t)$ depends on V_{ms}. This suggests that the range of controllability of v_{out} can be properly adjusted by using a center-tapped transformer phase-controlled rectifier shown in Figure 2.30. It is left as an exercise for the reader to prove that for this rectifier

$$\overline{v_{out}(t)} = \frac{N_2}{2\pi N_1}V_{ms}(1 + \cos\alpha) \tag{2.185}$$

and

$$v_{out}(t) \approx \overline{v_{out}(t)} \tag{2.186}$$

under the condition (2.173).

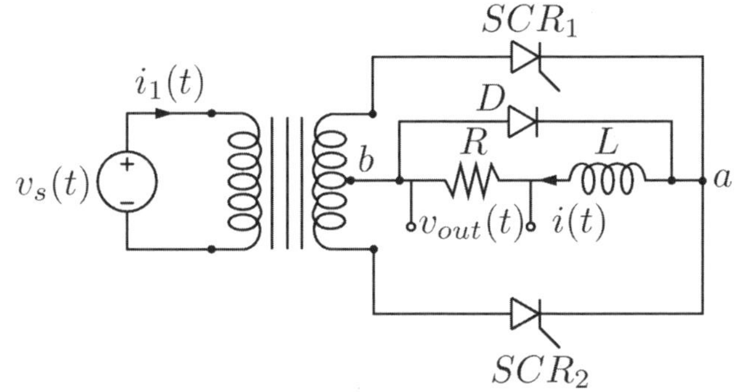

Fig. 2.30

Now we shall proceed to the discussion of the "six-pulse" bridge-controlled rectifier shown in Figure 2.31. The circuit of this rectifier is obtained from the circuit of the diode bridge rectifier shown in Figure 2.23 by replacing the six diodes by six SCRs and adding (when needed) the free-wheeling diode D. This freewheeling diode shall prevent the appearance of negative polarity voltage across the nodes 1 and 2. It turns out that such a voltage may appear only if the "firing angle" of the SCRs exceeds $\frac{\pi}{3}$, otherwise the voltage across the nodes 1 and 2 has only positive polarity. The first step in the analysis of this rectifier is to make the equivalent replacement of the three-phase voltage sources, the six SCRs and diode D by the voltage source $v_{eq}(t)$. This replacement results in the equivalent electric circuit shown in Figure 2.32. It is apparent that the equivalent voltage source $v_{eq}(t)$ must be equal at any instant of time t to the actual voltage between the nodes 1 and 2 in the original rectifier circuit shown in Figure 2.31. To find this voltage, the conduction pattern of the SCRs must be determined. To this end, we shall first recall that in the case of the three-phase diode bridge rectifier shown in Figure 2.23 one of the three diodes D_1, D_2 and D_3 is turned on as soon as the potential of the line connected to this diode assumes the highest positive value. Similarly, one of the three diodes D_4, D_5 and D_6 is turned on as soon as the potential of the line connected to this diode assumes the most negative value. This is not the case for the SCRs in the circuit shown in Figure 2.31. Indeed, one of the thyristors SCR_1, SCR_2 and SCR_3 is turned on while being connected to the line with the highest potential only if a triggering current is pulsed through its gate. Otherwise, this thyristor will be in the forward-blocking state and the previously conducting thyristor of this group will remain in the "on" state.

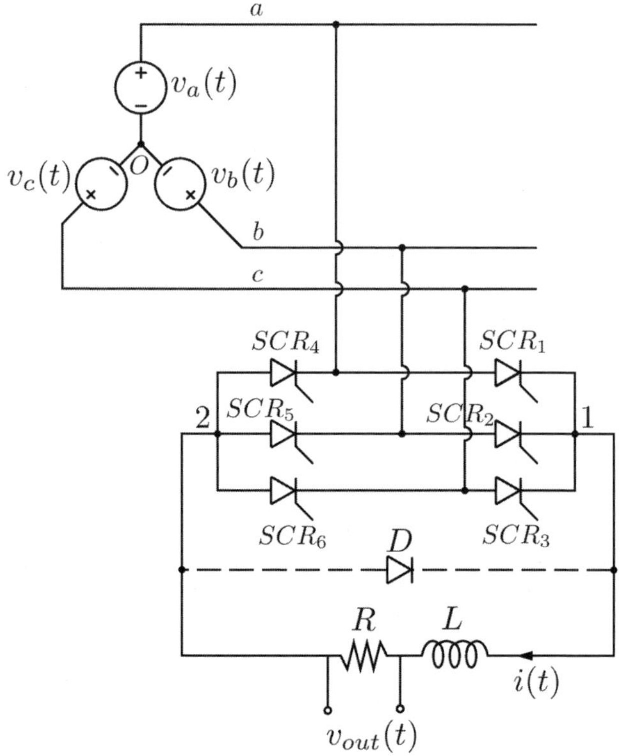

Fig. 2.31

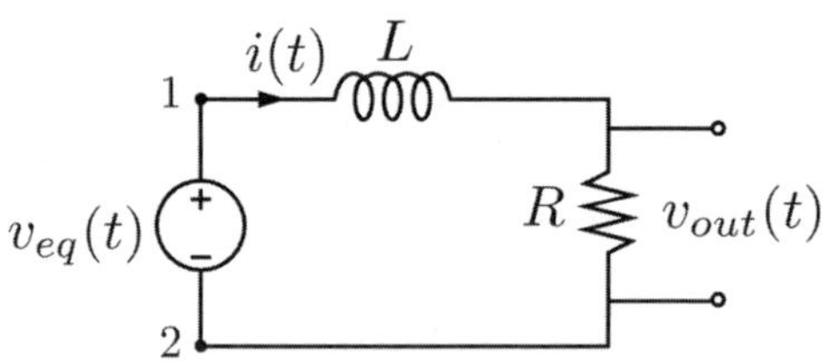

Fig. 2.32

Similarly, if one of the thyristors SCR_4, SCR_5 and SCR_6 is connected to the line with the most negative potential it will be turned on only after a triggering current is pulsed through its gate. Otherwise, this thyristor will be in the forward-blocking state and the previously conducting thyristor of this group will remain in the "on" state. In other words, for the rectifier

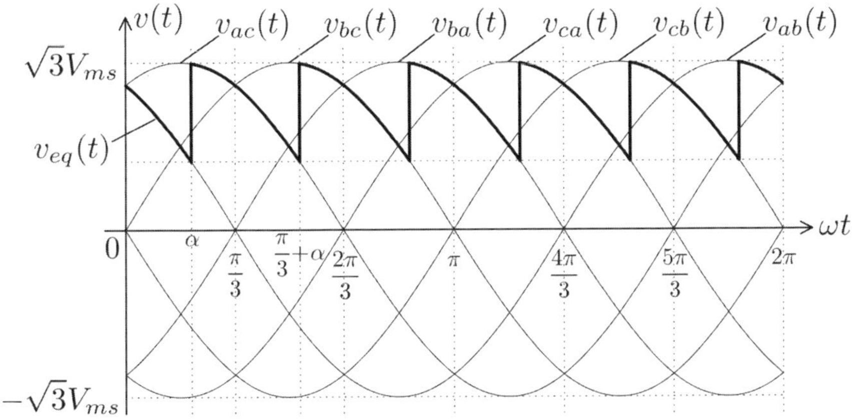

Fig. 2.33

shown in Figure 2.23, at any instant of time t only those two diodes conduct that connect the nodes 1 and 2 to the highest line voltage at time t. In the case of the rectifier shown in Figure 2.31, the highest line voltage at time t is applied across the nodes 1 and 2 only after the two SCRs connecting this voltage to the nodes 1 and 2 have been turned on by current pulses through their gates. Otherwise, the voltage between the nodes 1 and 2 is the line voltage provided by the two previously triggered SCRs. By using this remark, the equivalent voltage $v_{eq}(t)$ can be determined by using the plots of the line voltages as shown in Figure 2.33. It is seen from this figure that if the "firing angle" $\alpha = \omega t_0$ is less than $\frac{\pi}{3}$ then the equivalent voltage $v_{eq}(t)$ (i.e., voltage across the nodes 1 and 2) has positive polarity at any time instant and the freewheeling diode is not needed. If $\alpha > \frac{\pi}{3}$, then as is clear from Figure 2.33 the freewheeling diode D is needed to prevent the appearance of the negative polarity line voltage across the nodes 1 and 2.

Having determined $v_{eq}(t)$, we shall now proceed to the analysis of the steady state in the equivalent circuit shown in Figure 2.32 for the case $0 < \alpha < \frac{\pi}{3}$. We shall carry out this analysis for one period of $v_{eq}(t)$, namely, for the time interval

$$t_0 < t < t_0 + \frac{\pi}{3\omega}, \qquad \left(t_0 = \frac{\alpha}{\omega}\right), \tag{2.187}$$

and then periodically extend the found expression for $v_{out}(t)$ to other time intervals.

The KVL equation for the circuit in Figure 2.32 for the above time interval can be written as follows:

$$L\frac{di(t)}{dt}+Ri(t)=\sqrt{3}V_{ms}\cos\left(\omega t-\frac{\pi}{6}\right). \tag{2.188}$$

The steady-state solution of the above equation must satisfy the periodic boundary condition

$$i(t_0)=i\left(t_0+\frac{\pi}{3\omega}\right). \tag{2.189}$$

A general solution to equation (2.188) can be written as

$$i(t)=\frac{\sqrt{3}V_{ms}}{\sqrt{R^2+\omega^2L^2}}\cos\left(\omega t-\frac{\pi}{6}-\varphi\right)+Ae^{-\frac{R}{L}t}, \tag{2.190}$$

where, as before,

$$\tan\varphi=\frac{\omega L}{R} \tag{2.191}$$

and A is some constant. This constant is found from the boundary condition (2.189), which leads to

$$\frac{\sqrt{3}V_{ms}\cos\left(\omega t_0-\frac{\pi}{6}-\varphi\right)}{\sqrt{R^2+\omega^2L^2}}+Ae^{-\frac{Rt_0}{L}}=\frac{\sqrt{3}V_{ms}\cos\left(\omega t_0+\frac{\pi}{6}-\varphi\right)}{\sqrt{R^2+\omega^2L^2}}+Ae^{-\frac{Rt_0}{L}}e^{-\frac{\pi R}{3\omega L}}. \tag{2.192}$$

By using the simple identity

$$\cos\left(\omega t_0+\frac{\pi}{6}-\varphi\right)-\cos\left(\omega t_0-\frac{\pi}{6}-\varphi\right)=-\sin(\omega t_0-\varphi) \tag{2.193}$$

and by solving equation (2.192) for A, we find

$$A=-\frac{\sqrt{3}V_{ms}\sin(\omega t_0-\varphi)}{\left(1-e^{-\frac{\pi R}{3\omega L}}\right)\sqrt{R^2+\omega^2L^2}}e^{\frac{R}{L}t_0}. \tag{2.194}$$

By substituting the last formula into equation (2.190) and taking into account that

$$v_{out}(t)=Ri(t), \tag{2.195}$$

we derive

$$\boxed{\begin{aligned}v_{out}(t)=&\frac{\sqrt{3}V_{ms}R}{\sqrt{R^2+\omega^2L^2}}\cos\left(\omega t-\frac{\pi}{6}-\varphi\right)\\&-\frac{\sqrt{3}V_{ms}R\sin(\omega t_0-\varphi)}{\left(1-e^{-\frac{\pi R}{3\omega L}}\right)\sqrt{R^2+\omega^2L^2}}e^{-\frac{R}{L}(t-t_0)}.\end{aligned}} \tag{2.196}$$

By using the same line of reasoning as before, it can be shown that under the condition

$$\omega L \gg R \tag{2.197}$$

the output voltage $v_{out}(t)$ is almost constant in time and

$$\boxed{v_{out}(t) \approx \frac{3\sqrt{3}V_{ms}}{\pi} \cos \alpha.} \tag{2.198}$$

It can also be shown that

$$\boxed{\overline{v_{out}(t)} = \frac{3\sqrt{3}V_{ms}}{\pi} \cos \alpha.} \tag{2.199}$$

By varying α from 0 to $\frac{\pi}{3}$, the output voltage $v_{out}(t)$ can be continuously controlled within the range $\frac{3\sqrt{3}V_{ms}}{2\pi} < v_{out}(t) < \frac{3\sqrt{3}V_{ms}}{\pi}$. It is apparent from Figure 2.33 that this controllability is achieved at the expense of increasing the level of ripple in the equivalent voltage source $v_{eq}(t)$. Consequently, larger values of inductance L are needed to suppress this ripple and to achieve more or less constant value of output voltage $v_{out}(t)$.

It can be shown that by using the electric circuit shown in Figure 2.31 without the freewheeling diode D and for "firing angles" $\alpha > \frac{\pi}{3}$, the negative average value of voltage across the nodes 1 and 2 can be achieved, while the direction of current $i(t)$ due to the presence of the SCRs will remain the same. This means that, under the mentioned conditions, the above circuit operates as an inverter in the sense that power flow occurs from the dc side to the ac side. The detailed discussion of this matter is beyond the scope of this text.

Chapter 3

Inverters

3.1 Single-Phase Bridge Inverter

In this chapter, some basic principles of dc-to-ac energy conversion are presented. The power electronics circuits that accomplish this conversion are called inverters. The *voltage-source* inverters are discussed below. These inverters convert energy from fixed dc voltage sources into ac energy with voltages of desired and controllable frequencies and peak values. There are also *current-source* inverters where input dc sources maintain more or less constant currents. These inverters are not considered in this text. Their structures are somewhat similar to those of controlled bridge rectifiers studied at the end of the previous chapter. It is maybe for this reason that the term *inverter* is often used in literature for voltage-source inverters.

We begin with the discussion of the single-phase bridge inverter. The electric circuit of such inverter is shown in Figure 3.1. This circuit contains a dc voltage source V_0 as an input, four switches SW_1, SW_2, SW_3 and SW_4 on the four shoulders of the bridge and an LR branch with an output voltage $v_{out}(t)$ designated as the voltage across the terminals of the resistor R. The main challenge is to develop a proper strategy of switching that results in sinusoidal voltage $v_{out}(t)$ of desired frequency and peak value.

It is clear that such a switching strategy should periodically invert the polarity of the output voltage. This polarity inversion can be achieved by *periodically* repeating the following two steps of switching: step #1 of simultaneously turning SW_1 and SW_3 "on" and SW_2 and SW_4 "off"; and step #2 of simultaneously turning SW_1 and SW_3 "off" and SW_2 and SW_4 "on." It is apparent from Figure 3.1 that during the first step voltage V_0 of positive polarity appears across the nodes 1 and 2, while during the second step voltage V_0 of inverted (opposite) polarity appears across the nodes

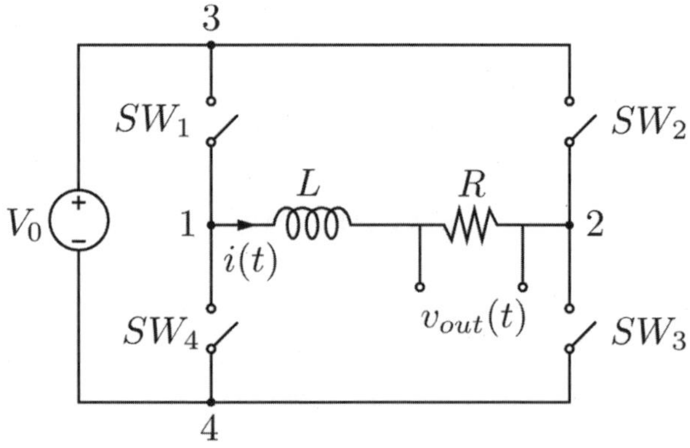

Fig. 3.1

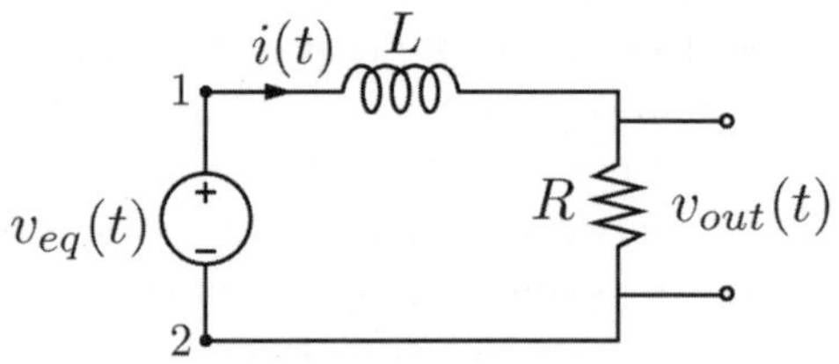

Fig. 3.2

1 and 2. This implies that for the above described switching the electric circuit shown in Figure 3.1 can be replaced by the equivalent circuit shown in Figure 3.2. In this figure, the equivalent voltage source $v_{eq}(t)$ is a periodic train (sequence) of rectangular voltage pulses of alternating polarity. The plot of $v_{eq}(t)$ is shown in Figure 3.3. It is apparent that $v_{eq}(t)$ is a function of half-wave symmetry,

$$v_{eq}(t) = -v_{eq}\left(t + \frac{T}{2}\right). \tag{3.1}$$

This means that at steady state the current $i(t)$ in the electric circuit shown in Figure 3.2 (as well as in the electric circuit shown in Figure 3.1) is a function of half-wave symmetry,

$$i(t) = -i\left(t + \frac{T}{2}\right). \tag{3.2}$$

By using the latter fact, we can formulate the steady-state analysis in the above electric circuit as the following boundary value problem with "an-

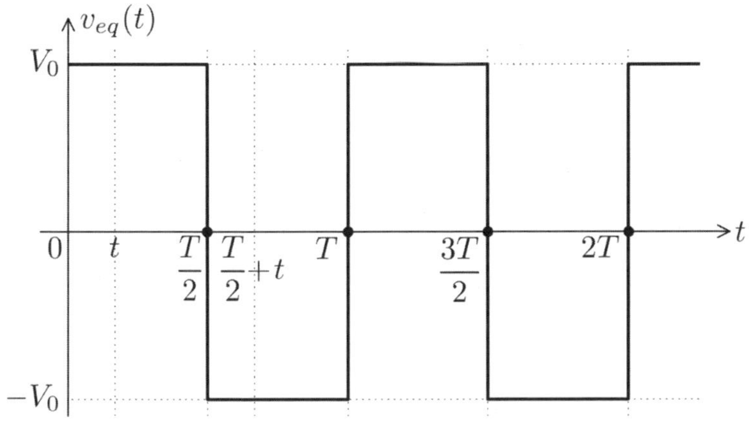

Fig. 3.3

tiperiodic" boundary condition:

$$L\frac{di(t)}{dt} + Ri(t) = V_0, \qquad \text{if } 0 < t < \frac{T}{2}, \tag{3.3}$$

$$i(0) = -i\left(\frac{T}{2}\right). \tag{3.4}$$

It is clear that the "antiperiodic" boundary condition (3.4) follows from formula (3.2) by setting t equal to zero. It is also clear that after solving the boundary value problem (3.3)-(3.4) for the time interval $[0, \frac{T}{2}]$, the current $i(t)$ can be extended to other time intervals by using the half-wave symmetry expressed by formula (3.2).

It is evident that a general solution to the differential equation (3.3) can be written in the form

$$i(t) = \frac{V_0}{R} + Ae^{-\frac{R}{L}t}, \qquad \left(0 < t < \frac{T}{2}\right), \tag{3.5}$$

where A is some constant that must be determined from the boundary condition (3.4). Indeed, this boundary condition leads to the following equation for A:

$$\frac{V_0}{R} + A = -\frac{V_0}{R} - Ae^{-\frac{RT}{2L}}. \tag{3.6}$$

From the last formula, we find

$$A\left(1 + e^{-\frac{RT}{2L}}\right) = -\frac{2V_0}{R}, \tag{3.7}$$

which results in

$$A = -\frac{2V_0}{R\left(1 + e^{-\frac{RT}{2L}}\right)}. \tag{3.8}$$

By substituting this expression for A into formula (3.5), we obtain

$$\boxed{i(t) = \frac{V_0}{R}\left[1 - \frac{2}{1 + e^{-\frac{RT}{2L}}}e^{-\frac{R}{L}t}\right], \qquad 0 \le t \le \frac{T}{2}.} \tag{3.9}$$

By taking into account that

$$v_{out}(t) = Ri(t), \tag{3.10}$$

we then find

$$\boxed{v_{out}(t) = V_0\left[1 - \frac{2}{1 + e^{-\frac{RT}{2L}}}e^{-\frac{R}{L}t}\right], \qquad 0 \le t \le \frac{T}{2}.} \tag{3.11}$$

Having found $v_{out}(t)$ for the time interval $[0, \frac{T}{2}]$, we can extend $v_{out}(t)$ to other time intervals through imposing the half-wave symmetry, i.e., by using the formula

$$v_{out}\left(t + \frac{T}{2}\right) = -v_{out}(t). \tag{3.12}$$

Next, we shall perform some simple analysis of formulas (3.9) and (3.11). From formula (3.9) we find that

$$i(0) = \frac{V_0}{R}\left[1 - \frac{2}{1 + e^{-\frac{RT}{2L}}}\right]. \tag{3.13}$$

Since

$$e^{-\frac{RT}{2L}} < 1, \tag{3.14}$$

we conclude from equation (3.13) that

$$i(0) < 0. \tag{3.15}$$

According to formula (3.4), the last inequality implies that

$$i\left(\frac{T}{2}\right) = -i(0) > 0. \tag{3.16}$$

From the last two inequalities and formula (3.10) follows that

$$v_{out}(0) < 0, \qquad \text{while} \qquad v_{out}\left(\frac{T}{2}\right) = -v_{out}(0) > 0. \tag{3.17}$$

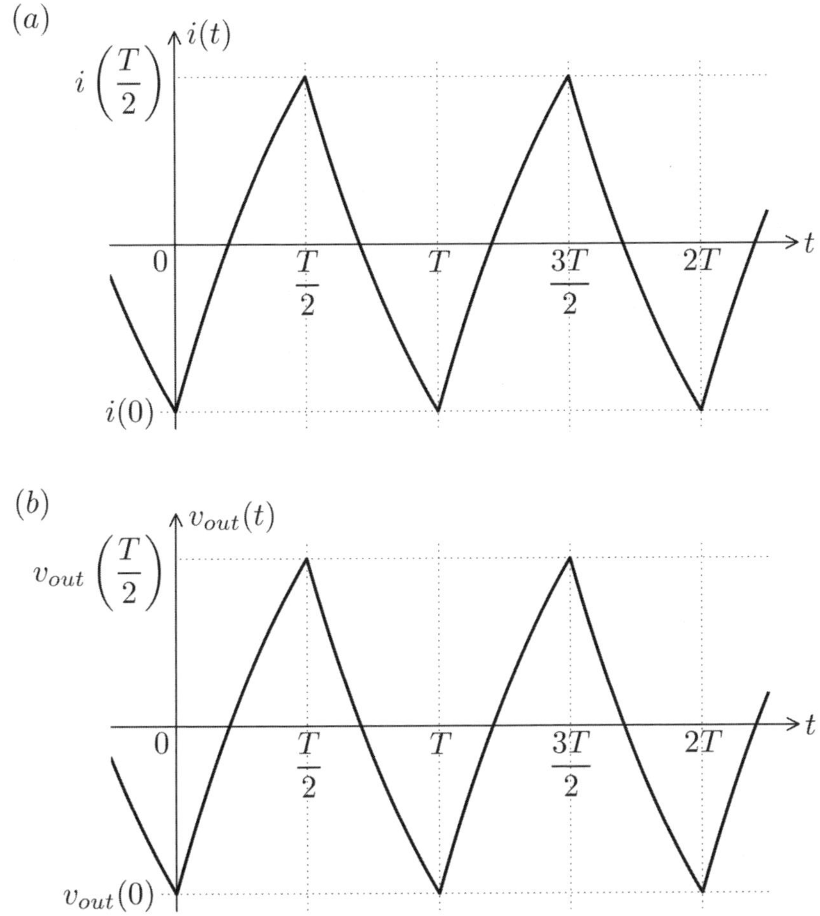

Fig. 3.4

By taking into account inequalities (3.15), (3.16) and (3.17) and using formulas (3.9) and (3.11), the plots of $i(t)$ and $v_{out}(t)$ can be constructed. These plots are shown in Figures 3.4a and 3.4b, respectively. It is apparent from Figure 3.4a that the current $i(t)$ changes its sign and, consequently, its direction of flow within each half-cycle. This fact has important implications concerning the design of the switches SW_1, SW_2, SW_3 and SW_4 shown in Figure 3.1. Indeed, single transistors are unidirectional (unilateral) switches. For instance, a BJT conducts an electric current from emit-

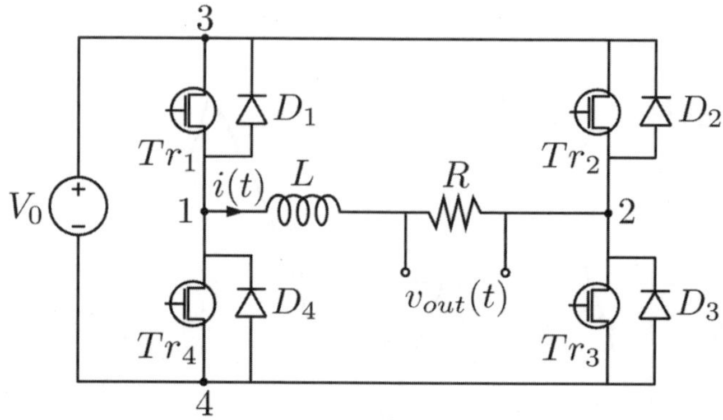

Fig. 3.5

ter to collector and, similarly, a power MOSFET (or IGBT) is designed to conduct an electric current from drain to source. This implies that a single transistor being in the "on" state cannot accommodate the flow of electric current in two opposite directions as needed in order to realize the current flow depicted in Figure 3.4a. It turns out that due to the presence of the inductor in the circuit shown in Figure 3.1, bidirectional (bilateral) switches SW_1, SW_2, SW_3 and SW_4 can be designed as parallel connections of power MOSFETs (or IGBTs) with freewheeling diodes. This is shown in Figure 3.5.

The operation of the electric circuit shown in this figure can be elucidated as follows. Immediately prior to the time instant $t = 0$, transistors Tr_2 and Tr_4 are in conducting ("on") states, while transistors Tr_1 and Tr_3 are in "off" states. This results in negative polarity voltage V_0 across the nodes 1 and 2 and in the current flow in the direction opposite to the one shown in Figure 3.5, which is consistent with Figure 3.4a. At time $t = 0$, transistors Tr_2 and Tr_4 are turned off, while transistors Tr_1 and Tr_3 are turned on, producing inversion of the polarity of the voltage V_0 across the nodes 1 and 2. However, transistors Tr_1 and Tr_3 cannot conduct the current $i(t)$ due to its direction at $t = 0$. To maintain the continuity of the current through the inductor L, diodes D_1 and D_3 are turned on and they form the closed path for freewheeling $i(t)$. Physically, diodes D_1 and D_3 are turned on because any (downward) disruption of continuity of $i(t)$ through L results in the induction of large voltage $L\frac{di(t)}{dt}$ of such polarity that will force these diodes into the conduction state. As soon as the current $i(t)$ is reduced to zero and its direction is reversed during the half-cycle $[0, \frac{T}{2}]$,

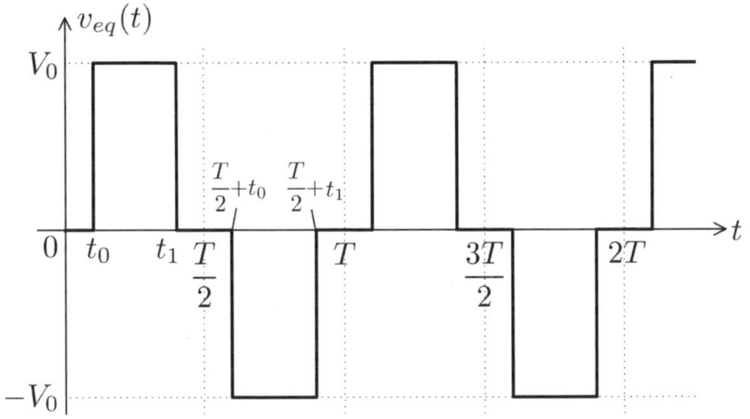

Fig. 3.6

transistors Tr_1 and Tr_3 start to conduct this current. At time $t = \frac{T}{2}$, transistors Tr_1 and Tr_3 are turned off, while transistors Tr_2 and Tr_4 are turned on. However, Tr_2 and Tr_4 cannot conduct the positive current $i(t)$ (with the direction shown in Figure 3.5). To maintain the continuity of the current through the inductor L, diodes D_2 and D_4 are turned on and form the closed path for the freewheeling current $i(t)$ until its direction is reversed. The described conduction pattern is periodically repeated.

It turns out that the bidirectional switches shown in Figure 3.5 can be used to control the width of rectangular pulses at each half-cycle. Particularly, the pattern of $v_{eq}(t)$ shown in Figure 3.6 can be produced through the appropriate switching. Indeed, immediately prior to the time instant $t = t_1$, transistors Tr_1 and Tr_3 are in "on" states, transistors Tr_2 and Tr_4 are in "off" states, and the current $i(t) > 0$, i.e., its direction coincides with the one shown in Figure 3.5. At time t_1, transistor Tr_3 is turned off. Due to the presence of the inductor, the continuity of $i(t)$ must be maintained. This is only possible if diode D_2 is turned on and freewheels the current $i(t)$ through the closed path formed by the LR branch, diode D_2 and transistor Tr_1. For this closed path, the voltage across the nodes 1 and 2 is equal to zero as shown in Figure 3.6. At time $\frac{T}{2} + t_0$, transistor Tr_1 is turned off, while transistors Tr_2 and Tr_4 are turned on. This switching action causes the voltage between nodes 1 and 2 to change to $-V_0$ as shown in Figure 3.6. Then, at time $\frac{T}{2} + t_1$, when $i(t) < 0$ and $i(t)$ has the direction opposite to the one shown in Figure 3.5, transistor Tr_4 is turned off, forcing diode D_1 to turn on, forming in this way the closed path for $i(t)$ through D_1, Tr_2

and the LR branch. This results in zero voltage across the nodes 1 and 2 as consistent with the plot of $v_{eq}(t)$ in Figure 3.6. The described switching pattern is periodically repeated.

The steady-state analysis of the electric circuit shown in Figure 3.2 with $v_{eq}(t)$ shown in Figure 3.6 can be carried out in a similar way as before. Namely, it is easy to see that electric current $i(t)$ during the half-cycle $[0, \frac{T}{2}]$ can be represented as follows:

$$i(t) = A_1 e^{-\frac{R}{L}t}, \qquad 0 \le t \le t_0, \tag{3.18}$$

$$i(t) = \frac{V_0}{R} + A_2 e^{-\frac{R}{L}t}, \qquad t_0 \le t \le t_1, \tag{3.19}$$

$$i(t) = A_3 e^{-\frac{R}{L}t}, \qquad t_1 \le t \le \frac{T}{2}, \tag{3.20}$$

where A_1, A_2 and A_3 are some constants. These constants are determined from the boundary condition (3.4) as well as from the following continuity conditions for electric current $i(t)$ at t_0 and t_1:

$$i(t_{0-}) = i(t_{0+}), \tag{3.21}$$

$$i(t_{1-}) = i(t_{1+}). \tag{3.22}$$

Having determined A_1, A_2 and A_3 from conditions (3.4), (3.21) and (3.22), we arrive at the final expressions for $i(t)$:

$$i(t) = -\frac{V_0}{R}\frac{e^{\frac{Rt_1}{L}} - e^{\frac{Rt_0}{L}}}{1 + e^{-\frac{RT}{2L}}} e^{-\frac{R}{L}\left(t+\frac{T}{2}\right)}, \qquad 0 \le t \le t_0, \tag{3.23}$$

$$i(t) = \frac{V_0}{R}\left[1 - \frac{e^{\frac{R}{L}\left(t_1-\frac{T}{2}\right)} + e^{\frac{Rt_0}{L}}}{1 + e^{-\frac{RT}{2L}}} e^{-\frac{R}{L}t}\right], \qquad t_0 \le t \le t_1, \tag{3.24}$$

$$i(t) = \frac{V_0}{R}\frac{e^{\frac{Rt_1}{L}} - e^{\frac{Rt_0}{L}}}{1 + e^{-\frac{RT}{2L}}} e^{-\frac{R}{L}t}, \qquad t_1 \le t \le \frac{T}{2}. \tag{3.25}$$

A plot of $i(t)$ based on formulas (3.23), (3.24) and (3.25) is shown in Figure 3.7. The plot of $v_{out}(t)$ is a scaled (by R) version of the plot for $i(t)$.

It is apparent from the above plot that $i(t)$ is a half-wave symmetric periodic function of t with period T. This period and, consequently, the fundamental frequency of $i(t)$ is controlled by the pattern of switching of SW_1, SW_2, SW_3 and SW_4 that can be chosen appropriately. It is also clear from the above plot as well as formulas (3.23), (3.24) and (3.25) that $i(t)$ is a piecewise exponential function of time t. This is true for the excitation of the electric circuit in Figure 3.2 by any sequence of rectangular pulses. It is also evident that the waveform of $i(t)$ shown in Figure 3.7 has more resemblance to a sinusoidal function than the waveform shown in Figure 3.4.

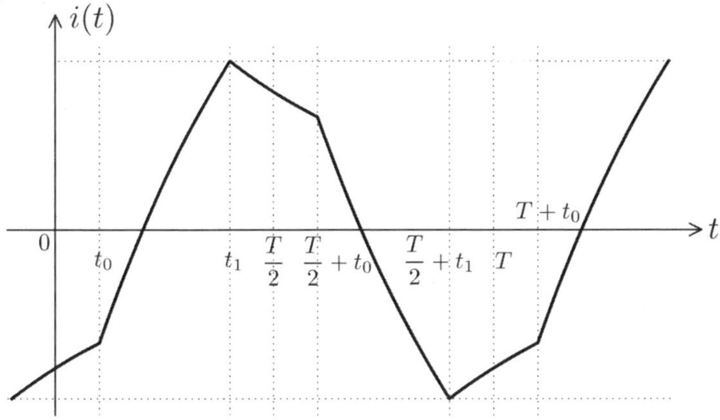

Fig. 3.7

This suggests that by choosing a more elaborate sequence of rectangular pulses for $v_{eq}(t)$ much better resemblance of $i(t)$ (and $v_{out}(t)$) with a sinusoidal function can be achieved. The latter can indeed be accomplished by using the pulse width modulation (PWM) technique discussed in the next section.

3.2 Pulse Width Modulation (PWM)

Pulse width modulation is used in power electronics to approximate a low frequency waveform by a sequence (train) of rectangular pulses whose width is properly modulated (by the usually low frequency waveform). A large number of PWM techniques have been developed over the years and discussed in literature (see, for instance, [23]). A common feature of most of these techniques is the suppression of low-order harmonics in the Fourier spectra of PWM voltages. This suppression is achieved at the expense of some amplification of higher-order harmonics in the Fourier spectra. However, these higher-order harmonics are suppressed in the output voltage $v_{out}(t)$ by an inductor in the LR branches of the inverters.

Next, we shall discuss below one simple version of PWM. In this version, each half-cycle $\frac{T}{2}$ is subdivided into k equal time intervals and the centers

t_i of these time intervals are specified by the formula

$$t_i = \frac{T}{2k}\left(i + \frac{1}{2}\right), \qquad (i = 0, 1, 2, ..., k-1). \tag{3.26}$$

This pulse width modulated voltage consists of a sequence of rectangular pulses centered at t_i and whose widths are sinusoidally modulated:

$$\Delta t_i = \frac{mT}{2k}\sin \omega t_i, \tag{3.27}$$

where as before

$$\omega = \frac{2\pi}{T}, \tag{3.28}$$

while m is called the modulation index (or depth of modulation) and usually

$$0 < m < 1. \tag{3.29}$$

It is shown below that by varying m the peak value of the sinusoidal output voltage can be controlled. An example of such pulse width modulated voltage is presented in Figure 3.8 for $k = 5$. For PWM to be effective, k is usually quite large,

$$k \gg 1. \tag{3.30}$$

Such sequences of width modulated rectangular pulses can be obtained by using specific switching patterns of transistors in the circuit shown in Figure 3.5. Some details of the realization of these switching patterns are discussed later in this section.

Now, we will be concerned with the study of the Fourier spectra of PWM voltages. It is clear from Figure 3.8 that $v_{eq}(t)$ has two types of symmetry: odd symmetry and half-wave symmetry. This implies (see Chapter 2 of Part I) that the Fourier series for $v_{eq}(t)$ contains only odd sine-type harmonics. Namely,

$$v_{eq}(t) = \sum_{n=1}^{\infty} V_{2n-1}\sin(2n-1)\omega t, \tag{3.31}$$

where

$$V_{2n-1} = \frac{4}{T}\int_0^{\frac{T}{2}} v_{eq}(t)\sin(2n-1)\omega t dt. \tag{3.32}$$

The immediate purpose of the subsequent discussion is the evaluation of the Fourier coefficients V_{2n-1}. To this end and by taking into account that

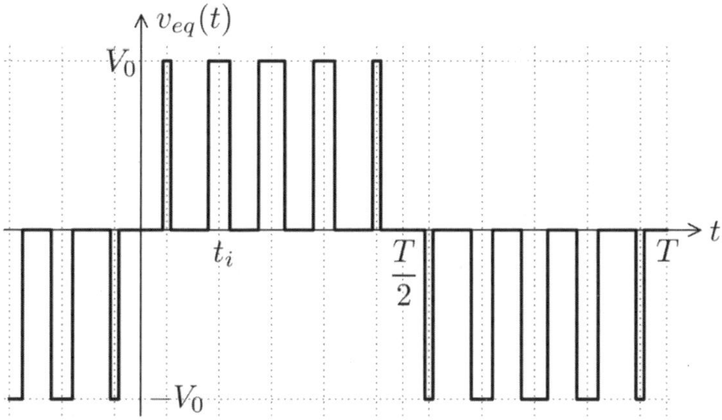

Fig. 3.8

$v_{eq}(t)$ is a sequence of k rectangular pulses of the same peak value V_0, the last formula can be written as follows:

$$V_{2n-1} = \frac{4V_0}{T} \sum_{i=0}^{k-1} \int_{\Delta t_i} \sin(2n-1)\omega t dt. \tag{3.33}$$

According to formulas (3.27) and (3.30), the widths of pulses Δt_i are quite small. For this reason, we shall use the following approximation:

$$\int_{\Delta t_i} \sin(2n-1)\omega t dt \approx \Delta t_i \sin(2n-1)\omega t_i. \tag{3.34}$$

This is the only approximation that is present in our derivation. It is evident that this approximation is quite accurate for small n when the period T_{2n-1} of $\sin(2n-1)\omega t$ is quite large in comparison with Δt_i:

$$T_{2n-1} = \frac{T}{2n-1} \gg \Delta t_i = \frac{mT}{2k} \sin \omega t_i. \tag{3.35}$$

This approximation may not be as accurate when T_{2n-1} and Δt_i are comparable. This suggests that our derivation will lead to accurate results for low-order sinusoidal harmonics in the Fourier series expansion (3.31).

Returning to our derivation and by substituting formula (3.34) into equation (3.33), we find

$$V_{2n-1} \approx \frac{4V_0}{T} \sum_{i=0}^{k-1} \Delta t_i \sin(2n-1)\omega t_i. \tag{3.36}$$

The last equation is further transformed by using expression (3.27) for Δt_i:

$$V_{2n-1} = \frac{2mV_0}{k} \sum_{i=0}^{k-1} \sin(2n-1)\omega t_i \sin \omega t_i. \tag{3.37}$$

By recalling the trigonometric identity

$$\sin\alpha \sin\beta = \frac{1}{2}\left[\cos(\alpha-\beta) - \cos(\alpha+\beta)\right], \tag{3.38}$$

the last formula can be transformed as follows:

$$V_{2n-1} = \frac{mV_0}{k}\left[\sum_{i=0}^{k-1} \cos(2n-2)\omega t_i - \sum_{i=0}^{k-1} \cos 2n\omega t_i\right]. \tag{3.39}$$

From formulas (3.26) and (3.28) follows that

$$\omega t_i = \frac{\pi}{2k}(2i+1) \tag{3.40}$$

and, consequently, the relation (3.39) can be written as

$$V_{2n-1} = \frac{mV_0}{k}\left[\sum_{i=0}^{k-1} \cos\frac{(n-1)\pi}{k}(2i+1) - \sum_{i=0}^{k-1} \cos\frac{n\pi}{k}(2i+1)\right]. \tag{3.41}$$

The subsequent derivation is based on the evaluation of the following two sums:

$$S_1 = \sum_{i=0}^{k-1} \cos\frac{(n-1)\pi}{k}(2i+1), \tag{3.42}$$

$$S_2 = \sum_{i=0}^{k-1} \cos\frac{n\pi}{k}(2i+1). \tag{3.43}$$

It is apparent that the expression for S_1 can be transformed as follows:

$$S_1 = \mathrm{Re}\left[\sum_{i=0}^{k-1} e^{j\frac{(n-1)\pi}{k}(2i+1)}\right] = \mathrm{Re}\left[e^{j\frac{(n-1)\pi}{k}} \sum_{i=0}^{k-1} e^{j\frac{(n-1)2\pi}{k}i}\right]. \tag{3.44}$$

By introducing the notation

$$q = e^{j\frac{(n-1)2\pi}{k}}, \tag{3.45}$$

we find that

$$\sum_{i=0}^{k-1} e^{j\frac{(n-1)2\pi}{k}i} = \sum_{i=0}^{k-1} q^i. \tag{3.46}$$

The last sum is a geometric series and, consequently,

$$\sum_{i=0}^{k-1} q^i = \frac{1-q^k}{1-q}. \tag{3.47}$$

Now, by combining formulas (3.45), (3.46) and (3.47), we derive

$$\sum_{i=0}^{k-1} e^{j\frac{(n-1)2\pi}{k}i} = \frac{1-e^{j(n-1)2\pi}}{1-e^{j\frac{(n-1)2\pi}{k}}}. \tag{3.48}$$

By using the last expression in formula (3.44), we arrive at

$$S_1 = \mathrm{Re}\left[e^{j\frac{(n-1)\pi}{k}} \frac{1-e^{j(n-1)2\pi}}{1-e^{j\frac{(n-1)2\pi}{k}}}\right]. \tag{3.49}$$

To conclude the evaluation of S_1, it is important to distinguish two cases.

Case a) when $n-1$ is not divisible by k, that is, for any natural number r we have

$$\boxed{n-1 \neq rk.} \tag{3.50}$$

The latter means that

$$q = e^{j\frac{(n-1)2\pi}{k}} \neq 1, \tag{3.51}$$

while, on the other hand,

$$e^{j(n-1)2\pi} = 1. \tag{3.52}$$

Thus, according to formula (3.49) we find

$$\boxed{S_1 = 0.} \tag{3.53}$$

Case b) deals with those "rare" instances when $n-1$ is divisible by k; this means that such a natural number r can be found that

$$\boxed{n-1 = rk.} \tag{3.54}$$

The latter implies that

$$q = e^{j\frac{(n-1)2\pi}{k}} = e^{jr2\pi} = 1. \tag{3.55}$$

In this case, the fraction in formula (3.49) is not defined. It turns out that in this case the sum S_1 can be evaluated differently. Namely, from formulas (3.55) and (3.46) we find

$$\sum_{i=0}^{k-1} e^{j\frac{(n-1)2\pi}{k}i} = k. \tag{3.56}$$

Furthermore, according to (3.54), we have

$$e^{j\frac{(n-1)\pi}{k}} = e^{jr\pi} = (-1)^r. \tag{3.57}$$

By using the last two formulas in the expression (3.44), we obtain

$$\boxed{S_1 = (-1)^r k.} \tag{3.58}$$

Thus, we have concluded the evaluation of S_1. The evaluation of sum S_2 can be now carried out based on the following observation: S_1 can be transformed into S_2 by replacing $n-1$ by n. This implies that as far as the value of S_2 is concerned, there are two distinct cases:

Case a) when

$$\boxed{n \neq rk} \tag{3.59}$$

and

$$\boxed{S_2 = 0,} \tag{3.60}$$

as well as

Case b) when

$$\boxed{n = rk} \tag{3.61}$$

and

$$\boxed{S_2 = (-1)^r k.} \tag{3.62}$$

It is easy to see that the condition (3.54) is equivalent to $2n-1 = 2rk+1$. Consequently, we conclude that

$$\text{if} \quad 2n-1 = 2rk+1, \quad \text{then} \quad S_1 = (-1)^r k. \tag{3.63}$$

Similarly, the condition (3.61) is equivalent to $2n-1 = 2rk-1$. Consequently,

$$\text{if} \quad 2n-1 = 2rk-1, \quad \text{then} \quad S_2 = (-1)^r k. \tag{3.64}$$

It is also easy to see from (3.50), (3.53), (3.59) and (3.60) that

$$S_1 = S_2 = 0, \quad \text{if} \quad 2n-1 \neq 2rk \pm 1. \tag{3.65}$$

Now, from formulas (3.41), (3.42), (3.43), (3.63), (3.64) and (3.65) we conclude that

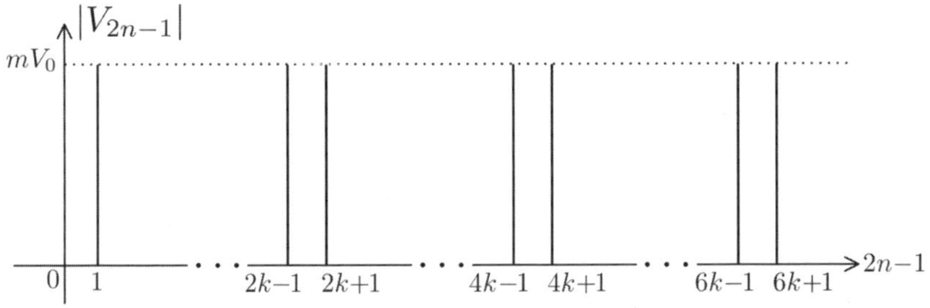

Fig. 3.9

$$\boxed{V_{2rk\pm 1} = (\pm)\,(-1)^r mV_0,} \tag{3.66}$$

while

$$\boxed{V_{2n-1} = 0 \quad \text{if} \quad 2n-1 \neq 2rk \pm 1.} \tag{3.67}$$

In particular, for $r = 0$ from (3.66) we obtain

$$\boxed{V_1 = mV_0.} \tag{3.68}$$

By using the last three formulas in equation (3.31), we find

$$\boxed{v_{eq}(t) = mV_0 \sin\omega t + mV_0 \sum_{r=1}^{\infty} (\pm)\,(-1)^r \sin(2rk \pm 1)\omega t.} \tag{3.69}$$

This means that the pulse width modulated voltage $v_{eq}(t)$ has a "sparse-twin" spectrum as illustrated in Figure 3.9. This spectrum is "sparse-twin" because "most" of the terms in the Fourier series expansion (3.31) are equal to zero, and those terms which are not equal to zero appear as pairs with equal peak values. It is worthwhile to mention again that the derivation of formula (3.69) has been based on approximation (3.34). For this reason, it is of interest to compare the "sparse-twin" spectrum shown in Figure 3.9 with the spectrum numerically computed without using this approximation. For the purpose of this comparison, such numerically computed spectra are shown in Figures 3.10a and 3.10b for $k = 10$ and $k = 50$, respectively, where $m = 0.3$.

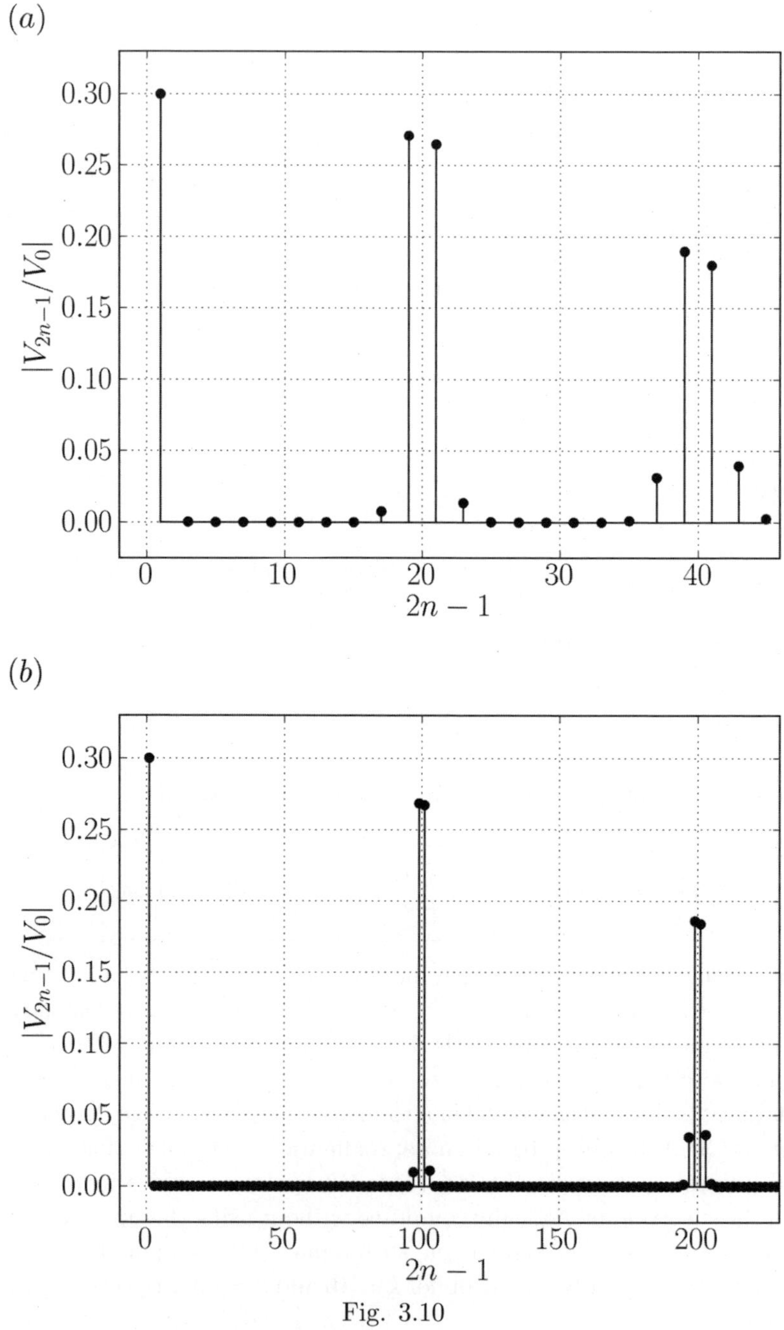
(a)
0.30
0.25
0.20
0.15
0.10
0.05
0.00
$|V_{2n-1}/V_0|$
0
10
20
30
40
$2n-1$
(b)
0.30
0.25
0.20
0.15
0.10
0.05
0.00
$|V_{2n-1}/V_0|$
0
50
100
150
200
$2n-1$

Fig. 3.10

The numerically computed spectrum for $k = 10$ and $m = 0.3$ is given in Table 1.

Table 1

$2n-1$	V_{2n-1}/V_0	$2n-1$	V_{2n-1}/V_0
1	**0.299917**	17	$7.70548{\cdot}10^{-3}$
3	$2.49512{\cdot}10^{-4}$	19	**0.270925**
5	$4.80463{\cdot}10^{-7}$	21	**−0.264744**
7	$1.19418{\cdot}10^{-9}$	23	$-1.36372 \cdot 10^{-2}$
9	$3.36710{\cdot}10^{-12}$	25	$-2.84004 \cdot 10^{-4}$
11	$1.67320{\cdot}10^{-11}$	27	$-3.75066 \cdot 10^{-6}$
13	$4.85871{\cdot}10^{-8}$	29	$-3.76500 \cdot 10^{-8}$
15	$3.82028{\cdot}10^{-5}$	31	$-6.38020 \cdot 10^{-8}$

Next, we shall use formula (3.69) in the analysis of the electric circuit shown in Figure 3.2. We shall treat each term in (3.69) as an ac voltage source of frequency $(2rk \pm 1)\omega$ with $r = 0, 1, \ldots$, and by using the superposition principle and ac steady-state analysis, we derive the following expression for the current $i(t)$:

$$i(t) = \frac{mV_0}{\sqrt{R^2 + \omega^2 L^2}} \sin(\omega t - \varphi) + mV_0 \sum_{r=1}^{\infty} \frac{(\pm)\,(-1)^r}{\sqrt{R^2 + (2rk \pm 1)^2 \omega^2 L^2}} \sin\left[(2rk \pm 1)\omega t + \varphi_r^{\pm}\right], \tag{3.70}$$

where

$$\tan\varphi = \frac{\omega L}{R}, \tag{3.71}$$

$$\tan\varphi_r^{\pm} = \frac{(2rk \pm 1)\omega L}{R}. \tag{3.72}$$

For sufficiently large number k of pulses (see (3.30)), we have

$$(2rk \pm 1)\omega L \gg \omega L \quad \text{for all} \quad r \geq 1, \tag{3.73}$$

and all terms in the sum in formula (3.70) are small and can be neglected. This leads to

$$i(t) \approx \frac{mV_0}{\sqrt{R^2 + \omega^2 L^2}} \sin(\omega t - \varphi) \tag{3.74}$$

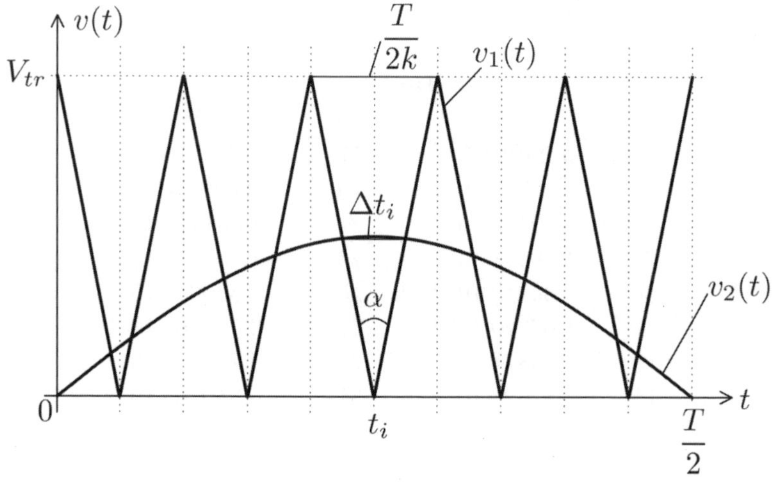

Fig. 3.11

and

$$\boxed{v_{out}(t) \approx \frac{mV_0R}{\sqrt{R^2+\omega^2L^2}}\sin(\omega t-\varphi).} \tag{3.75}$$

Thus, PWM leads to practically sinusoidal output voltage $v_{out}(t)$, and its peak value can be controlled by varying the modulation index m.

Next, we shall briefly discuss how PWM voltages can be generated. Both analog and digital techniques can be used for this purpose. One analog technique is to generate a triangular (low-voltage) waveform $v_1(t)$ with the frequency which is $2k$ times the inverter output frequency as well as to generate a low-voltage sinusoidal (modulating) function $v_2(t)$ (see Figure 3.11):

$$v_1(t) = v_{tr}(t), \tag{3.76}$$

$$v_2(t) = V_m \sin \omega t. \tag{3.77}$$

These low-voltage (and low-power) waveforms can be generated by using operational amplifiers, for instance. The difference of these waveforms

$$v_G(t) = v_2(t) - v_1(t) \tag{3.78}$$

can be used as a voltage controlling the switching of the transistors in the inverter circuit. For instance, this voltage $v_G(t)$ can be applied as a gate voltage in the IGBTs which are frequently used in inverters. These transistors will be in "on" states and produce rectangular pulses with time durations equal to the time intervals where

$$v_G > 0. \tag{3.79}$$

According to Figure 3.11, these time intervals Δt_i can be computed as

$$\Delta t_i \approx \alpha V_m \sin \omega t_i. \tag{3.80}$$

On the other hand,

$$\alpha \approx \frac{T/(2k)}{V_{tr}}. \tag{3.81}$$

By substituting the last formula in (3.80), we find

$$\Delta t_i \approx \frac{V_m}{V_{tr}} \frac{T}{2k} \sin \omega t_i. \tag{3.82}$$

It is apparent that the last formula coincides with formula (3.27) when the modulation index is defined as

$$\boxed{m = \frac{V_m}{V_{tr}}.} \tag{3.83}$$

Thus, by controlling the ratio of peak values of the sinusoidal and triangular waveforms, the modulation index m and, consequently, the peak value of the sinusoidal output voltage $v_{out}(t)$ (see (3.75)) can be controlled.

The presented discussion of pulse width modulation is based on the frequency-domain technique and leads to the formulas (3.70), (3.74) and (3.75) which are approximate in nature. The origin of the approximate nature of these formulas can be traced back to equation (3.34). It turns out that *exact* and *explicit* analytical expressions for $i(t)$ and $v_{out}(t)$ can be derived by using the time-domain technique, and this actually can be done when $v_{eq}(t)$ is any periodic sequence (train) of rectangular voltage pulses with half-wave symmetry. Consider the positive half-cycle $0 < t < \frac{T}{2}$. For this time interval $v_{eq}(t)$ can be represented by the formula

$$v_{eq}(t) = \begin{cases} 0, & \text{if } t_{2j} < t < t_{2j+1}, \\ V_0, & \text{if } t_{2j+1} < t < t_{2j+2}, \end{cases} \tag{3.84}$$

where $j = 0, 1, ..., k$ and

$$t_0 = 0, \qquad t_{2k+1} = \frac{T}{2}. \tag{3.85}$$

It is clear that the current $i(t)$ in the electric circuit shown in Figure 3.2 is given by the equations

$$i(t) = A_{2j+1} e^{-\frac{R}{L}t}, \qquad t_{2j} < t < t_{2j+1}, \tag{3.86}$$

$$i(t) = \frac{V_0}{R} + A_{2j+2}e^{-\frac{R}{L}t}, \qquad t_{2j+1} < t < t_{2j+2}, \tag{3.87}$$

where constants A_{2j+1} and A_{2j+2} must be determined from the continuity of electric current $i(t)$ at times t_{2j} and t_{2j+1} as well as from the "antiperiodic" boundary condition (3.4). This leads, respectively, to the following simultaneous equations:

$$A_2 - A_1 = -\frac{V_0}{R}e^{\frac{Rt_1}{L}}, \tag{3.88}$$

$$A_3 - A_2 = \frac{V_0}{R}e^{\frac{Rt_2}{L}}, \tag{3.89}$$

$$\vdots$$

$$A_{2j} - A_{2j-1} = -\frac{V_0}{R}e^{\frac{Rt_{2j-1}}{L}}, \tag{3.90}$$

$$A_{2j+1} - A_{2j} = \frac{V_0}{R}e^{\frac{Rt_{2j}}{L}}, \tag{3.91}$$

$$\vdots$$

$$A_{2k+1} - A_{2k} = \frac{V_0}{R}e^{\frac{Rt_{2k}}{L}} \tag{3.92}$$

and

$$A_1 + A_{2k+1}e^{-\frac{RT}{2L}} = 0. \tag{3.93}$$

By summing up all equations from (3.88) to (3.92), we find

$$A_{2k+1} - A_1 = \frac{V_0}{R}\sum_{j=1}^{2k}(-1)^j e^{\frac{Rt_j}{L}}. \tag{3.94}$$

By solving the two simultaneous equations (3.93) and (3.94), we derive

$$\boxed{A_{2k+1} = \frac{V_0}{R}\frac{\sum_{j=1}^{2k}(-1)^j e^{\frac{Rt_j}{L}}}{1+e^{-\frac{RT}{2L}}},} \tag{3.95}$$

$$\boxed{A_1 = -\frac{V_0}{R}\frac{\sum_{j=1}^{2k}(-1)^j e^{\frac{Rt_j}{L}}}{1+e^{-\frac{RT}{2L}}}e^{-\frac{RT}{2L}}.} \tag{3.96}$$

Having found A_1, all other A-coefficients can be computed by using the formula

$$\boxed{A_j = A_1 + \frac{V_0}{R}\sum_{n=1}^{j-1}(-1)^n e^{\frac{Rt_n}{L}}.} \tag{3.97}$$

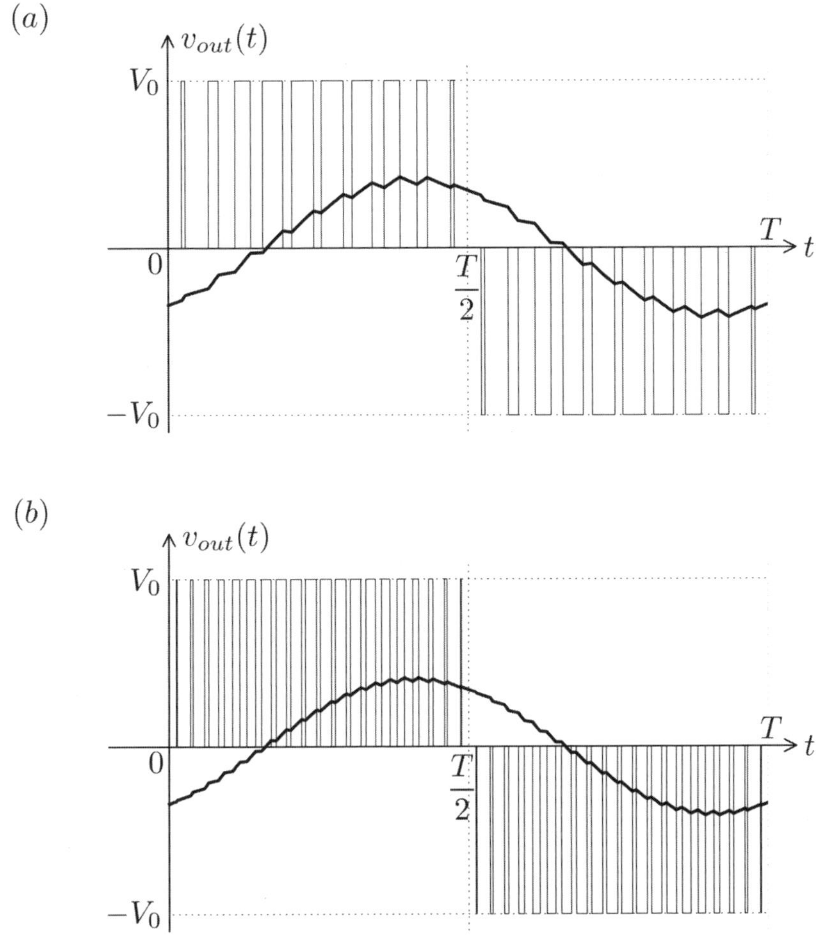

Fig. 3.12

Thus, by means of the last three formulas as well as equations (3.86) and (3.87), we can compute the current $i(t)$ as well as the output voltage $v_{out}(t)$. It is clear that this voltage is a function of V_0, R, L and t_1, t_2, ..., t_{2k}:

$$v_{out}(t) = F(t, V_0, R, L, t_1, t_2, ..., t_{2k}). \tag{3.98}$$

Examples of $v_{out}(t)$ computed by using the presented formulas are shown in Figure 3.12 for $k = 10$ (Figure 3.12a) and $k = 20$ (Figure 3.12b).

It is interesting to point out that, in the time domain, the problem of pulse width modulation can be stated as the following optimization problem: find times t_1, t_2, ..., t_{2k} (and possibly R and L) such that the function

$v_{out}(t)$ in (3.98) gives the best (in some sense) approximation to the desired function. For inverters, the desired function is naturally chosen as

$$\tilde{v}_{out}(t) = V_m \sin(\omega t - \varphi). \tag{3.99}$$

In particular, the least square approximation can be chosen for the selection of t_j. In this case, the problem is reduced to the minimization of the following integral functional:

$$\min_{(t_1,\ldots,t_{2k})} \int_0^T |F(t, V_0, R, L, t_1, t_2, ..., t_{2k}) - V_m \sin(\omega t - \varphi)|^2 dt. \tag{3.100}$$

This problem can be handled by using existing minimization techniques. Further discussion of this matter is beyond the scope of this text.

3.3 Three-Phase Inverters; AC-to-AC Converters and AC Motor Drives

Three-phase inverters are widely used in ac motor drives as well as in uninterruptible ac power supplies for three-phase loads. Such inverters are designed to produce sinusoidal three-phase ac outputs whose frequencies and voltage peak values can both be controlled. It is possible to design three-phase inverters as "parallel" connections of three single-phase inverters discussed in the previous sections. In such designs, each of these three inverters produces an ac output voltage of the same peak value and frequency, but phase-shifted in time by 120° with respect to one another. The latter can be achieved by using three identical reference sinusoidal voltages $v_2(t)$ in pulse width modulation (see Figure 3.11) but phase-shifted in time by 120°. Although the described design is conceptually simple, it often requires three-phase transformers as a link between such inverters and three-phase loads as well as twelve switches. There exists another conceptual design in which a three-phase bridge with six switches is used. This design is illustrated by Figure 3.13. In this figure, switches are assumed to be bilateral (bidirectional). Such switches are designed in the same way as discussed in the first section of this chapter, namely by using transistors connected in parallel with freewheeling diodes. Different patterns of periodical switching are possible in the circuit shown in Figure 3.13. First, we discuss a simple (i.e., without pulse width modulation) pattern of switching that is repeated during each time period T. As before, the choice of T determines the frequency of output ac three-phase voltages. In this simple pattern, the six switches continuously conduct during different time

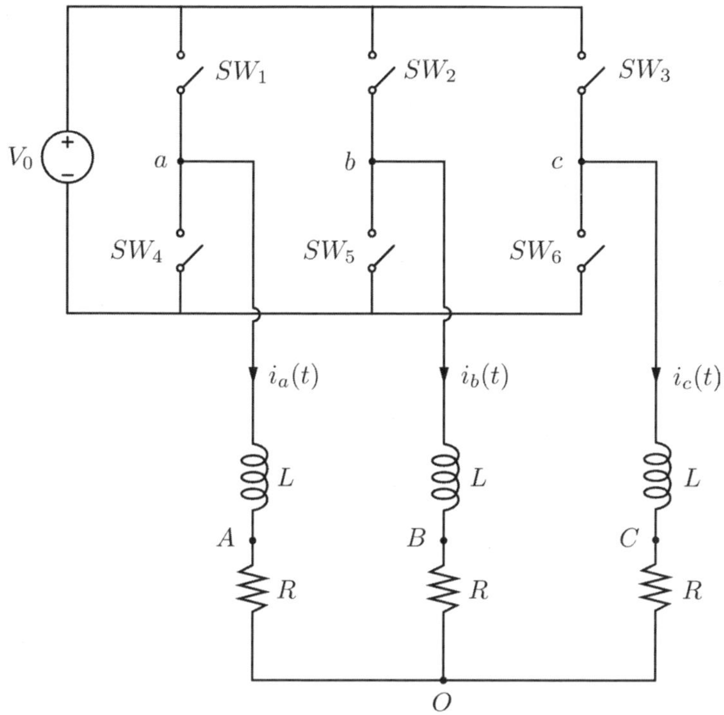

Fig. 3.13

intervals of duration $\frac{T}{2}$ and at each instant of time there are only three conducting switches. This switching pattern is fully specified by Table 2. Next, we shall find voltages at nodes a, b and c (see Figure 3.13) at each time interval in Table 2. It is apparent that during the time interval $(0, \frac{T}{6})$ when switches SW_1, SW_3 and SW_5 are in the "on" (conducting) state, phases a and c are connected in parallel and then in series with the phase b. This is illustrated by Figure 3.14. Since it is assumed that all three phases have identical inductances and resistances, it is easy to conclude from Figure 3.14 that

$$v_{aO}(t) = \frac{V_0}{3}, \qquad v_{bO}(t) = -\frac{2V_0}{3}, \qquad v_{cO}(t) = \frac{V_0}{3} \tag{3.101}$$

during the time interval $(0, \frac{T}{6})$.

Next, we consider the time interval $(\frac{T}{6}, \frac{T}{3})$ during which (according to Table 2) switches SW_1, SW_5 and SW_6 are in the "on" (conducting) state. This implies that during the above time interval phases b and c are

Table 2

Time Interval	Conducting Switches
$\left(0, \frac{T}{6}\right)$	SW_1, SW_3, SW_5
$\left(\frac{T}{6}, \frac{T}{3}\right)$	SW_1, SW_5, SW_6
$\left(\frac{T}{3}, \frac{T}{2}\right)$	SW_1, SW_2, SW_6
$\left(\frac{T}{2}, \frac{2T}{3}\right)$	SW_2, SW_4, SW_6
$\left(\frac{2T}{3}, \frac{5T}{6}\right)$	SW_2, SW_3, SW_4
$\left(\frac{5T}{6}, T\right)$	SW_3, SW_4, SW_5

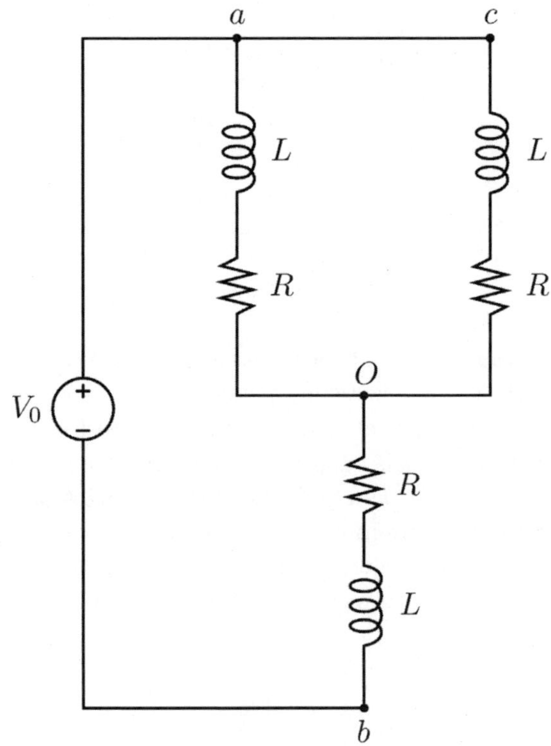

Fig. 3.14

connected in parallel and then in series with phase a. This is illustrated by Figure 3.15. Since all three phases have identical parameters, it is easy to

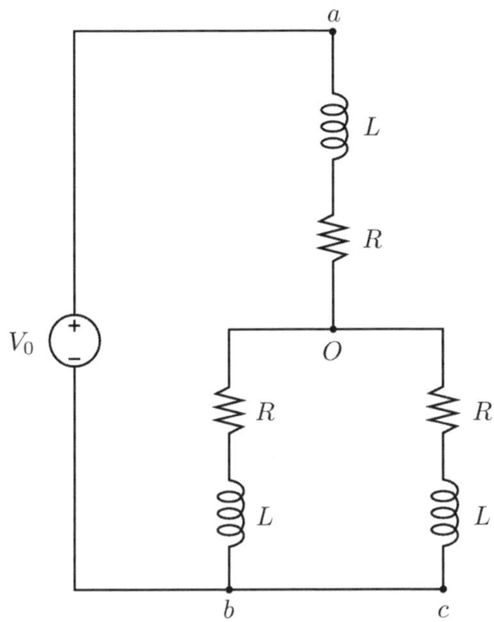

Fig. 3.15

conclude from Figure 3.15 that

$$v_{aO}(t) = \frac{2V_0}{3}, \qquad v_{bO}(t) = -\frac{V_0}{3}, \qquad v_{cO}(t) = -\frac{V_0}{3}. \tag{3.102}$$

By repeating the same line of reasoning as before, it is easy to find the values of voltages $v_{aO}(t)$, $v_{bO}(t)$ and $v_{cO}(t)$ in all other time intervals. These values are given in Table 3. These values are also graphically represented in Figure 3.16. It is evident from this figure that voltages $v_{aO}(t)$, $v_{bO}(t)$ and $v_{cO}(t)$ are identical but progressively shifted in time by $\frac{T}{3}$. Namely,

$$v_{bO}(t) = v_{aO}\left(t - \frac{T}{3}\right), \tag{3.103}$$

$$v_{cO}(t) = v_{aO}\left(t - \frac{2T}{3}\right). \tag{3.104}$$

It is also apparent that at any instant of time t, we have

$$v_{aO}(t) + v_{bO}(t) + v_{cO}(t) = 0. \tag{3.105}$$

By using the time-domain technique and the same reasoning as in the derivation of formulas (3.95), (3.96) and (3.97), the explicit analytical exponential-type expressions for $i_a(t)$ and $v_{AO}(t) = Ri_a(t)$ can be obtained. A typical graph of $v_{AO}(t)$ computed this way is shown in Figure 3.17. It is clear that $v_{BO}(t)$ and $v_{CO}(t)$ can be obtained from the graph in Figure 3.17 by shifting (translating) it in time by $\frac{T}{3}$ and $\frac{2T}{3}$, respectively. It is noticeable that $v_{AO}(t)$ is somewhat close to a sinusoidal waveform. This has been achieved by using only two-level (two-step) approximations for positive (or negative) half-cycles of phase voltages (see Figure 3.16). The closeness to sinusoidal waveforms can be improved by using multilevel inverters with a larger number of switches. However, three-phase inverters with switching strategies based on pulse width modulation and having the advantage of simultaneous control of frequency and peak value of output voltages are preferable nowadays.

The pulse width modulation switching in three-phase inverters can be accomplished in many different ways. One of these ways is very similar to the switching discussed in the previous section and illustrated by Figure 3.11. It is based on comparison of the triangular waveform $v_1(t)$ with three reference sinusoidal waveforms $v_2^{(a)}(t)$, $v_2^{(b)}(t)$ and $v_2^{(c)}(t)$ which are identical but shifted in time with respect to one another by $\frac{T}{3}$ (or 120°). Then, the switches SW_1 and SW_4 in the bridge branch with node a (see Figure 3.13) are controlled in the following way. If the sinusoid $v_2^{(a)}(t)$ is larger than the triangular waveform, then switch SW_1 is closed while switch SW_4 is open. On the other hand, if the sinusoid $v_2^{(a)}(t)$ is less than the triangular waveform, then the switch SW_1 is open, while the switch SW_4 is closed. The switches SW_2 and SW_5, SW_3 and SW_6 are controlled in the similar way by comparing the triangular waveform with the sinusoids $v_2^{(b)}(t)$ and $v_2^{(c)}(t)$, respectively. The described pattern of switching will result in the

Table 3

Time Interval	$v_{aO}(t)$	$v_{bO}(t)$	$v_{cO}(t)$
$\left(0, \frac{T}{6}\right)$	$\frac{V_0}{3}$	$-\frac{2V_0}{3}$	$\frac{V_0}{3}$
$\left(\frac{T}{6}, \frac{T}{3}\right)$	$\frac{2V_0}{3}$	$-\frac{V_0}{3}$	$-\frac{V_0}{3}$
$\left(\frac{T}{3}, \frac{T}{2}\right)$	$\frac{V_0}{3}$	$\frac{V_0}{3}$	$-\frac{2V_0}{3}$
$\left(\frac{T}{2}, \frac{2T}{3}\right)$	$-\frac{V_0}{3}$	$\frac{2V_0}{3}$	$-\frac{V_0}{3}$
$\left(\frac{2T}{3}, \frac{5T}{6}\right)$	$-\frac{2V_0}{3}$	$\frac{V_0}{3}$	$\frac{V_0}{3}$
$\left(\frac{5T}{6}, T\right)$	$-\frac{V_0}{3}$	$-\frac{V_0}{3}$	$\frac{2V_0}{3}$

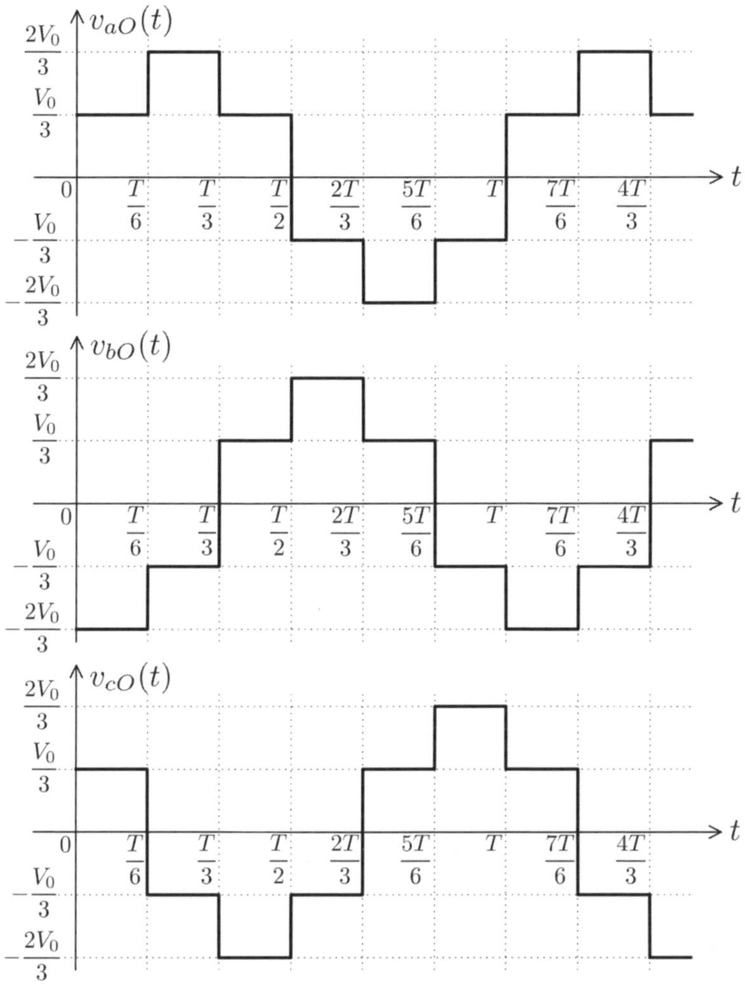

Fig. 3.16

trains of rectangular pulses for the potentials of nodes a, b and c. From these trains of rectangular pulses, the trains of rectangular pulses for line voltages $v_{ab}(t)$, $v_{ca}(t)$ and $v_{bc}(t)$ can be constructed and their Fourier analysis can be performed in a similar way as discussed in the previous section. This analysis will reveal that lower-order harmonics (except the fundamental) in the Fourier spectra are suppressed, while the higher-order harmonics are suppressed by phase inductances (see Figure 3.13).

It is worthwhile to point out that the problem of pulse width modulation

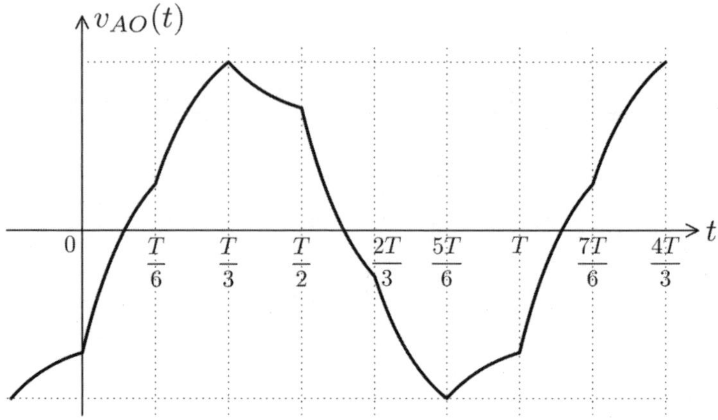

Fig. 3.17

can also be treated in the time domain in a way similar to that which was discussed at the end of the last section. In the time domain, we deal with the solution of the following Kirchhoff equations (see Figure 3.13):

$$L\frac{di_a(t)}{dt} + Ri_a(t) = v_a(t) - v_O(t), \tag{3.106}$$

$$L\frac{di_b(t)}{dt} + Ri_b(t) = v_a\left(t - \frac{T}{3}\right) - v_O(t), \tag{3.107}$$

$$L\frac{di_c(t)}{dt} + Ri_c(t) = v_a\left(t - \frac{2T}{3}\right) - v_O(t), \tag{3.108}$$

$$i_a(t) + i_b(t) + i_c(t) = 0, \tag{3.109}$$

where $v_a(t)$, $v_a\left(t - \frac{T}{3}\right)$ and $v_a\left(t - \frac{2T}{3}\right)$ are the potentials of nodes a, b and c, respectively, while $v_O(t)$ is the potential of node O.

By summing up equations (3.106), (3.107) and (3.108), we find

$$L\frac{d}{dt}\left[i_a(t) + i_b(t) + i_c(t)\right] + R\left[i_a(t) + i_b(t) + i_c(t)\right]$$
$$= v_a(t) + v_a\left(t - \frac{T}{3}\right) + v_a\left(t - \frac{2T}{3}\right) - 3v_O(t). \tag{3.110}$$

By taking into account formula (3.109) in the last equation, we obtain

$$v_O(t) = \frac{1}{3}\left[v_a(t) + v_a\left(t - \frac{T}{3}\right) + v_a\left(t - \frac{2T}{3}\right)\right]. \tag{3.111}$$

By substituting the last expression into formulas (3.106), (3.107) and (3.108) we end up with differential equations whose right-hand sides are expressed in terms of $v_a(t)$ and its shifts. It is easy to see that these equations

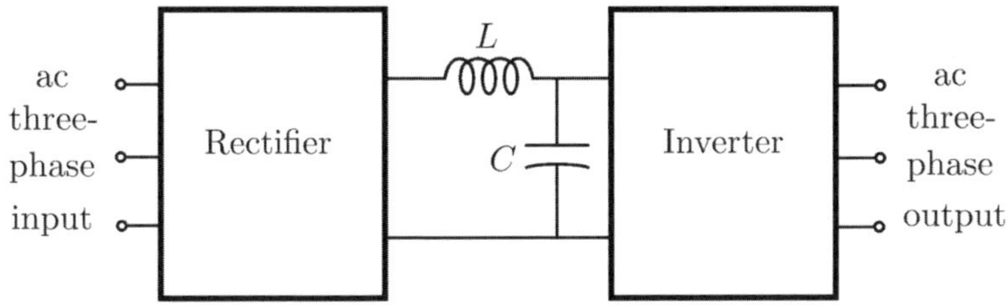

Fig. 3.18

exhibit (as expected) symmetry under translation by $\frac{T}{3}$. More importantly, these equations can be analytically solved for any train of rectangular pulses in $v_a(t)$ using the technique described at the end of the previous section and the solutions are exponential functions of switching times. By using these analytical solutions, the problem of pulse width modulation can be stated as the optimization problem of finding such switching times that will provide (in some chosen sense) the best approximation for desired ac three-phase voltages with respect to nodes A, B and C (see Figure 3.13).

At the end of this section, we shall briefly discuss ac-to-ac converters and their application in ac motor drives. One very widely used way to construct three-phase ac-to-ac converters is by cascading three-phase rectifiers with three-phase inverters. Such a cascading is illustrated by Figure 3.18. In this figure, there is an LC link between the rectifier and inverter, and such cascades are often called *dc link converters*. In such converters usually three-phase diode rectifiers are used, while the controllability of peak value of ac output voltage (and its frequency) is achieved by using PWM inverters. The link capacitors are used to maintain more or less constant input voltages for the inverters because voltages across capacitors are continuous functions of time and cannot change abruptly. The link inductors are used to suppress higher-order (high-frequency) harmonics generated by pulse width modulation in the inverters and, in this way, to isolate the rectifiers and power systems from their detrimental effects. In some cases, the link inductors are not used because they increase the overall size, weight and cost of converters and may negatively affect the input voltages of inverters.

Ac-to-ac converters are used in ac motor drives for frequency control of speed of induction and synchronous motors. First, we briefly consider the case of induction motors. The electromagnetic torque-speed characteristics of such motors for some fixed frequency f are shown in Figure 3.19 by the continuous bold line 1 (consult on this matter Chapter 6 of Part II). It is

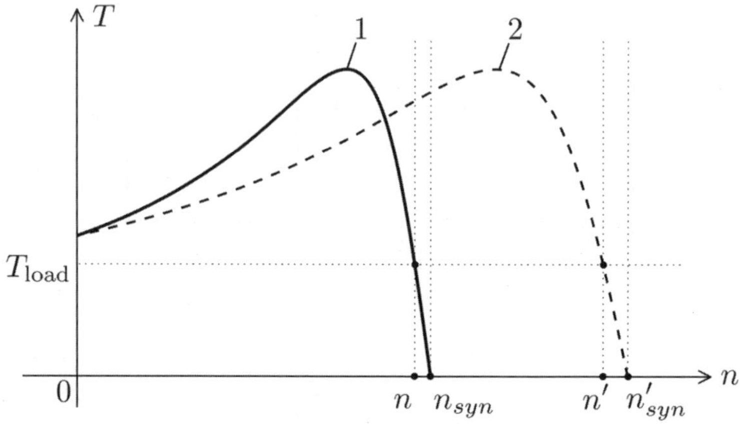

Fig. 3.19

clear from this figure that the actual rotor speed n of the induction motor is attained when the electromagnetic torque is equal to the load torque. It is also clear that this speed is very close to the synchronous (i.e., stator magnetic field) speed n_{syn} which is controlled by the frequency f of the stator ac voltage according to the formula

$$n \approx n_{syn} = \frac{120f}{p}, \tag{3.112}$$

where p is the number of poles of the stator winding. As the frequency of the power supply to the stator winding of the induction motor is increased to some value f', this results in the increase of synchronous speed and in the modification of the torque-speed characteristics as shown in Figure 3.19 by the dashed line 2. This, in turn, results in the increase of the mechanical rotor speed of the induction motor

$$n' \approx n'_{syn} = \frac{120f'}{p}. \tag{3.113}$$

The presented discussion clearly reveals the mechanisms of frequency control of speed of induction motors. In the case of synchronous motors (and, specifically, in the case of synchronous motors with permanent magnets on rotors) the rotor speed coincides with the synchronous speed,

$$n = n_{syn} = \frac{120f}{p}, \tag{3.114}$$

which implies the exact controllability of speed of synchronous motors by means of proper frequency variation. This *exact* controllability can be of importance in some applications as is the case, for instance, of spindle motors of hard disk drives.

In the case of ac motors, the electromagnetic torques are determined by stator magnetic fields. These fields are (roughly) proportional to the magnetic flux linkages of the stator windings which, in turn, are determined by the voltages applied to the stator windings according to the formula

$$v(t) = \frac{d\psi(t)}{dt}, \tag{3.115}$$

where $\psi(t)$ stands for stator winding flux linkages.

In the case when

$$v(t) = V_m \cos \omega t, \tag{3.116}$$

from formula (3.115) we find

$$\psi(t) = \frac{V_m}{\omega} \sin \omega t \tag{3.117}$$

and

$$\psi_m = \frac{V_m}{\omega}. \tag{3.118}$$

It is clear from the last formula that in order to maintain electromagnetic torque more or less constant as the frequency is varied in order to control the motor speed, the ratio in the right-hand side of formula (3.118) must be maintained constant:

$$\boxed{\frac{V_m}{\omega} = const.} \tag{3.119}$$

This is usually called the *constant volts per hertz criterion.*

Modern ac-to-ac (dc link) converters with pulse width modulation in inverters usually meet this constant volts per hertz requirement because the PWM techniques allow for efficient and simultaneous control of frequency and peak value of ac three-phase output voltages.

Chapter 4

DC-to-DC Converters (Choppers)

4.1 Buck Converter

In this chapter, dc-to-dc converters are discussed. These converters find many applications in various areas of technology where several different levels of dc voltage are (simultaneously) required. Integrated circuits and electronic circuits, in general, are examples of such applications. There are many different designs of choppers. In this chapter, we discuss only the most basic and simplest versions of dc-to-dc converters with direct electric as well as magnetic coupling connections between input and output terminals.

We start with the study of the buck (step-down) chopper whose electric circuit is shown in Figure 4.1. This circuit contains five basic elements: transistor Tr, freewheeling diode D, inductor L, capacitor C and resistor R. The same five elements are used in the designs of boost and buck-boost choppers discussed in subsequent sections. However, these elements are differently interconnected in those choppers, resulting in their different performance.

The operation of the circuit shown in Figure 4.1 can be described as follows. Transistor Tr is periodically turned "on" and "off." When the transistor is "on," the diode is reverse biased and "off," and voltage V_0 is applied across terminals a and b. When the transistor is "off," the freewheeling diode is automatically turned "on" to maintain the continuity of the current $i_L(t)$ through the inductor. This implies that when the transistor is "off," zero voltage appears across terminals a and b. Thus, the voltage source V_0, the transistor and the diode can be replaced by the equivalent voltage source $v_{eq}(t)$ as shown in Figures 4.2a and 4.2b, where T is the period of repeated switching, while D is the fraction of this period when the transistor is "on." This quantity is termed the "duty factor." It

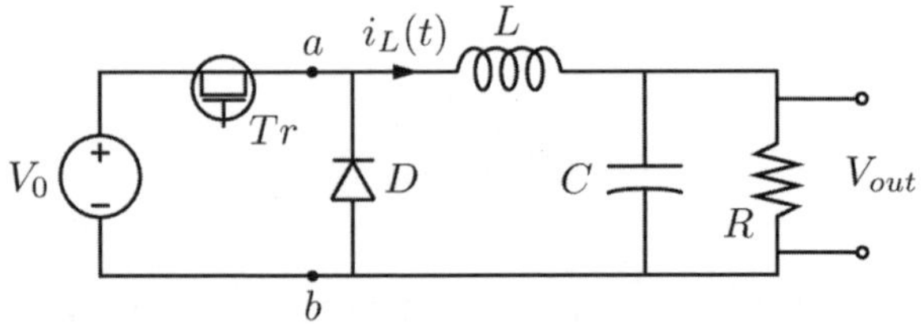

Fig. 4.1

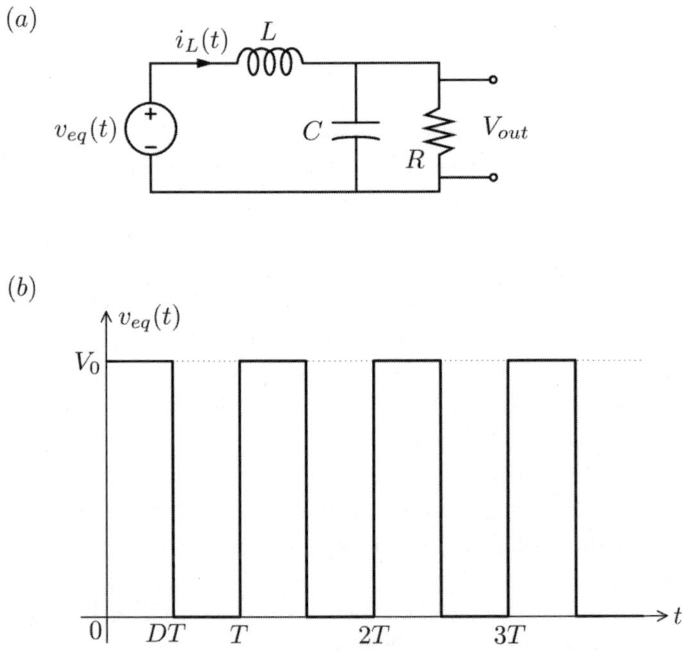

Fig. 4.2

is evident from Figure 4.2b that as a result of periodic switching the input dc voltage is "chopped" and transformed into a periodic sequence (train) of identical rectangular pulses. This explains why dc-to-dc converters are called choppers.

It turns out that the circuit shown in Figure 4.1 has two distinct modes of operation: "*continuous*" mode when the current $i_L(t)$ is always strictly

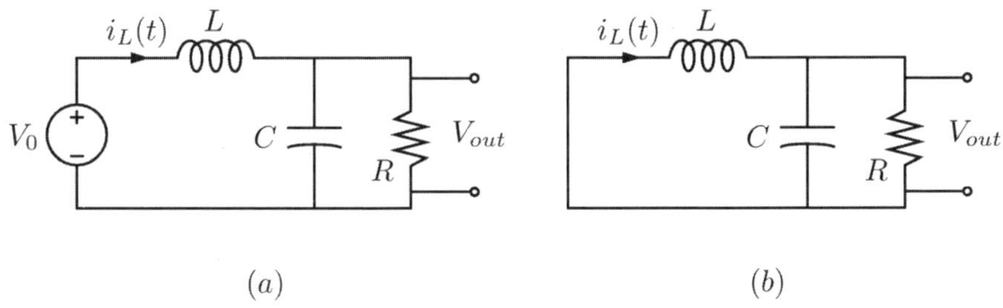

Fig. 4.3

positive,

$$i_L(t) > 0, \tag{4.1}$$

and the "*discontinuous*" mode when the current $i_L(t)$ reaches zero during the time interval when the transistor is "off" and remains equal to zero until the transistor is turned "on" again. This zero value is maintained because the flow of the current $i_L(t)$ in the direction opposite to the one shown in Figure 4.1 is prohibited by the diode. Thus, the discontinuous mode of operation is characterized by the formula

$$i_L(t) \geq 0, \tag{4.2}$$

and the equality in this formula is realized during some time interval when the transistor is "off."

It turns out that the performance of the chopper is quite different for these two modes of operation. For this reason, these two modes are discussed below separately. We shall first start with the discussion of the continuous mode of operation and we assume that the capacitance C is large enough that the ripple of the voltage across this capacitor (as well as across the resistor R) is negligible. In other words, it is assumed that

$$\boxed{v_C(t) = V_C = const > 0.} \tag{4.3}$$

If this is not the case, then the chopper is not properly designed because the output voltage V_{out} cannot be maintained constant. The assumption (4.3) is used throughout this chapter in the analysis of all dc-to-dc converters.

Figure 4.2b suggests that the circuit in Figure 4.2a can be transformed into two circuits shown in Figures 4.3a and 4.3b for the time intervals $(0, DT)$ and (DT, T), respectively. First, consider the time interval

$$0 < t < DT. \tag{4.4}$$

By using KVL for the circuit shown in Figure 4.3a, we find

$$L\frac{di_L(t)}{dt} + V_C = V_0, \tag{4.5}$$

which implies that

$$\frac{di_L(t)}{dt} = \frac{V_0 - V_C}{L} = const. \tag{4.6}$$

Similarly, for the time interval

$$DT < t < T, \tag{4.7}$$

from the circuit shown in Figure 4.3b we conclude that

$$L\frac{di_L(t)}{dt} + V_C = 0 \tag{4.8}$$

and

$$\frac{di_L(t)}{dt} = -\frac{V_C}{L} < 0. \tag{4.9}$$

Since we consider the steady-state performance of the buck chopper, the values of $i_L(t)$ at the beginning and at the end of one period must be the same. This implies that

$$i_L(0) = i_L(T). \tag{4.10}$$

The latter is only possible if the right-hand side of formula (4.6) is positive; otherwise, $i_L(t)$ would be monotonically decreasing throughout the period $[0, T]$ (see (4.9)) and the equality (4.10) is not possible. Thus, we conclude that

$$\frac{di_L(t)}{dt} = \frac{V_0 - V_C}{L} > 0 \qquad \text{and} \qquad V_0 > V_C. \tag{4.11}$$

The last inequality indicates that the chopper shown in Figure 4.1 is the step-down (buck) chopper. Next, by integrating equations (4.6) and (4.9), we find

$$i_L(t) = I_{min} + \frac{V_0 - V_C}{L}t, \qquad 0 < t < DT, \tag{4.12}$$

$$i_L(t) = I_{max} - \frac{V_C}{L}(t - DT), \qquad DT < t < T. \tag{4.13}$$

The plot of $i_L(t)$ is shown in Figure 4.4. From the last two formulas as well as Figure 4.4, we find

$$I_{max} = I_{min} + \frac{V_0 - V_C}{L}DT, \tag{4.14}$$

$$I_{min} = I_{max} - \frac{V_C}{L}(1 - D)T, \tag{4.15}$$

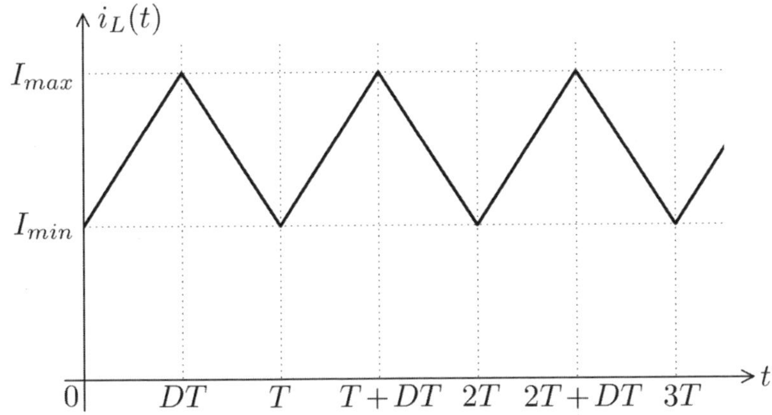

Fig. 4.4

which is equivalent to

$$I_{max} - I_{min} = \frac{V_0 - V_C}{L} DT, \tag{4.16}$$

$$I_{max} - I_{min} = \frac{V_C}{L}(1 - D)T. \tag{4.17}$$

From the last two formulas follows that

$$\frac{V_0 - V_C}{L} DT = \frac{V_C}{L}(1 - D)T, \tag{4.18}$$

which leads to

$$\boxed{V_{out} = V_C = DV_0.} \tag{4.19}$$

It is clear from the last relation that the circuit shown in Figure 4.1 is indeed a step-down (buck) converter and that the output voltage can be fully controlled by varying the duty factor D. The desired variations of the duty factor can be accomplished by using the pulse width modulation technique similar to the one discussed in the previous chapter. Namely, the switching of the transistor can be controlled by the difference between a reference dc voltage $v_2(t)$ and a triangular carrier waveform voltage $v_1(t)$ (see Figure 4.5). It is clear that by changing the level of $v_2(t)$ the width of the rectangular pulses (and duty factor D) can be effectively controlled. The input-output (transfer) relation has been derived under the assumption of the continuous mode of operation, that is, when the strict inequality (4.1) is valid. This inequality is valid if and only if

$$I_{min} > 0. \tag{4.20}$$

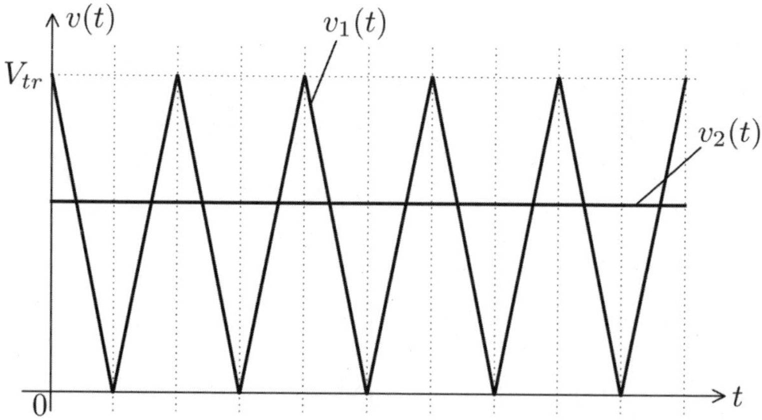

Fig. 4.5

Next, we shall discuss under what constraints on inductance L or duty factor D the last inequality is valid. To find these constraints, we shall carry out the analysis of electric currents in the chopper. According to KCL, we have

$$i_C(t) = i_L(t) - i_R(t). \tag{4.21}$$

It is apparent that

$$i_R(t) = \frac{V_C}{R} = \frac{DV_0}{R} \tag{4.22}$$

and

$$i_C(t) = C\frac{dv_C(t)}{dt}. \tag{4.23}$$

By substituting the last two formulas into equation (4.21), we find

$$C\frac{dv_C(t)}{dt} = i_L(t) - \frac{DV_0}{R}. \tag{4.24}$$

Now, we shall integrate both sides of the last equation over one period T. This yields

$$C\int_0^T \frac{dv_C(t)}{dt}dt = \int_0^T i_L(t)dt - \frac{DV_0}{R}T. \tag{4.25}$$

It is easy to see that

$$\int_0^T \frac{dv_C(t)}{dt}dt = v_C(T) - v_C(0) = 0, \tag{4.26}$$

because at the steady state $v_C(t)$ (as all other quantities) is a periodic function of time with period T. Furthermore, the second integral in formula (4.25) is equal to the area under the plot of $i_L(t)$ for one period T. From this fact and Figure 4.4, we find

$$\int_0^T i_L(t)dt = I_{min}T + \frac{I_{max} - I_{min}}{2}T, \tag{4.27}$$

or

$$\int_0^T i_L(t)dt = \frac{I_{max} + I_{min}}{2}T. \tag{4.28}$$

By using formulas (4.26) and (4.28) in equation (4.25), we derive

$$I_{max} + I_{min} = \frac{2DV_0}{R}. \tag{4.29}$$

On the other hand, from formulas (4.17) and (4.19) follows that

$$I_{max} - I_{min} = \frac{(1-D)DT}{L}V_0. \tag{4.30}$$

By adding equations (4.29) and (4.30), we arrive at

$$\boxed{I_{max} = DV_0\left(\frac{1}{R} + \frac{(1-D)T}{2L}\right).} \tag{4.31}$$

Similarly, by subtracting equations (4.29) and (4.30), we find

$$\boxed{I_{min} = DV_0\left(\frac{1}{R} - \frac{(1-D)T}{2L}\right).} \tag{4.32}$$

From the last formula we conclude that the inequality (4.20) holds and, consequently, the continuous mode of operation is realized if

$$\frac{1}{R} - \frac{(1-D)T}{2L} > 0. \tag{4.33}$$

It is easy to see that the last inequality is fulfilled for any value of duty factor D if the inductance L in the circuit shown in Figure 4.1 is sufficiently large, namely if

$$\boxed{L > \tilde{L} = \frac{RT}{2}.} \tag{4.34}$$

On the other hand, if the inductance L in the converter circuit is smaller than $\tilde{L}$, then the constraint can be imposed on the duty factor D to achieve

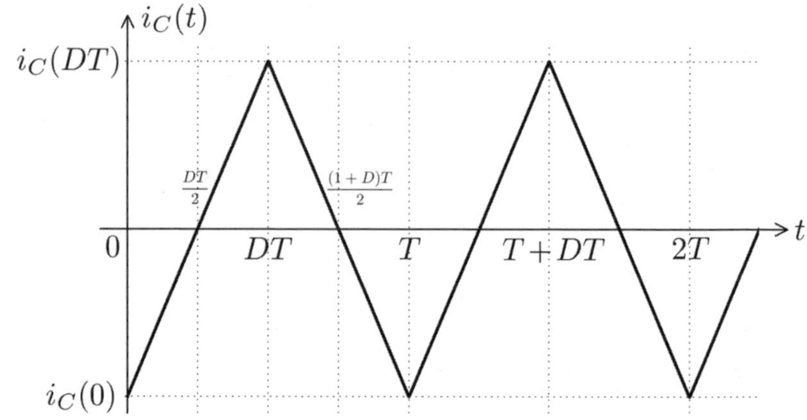

Fig. 4.6

the inequality (4.33) and, consequently, to guarantee the continuous mode of operation. From inequality (4.33), we find

$$\boxed{D > \tilde{D} = 1 - \frac{2L}{RT}.} \tag{4.35}$$

Formulas (4.31) and (4.32) can be used to fully describe the currents in the converter circuit. Indeed, by substituting these formulas into equations (4.12) and (4.13), we find explicit expressions for $i_L(t)$ for time intervals when the transistor is "on" and "off," respectively. The value of the current $i_R(t)$ is given by formula (4.22). Now, by using formulas (4.12), (4.13) and (4.22) in equation (4.21), we end up with the following expressions for the current $i_C(t)$ through the capacitor:

$$i_C(t) = I_{min} + \frac{(1-D)V_0}{L}t - \frac{DV_0}{R}, \qquad \text{if } 0 < t < DT, \tag{4.36}$$

$$i_C(t) = I_{max} - \frac{DV_0}{L}(t - DT) - \frac{DV_0}{R}, \qquad \text{if } DT < t < T. \tag{4.37}$$

The plot of $i_C(t)$ is presented in Figure 4.6, where

$$i_C(0) = -i_C(DT) = -\frac{V_0 D(1-D)T}{2L}, \tag{4.38}$$

as can be found from formulas (4.31), (4.32), (4.36) and (4.37).

By using the last figure, the ripple ΔQ of the electric charge of capacitor C can be computed as follows:

$$\Delta Q = \int_{\frac{DT}{2}}^{\frac{(1+D)T}{2}} i_C(t)dt = \frac{i_C(DT)}{2}\frac{T}{2} = \frac{V_0 D(1-D)T^2}{8L}. \tag{4.39}$$

From the last formula, the ripple ΔV_C of the voltage across the capacitor C can be evaluated as

$$\Delta V_C = \frac{\Delta Q}{C} = \frac{V_0 D(1-D)T^2}{8LC} = \frac{V_0 D(1-D)}{8f^2 LC}, \tag{4.40}$$

where $f = \frac{1}{T}$ is the frequency of switching. By taking into account formula (4.19), we further derive

$$\boxed{\frac{\Delta V_C}{V_C} = \frac{1-D}{8f^2 LC}.} \tag{4.41}$$

The last formula clearly reveals that the level of ripple in V_C and, consequently, in the output voltage V_{out} can be effectively suppressed by the proper choice of energy storage elements L and C as well as by increasing the switching frequency f. It is also evident from the last formula that there exists a trade-off between the switching frequency and the values of the energy storage elements.

Now, we proceed to the discussion of the *discontinuous* mode of operation of the buck chopper. In this mode of operation, the current $i_L(t)$ through the inductor is strictly positive only during the portion $D_1 T$ (with $D_1 > D$) of the switching cycle, i.e.,

$$i_L(t) > 0, \qquad \text{if } 0 < t < D_1 T, \tag{4.42}$$

and this current is equal to zero during the rest of the cycle, that is,

$$i_L(t) = 0, \qquad \text{if } D_1 T < t < T. \tag{4.43}$$

By using the same reasoning as before (see the derivation of equations (4.6) and (4.9)), we find

$$\frac{di_L(t)}{dt} = \frac{V_0 - V_C}{L}, \qquad \text{if } 0 < t < DT, \tag{4.44}$$

$$\frac{di_L(t)}{dt} = -\frac{V_C}{L}, \qquad \text{if } DT < t < D_1 T. \tag{4.45}$$

By integrating the last two equations, we obtain

$$i_L(t) = \frac{V_0 - V_C}{L} t, \qquad \text{if } 0 < t < DT, \tag{4.46}$$

$$i_L(t) = I_{max} - \frac{V_C}{L}(t - DT), \qquad \text{if } DT < t < D_1 T. \tag{4.47}$$

By using the last two equations and formula (4.43), the plot of $i_L(t)$ can be constructed as shown in Figure 4.7. It is clear from the last two formulas

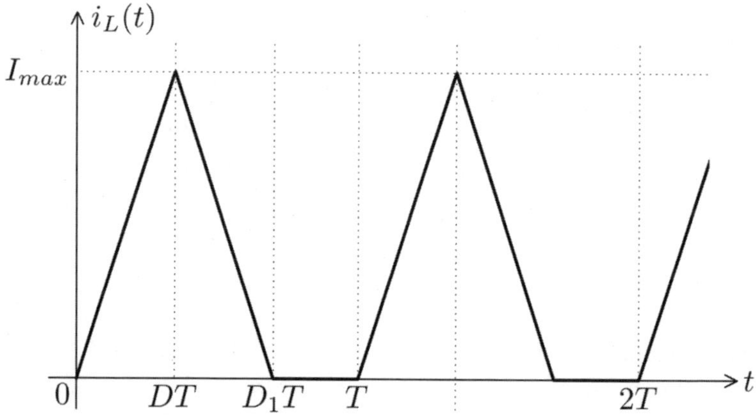

Fig. 4.7

as well as from the last figure that

$$I_{max} = \frac{V_0 - V_C}{L} DT, \tag{4.48}$$

$$I_{max} = \frac{V_C}{L}(D_1 - D)T. \tag{4.49}$$

These are two equations with respect to three unknowns I_{max}, V_C and D_1. Thus, we need an additional equation that can be derived by using the power balance condition that the average per period T input power P_{in} supplied by the input voltage source is equal to the average output power P_{out} dissipated across the resistor R. Power P_{in} can be computed as follows:

$$P_{in} = \frac{1}{T}\int_0^{DT} V_0 i_L(t)dt = \frac{V_0}{T}\int_0^{DT} i_L(t)dt = \frac{V_0 I_{max} D}{2}. \tag{4.50}$$

On the other hand,

$$P_{out} = \frac{V_C^2}{R}. \tag{4.51}$$

Since

$$P_{in} = P_{out}, \tag{4.52}$$

from formulas (4.50) and (4.51) we derive

$$\frac{V_0 I_{max} D}{2} = \frac{V_C^2}{R}. \tag{4.53}$$

Now, we have three equations (4.48), (4.49) and (4.53) with respect to three unknowns I_{max}, V_C and D_1. Our immediate goal is to find the expression for the output voltage $V_{out} = V_C$. To this end, we substitute formula (4.48) into the last equation to arrive at

$$\frac{V_0(V_0 - V_C)D^2T}{2L} = \frac{V_C^2}{R}. \tag{4.54}$$

This relation can be further transformed as

$$V_C^2 + \frac{D^2RT}{2L}V_0V_C - \frac{D^2RT}{2L}V_0^2 = 0. \tag{4.55}$$

We next introduce the non-dimensional parameter

$$\boxed{k = \frac{D^2RT}{4L}} \tag{4.56}$$

and write the quadratic equation (4.55) in the form

$$\left(\frac{V_C}{V_0}\right)^2 + 2k\left(\frac{V_C}{V_0}\right) - 2k = 0. \tag{4.57}$$

By solving the last equation and taking into account that $\frac{V_C}{V_0}$ is positive, we find

$$\frac{V_C}{V_0} = -k + \sqrt{k^2 + 2k}, \tag{4.58}$$

or

$$\boxed{\frac{V_C}{V_0} = k\left(\sqrt{1 + \frac{2}{k}} - 1\right).} \tag{4.59}$$

It is apparent from the last formula that the performance of the buck chopper in the discontinuous mode of operation is quite different from its performance in the continuous mode. Indeed, in the discontinuous mode, the output voltage depends not only on D as in the case of the continuous mode but on R, L and T as well, as it is evident from formulas (4.56) and (4.59). If the inductance L in the converter circuit is smaller than $\tilde{L}$ (see formula (4.34)), then depending on the value of D the continuous or discontinuous modes of operation can be realized. This is illustrated by Figure 4.8, where line 1 corresponds to the continuous mode (see formula (4.19)), while curve 2 computed by using equation (4.59) corresponds to the discontinuous mode. The value $\tilde{D}$ is defined by formula (4.35).

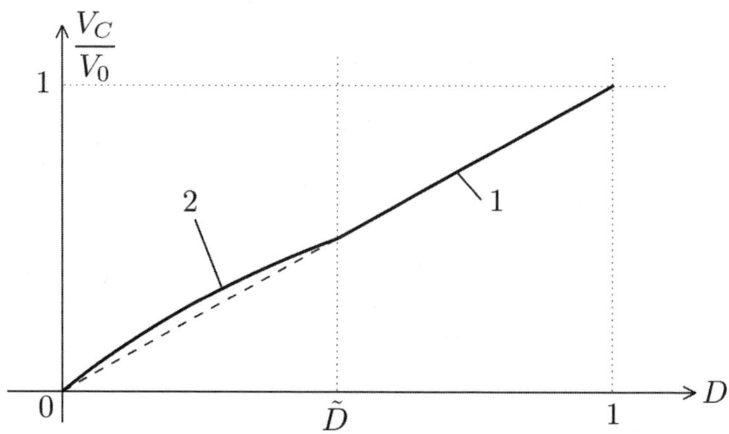

Fig. 4.8

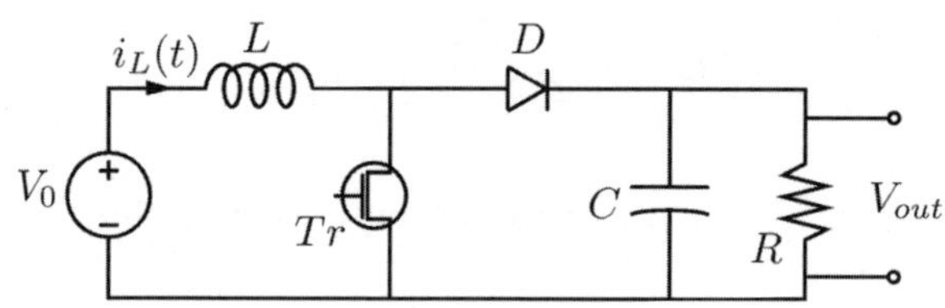

Fig. 4.9

4.2 Boost Converter

In this section, we shall discuss the boost (step-up) converter in which the controllable output dc voltage is larger than the input dc voltage. The electric circuit of this converter is shown in Figure 4.9. The operation of this circuit can be described as follows. Transistor Tr is periodically turned "on" and "off." Consider one period $[0, T]$ of this switching. When the transistor is "on" during the time interval

$$0 < t < DT, \tag{4.60}$$

the diode is reverse biased and in the "off" state. This implies that during this time interval the circuit shown in Figure 4.9 is reduced to the circuit shown in Figure 4.10a. When the transistor is "off" during the time interval

$$DT < t < T, \tag{4.61}$$

the freewheeling diode is turned "on" to maintain the continuity of the current $i_L(t)$ through the inductor. This implies that for this time interval

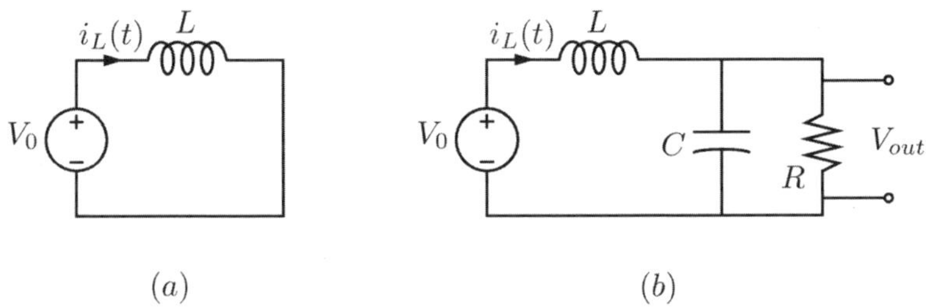

Fig. 4.10

the circuit shown in Figure 4.9 is reduced to the one shown in Figure 4.10b. The boost chopper, similar to the buck chopper, has two distinct modes of operation: *continuous* mode and *discontinuous* mode. We shall first discuss the continuous mode of operation when the inequality (4.1) is valid for any instant of time. It will be assumed in our discussion that the capacitance C in the electric circuit shown in Figure 4.9 is large enough that the ripple of the voltage across the capacitor is negligible and the relation (4.3) is quite accurate.

By using KVL for the circuit shown in Figure 4.10a, we find that

$$L\frac{di_L(t)}{dt} = V_0, \tag{4.62}$$

which implies that

$$\frac{di_L(t)}{dt} = \frac{V_0}{L} > 0, \qquad \text{if } 0 < t < DT. \tag{4.63}$$

Similarly, by using KVL for the circuit shown in Figure 4.10b, we conclude that

$$L\frac{di_L(t)}{dt} + V_C = V_0, \tag{4.64}$$

which leads to

$$\frac{di_L(t)}{dt} = \frac{V_0 - V_C}{L}, \qquad \text{if } DT < t < T. \tag{4.65}$$

Since we are interested in the steady-state performance of the boost chopper, from formulas (4.63) and (4.65) we conclude that

$$\frac{V_0 - V_C}{L} < 0. \tag{4.66}$$

Otherwise, $i_L(t)$ would be a monotonically increasing function of time throughout the entire period $[0, T]$, which is not possible at the steady state

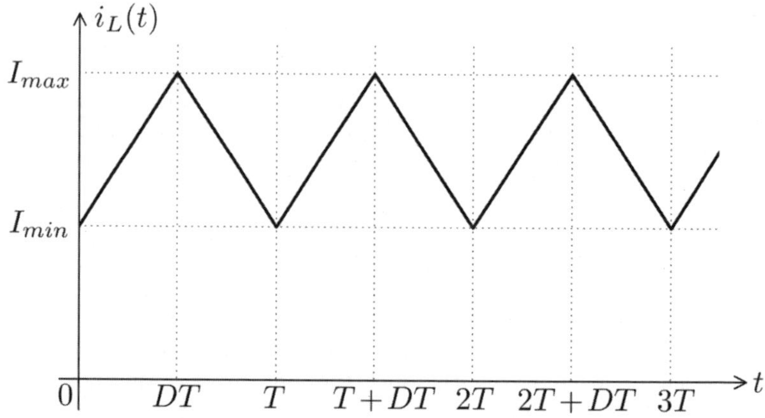

Fig. 4.11

when the equality (4.10) must be satisfied. The inequality (4.66) implies that

$$V_C > V_0 \tag{4.67}$$

and the chopper shown in Figure 4.9 is the boost (step-up) converter.

Next, by integrating equations (4.63) and (4.65), we obtain

$$i_L(t) = I_{min} + \frac{V_0}{L}t, \qquad \text{if } 0 < t < DT, \tag{4.68}$$

$$i_L(t) = I_{max} + \frac{V_0 - V_C}{L}(t - DT), \qquad \text{if } DT < t < T. \tag{4.69}$$

The plot of $i_L(t)$ is shown in Figure 4.11. From the last two formulas as well as Figure 4.11 we derive

$$I_{max} - I_{min} = \frac{V_0}{L}DT, \tag{4.70}$$

$$I_{max} - I_{min} = \frac{V_C - V_0}{L}(1 - D)T. \tag{4.71}$$

From relations (4.70) and (4.71) follows that

$$\frac{V_0}{L}DT = \frac{V_C - V_0}{L}(1 - D)T, \tag{4.72}$$

which is equivalent to

$$V_0 D = (V_C - V_0)(1 - D). \tag{4.73}$$

From the last formula we find the following transfer (input-output) relation:

$$\boxed{V_{out} = V_C = \frac{V_0}{1 - D}}. \tag{4.74}$$

It is clear from the last formula that the circuit shown in Figure 4.9 is the boost chopper and that by controlling the duty factor D the output voltage can be effectively controlled. It is also clear from the last formula that

$$\lim_{D\to 1} V_C = \infty. \tag{4.75}$$

The latter is clearly impossible. This unrealistic performance of the circuit shown in Figure 4.9 is due to the idealization of the inductor in this circuit when its resistance is entirely neglected. The detailed analysis, which accounts for this resistance, suggests that formula (4.74) may be valid for the following range of variation of the duty factor:

$$0 \le D \le 0.8. \tag{4.76}$$

The latter implies that

$$1 \le \frac{V_{out}}{V_0} \le 5. \tag{4.77}$$

The input-output (transfer) relation (4.74) has been derived under the tacit assumption that

$$I_{min} > 0, \tag{4.78}$$

which must be satisfied for the continuous mode of operation. Next, we shall discuss under what constraints on inductance L or duty factor D the last inequality is valid. To do this, we shall derive the explicit expression for I_{min}. To this end, we shall invoke the principle of power balance which implies that the average per period T input power P_{in} supplied to the converter by the voltage source V_0 is equal to the average output power P_{out} dissipated across the resistor R:

$$P_{in} = P_{out}. \tag{4.79}$$

Power P_{in} can be computed as follows:

$$P_{in} = \frac{1}{T}\int_0^T V_0 i_L(t)dt = \frac{V_0}{T}\int_0^T i_L(t)dt. \tag{4.80}$$

The last integral is equal to the area under the plot of $i_L(t)$ for one period T. From this fact and Figure 4.11, we find

$$\int_0^T i_L(t)dt = I_{min}T + \frac{I_{max} - I_{min}}{2}T = \frac{I_{max} + I_{min}}{2}T. \tag{4.81}$$

By using the last formula in equation (4.80), we obtain

$$P_{in} = \frac{V_0(I_{max} + I_{min})}{2}. \tag{4.82}$$

On the other hand, according to formula (4.74), we derive

$$P_{out} = \frac{V_C^2}{R} = \frac{V_0^2}{(1-D)^2 R}. \tag{4.83}$$

By substituting the last two formulas into equation (4.79), we arrive at

$$\frac{V_0(I_{max} + I_{min})}{2} = \frac{V_0^2}{(1-D)^2 R}, \tag{4.84}$$

which is tantamount to

$$I_{max} + I_{min} = \frac{2V_0}{(1-D)^2 R}. \tag{4.85}$$

Now, by adding equations (4.85) and (4.70), we find

$$\boxed{I_{max} = V_0 \left[\frac{1}{(1-D)^2 R} + \frac{DT}{2L} \right].} \tag{4.86}$$

Similarly, by subtracting equation (4.70) from (4.85), we obtain

$$\boxed{I_{min} = V_0 \left[\frac{1}{(1-D)^2 R} - \frac{DT}{2L} \right].} \tag{4.87}$$

From the last formula we conclude that the inequality (4.78) holds and the continuous mode of operation is realized if

$$\boxed{\frac{1}{(1-D)^2 R} - \frac{DT}{2L} > 0.} \tag{4.88}$$

It is easy to see that the last inequality is fulfilled for any value of duty factor D if the inductance L in the circuit shown in Figure 4.9 is sufficiently large, namely if

$$\boxed{L > \tilde{L} = \frac{RT}{2} \max_{0 \le D \le 0.8} \left[D(1-D)^2 \right].} \tag{4.89}$$

On the other hand, if the inductance L in the converter circuit is smaller than $\tilde{L}$, then the constraint can be imposed on the duty factor D to achieve the inequality (4.88) and, consequently, to guarantee the continuous mode of operation. Indeed, from inequality (4.88) we find

$$D(1-D)^2 < \frac{2L}{RT}. \tag{4.90}$$

To find the values of D for which the inequality (4.90) is valid, consider the function

$$f(D) = D(1-D)^2. \tag{4.91}$$

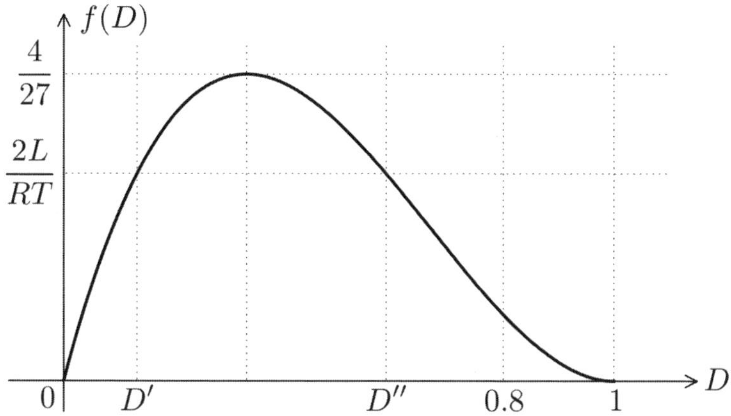

Fig. 4.12

It is clear that this function is positive for $0 < D < 1$ and that $f(0) = f(1) = 0$. Next, it is apparent that

$$f'(D) = (1 - 3D)(1 - D). \tag{4.92}$$

The latter implies that $f'(D) = 0$ for $D = \frac{1}{3}$ and that $f(D)$ achieves its maximum value for this value of D. This maximum value is

$$\max_{0 \le D \le 0.8} \left[D(1 - D)^2 \right] = f\left(\frac{1}{3}\right) = \frac{4}{27}. \tag{4.93}$$

Now, the plot of $f(D)$ can be easily constructed and it is represented by the continuous line in Figure 4.12. By using this figure, it is easy to conclude that the inequality (4.90) is valid and, consequently, the continuous mode of operation of the boost chopper is realized for the values of D that are within the intervals

$$0 < D < D', \tag{4.94}$$

$$D'' < D < 0.8, \tag{4.95}$$

where D' and D'' are two real roots of the cubic equation

$$f(D) = D(1 - D)^2 = \frac{2L}{RT}. \tag{4.96}$$

It is also clear from Figure 4.12 that if

$$\frac{2L}{RT} > \frac{4}{27}, \tag{4.97}$$

then the continuous mode of operation is realized for any D. The last inequality can also be written as

$$\boxed{L > \tilde{L} = \frac{2}{27} RT.} \tag{4.98}$$

By taking into account formula (4.93), it is easy to see that the inequalities (4.89) and (4.98) are identical.

Now, we shall proceed to the discussion of the *discontinuous* mode of operation of the boost chopper. In this mode of operation, the current $i_L(t)$ through the inductor is strictly positive only during some portion $D_1 T$ (with $D_1 > D$) of the switching cycle, that is,

$$i_L(t) > 0, \qquad \text{if } 0 < t < D_1 T, \tag{4.99}$$

and the current $i_L(t)$ is equal to zero during the rest of the switching cycle, that is,

$$i_L(t) = 0, \qquad \text{if } D_1 T < t < T. \tag{4.100}$$

This zero value of the current $i_L(t)$ is maintained because the flow of the current $i_L(t)$ in the direction opposite to the one shown in Figure 4.9 is prohibited by the diode.

By using the same reasoning as before (see the derivation of equation (4.63) and (4.65)-(4.66)), we find

$$\frac{di_L(t)}{dt} = \frac{V_0}{L} > 0, \qquad \text{if } 0 < t < DT, \tag{4.101}$$

$$\frac{di_L(t)}{dt} = \frac{V_0 - V_C}{L} < 0, \qquad \text{if } DT < t < D_1 T. \tag{4.102}$$

By integrating the last two equations, we derive that

$$i_L(t) = \frac{V_0}{L} t, \qquad \text{if } 0 < t < DT, \tag{4.103}$$

$$i_L(t) = I_{max} + \frac{V_0 - V_C}{L}(t - DT), \qquad \text{if } DT < t < D_1 T. \tag{4.104}$$

By using the last two equations and formula (4.100), $i_L(t)$ can be plotted as shown in Figure 4.13. It is apparent from the last two formulas as well as from Figure 4.13 that

$$I_{max} = \frac{V_0}{L} DT, \tag{4.105}$$

$$I_{max} = \frac{V_C - V_0}{L}(D_1 - D)T. \tag{4.106}$$

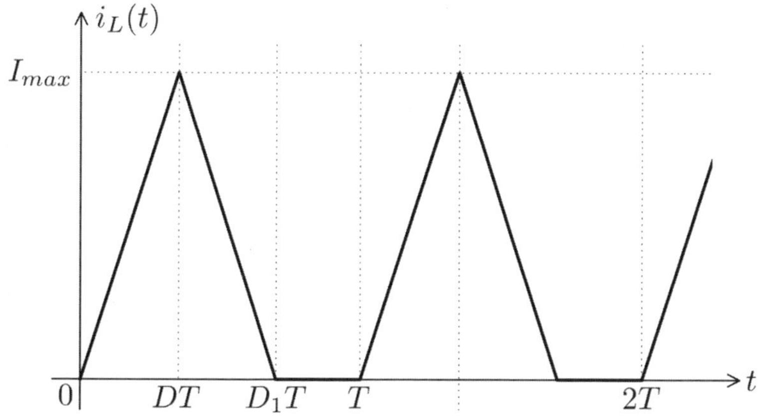

Fig. 4.13

These are two equations with respect to three unknowns I_{max}, V_C and D_1. An additional equation is needed and this equation can be derived by using the power balance condition

$$P_{in} = P_{out}, \tag{4.107}$$

where P_{in} and P_{out} have the same meaning as before.

It is clear that

$$P_{in} = \frac{1}{T}\int_0^T V_0 i_L(t)dt = \frac{V_0}{T}\int_0^{D_1T} i_L(t)dt. \tag{4.108}$$

The last integral can be evaluated by using Figure 4.13 as follows:

$$\int_0^{D_1T} i_L(t)dt = \frac{I_{max}D_1T}{2}. \tag{4.109}$$

From the last two equations we find

$$P_{in} = \frac{V_0 I_{max} D_1}{2}. \tag{4.110}$$

On the other hand,

$$P_{out} = \frac{V_C^2}{R}. \tag{4.111}$$

From formulas (4.107), (4.110) and (4.111) we find

$$\frac{V_0 I_{max} D_1}{2} = \frac{V_C^2}{R}. \tag{4.112}$$

Thus, now we have three equations (4.105), (4.106) and (4.112) for three unknowns I_{max}, V_C and D_1. Our immediate goal is to find the expression

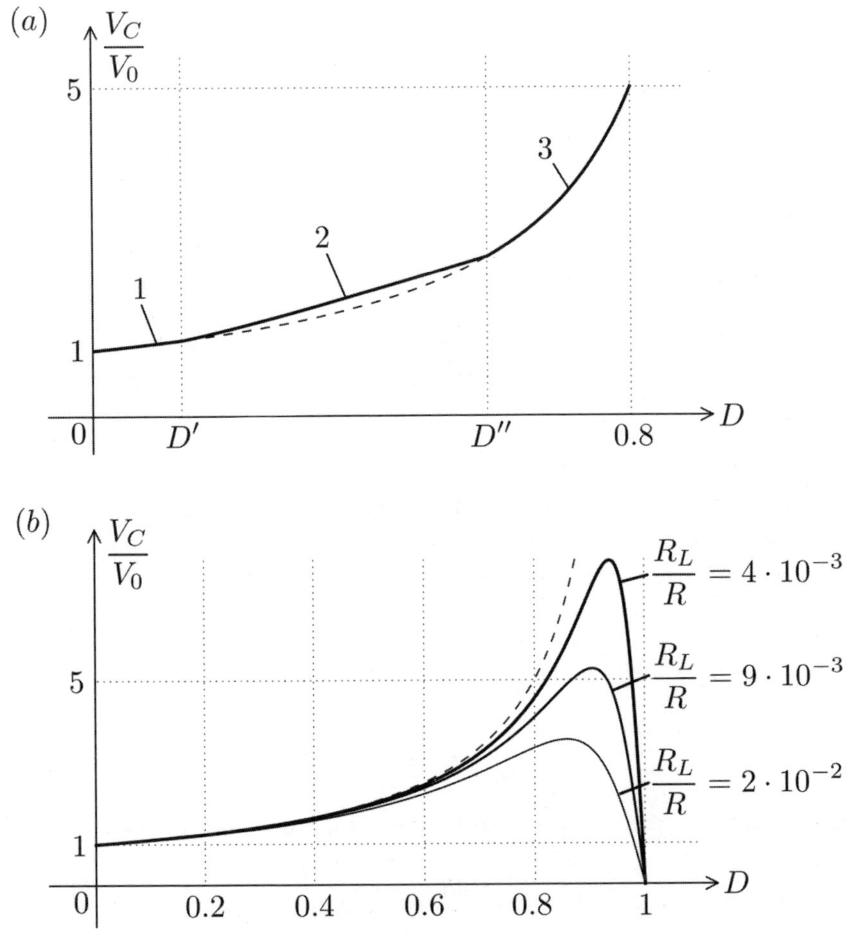

Fig. 4.14

for the output voltage $V_{out} = V_C$. To this end, we shall first derive the formula for D_1 by using equations (4.105) and (4.106). It is clear from these equations that

$$\frac{V_0}{L}DT = \frac{V_C - V_0}{L}(D_1 - D)T. \tag{4.113}$$

By using simple algebraic transformation, we find from the last relation that

$$D_1 = \frac{V_C}{V_C - V_0}D. \tag{4.114}$$

Next, by substituting formulas (4.105) and (4.114) into equation (4.112), we end up with

$$\frac{V_0^2 V_C D^2 T}{2L(V_C - V_0)} = \frac{V_C^2}{R}, \tag{4.115}$$

which leads to

$$V_0^2 \frac{D^2 RT}{2L} = (V_C - V_0)V_C. \tag{4.116}$$

By introducing the non-dimensional parameter

$$\boxed{k = \frac{D^2 RT}{4L},} \tag{4.117}$$

equation (4.116) can be written as the following quadratic equation:

$$\left(\frac{V_C}{V_0}\right)^2 - \frac{V_C}{V_0} - 2k = 0. \tag{4.118}$$

By solving the last equation and taking into account that $\frac{V_C}{V_0}$ is positive, we find

$$\boxed{\frac{V_C}{V_0} = \frac{1}{2}\left(1 + \sqrt{1+8k}\right).} \tag{4.119}$$

It is apparent from the last formula that the performance of the boost chopper in the discontinuous mode of operation is quite different from its performance in the continuous mode. Indeed, in the continuous mode the output voltage depends only on the duty factor D (see formula (4.74)), while in the discontinuous mode the output voltage depends on k, which is a function of D, R, L and T. If the inductance L in the converter circuit is smaller than $\tilde{L}$ given by formula (4.98), then depending on the value of duty factor D the continuous or discontinuous mode of operation can be realized. This is illustrated by Figure 4.14a, in which the lines marked "1" and "3" correspond to continuous mode (see formulas (4.94) and (4.95)) while the line marked "2" corresponds to the discontinuous mode. It must be remarked that the curves "1" and "3" are computed by using formula (4.74), while the curve "2" is computed by using formulas (4.117) and (4.119). The values D' and D'' are computed by solving cubic equation (4.96).

It was mentioned before that the transfer relation (4.74) for continuous mode of operation was derived by neglecting the small resistance R_L of the inductor and this is the reason why formula (4.74) fails for duty factors D close to one. The transfer relations that account for small values of R_L are

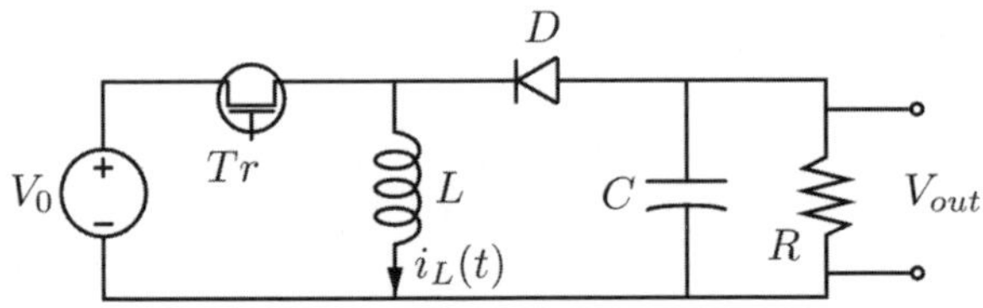

Fig. 4.15

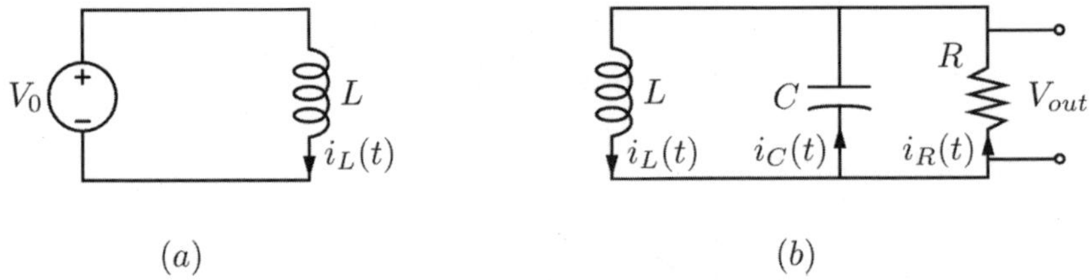

Fig. 4.16

shown in Figure 4.14b for different values of ratio R_L/R. These transfer relations have been computed by using the formula given in the problem 52. This formula as well as the curves in Figure 4.14b suggest that the small parameter R_L may have a strong effect on the performance of the boost converter.

4.3 Buck-Boost Converter

In this section, we shall discuss the buck-boost converter whose controllable output dc voltage can be below or above the input dc voltage. The electric circuit of this converter is shown in Figure 4.15. The operation of this circuit can be described as follows. Transistor Tr is periodically turned "on" and "off." Consider one period $[0, T]$ of this switching. When the transistor is "on" during the time interval

$$0 < t < DT, \tag{4.120}$$

the diode is reverse biased and in the "off" state. This means that during this time interval the circuit shown in Figure 4.15 is reduced to the circuit shown in Figure 4.16a. On the other hand, when the transistor is "off" during the time interval

$$DT < t < T, \tag{4.121}$$

the freewheeling diode is turned "on" to maintain the continuity of the current $i_L(t)$ through the inductor. This means that for this time interval the circuit shown in Figure 4.15 is reduced to the one shown in Figure 4.16b. The buck-boost chopper has two distinct modes of operation: *continuous* mode and *discontinuous* mode. We shall first discuss the *continuous* mode when the inequality (4.1) is valid for any instant of time. It will be assumed throughout our discussion that the capacitance C in the electric circuit shown in Figure 4.15 is sufficiently large that the ripple of the voltage across the capacitor can be neglected and the relation (4.3) is quite accurate.

By using KVL for the circuit shown in Figure 4.16a, we find that

$$L\frac{di_L(t)}{dt} = V_0 \tag{4.122}$$

and

$$\frac{di_L(t)}{dt} = \frac{V_0}{L} > 0, \qquad \text{if } 0 < t < DT. \tag{4.123}$$

Similarly, by using KVL for the circuit shown in Figure 4.16b, we arrive at

$$L\frac{di_L(t)}{dt} + V_C = 0, \tag{4.124}$$

which leads to

$$\frac{di_L(t)}{dt} = -\frac{V_C}{L} < 0, \qquad \text{if } DT < t < T. \tag{4.125}$$

By integrating equations (4.123) and (4.125) we respectively obtain

$$i_L(t) = I_{min} + \frac{V_0}{L}t, \qquad \text{if } 0 < t < DT, \tag{4.126}$$

$$i_L(t) = I_{max} - \frac{V_C}{L}(t - DT), \qquad \text{if } DT < t < T. \tag{4.127}$$

The last two equations lead to the plot of $i_L(t)$ shown in Figure 4.17. From the last two equations as well as Figure 4.17, we find

$$I_{max} - I_{min} = \frac{V_0}{L}DT, \tag{4.128}$$

$$I_{max} - I_{min} = \frac{V_C}{L}(1 - D)T. \tag{4.129}$$

It is apparent from the last two formulas that

$$\frac{V_0}{L}DT = \frac{V_C}{L}(1 - D)T, \tag{4.130}$$

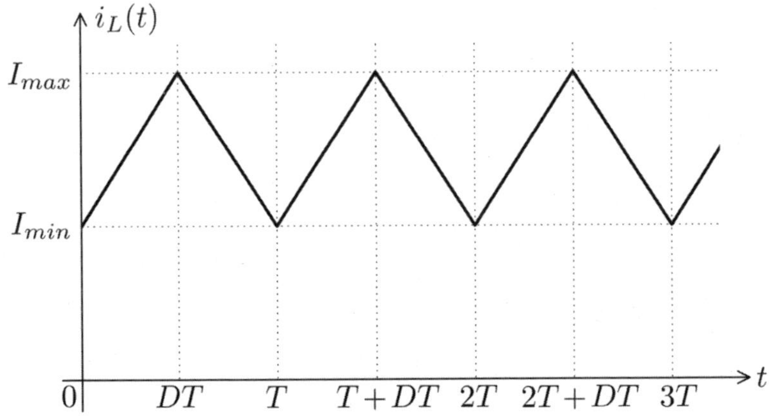

Fig. 4.17

which leads to the following transfer (input-output) relation:

$$\boxed{V_{out} = V_C = \frac{D}{1-D}V_0.} \tag{4.131}$$

It is clear from the last formula that

$$\lim_{D\to 1} V_C = \infty. \tag{4.132}$$

As in the case of the boost chopper, this unrealistic performance is due to the idealization of the inductor in the circuit in Figure 4.15 when its finite resistance is completely neglected. The detailed analysis, which accounts for this resistance, suggests that formula (4.131) is valid for the following range of variation of the duty factor:

$$0 \le D \le 0.8. \tag{4.133}$$

Next, it is easy to see that $F(D) = \frac{D}{1-D}$ is a monotonically increasing function of D. Indeed,

$$F'(D) = \frac{1}{(1-D)^2} > 0. \tag{4.134}$$

Furthermore,

$$F(0) = 0 \qquad \text{and} \qquad F(0.5) = 1. \tag{4.135}$$

This means that

$$F(D) < 1, \qquad \text{if } 0 \le D < 0.5, \tag{4.136}$$

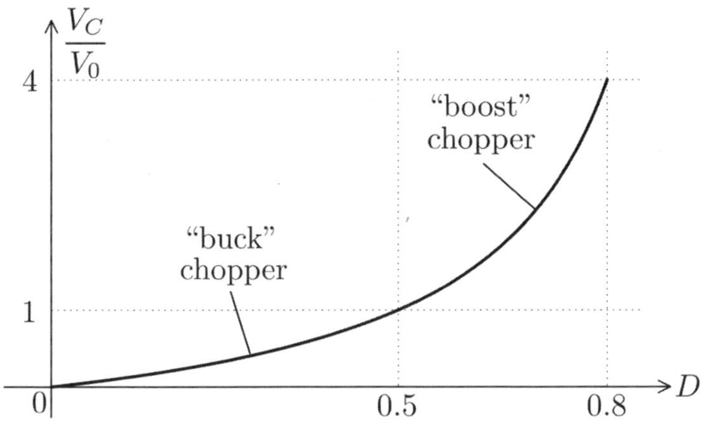

Fig. 4.18

and, consequently, according to the transfer relation (4.131)

$$0 \le V_C = V_{out} < V_0, \qquad \text{if } 0 \le D < 0.5. \tag{4.137}$$

This implies that the circuit shown in Figure 4.15 operates as a buck (step-down) chopper when the duty factor is varied between 0 and 0.5.

On the other hand, it is clear that

$$F(D) > 1, \qquad \text{if } 0.5 < D \le 0.8, \tag{4.138}$$

and according to formula (4.131)

$$V_0 < V_C = V_{out} \le 4V_0. \tag{4.139}$$

Thus, the circuit shown in Figure 4.15 operates as a boost (step-up) chopper when the duty factor D is varied between 0.5 and 0.8.

The transfer relation (4.131) is illustrated by Figure 4.18. This transfer relation has been derived under the assumption that

$$I_{min} > 0, \tag{4.140}$$

which is satisfied when the continuous mode of operation is realized. Now, we shall discuss under what constraints on inductance L and duty factor D the last inequality is actually valid. To do this, we shall first find the explicit expression for I_{min}. To this end, we shall invoke the principle of power balance

$$P_{in} = P_{out}, \tag{4.141}$$

where P_{in} and P_{out} have the same meaning as before.

It is clear that

$$P_{in} = \frac{1}{T}\int_0^{DT} V_0 i_L(t)dt = \frac{V_0}{T}\int_0^{DT} i_L(t)dt. \tag{4.142}$$

It is apparent from Figure 4.17 that

$$\int_0^{DT} i_L(t)dt = \frac{I_{max} + I_{min}}{2} DT. \tag{4.143}$$

From the last two formulas follows that

$$P_{in} = \frac{I_{max} + I_{min}}{2} V_0 D. \tag{4.144}$$

On the other hand, from the relation (4.131) we find

$$P_{out} = \frac{V_C^2}{R} = \frac{D^2 V_0^2}{(1-D)^2 R}. \tag{4.145}$$

By using the last two formulas in equation (4.141), we end up with

$$I_{max} + I_{min} = V_0 \frac{2D}{(1-D)^2 R}. \tag{4.146}$$

By adding equations (4.128) and (4.146), we obtain

$$\boxed{I_{max} = V_0 D\left[\frac{1}{(1-D)^2 R} + \frac{T}{2L}\right].} \tag{4.147}$$

On the other hand, by subtracting equation (4.128) from (4.146), we arrive at

$$\boxed{I_{min} = V_0 D\left[\frac{1}{(1-D)^2 R} - \frac{T}{2L}\right].} \tag{4.148}$$

From the last formula can be inferred that the inequality (4.140) holds and the continuous mode of operation is realized if

$$\boxed{\frac{1}{(1-D)^2 R} - \frac{T}{2L} > 0.} \tag{4.149}$$

It is easy to see that the last inequality is valid for any value of duty factor D if the inductance L is sufficiently large, namely if

$$\boxed{L > \tilde{L} = \frac{RT}{2}.} \tag{4.150}$$

If the inductance L in the converter circuit is smaller than $\tilde{L}$, then constraints can be imposed on the duty factor D to fulfill the inequality (4.149) and in this way to guarantee the continuous mode of operation. Indeed, from inequality (4.149) we find that

$$\frac{2L}{RT} > (1-D)^2, \tag{4.151}$$

which implies that

$$\boxed{D > \tilde{D} = 1 - \sqrt{\frac{2L}{RT}}.} \tag{4.152}$$

Thus, if the duty factor is chosen to satisfy the condition (4.152), then the continuous mode of operation is realized. Furthermore, it is clear that the last inequality is valid for any D if the inductance L satisfies the condition (4.150).

Now, we shall turn to the discussion of the *discontinuous* mode of operation of the buck-boost chopper. In this mode of operation, the current $i_L(t)$ through the inductor is strictly positive only during some portion D_1T (with $D_1 > D$) of the switching cycle,

$$i_L(t) > 0, \qquad \text{if } 0 < t < D_1T, \tag{4.153}$$

and the current $i_L(t)$ is equal to zero during the rest of the switching cycle,

$$i_L(t) = 0, \qquad \text{if } D_1T < t < T. \tag{4.154}$$

This zero value of the current $i_L(t)$ is maintained because the flow of $i_L(t)$ in the direction opposite to the one shown in Figure 4.15 is prohibited by the diode.

Next, by using the same reasoning as before (see the derivation of equations (4.123) and (4.125)), we find

$$\frac{di_L(t)}{dt} = \frac{V_0}{L} > 0, \qquad \text{if } 0 < t < DT, \tag{4.155}$$

$$\frac{di_L(t)}{dt} = -\frac{V_C}{L} < 0, \qquad \text{if } DT < t < D_1T. \tag{4.156}$$

By integrating equations (4.155) and (4.156), we obtain

$$i_L(t) = \frac{V_0}{L}t, \qquad \text{if } 0 < t < DT, \tag{4.157}$$

$$i_L(t) = I_{max} - \frac{V_C}{L}(t - DT), \qquad \text{if } DT < t < D_1T. \tag{4.158}$$

By using the last equations and formula (4.154), function $i_L(t)$ can be plotted as shown in Figure 4.19. It is apparent from formulas (4.157) and

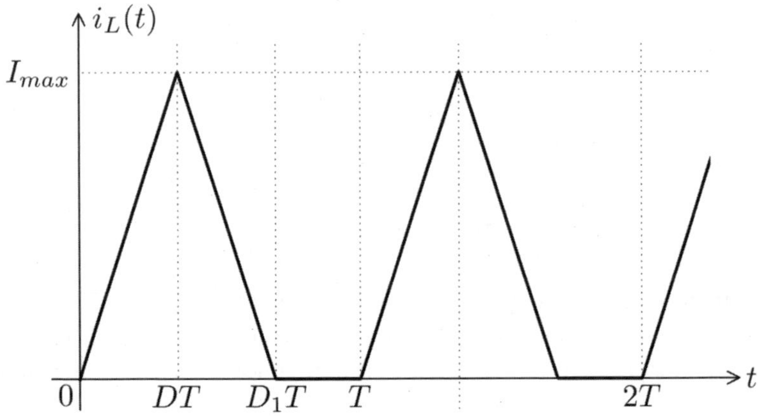

Fig. 4.19

(4.158) as well as from Figure 4.19 that

$$I_{max} = \frac{V_0}{L} DT, \tag{4.159}$$

$$I_{max} = \frac{V_C}{L}(D_1 - D)T. \tag{4.160}$$

These are two equations with respect to three unknowns I_{max}, V_C and D_1. Thus, an additional equation is needed, and this equation can be derived from the power balance condition

$$P_{in} = P_{out}. \tag{4.161}$$

It is clear that

$$P_{in} = \frac{V_0}{T} \int_0^{DT} i_L(t) dt. \tag{4.162}$$

The last integral can be evaluated by using Figure 4.19 as follows:

$$\int_0^{DT} i_L(t) dt = \frac{I_{max}}{2} DT. \tag{4.163}$$

From the last two equations follows that

$$P_{in} = \frac{V_0 I_{max} D}{2}. \tag{4.164}$$

On the other hand, we have

$$P_{out} = \frac{V_C^2}{R}. \tag{4.165}$$

By using formulas (4.164) and (4.165) in equation (4.161), we end up with

$$\frac{V_0 I_{max} D}{2} = \frac{V_C^2}{R}. \tag{4.166}$$

As a result, we have now three equations (4.159), (4.160) and (4.166) for three unknowns I_{max}, V_C and D_1. Our immediate goal is to find the expression for the output voltage $V_{out} = V_C$. This can be accomplished by substituting formula (4.159) into equation (4.166), which leads to

$$\frac{V_0^2 D^2 T}{2L} = \frac{V_C^2}{R}. \tag{4.167}$$

By introducing, as before, the non-dimensional parameter

$$\boxed{k = \frac{D^2 RT}{4L},} \tag{4.168}$$

from equation (4.167) we finally derive

$$\boxed{V_{out} = V_C = V_0\sqrt{2k}.} \tag{4.169}$$

The last two formulas can be effectively used to predict the output voltage V_{out} in the case of the discontinuous mode of operation.

If the inductance L in the converter circuit is less than $\tilde{L}$ given by formula (4.150), then depending on the value of the duty factor D the continuous or discontinuous mode of operation can be realized. This is illustrated by Figure 4.20 where the curve marked "1" corresponds to the continuous mode of operation, while the curve marked "2" corresponds to the discontinuous mode of operation. Curves "1" and "2" are computed by using formulas (4.131) and (4.169), respectively. The value of $\tilde{D}$ is found by using equation (4.152).

4.4 Flyback and Forward Converters

In this section, we discuss "*flyback*" and "*forward*" dc-to-dc converters. In these converters, the input and output terminals are electrically isolated. This is achieved by using magnetic coupling between these terminals. For this reason, such choppers are often called indirect converters. These types of converters are often used in various switching power supplies.

We begin with the discussion of the flyback converter. The electric circuit of this converter is shown in Figure 4.21. Similar to the converters discussed in the previous sections, the operation of the flyback converter

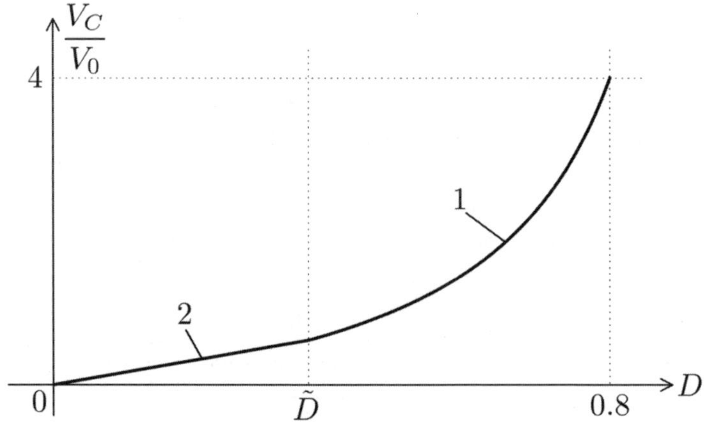

Fig. 4.20

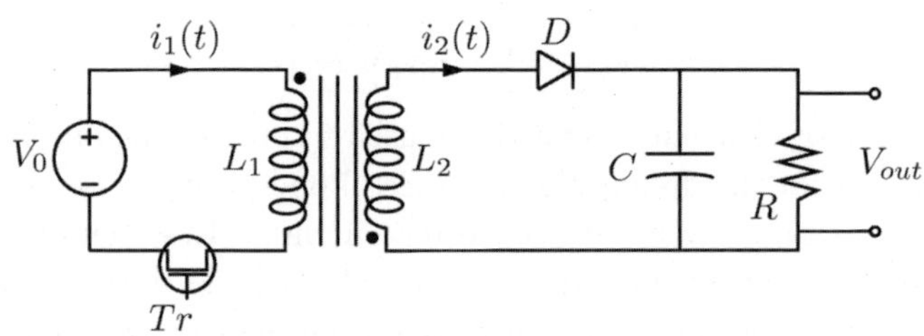

Fig. 4.21

is based on periodic switching of transistor Tr. Before proceeding to the analysis of the circuit shown in Figure 4.21 and the derivation of the transfer (input-output) relation, it is worthwhile to discuss two aspects important for understanding the performance of the flyback converter.

The *first* aspect is related to the *dot convention.* The essence of the dot convention can be stated as follows: a monotonically *increasing* (in time) current entering the dotted terminal of one coil results in such induced voltage across the terminals of the other coil that its dotted terminal is at *positive* potential, i.e., higher potential than the undotted terminal. This dot convention also implies that, vice versa, a monotonically *decreasing* (in time) current entering the dotted terminal of one coil results in such induced voltage across the terminals of the other coil that its dotted terminal is at negative potential. The dot convention is introduced to avoid the ambiguity in voltage polarities which may exist due to different coil

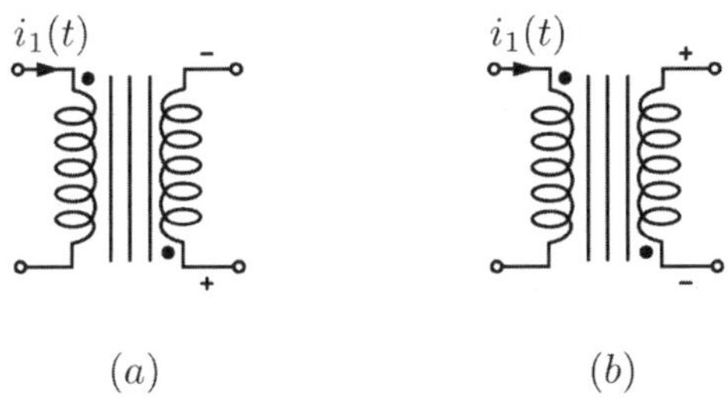

Fig. 4.22

winding directions. This ambiguity is usually removed by the preliminary calibration of magnetically coupled coils for the purpose of identification and assignment of dotted terminals. This calibration can be accomplished experimentally without any prior knowledge of relative winding directions of the coils. This calibration can also be accomplished theoretically by using the right-hand rule. It is clear that the desired assignment of dotted terminals can always be achieved by the appropriate choice of coil winding directions. In circuit analysis, it is tacitly assumed that the proper calibration has been performed and resulted in the assigned dot notations. The essence of the dot convention is illustrated by Figure 4.22a in the case when

$$\frac{di_1(t)}{dt} > 0, \tag{4.170}$$

and by Figure 4.22b in the case when

$$\frac{di_1(t)}{dt} < 0. \tag{4.171}$$

The *second* aspect is related to the generalization of the principle of continuity of electric current through an inductor. This continuity is always valid for a single inductor and it follows from the principle of continuity of energy stored in the magnetic field of the inductor. In the case of a single inductor, this magnetic energy is given by the formula

$$w_m(t) = \frac{Li^2(t)}{2} \tag{4.172}$$

and its continuity implies the continuity of $i(t)$. In the case of two coupled inductors, the continuity of electric current through each inductor may not be preserved. However, the more fundamental principle of the continuity in time of magnetic energy cannot be violated and must always be preserved. Otherwise, it leads to the possibility of infinite power sources, which are

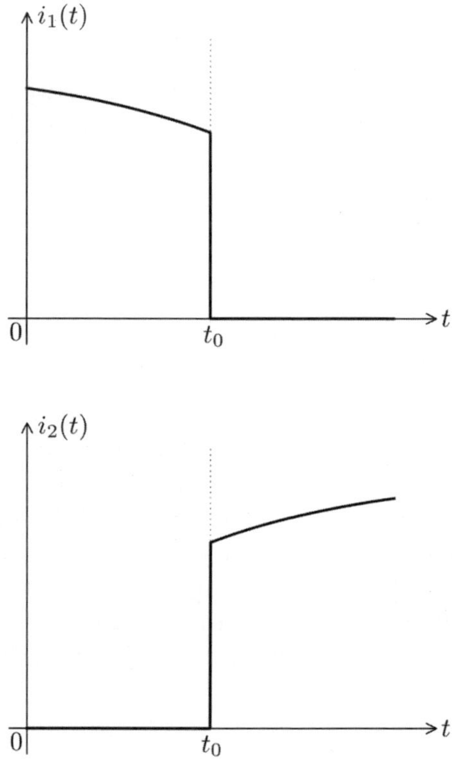

Fig. 4.23

not possible and have never been observed. In the case of two coupled inductors, the magnetic energy is given by the formula

$$w_m(t) = \frac{L_1 i_1^2(t)}{2} + \frac{L_2 i_2^2(t)}{2} + M i_1(t) i_2(t). \tag{4.173}$$

Now, to illustrate how the currents through individual inductors may be discontinuous while the continuity of magnetic energy is preserved, consider the following situation, which is quite relevant to the performance of the flyback converter. Let us assume that the current $i_1(t)$ through the first inductor is not equal to zero during some time interval $(0, t_0)$, while the current $i_2(t)$ through the second inductor is equal to zero during the same interval (see Figure 4.23). Furthermore, we assume that at time t_0 the current $i_1(t)$ is suddenly reduced to zero (as a result of some switching, for instance). Then, the current $i_2(t)$ must be suddenly increased to maintain the continuity of magnetic energy. Such sudden changes are illustrated in

Figure 4.23. Indeed, according to the principle of continuity of magnetic energy, we have

$$w_m(t_{0-}) = w_m(t_{0+}). \tag{4.174}$$

By using formula (4.173), we find that

$$w_m(t_{0-}) = \frac{L_1 i_1^2(t_{0-})}{2} \tag{4.175}$$

and

$$w_m(t_{0+}) = \frac{L_2 i_2^2(t_{0+})}{2}. \tag{4.176}$$

By substituting the last two formulas into equation (4.174), we easily find that sudden changes in currents $i_1(t)$ and $i_2(t)$ are related by the formula

$$\boxed{\frac{i_2(t_{0+})}{i_1(t_{0-})} = \sqrt{\frac{L_1}{L_2}}.} \tag{4.177}$$

Now, we are equipped to proceed to the analysis of the electric circuit of the flyback converter shown in Figure 4.21. As before, it will be assumed in this analysis that the capacitance C is large enough to result in negligible ripples of voltage across the capacitor. In other words, it will be assumed that

$$v_C(t) = V_C = const > 0. \tag{4.178}$$

As mentioned before, the transistor Tr is periodically turned "on" and "off." Consider one period $[0, T]$ of this switching. When the transistor is "on" during the time interval

$$0 < t < DT, \tag{4.179}$$

according to KVL we find

$$L_1 \frac{di_1(t)}{dt} = V_0 \tag{4.180}$$

and

$$\frac{di_1(t)}{dt} = \frac{V_0}{L_1} > 0. \tag{4.181}$$

This implies that the monotonically increasing in time current $i_1(t)$ enters the dotted terminal of the first coil. According to the dot convention, this results in such induced voltage in the second coil that its dotted terminal has a positive potential while its other terminal has a negative potential.

This means that the diode D is reverse biased and in the "off" state. Consequently,

$$i_2(t) = 0. \tag{4.182}$$

Now, by integrating equation (4.181), we derive

$$i_1(t) = I_{min}^{(1)} + \frac{V_0}{L_1}t, \tag{4.183}$$

and

$$\boxed{I_{max}^{(1)} - I_{min}^{(1)} = \frac{V_0}{L_1}DT.} \tag{4.184}$$

Next, we consider the time interval

$$DT < t < T, \tag{4.185}$$

when the transistor Tr is turned "off." During the turning-off process, the current $i_1(t)$ is monotonically decreased from its value $I_{max}^{(1)}$ immediately before the switching to zero immediately after switching:

$$i_1(t) = 0. \tag{4.186}$$

According to the dot convention, this monotonic decrease in time of $i_1(t)$ results in such induced voltage across the terminals of the second coil that its dotted terminal is at negative potential, while its other terminal has positive potential. This means that the diode D is forward biased and turned "on" during the switching. This diode will remain in the "on" state after switching to maintain the continuity of magnetic energy.

According to KVL, we find

$$L_2\frac{di_2(t)}{dt} + V_C = 0, \tag{4.187}$$

$$\frac{di_2(t)}{dt} = -\frac{V_C}{L_2} < 0. \tag{4.188}$$

By integrating the last equation from DT to t, we derive

$$i_2(t) = I_{max}^{(2)} - \frac{V_C}{L_2}(t - DT) \tag{4.189}$$

and

$$\boxed{I_{max}^{(2)} - I_{min}^{(2)} = \frac{V_C}{L_2}(1 - D)T.} \tag{4.190}$$

When the transistor Tr is turned "on" again at time T, this results in monotonically increasing current $i_1(t)$ which induces the voltage across the

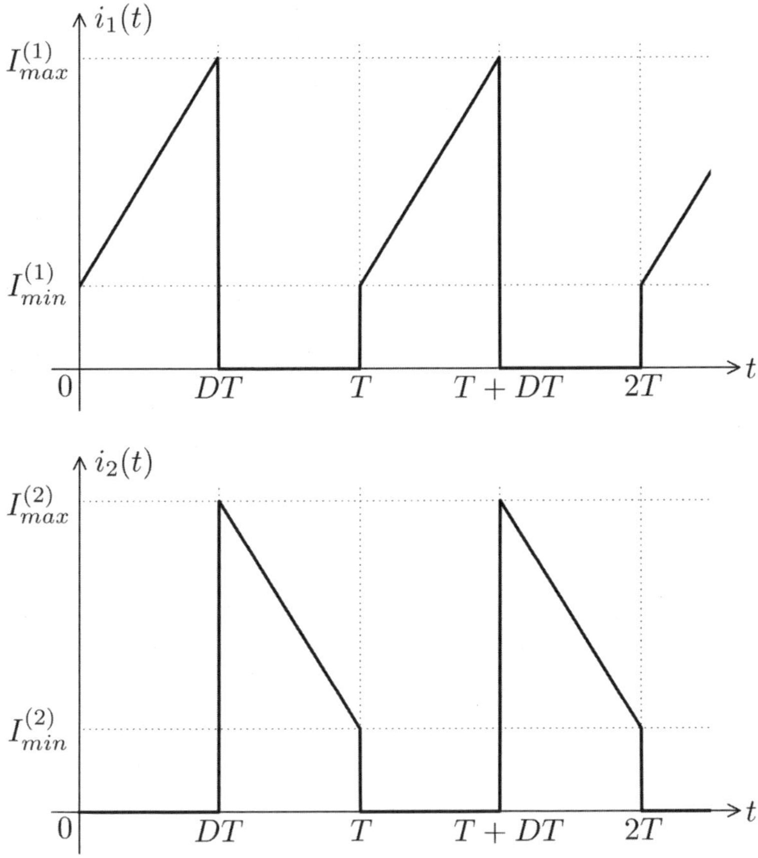

Fig. 4.24

terminals of the second coil of such polarity that the diode D is reverse biased and turned "off." This implies that during the second period (as well as during all subsequent periods) the time variations of $i_1(t)$ and $i_2(t)$ are periodically repeated. These time variations are illustrated by Figure 4.24. Next, we shall relate currents $I^{(1)}_{max}$ and $I^{(1)}_{min}$ to $I^{(2)}_{max}$ and $I^{(2)}_{min}$, respectively, by using the principle of continuity of magnetic energy at time instants $t = DT$ and $t = T$. According to this principle, we have

$$w_m(DT_-) = w_m(DT_+), \tag{4.191}$$

which leads to

$$\frac{L_1\left(I^{(1)}_{max}\right)^2}{2} = \frac{L_2\left(I^{(2)}_{max}\right)^2}{2}. \tag{4.192}$$

From the last formula, we derive

$$I_{max}^{(2)} = \sqrt{\frac{L_1}{L_2}} I_{max}^{(1)}. \tag{4.193}$$

Similarly, we find

$$w_m(T_-) = w_m(T_+), \tag{4.194}$$

which according to Figure 4.24 implies that

$$\frac{L_2 \left(I_{min}^{(2)}\right)^2}{2} = \frac{L_1 \left(I_{min}^{(1)}\right)^2}{2}. \tag{4.195}$$

From the last formula, we derive

$$I_{min}^{(2)} = \sqrt{\frac{L_1}{L_2}} I_{min}^{(1)}. \tag{4.196}$$

By substituting formulas (4.193) and (4.196) into equation (4.190), we obtain

$$\sqrt{\frac{L_1}{L_2}} \left(I_{max}^{(1)} - I_{min}^{(1)}\right) = \frac{V_C}{L_2}(1-D)T. \tag{4.197}$$

Now, by substituting formula (4.184) into the last equation, we end up with

$$\sqrt{\frac{L_1}{L_2}} \frac{V_0}{L_1} DT = \frac{V_C}{L_2}(1-D)T, \tag{4.198}$$

which can be further transformed to result in

$$\boxed{V_C = \sqrt{\frac{L_2}{L_1}} \frac{D}{1-D} V_0.} \tag{4.199}$$

When two coils are wound around the same leg of a ferromagnetic core, then the following formulas for inductances are valid (see Chapter 3 of Part I):

$$L_1 = \frac{N_1^2}{\mathcal{R}_{me}}, \qquad L_2 = \frac{N_2^2}{\mathcal{R}_{me}}. \tag{4.200}$$

Usually, coils are wound around a one-leg (toroidal) core. In this case, the equivalent magnetic reluctance is given by the formula

$$\mathcal{R}_{me} = \frac{\ell}{\mu A}, \tag{4.201}$$

where A and ℓ are the cross-sectional area and average length of the core, respectively. From formulas (4.200), we find

$$\sqrt{\frac{L_2}{L_1}} = \frac{N_2}{N_1}, \tag{4.202}$$

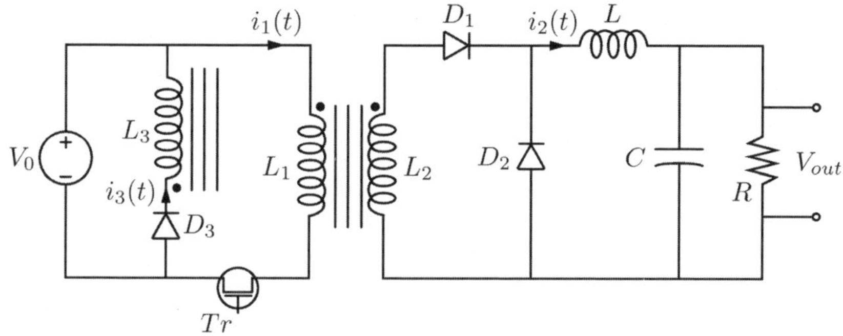

Fig. 4.25

and the formula (4.199) can be written as

$$\boxed{V_{out} = V_C = \frac{N_2}{N_1}\frac{D}{1-D}V_0.} \tag{4.203}$$

It is apparent from the last formula that the dependence of the output voltage on the duty factor D is the same as in the case of the buck-boost converter. For this reason, this converter is often called a *buck-boost-derived* converter. It is also apparent from the last formula that by using a large ratio of N_2 to N_1 high output voltages V_{out} can be achieved. The value of this high voltage is controlled by the switching of the transistor (i.e., controlling D) on the low-voltage side of the converter.

Now, we consider the forward converter. This converter has three coils which are wound around the same ferromagnetic core. The electric circuit of this converter is shown in Figure 4.25. As before, the operation of this converter is based on periodic switching of the transistor Tr. Consider one period $[0, T]$ of this switching. When the transistor is "on" during the time interval

$$0 < t < DT, \tag{4.204}$$

diode D_3 is reverse biased and, consequently,

$$i_3(t) = 0. \tag{4.205}$$

By using KVL, we also find that

$$L_1\frac{di_1(t)}{dt} = V_0 \tag{4.206}$$

and

$$\frac{di_1(t)}{dt} = \frac{V_0}{L_1} > 0. \tag{4.207}$$

This implies that the monotonically increasing in time current $i_1(t)$ enters the dotted terminal of the first coil. According to the dot convention, this results in such induced voltage in the second coil that its dotted terminal has a high (positive) potential, while its other terminal has a low (negative) potential. This means that the diode D_1 is forward biased and "on," while the diode D_2 is reverse biased and "off." Now, by using KVL, we find

$$L\frac{di_2(t)}{dt} + V_C = V_2, \tag{4.208}$$

where V_C is the voltage across the capacitor which, as before, is assumed to be constant and positive (see (4.178)), while V_2 is the induced voltage across the terminals of the second coil. To find this voltage, we remark that if the leakage flux is neglected then all turns of the first and second coils are linked by the same flux $\Phi(t)$. Consequently,

$$V_0 = N_1\frac{d\Phi(t)}{dt}, \tag{4.209}$$

while

$$V_2 = N_2\frac{d\Phi(t)}{dt}. \tag{4.210}$$

From the last two formulas we find

$$V_2 = \frac{N_2}{N_1}V_0. \tag{4.211}$$

By substituting formula (4.211) into equation (4.208), we end up with

$$\frac{di_2(t)}{dt} = \frac{\frac{N_2}{N_1}V_0 - V_C}{L}. \tag{4.212}$$

Now, consider the time interval

$$DT < t < T \tag{4.213}$$

when the transistor is turned off. This turning-off results in a forced sudden monotonic decrease in time of current $i_1(t)$ from some positive value before the switching to zero value after switching. According to the dot convention, this monotonic decrease in time of $i_1(t)$ results in such induced voltage across the terminals of the second coil that its dotted terminal is at negative potential while its other terminal has positive potential. This implies that the diode D_1 is reverse biased and "off," while the diode D_2 is forward biased and "on."

By using KVL, we find

$$L\frac{di_2(t)}{dt} + V_C = 0, \tag{4.214}$$

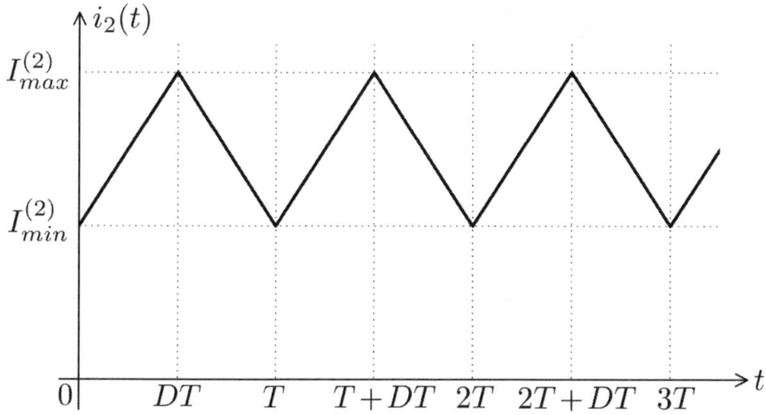

Fig. 4.26

and

$$\frac{di_2(t)}{dt} = -\frac{V_C}{L} < 0. \tag{4.215}$$

Since we consider the steady-state (periodically repeated) performance of the forward converter, the last inequality implies that the right-hand side of equation (4.212) must be positive,

$$\frac{di_2(t)}{dt} = \frac{\frac{N_2}{N_1}V_0 - V_C}{L} > 0. \tag{4.216}$$

By integrating equations (4.216) and (4.215) we respectively find

$$i_2(t) = I^{(2)}_{min} + \frac{\frac{N_2}{N_1}V_0 - V_C}{L}t, \qquad \text{if } 0 < t < DT, \tag{4.217}$$

$$i_2(t) = I^{(2)}_{max} - \frac{V_C}{L}(t - DT), \qquad \text{if } DT < t < T. \tag{4.218}$$

A typical plot of $i_2(t)$ is shown in Figure 4.26. From equations (4.217) and (4.218) we respectively derive

$$I^{(2)}_{max} - I^{(2)}_{min} = \frac{\frac{N_2}{N_1}V_0 - V_C}{L}DT, \tag{4.219}$$

$$I^{(2)}_{max} - I^{(2)}_{min} = \frac{V_C}{L}(1 - D)T. \tag{4.220}$$

It is easy to conclude from the last two formulas that

$$\frac{\frac{N_2}{N_1}V_0 - V_C}{L}DT = \frac{V_C}{L}(1 - D)T, \tag{4.221}$$

which after simple transformations leads to the following transfer (input-output) relation:

$$\boxed{V_{out} = V_C = \frac{N_2}{N_1} D V_0.} \tag{4.222}$$

It is apparent from the last formula that the dependence of the output voltage on the duty factor D is the same as in the case of the buck converter. This explains why this converter is often called a *buck-derived* converter. It is also apparent from the last formula that by using a large ratio of N_2 to N_1 high output voltages V_{out} can be achieved. The value of high output voltage is controlled by the switching of the transistor Tr on the low-voltage side, i.e., by controlling the duty factor D.

It turns out that there are some constraints on the range of variation of D. To find these constraints, we consider the intended function of the third coil in the operation of the forward converter. This function is to "catch" and "extinguish" the magnetic flux in the ferromagnetic core during the time interval when the transistor Tr is "off." For this reason, this winding is sometimes called the "catch winding." It would be better to call it the "flux-resetting" winding. If this magnetic flux is removed by the end of period $[0, T]$, then at the beginning of the next period the magnetic conditions of the core will be the same as at the beginning of the previous period, i.e., flux resetting is achieved. Otherwise, the build-up of magnetic flux in the core will occur and the converter cannot function properly.

During the turning-off of the transistor Tr, we find in accordance with the dot convention that the voltage across the terminals of the third coil is induced with such polarity that the dotted terminal is at appreciable negative potential. This implies that the diode D_3 is forward biased and the flow of the current $i_3(t)$ commences. The value of this current at the moment of switching is such that the continuity of magnetic energy is maintained. To find the flux resetting conditions, it is convenient to write the KVL equation for the loop formed by the input voltage source and the third coil in the form

$$N_3 \frac{d\Phi(t)}{dt} = -V_0. \tag{4.223}$$

The negative sign in the last equation reflects the fact that the direction of the current $i_3(t)$ is such that instantaneous input power is negative. This means that the energy stored in the magnetic field of the core during the time interval when the transistor Tr is "on" is being returned to the source during the time interval when the transistor Tr is "off." This energy return implies the reduction in core magnetic field and in core magnetic flux, which is consistent with the negative sign in the last equation. By integrating the last equation from DT to t, we find

$$\Phi(t) = \Phi(DT) - \frac{V_0}{N_3}(t - DT). \tag{4.224}$$

From formula (4.209) follows that

$$\Phi(DT) = \frac{V_0}{N_1}DT. \tag{4.225}$$

By combining the last two formulas, we obtain

$$\Phi(t) = \frac{V_0}{N_1}DT - \frac{V_0}{N_3}(t - DT). \tag{4.226}$$

The flux resetting will occur if

$$\Phi(t_0) = 0 \qquad \text{for } t_0 < T. \tag{4.227}$$

This means according to equation (4.226) that

$$\frac{V_0}{N_1}DT = \frac{V_0}{N_3}(t_0 - DT). \tag{4.228}$$

Since $t_0 < T$, from the last formula we obtain

$$\frac{DT}{N_1} < \frac{(1 - D)T}{N_3}, \tag{4.229}$$

which after simple transformations leads to the inequality

$$\boxed{D < \frac{N_1}{N_1 + N_3}.} \tag{4.230}$$

Thus, the flux resetting occurs only if the duty factor does not exceed the right-hand side in the last formula. In the somewhat typical case when $N_1 = N_3$, the flux resetting condition is

$$\boxed{D < 0.5.} \tag{4.231}$$

This concludes the discussion of the forward converter.

Problems

(1) What is the subject of power electronics? What are the main types of power converters?
(2) What is the function of energy storage elements in power converters? Explain the trade-off between switching speed and overall size, weight and cost of power converters.
(3) What are the main engineering applications of power electronics?
(4) Give a concise summary of the basic facts and mathematical relations of the drift-diffusion model for mobile carrier transport in semiconductors.
(5) Give a concise summary of the basic facts related to the physics of p-n junctions at equilibrium.
(6) By using the drift-diffusion model derive the Shockley equation (1.72). What is the physical origin of reverse (negative) saturation current I_s?
(7) What is a unique design feature of power diodes resulting in increase of their breakdown voltage?
(8) Describe the design and the principle of operation of the n^+pn BJT. Explain why the base in BJT devices is narrow.
(9) Explain how the BJT can be used as a current-controlled switch in the common emitter configuration. What are the advantages and disadvantages of the BJT as a switch?
(10) Explain the design and the principle of operation of the thyristor (SCR) by using the two-transistor model. Draw and explain the (idealized) I-V curve for the thyristor.
(11) Describe the design and the principle of operation of the MOSFET and how it can be used as a voltage-controlled switch. What are the advantages and disadvantages of the MOSFET as a switch in comparison with the BJT?

(12) What are the unique design features of the power MOSFET?

(13) Explain the design and principle of operation of the IGBT. What is the main advantage of the IGBT in comparison with the power MOSFET?

(14) Explain what snubber circuits are and give examples of such circuits.

(15) Explain what resonant switches are and how they can be used for "soft" switching of semiconductor devices. Give examples of resonant switches used for zero-current switching (ZCS) and zero-voltage switching (ZVS).

(16) Carry out the analysis of the single-phase rectifier with RL load (see Figure 2.1a or 2.1b) by using the frequency-domain technique.

(17) Explain how the performance of the center-tapped transformer rectifier shown in Figure 2.6 is different from the performance of the rectifier shown in Figure 2.1a.

(18) What is the physical mechanism of suppressing ripple in the output voltage of the rectifier shown in Figure 2.9? Explain the physics of operation of this rectifier.

(19) Carry out the analysis of the center-tapped transformer rectifier shown in Figure 2.15 by using the time-domain technique.

(20) Suppose you want to build a power converter to provide 10 V dc to a device by using an available single-phase ac voltage of 120 V rms. What turns ratio should you use in the center-tapped transformer rectifier shown in Figure 2.6 to achieve this?

(21) Analyze the single-phase bridge rectifier with RLC load shown in Figure 2.17 by using the frequency-domain technique. Explain how the suppression of ripple is improved by the presence of two energy storage elements (inductor and capacitor).

(22) Suppose you have a three-phase power system from which three dc voltage supplies are being powered. One dc supply employs a three-phase diode bridge rectifier (Figure 2.23), another uses a three-phase, three-pulse diode rectifier (Figure 2.20), and the remaining dc supply uses a single-phase bridge rectifier (Figure 2.1) connected to one of the three phases. By using just oscilloscope measurements across the terminals of the RL branches, how might it be possible to determine which rectification scheme is used?

(23) Analyze the three-phase rectifier shown in Figure 2.20 by using the frequency-domain technique.

(24) Analyze the three-phase bridge rectifier (Figure 2.23) by using the frequency-domain technique.
(25) Derive the output voltage expression for the twelve-pulse three-phase diode rectifier shown in Figure 2.26 in the case when the ripple is small.
(26) Explain the function of the freewheeling diode in the phase-controlled rectifier shown in Figure 2.27.
(27) Analyze the center-tapped transformer phase-controlled rectifier shown in Figure 2.30.
(28) Carry out the analysis of the rectifier shown in Figure 2.31 by using the frequency-domain technique.
(29) Suppose you have a three-phase power system from which two dc voltage supplies are being powered. One dc supply employs a three-phase diode bridge rectifier (Figure 2.23). The other dc supply uses a three-phase SCR rectifier (Figure 2.31). By using just oscilloscope measurements across the terminals of the RL branches, how might it be possible to determine which rectification scheme is used?
(30) Draw the circuit of the single-phase bridge inverter and explain how by the appropriate switching the polarity across the terminals of the RL branch can be periodically inverted.
(31) Explain why bidirectional (bilateral) switches are needed for the operation of bridge inverters and how these switches are designed.
(32) Explain how the bidirectional switches shown in Figure 3.5 can be used to control the width of rectangular pulses in single-phase bridge inverters.
(33) What is pulse width modulation (PWM) and what are the generic features of Fourier spectra of PWM voltages?
(34) Describe the main steps in the derivation of formula (3.69) for the Fourier series expansion of PWM voltages. What is the significance of the depth of modulation?
(35) What are the main functions of the inductor in the single-phase inverter (see Figure 3.5) with PWM?
(36) How can PWM voltages be generated? (Explain how switches are controlled to achieve PWM voltages.)
(37) Explain how the time-domain technique can be used to analyze single-phase inverters with PWM. Derive formulas (3.95), (3.96) and (3.97).
(38) How can the problem of PWM be stated as an optimization problem in the time domain?

(39) Draw the circuit of the three-phase bridge inverter and explain the pattern of switching that produces the voltage waveforms shown in Figure 3.16.
(40) Describe a pattern of switching that results in three-phase PWM voltages.
(41) Explain how ac-to-ac converters can be designed by cascading three-phase rectifiers with three-phase inverters.
(42) Explain how ac-to-ac converters can be used in ac motor drives for frequency control of speed of induction and synchronous motors.
(43) Explain what the "constant volts per hertz" criterion is in ac motor drives and why it is needed.
(44) What are the major types of dc-to-dc converters (choppers)?
(45) What are two distinct regimes of chopper operation?
(46) What is the main assumption made in the analysis of choppers?
(47) Suppose you are considering to build a buck chopper with parameters $L = 20$ mH and $R = 50\ \Omega$ with the switching rate of 1 kHz. You want to output a dc voltage that is one-fourth of the input dc value. In what mode of operation will the chopper operate and what duty factor is necessary to achieve the desired output voltage?
(48) Suppose you want to build a power converter to provide 10 V dc to a device by using an available single-phase ac voltage of 120 V rms. You use a single-phase bridge rectifier to convert the ac voltage to dc. Design a buck chopper circuit cascaded with the rectifier to achieve the desired dc voltage. How does such a device compare to the center-tapped transformer-based design considered in question 20?
(49) Explain how the duty factor can be controlled in choppers.
(50) Formula (4.19) has been derived by neglecting the resistance R_L of the inductor. Demonstrate that in the case when this resistance is not neglected but $R_L T \ll L$, formula (4.19) is replaced by the following:

$$V_{out} = V_C = V_0 \frac{D}{1 + \frac{R_L}{R}}.$$

(Hint: use two-term Taylor expansions for exponentials.)
(51) Perform the ripple analysis (i.e., derive the formula for $\Delta V_C / V_C$) for the boost converter in the case of continuous current mode of operation.

(52) Demonstrate that in the case when the resistance R_L of the inductor is not neglected but $R_L T \ll L$, formula (4.74) is replaced by the following:

$$V_{out} = V_C = V_0 \frac{1 - D}{(1 - D)^2 + \frac{R_L}{R}}.$$

(53) Perform the ripple analysis (i.e., derive the formula for $\Delta V_C / V_C$) for the buck-boost converter in the case of continuous current mode of operation.

(54) Demonstrate that in the case when the resistance R_L of the inductor is not neglected but $R_L T \ll L$, formula (4.131) is replaced by the following:

$$V_{out} = V_C = V_0 \frac{D(1 - D)}{(1 - D)^2 + \frac{R_L}{R}}.$$

(55) Produce the plots of V_{out}/V_0 as a function of D for various small values of ratio R_L/R.

(56) Draw the circuit of the flyback converter and explain its principle of operation.

(57) Formula (4.199) has been derived for the flyback converter under the tacit assumption that $I^{(1)}_{min} > 0$ (and, consequently, $I^{(2)}_{min} > 0$). Derive the condition (i.e., inequalities) for L_2 (or L_1) under which this assumption (and, consequently, formula (4.199)) is valid.

(58) Perform the analysis of the flyback converter in the case when the small resistances of the two inductors are taken into account.

(59) Draw the circuit and explain the principle of operation of the forward converter. What is the purpose of the "flux resetting" winding with L_3 (see Figure 4.25)? Explain the reason for the limitation on the range of duty factor values (see formula (4.231)).

(60) Compare and contrast the operation of the flyback and forward converters. Explain when you might want to use one converter instead of the other.

Bibliography

[1] A. Ahmed, *Power Electronics for Technology*, Prentice Hall, 1999.

[2] V. Ajjarapu, *Computational Techniques for Voltage Stability Assessment and Control*, Springer, 2006.

[3] A. Arapostathis, S. Sastry and P. Varaiya, "Analysis of Power Flow Equations," *Int. J. Electr. Power and Energy Systems*, vol. 3, pp. 115-126, 1981.

[4] B. J. Baliga, *Power Semiconductor Devices*, PWS Publishing Company, 1996.

[5] B. J. Baliga, *Fundamentals of Power Semiconductor Devices*, Springer, 2008.

[6] A. R. Bergen and V. Vittal, *Power System Analysis* (second edition), Prentice Hall, 1986.

[7] G. Bertotti, "Physical Interpretation of Eddy Current Losses in Ferromagnetic Materials," *Journal of Applied Physics*, vol. 57, pp. 2118-2126, 1985.

[8] G. Bertotti, "General Properties of Power Losses in Soft Ferromagnetic Materials," *IEEE Transactions on Magnetics*, vol. 24, pp. 621-630, 1988.

[9] G. Bertotti, *Hysteresis in Magnetism (for physicists, material scientists and engineers)*, Academic Press, 1998.

[10] B. K. Bose, *Modern Power Electronics and AC Drives*, Prentice Hall, 2002.

[11] J. R. Brauer, *Magnetic Actuators and Sensors, 2nd Edition*, Wiley-IEEE Press, 2014.

[12] D. Brown and E. P. Hamilton III, *Electromechanical Energy Conversion*, Macmillan Publishing Company, 1984.

[13] E. E. Buchanan and E. J. Miller, "Resonant Switching Power Conversion Technique," *IEEE Power Electronics Specialists Conference*, 1975 Record, pp. 188-193.

[14] S. J. Chapman, *Electric Machinery Fundamentals (5th edition)*, McGraw-Hill, 2011.

[15] V. Del Toro, *Basic Electric Machines*, Prentice Hall, Englewood Cliffs, 1990.

[16] V. Del Toro, *Electric Power Systems*, Prentice Hall, 1992.

[17] F. P. Emad and I. D. Mayergoyz, "Computation of Lumped Parameters of Doubly-Fed Machines," *IEEE Transactions on Magnetics*, vol. 22, no. 5, pp. 1049-1051, 1986.

[18] F. Fiorillo, *Measurement and Characterization of Magnetic Materials*, Elsevier, 2005.

[19] M. J. Fisher, *Power Electronics*, PWS-Kent Publishing Company, 1991.

[20] E. P. Furlani, *Permanent Magnet and Electromechanical Devices (materials, analysis and applications)*, Academic Press, 2001.

[21] A. E. Fitzgerald, C. Kingsley, Jr. and S. D. Umans, *Electric Machinery (fifth edition)*, McGraw-Hill, 1990.

[22] B. R. Gungor, *Power Systems*, Harcourt Brace Jovanovich Publishers, 1988.

[23] D. G. Holmes and T. A. Lipo, *Pulse Width Modulation for Power Converters (Principles and Practice)*, John Wiley and Sons, 2003.

[24] J. G. Kassakian, M. F. Schlecht, G. C. Verghese, *Principles of Power Electronics*, Addison-Wesley, 1991.

[25] E. W. Kimbark, *Power System Stability (volume III: Synchronous Machines)*, John Wiley and Sons, 1956.

[26] E. W. Kimbark, *Direct Current Transmission*, Wiley-Interscience, 1971.

[27] J. L. Kirtley, *Electric Power Principles (Sources, Conversion, Distribution and Use)*, Wiley, 2011.

[28] A. J. Korsak, "On the Question of Uniqueness of Stable Load Flow Solutions," *IEEE Trans. Power Appar. & Systems*, vol. PAS-91, pp. 1093-1100, 1972.

[29] P. T. Krein, *Elements of Power Electronics*, Oxford University Press, 1998.

[30] F. C. Lee, *High Frequency Resonant, Quasi-Resonant and Multi-Resonant Converters*, Virginia Power Electronics Center, 1989.

[31] I. D. Mayergoyz, *Iterative Techniques for Calculation of Stationary Fields in Nonlinear and Anisotropic Media*, Naukova Dumka, Kyiv, 1979.

[32] I. D. Mayergoyz and F. P. Emad, "A New Method for the Calculation of Magnetic Fields in AC Machines," *IEEE Transactions on Magnetics*, vol. 22, no. 5, pp. 1046-1048, 1986.

[33] I. D. Mayergoyz, F. P. Emad, M. A. Sherif, "Electromagnetic Field Analysis of Unbalanced Regimes of Synchronous Machines," *Journal of Applied Physics*, vol. 63, no. 8, pp. 3188-3190, 1988.

[34] I. D. Mayergoyz, *Mathematical Models of Hysteresis*, Springer, 1991.

[35] I. D. Mayergoyz and W. Lawson, *Basic Electric Circuit Theory (A one-semester text)*, Academic Press, 1997.

[36] I. Mayergoyz, *Nonlinear Diffusion of Electromagnetic Fields (with applications to eddy currents and superconductivity)*, Academic Press, 1998.

[37] I. D. Mayergoyz and C. Serpico, "Nonlinear Diffusion of Electromagnetic Fields and Excess Eddy Current Losses," *Journal of Applied Physics*, vol. 85, no. 8, pp. 4910-4912, 1999.

[38] I. Mayergoyz, *Mathematical Models of Hysteresis and Their Applications*, Elsevier-Academic Press, 2003.

[39] I. Mayergoyz and C. Tse, *Spin-Stand Microscopy of Hard Disk Data*, Elsevier, 2007.

[40] E. J. Miller, "Resonant Switching Power Conversion," *IEEE Power Electronics Specialists Conference*, 1976 Record, pp. 206-211.

[41] N. Mohan, *Electric Power Systems (A First Course)*, John Wiley and Sons, 2012.

[42] N. Mohan, *Electric Machines and Drives (A First Course)*, John Wiley and Sons, 2012.

[43] N. Mohan, T. M. Undeland and W. P. Robbins, *Power Electronics (Converters, Applications and Design)*, John Wiley and Sons, 1995.
[44] M. A. Pai, *Power System Stability (Analysis by the Direct Method of Lyapunov)*, North-Holland, 1981.
[45] M. A. Pai, *Energy Function Analysis for Power System Stability*, Kluwer Academic Publisher, 1989.
[46] A. G. Phadke and J. S. Thorp, *Synchronized Phasor Measurements and Their Applications*, Springer, 2008.
[47] H. Saadat, *Power System Analysis (second edition)*, McGraw-Hill, 2002.
[48] P. W. Sauer and M. A. Pai, *Power System Dynamics and Stability*, Prentice Hall, 1998.
[49] C. Serpico, C. Visone, I. D. Mayergoyz, V. Basso and G. Miano, "Eddy Current Losses in Ferromagnetic Laminations," *Journal of Applied Physics*, vol. 87, no. 9, pp. 6923-6925, 2000.
[50] M. Shur, *Physics of Semiconductor Devices*, Prentice Hall, 1990.
[51] G. R. Slemon, *Electric Machines and Drives*, Addison-Wesley, 1992.
[52] R. Stein and W. T. Hunt, Jr., *Electric Power System Components (Transformer and Rotating Machines)*, Van Nostrand Reinhold Company, 1979.
[53] W. D. Stevenson, Jr., *Elements of Power System Analysis* (fourth edition), McGraw-Hill, 1982.
[54] R. D. Strattan and F. J. Young, "Iron Losses in Elliptically Rotating Fields," *Journal of Applied Physics*, vol. 33, no. 3, pp. 1285-1286, 1962.
[55] B. G. Streetman and S. Banerjee, *Solid State Electronic Devices (6th Edition)*, Prentice Hall, 2005.
[56] C. J. Tavora, O. J. M. Smith, "Equilibrium Analysis of Power Systems," *IEEE Trans. Power Appar. & Systems*, vol. PAS-91, pp. 1131-1137, 1972.
[57] A. M. Trzynadlowski, *Introduction to Modern Power Electronics*, John Wiley and Sons, 1998.
[58] A. I. Voldek, *Electric Machines*, Russian Publisher "Energya," 1978.
[59] D. C. White and H. H. Woodson, *Electromechanical Energy Conversion*, John Wiley and Sons, 1959.
[60] T. Wildi, *Electrical Machines, Drives, and Power Systems* (fourth edition), Prentice Hall, 2000.
[61] F. J. Young and H. L. Schenk, "Iron Losses Due to Elliptically Polarized Magnetic Fields," *Journal of Applied Physics*, vol. 37, no. 3, pp. 1210-1211, 1966.

Index